The Nonlinear Theory of Elastic Shells

One Spatial Dimension

#17297434

The Nonlinear Theory of Elastic Shells

One Spatial Dimension

A. Libai
Department of Aeronautical Engineering
Technion, Israel Institute of Technology
Haifa, Israel

J. G. Simmonds
Department of Applied Mathematics
University of Virginia
Charlottesville, Virginia

ACADEMIC PRESS, INC.
Harcourt Brace Jovanovich, Publishers

Boston San Diego New York
Berkeley London Sydney
Tokyo Toronto

ACADEMIC PRESS, INC.
1250 Sixth Avenue, San Diego, CA 92101

United Kingdom Edition published by
ACADEMIC PRESS INC. (LONDON) LTD.
24–28 Oval Road, London NW1 7DX

Library of Congress Cataloging-in-Publication Data

Libai, A. (Avinoam), 1929-
The nonlinear theory of elastic shells of one spatial dimension.

Includes bibliographies and index.
1. Shells (Engineering) 2. Elasticity.
3. Nonlinear theories. I. Simmonds, James G.
II. Title.
TA660.S5L457 1988 624.1'7762 87-33497
ISBN 0-12-447940-5

88 89 90 91 9 8 7 6 5 4 3 2 1
Printed in the United States of America

To our beloved families

Contents

An asterisk (*) indicates a section that may be omitted without loss of continuity

Contents

Contents

Preface

This book is a greatly expanded version of parts of our monograph, "Nonlinear Elastic Shell Theory," that appeared on pp. 271-371 of Volume 23 of *Advances in Applied Mechanics* (J. W. Hutchinson & T. Y. Wu, eds.), Academic Press, 1983. We have added several new chapters and rewritten or supplemented others extensively. Discussions of stability and variational principles, which were omitted from the monograph for lack of space, are included here. Furthermore, some material on load potentials, nonlinear constitutive laws, thermodynamics (as applied to shells), and boundary conditions is original. Space, time, and our proclivities have prevented us from discussing numerical solutions of shell equations, an important area that needs to be surveyed, summarized, and streamlined. The state of the art may be gleaned from books by Hinton & Owen,[1] Hughes & Hinton,[2] and Bushnell.[3]

This book has two main goals: to lay a foundation for the nonlinear theory of thermoelastic shells undergoing large rotations and large strains and to present, early on, relatively simple equations for practical application. We have tried to write for those who know some continuum mechanics but little shell theory, but we think that experts will find here much that is novel—in presentation if not in content. Thus, after an introductory chapter that describes what we mean by a nonlinear elastic shell and spells out our approach, we expose, in the next chapter, the bedrock of 3-dimensional continuum mechanics. On this foundation we build an ascending stairway of four chapters. Our hope is to raise gradually the reader's understanding of shell theory by developing, for rods and special shells, concrete results that are useful in themselves. There are some topics, however, that we think are best treated as special cases of, or as approximations to, the general shell equations. It is our intention to include these in a second volume—the descending portion of the stairway.

To aid the reader, we have posed and solved, in various places, simple problems to illustrate general principles, have used an asterisk (*) to indicate those sections that can be omitted without loss of continuity, and have

[1]"Finite Element Software for Plates and Shells," Pineridge Press, Swansea, U. K., 1984.

[2]"Finite Element Methods for Plate and Shell Structures," vols. 1 and 2, Pineridge Press, Swansea, U. K., 1986.

[3]"Computerized Buckling Analysis of Shells," Nijhoff, Dordrecht, The Netherlands, 1985.

summarized our notational scheme and listed the important symbols we use in an appendix at the end of the book.

The chapters on "birods", "beamshells","axishells", and "unishells" also illustrate, in situations of increasing complexity, our *mixed* approach to shell theory, namely, to use those equations of 3-dimensional continuum mechanics that are independent of material properties to derive corresponding rod or shell equations, but to *postulate* the form of those rod or shell equations that depend unavoidably on material properties.

We might take as an epigraph for this book Truesdell's dictum (1983):[4]

> In mathematical practice today it is, unfortunately, often forgotten that to derive basic equations is even so much a mathematician's duty as to study their properties

to which we would add Koiter's observation (1969)[5] that

> Flexible bodies like thin shells require a flexible approach.

In obedience to these mentors, we have tried to derive the equations of shell theory with care and to cast them in many different forms, knowing from experience that a set of dependent variables or a reference frame that works for one class of problems may be unsuitable for another. We also acknowledge the profound influence of Eric Reissner, as our many citations to his papers attest. He was the teacher of one of us (JGS), and his lessons have not been forgotten.

In assembling material for this book we have encountered many gaps. In most cases, we have resisted the temptation—with its potential for distraction and frustration—to try to fill them. Here and there, therefore, the reader will find open questions and problems.

Parts of this book were written while Professor Libai was visiting the Department of Applied Mathematics at the University of Virginia, on sabbatical leave from the Department of Aeronautical Engineering, the Technion, Israel Institute of Technology. We thank the Center for Advanced Studies at the University of Virginia for its partial support of Professor Libai's visit. We also thank the Fund for the Promotion of Research at the Technion and the National Science Foundation for supporting much of our research. Our colleagues, Ernst Axelrad and Hubertus Weinitschke, and our students, Dawn Fisher, Jim Fulton, and Dror Rubinstein, have made useful suggestions in various chapters. Rohn England, another student, has read the entire manuscript with an eye to both syntax and semantics. Tom Tartaglino and Judith Trott prepared all of the figures, except for 5.10-5.12, which Jim Fulton prepared using the *troff* preprocessors *grap* and *pic*. Thanks to all!

---

[4]The influence of elasticity on analysis: the classical heritage. *Bull. Amer. Math. Soc.* **9**, 293-310.

[5]Foundations and basic equations of shell theory: A survey of Recent Progress. *In* "Theory of Thin Shells," Proc. I.U.T.A.M. Sympos., Copenhagen, 1967 (F. I. Niordson, ed.), pp. 93-105. Springer-Verlag, Berlin, etc.

Preface

We typeset the manuscript ourselves using the *ditroff* text processing system with 4.2 BSD UNIX on a VAX 11/780, run jointly by the Departments of Computer Science and Applied Mathematics at the University of Virginia. Thus, all errors—substantial or typographical—rest squarely on us.

Chapter I

Introduction

A. What Is a Shell?

A *shell* is a curved, thin-walled structure. A quantitative definition will be given in later chapters. Two important degenerate classes of shells are *plates* (shells which are flat when undeformed) and *membranes* (shells whose walls offer no resistance to bending). Shells may be made of a single inhomogeneous or anisotropic material or may be made of layers of different materials. The primary function of a shell may be to transfer loads from one of its edges to another, to support a surface load, to provide a covering, to contain a fluid, to please the eye or ear, or a combination of these. Shells occur in nature and as artifacts. Aortic valves, automobile hoods, balloons, beer cans, bellows, bells, bladders, bowls, contact lenses, crab carapaces, domes, ducts, egg coverings, footballs, funnels, inner tubes, light bulb casings, loudspeaker cones, manhole covers, parachutes, peanut hulls, Ping-Pong balls, panels, pipes, pressure vessels, silos, skulls, straws, tents, tires, trumpets, umbrellas, vaults, wine glasses, and woks are all shells. The aim of shell theory is to describe the static or dynamic behavior of structures like the above by equations that involve no more than one or two *spatial* variables.

B. Elastic Shells and Nonlinear Behavior

If a shell is prevented from moving as a rigid body, then it is *elastic* if, upon application and removal of a sufficiently small load, it tends to return to its initial shape.

The *behavior* of an elastic shell is said to be *nonlinear* if, under static conditions, the deflection of any point of the shell is *not* proportional to the magnitude of an applied load. Two sources of nonlinearity are often distinguished: *geometric* and *material*. One speaks of a geometrically nonlinear shell theory if the strain-displacement relations are nonlinear but the stress-strain relations linear. Most papers on nonlinear shell theory make this assumption and for good reason: traditional engineering materials such as steel and aluminum remain elastic only if the principal strains are relatively small, say less than half of a percent. While there have been attempts to formulate shell theories in which the strain-displacement relations are linear but the stress-strain relations nonlinear (e.g., Zerna, 1960), the consistency of this assumption, at least in 3-dimensional elasticity, has been challenged by Bharatha & Levinson (1977).

Our aim in this book is to present a shell theory that, in its general form, is both geometrically and materially nonlinear. Thus, we intend to present equations that can be applied to biologic or rubber-like shells undergoing large deformations as well as to more traditional shells.

C. Approaches to Shell Theory

An *exact* 2-dimensional theory of shells does not exist: no matter how thin, a shell remains a 3-dimensional continuum. Indeed, the response of a shell to external loads—mechanical or thermal—depends critically on its thickness. Different approaches to shell theory may be classified according to how the reality of 3-dimensions is handled.

At one extreme lies the *direct* approach which, from the start, denies that a shell is 3-dimensional. This approach, pioneered by the Cosserat brothers (1908, 1909), proposed independently in a long-overlooked paper by Weatherburn (1925), revived independently by Ericksen & Truesdell (1958) and Günter (1958), and brought to perfection by Green, Naghdi, & Wainwright (1965) and Cohen & DeSilva (1966), starts by modeling a shell as a 2-dimensional "oriented" continuum, represented by a surface with one or more attached *directors* (vectors) and endowed with pointwise properties such as mass, bending stiffness, etc. Next, 2-dimensional laws of conservation of mass; balance of linear, rotational, and "director" momentum; and thermodynamics are *postulated*. Finally, the addition of a 2-dimensional free-energy density yields a complete thermoelastic theory of shells.

A limitation of the direct method is its apparent isolation from 3-dimensional continuum mechanics. To remedy this, Green, Laws, & Naghdi (1968) assume for the deformed position vector of a 3-dimensional, shell-like continuum an infinite series expansion in a thickness coordinate and identify the directors of the direct approach with the coefficients in this expansion. See the treatise by Naghdi (1972) for a complete exposition.

At the other end of the spectrum lies the *derived* approach to shell theory. Here, one starts from the 3-dimensional equations of thermoelasticity and attempts, by formal, asymptotic methods, to exploit the special geometry and loading that characterize those bodies we call shells. The end results are "interior" equations that describe the behavior of shells away from edges and other geometric or load discontinuities. Contributors to this approach (who confine themselves to isothermal, linear theory) include Goodier (1938), Goldenveiser (1945, 1962, 1963, 1966, 1969a, 1969b), Green (1962), Johnson & Reissner (1958), Reissner (1960, 1963, 1969), Reiss (1960, 1962), Johnson (1963), and Cicala (1966).

A drawback of the derived approach, aside from its tediousness, is that we must know at the edges of the shell the distribution of the applied stress or displacements over the thickness. As Koiter (1970) has emphasized, we never know the stress distribution precisely, except at a free edge. Another drawback of the derived approach is that, because the thickness of the shell is always incorporated in the expansion parameter, one set of uniformly valid interior (i.e., shell) equations does not emerge. Instead, there is one set of equations for a membrane state, another for an inextensional bending state, another for a "simple" edge-effect, another for a "degenerate" edge-effect, and, if one is dealing, say, with an infinite cylindrical shell subject to self-equilibrating edge loads, still another set of equations is needed to recover the "semi-membrane" theory of Vlasov (see Novozhilov, 1970, Section 49).

In contrast, the approach of John (1965a, 1965b, 1969, 1971) to shell theory is remarkable. His analysis is nonlinear, his error estimates are rigorous rather than formal, and in his 1969 and 1971 papers, he obtains uniformly valid interior shell equations (at the expense, however, of introducing cubic terms in the 3-dimensional strain-energy density).[1] John's work has been extended to shells of variable thickness under distributed surface loads by Berger (1973).

D. The Approach of this Book to Shell Theory

Our approach is a *mixed one*, based on the following dichotomy. The equations of 3-dimensional continuum mechanics fall into two groups: the generic equations, which are independent of the material properties of a body, namely conservation of mass, balance of linear and rotational momentum, heat flow, and the Clausius-Duhem inequality (the Second Law of Thermodynamics); and those which are not, namely Conservation of Energy (the First Law of Thermodynamics) and the constitutive relations. The second group of equations serve to join the mechanical and thermal variables which never appear together in the generic equations. The internal energy that appears in the Conservation of Energy and the constitutive relations must be determined experimentally; hence it is unavoidable that these relations be approximate.

To derive a nonlinear shell theory, we descend from the generic 3-dimensional equations, via a weighted integration through the thickness, to obtain generic 2-dimensional equations of balance of momentum, heat flow, and an entropy inequality. Conservation of mass is satisfied identically because we use a material (Lagrangian) formulation. On the other hand, because Conservation of Energy and the constitutive relations in 3-dimensions are not exact, we do not use them, but instead *postulate* analogous 2-dimensional relations. That is, *we throw all approximations in shell theory into those field equations which by their very nature are approximate.* The reason for calling our approach "mixed" is now clear: the exact parts of our shell equations follow from 3-

[1]Koiter & Simmonds (1973) have shown how to obtain John's results using only the standard quadratic strain-energy density.

dimensional continuum mechanics; the approximate parts are postulated, *ab initio*.

In more detail, our procedure begins with the 3-dimensional equations of balance of linear and rotational momentum and the Clausius-Duhem inequality for a body, written in *integral-impulse form* and referred to the reference shape of the body. [Thus, impulsive or concentrated loads and discontinuous unknowns are incorporated simply and naturally, and the artifice of delta "functions" is never needed. See Truesdell & Toupin (1960, p. 232, footnote #4) or Truesdell (1984a, p. 33-34, footnote #19 and p. 38) for a history and an elaboration of this point.] By specializing these equations to a shell-like body, we obtain analogous, exact 2-dimensional equations for shells, together with definitions of stress resultants, stress couples, a deformed position vector, a spin vector, and several thermal variables, all expressed as certain *weighted* integrals through the thickness of the 3-dimensional body in its reference shape. We emphasize that the descent from 3-dimensions to 2 involves *no series expansions in the thickness coordinate.*

Next, assuming sufficient smoothness of various fields in space and time, we obtain differential equations of motion and a differential entropy inequality. From the differential equations, we derive a Mechanical Work Identity.[2] In the process, definitions of 2-dimensional strains fall out automatically.

At this point, if we assume isothermal deformations, the machinery is in place to develop a complete *Mechanical Theory of Shells*. In particular, an *elastic* theory may be defined by assuming the existence of a strain-energy density that depends only on the strains delivered by the Mechanical Work Identity. We develop this theory extensively in Chapters III-V.

If the temperature is not constant, we must couple the mechanical variables with the 2-dimensional thermal variables; the latter include heat flux, the entropy resultant, and an entropy flux vector. This is done by postulating a 2-dimensional Law of Conservation of Energy. A thermoelastic shell is then defined as one in which the internal energy depends on the present values of the "state" variables (the strains plus some of the thermal variables) and their spatial gradients. By arguments in the spirit of Coleman & Noll (1963), we deduce from the shell version of the Clausius-Duhem inequality suitable forms of, and restrictions on, the constitutive relations. As one application of our results, we use the ideas of Duhem (1911),[3] Ericksen (1966a,b), Koiter (1969, 1971), and Gurtin (1973, 1975) to cast the problem of stability of an equilibrium configuration in a thermodynamic setting.

---

[2]Antman & Osborn (1979) have shown that the closely related Virtual Work Principle can, in 3-dimensions, be obtained *directly* from the integral form of balance of linear and rotational momentum. This is a most satisfying result—to obtain the Virtual Work Principle via differential equations requires more smoothness than is necessary. A similar observation was made by Carey & Dinh (1986).

[3]See Truesdell (1984b, pp. 38-44) for a summary of Duhem's long-neglected ideas on the stability of deformable, heat conducting bodies.

E. Outline of the Book

We are convinced that the best path to understanding general shell theory takes the reader upward through stations of increasing complexity, each station offering a view of interest in its own right. Thus, from our base in 3-dimensional continuum mechanics, we first derive the 1-dimensional longitudinal equation of motion for a rod of bisymmetric cross-section, or *birod* for short. Though a birod is *not* a shell, we think it ideal for revealing, in as simple a context as possible, many of the features of the construction of general shell theory. Furthermore, because the longitudinal motion of a birod is discussed in many books on strength of materials or applied mathematics—see Lin & Segel (1974) for a particularly good discussion—the analysis should build on the reader's experience.

The chapter on birods is followed by one on *beamshells*—infinite general cylindrical shells in which every particle moves in a fixed plane perpendicular to the generators of the cylindrical shell in its (fixed) reference shape. Mathematically, the governing equations are identical to those for the plane motion of a curved beam. The nonlinear theory of curved beams, of course, has a large literature of its own. See, for example, the review by Gorski (1976).

Next, we take up axisymmetric shells of revolution undergoing torsionless, axisymmetric deformation. We call these *axishells* for short. While there are many similarities with beamshells, here we first encounter the effects of double curvature. (It is non-zero Gaussian curvature that separates the shells from the plates—shells of non-zero mean curvature but zero Gaussian curvature, such as beamshells, have their own peculiar behaviors). There is a vast literature on axishells, and this chapter is the longest in the book.

Finally, we give a brief treatment of shells capable of exhibiting 1-dimensional static strain fields, i.e., general helicoidal shells or *unishells* (Simmonds, 1979, 1984), beginning with the important special case of bent, curved tubes.

The great complexity of the equations of general shell theory has spurred major efforts at simplification. There are three major lines. First, the general equations can be restricted to shells of special geometry subject to special loads and boundary/initial conditions. The theories treated in this book are of this type. Second, there are approximate theories of *general shells.* These theories include quasi-shallow theory, the independent extensions by Libai (1962) and Koiter (1966) of conventional shallow shell theory due to Donnell (1933), Marguerre (1938), Mushtari (1938), and Vlasov (1944); membrane theory; and the theory of (membrane-)inextensional bending. We shall touch on the form of these theories for beamshells and axishells; we intend a general treatment in a future volume.

The third line of simplification is to discretize the general shell equations, thereby reducing their solution to a nonlinear algebraic problem. A fruitful approach is via finite elements, proposed by Courant (1943) before advanced computers made them feasible, but first popularized by Turner, Clough, Martin & Topp (1956). Though the finite element method is powerful, there are pitfalls for the naive user of the large computer codes embodying these methods. The literature on finite elements methods for shells is large, but we make no attempt to survey it.

In the chapters on beamshells and axishells we devote considerable attention to, and include new results on, load potentials, constitutive relations (especially for rubber-like shells), variational principles, nonlinear membrane theory, and thermodynamics. We also discuss, but not exhaustively, stability with particular emphasis on the mechanical (isothermal) theory.

We close this Introduction by noting that the ascending approach to shell theory allows us to elaborate on a concept just once in its simplest context. Thus, shock relations, mechanical boundary conditions, the Virtual Work Principle, variational formulations, and some basic techniques from thermodynamics are motivated and developed in the chapter on birods, while an extended discussion on load potentials, nonlinear stress-strain relations, and thermodynamic stability is to be found in the chapter on beamshells. Of course, there are certain topics such as the finite rotation vector, the Kirchhoff boundary conditions and their modification, and the full implication of the intrinsic form of the equations of motion that can only be adequately addressed within the theory of general shells. It is our plan to do so in a future volume.

References

[A letter in brackets indicates the section where the reference appears.]

Antman, S. S., and Osborn, J. E. (1979). The principle of virtual work and integral laws of motion. *Arch. Rat. Mech. Anal.* **69**, 232-261. [D]

Berger, N. (1973). Estimates for stress derivatives and error in interior equations for shells of variable thickness with applied forces. *SIAM J. Appl. Math.* **24**, 97-120. [C]

Bharatha, S., and Levinson, M. (1977). On physically nonlinear elasticity. *J. Elasticity* **7**, 307-324. [B]

Carey, G. F., and Dinh, H. T. (1986). Conservation principles and variational problems. *Acta Mech.* **58**, 93-97. [D]

Cicala, P. (1966). "Systematic Approximation Approach to Linear Shell Theory." Levrotto and Bella, Torino. [C]

Cohen, H., and DeSilva, C. N. (1966). Nonlinear theory of elastic surfaces. *J. Math. Phys.* **7**, 246-253. [C]

Coleman, B. D., and Noll, W. (1963). The thermodynamics of elastic materials with heat conduction and viscosity. *Arch. Rat. Mech. Anal.* **51**, 1-53. [D]

Cosserat, E., and Cosserat, F. (1908). Sur la théorie des corps minces. *Compt. Rend.* **146**, 169-172. [C]

Cosserat, E., and Cosserat, F. (1909). "Théoric des Corps Déformables." Herrmann, Paris. [C]

Courant, R. (1943). Variational methods for the solution of problems of equilibrium and vibration. *Bull. Am. Math. Soc.* **49**, 1-23. [E]

Donnell, L. H. (1933). Stability of thin-walled tubes under torsion. *NACA Rept.* 479. [E]
Duhem, P. (1911). "Traité d'Energétique ou de Thermodynamique Générale," 2 vols. Gauthier-Villars, Paris [D]
Ericksen, J. L. (1966a). A thermo-kinetic view of elastic stability theory. *Int. J. Solids & Struct.* **2**, 573-580. [D]
Ericksen, J. L. (1966b). Thermoelastic stability. *In* "Proc. 5th U.S. Nat. Cong. Appl. Mech.," pp. 187-193. ASME, New York. [D]
Ericksen, J. L., and Truesdell, C. (1958). Exact theory of stress and strain in rods and shells. *Arch. Rat. Mech. Anal.* **1**, 295-323. [C]
Goldenveiser, A.L. (1945). A qualitative investigation of the state of stress in a thin shell (in Russian). *P.M.M.* **9**, 463-478. [C]
Goldenveiser, A. L. (1962). Derivation of an approximate theory of bending of a plate by the methods of asymptotic integration of the equations of the theory of elasticity. *P.M.M.* **26**, 1000-1025. [C]
Goldenveiser, A. L. (1963). Derivation of an approximate theory of shells by means of asymptotic integration of the equations of the theory of elasticity. *P.M.M.* **27**, 903-924. [C]
Goldenveiser, A. L. (1966). The principles of reducing three-dimensional problems of elasticity to two-dimensional problems of the theory of plates and shells. *In* "Proc. 11th Int. Cong., Munich, 1964" (H. Gortler, ed.), pp. 306-311. Springer-Verlag, Berlin and New York. [C]
Goldenveiser, A. L. (1969a). On two-dimensional equations of the general linear theory of thin elastic shells. *In* "Problems of Hydrodynamics and Continuum Mechanics" (The Sedov Anniversary Volume), pp. 334-351. SIAM, Philadelphia. [C]
Goldenveiser, A. L. (1969b). Problems in the rigorous deduction of the theory of thin elastic shells. *In* "Theory of Thin Shells," Proc. I.U.T.A.M. Sympos., Copenhagen, 1967 (F.I. Niordson, ed.), pp. 31-38. Springer-Verlag, Berlin, etc. [C]
Goodier, J. N. (1938). On the problem of a beam and a plate in the theory of elasticity. *Trans. Roy. Soc. Canada.* **32**, 65-88. [C]
Gorski, W. (1976). A review of the literature and a bibliography of finite elastic deflections of bars. *Civil Engr. Trans. I. E. Australia* **CE 18**, 74-85. [E]
Green, A. E. (1962). On the linear theory of thin elastic shells. *Proc. Roy. Soc. London* **A226**, 143-160. [C]
Green, A. E., Laws, N., and Naghdi, P. M. (1968). Rods, plates, and shells. *Proc. Camb. Phil. Soc.* **64**, 895-913. [C]
Green, A. E., Naghdi, P.M., and Wainwright, W. L. (1965). A general theory of a Cosserat surface. *Arch. Rat. Mech. Anal.* **20**, 287-308. [C]
Günter, W. (1958). Zur Statik und Kinematik des Cosseratchen Kontinuums. *Abh. Braunschweig. Wiss. Ges,* **10**, 195-213. [C]
Gurtin, M. E. (1973). Thermodynamics and the energy criterion for stability. *Arch. Rat. Mech. Anal.* **52**, 93-103. [D]
Gurtin, M. E. (1975). Thermodynamics and stability. *Arch. Rat. Mech. Anal.* **59**, 63-96. [D]

John, F. (1965a). A priori estimates applied to nonlinear shell theory. *Proc. Sympos. Appl. Math.* **17,** pp. 102-110. Am. Math Soc. [C]
John, F. (1965b). Estimates for the derivatives of the stresses in a thin shell and interior shell equations. *Comm. Pure Appl. Math.* **18,** 235-267. [C]
John, F. (1969). Refined interior shell equations. *In* "Theory of Thin Shells," Proc. of the I.U.T.A.M. Symposium, Copenhagen, 1967 (F.I. Niordson, ed.), pp. 1-14. Springer-Verlag, Berlin, etc. [C]
John, F. (1971). Refined interior equations for thin elastic shells. *Comm. Pure Appl. Math.* **24,** 583-615. [C]
Johnson, M. W. (1963). A boundary layer theory for unsymmetric deformation of circular cylindrical elastic shells. *J. Math. & Phys.* **42,** 167-188. [C]
Johnson, M. W., and Reissner, E. (1958). On the foundations of the theory of thin elastic shells. *J. Math & Phys.* **37,** 375-392. [C]
Koiter, W. T. (1966). On the nonlinear theory of thin elastic shells. *Proc. Kon. Ned. Ak. Wet.* **B69,** 1-54. [E]
Koiter, W. T. (1969). On the thermodynamic background of elastic stability theory. *In* "Problems of Hydrodynamics and Continuum Mechanics" (The Sedov Anniversary Volume), pp. 423-433. SIAM, Philadelphia. [D]
Koiter, W. T. (1970). On the mathematical foundation of shell theory. *Actes. Congr. Int. Math.* **3,** 123-130. [C]
Koiter, W. T. (1971). Thermodynamics of elastic stability. *In* "Proc. 3rd Canadian Cong. Appl. Mech., Calgary," pp. 29-37. [D]
Koiter, W. T., and Simmonds, J. G. (1973). Foundations of shell theory. *In* "Proc. 13th Int. Cong. Theor. Appl. Mech., Moscow, 1972" (E. Becker & G. K. Mikhailov, eds.), pp. 150-176. Springer-Verlag, Berlin and New York. [C]
Libai, A. (1962). On the nonlinear elastokinetics of shells and beams. *J. Aerospace Sci.* **29,** 1190-1195, 1209. [E]
Lin, C. C., and Segel, L. A. (1974). "Mathematics Applied to Deterministic Problems in the Natural Sciences." Macmillan, New York. [E]
Marguerre, K. (1938). Zur Theorie der gekrümmten Platte grosser Formänderung. *In* "Proc. 5th Int. Cong. Appl. Mech.," pp. 93-101. Wiley, New York. [E]
Mushtari, K. M. (1938). Certain generalizations of the theory of thin shells (in Russian). *Izv. Fiz. Mat. Ob-va pri Kaz. Un-te.* **11,** 71-150. [E]
Naghdi, P. M. (1972). The theory of plates and shells. *In* "Encyclopedia of Physics", 2nd ed. (S. Flügge, ed.), vol. VIa/2. Springer. [C]
Novozhilov, V. V. (1970). "Thin Shell Theory," 2nd ed. Wolters-Noordhoff, Groningen. [C]
Reiss, E. L. (1960). A theory for the small rotationally symmetric deformations of cylindrical shells. *Comm. Pure Appl. Math.* **13,** 531-550. [C]
Reiss, E. L. (1962). On the theory of cylindrical shells. *Q. J. Mech. Appl. Math.* **15,** 325-338. [C]
Reissner, E. (1960). On some problems in shell theory. *In* "Proc. 1st Sympos. Naval Structural Mech.," pp. 74-114. Pergamon, Oxford, etc. [C]
Reissner, E. (1963). On the derivation of the theory of thin elastic shells. *J.*

Math. & Phys. **42,** 263-277. [C]

Reissner, E. (1969). On the foundations of generalized linear shell theory. *In* "Theory of Thin Shells," Proc. I.U.T.A.M. Sympos., Copenhagen, 1967 (F. I. Niordson, ed.), pp. 15-30. Springer-Verlag, Berlin, etc. [C]

Simmonds, J. G. (1979). Surfaces with metric and curvature tensors that depend on one coordinate only are general helicoids. *Q. Appl. Math.* **37,** 82-85. [E]

Simmonds, J. G. (1984). General helicoidal shells undergoing large, one-dimensional strains or large inextensional deformations. *Int. J. Solids & Struct.* **20,** 13-30. [E]

Truesdell, C. A. (1984a). "An Idiot's Fugitive Essays on Science." Springer-Verlag, New York, etc. [D]

Truesdell, C. A. (1984b). "Rational Thermodynamics," 2nd ed. Springer-Verlag, New York, etc. [D]

Truesdell, C. A., and Toupin, R. A. (1960). The classical field theories. *In* "Encyclopedia of Physics" (S. Flügge, ed.), vol. III/1. Springer-Verlag, Berlin and New York. [D]

Turner, M. J., Clough, R. W., Martin, H. C., and Topp, L. J. (1956). Stiffness and deflection analysis of complex structures. *J. Aerospace Sci.* **23,** 805-823,854. [E]

Vlasov, V. V. (1944). The fundamental differential equations of the general theory of elastic shells. *NACA* T.M. 1241. [E]

Weatherburn, C. E. (1925). On small deformation of surfaces and of thin elastic shells. *Quart. J. Math.* **50,** 272-296. [C]

Zerna, W. (1960). Über eine nichtlineare allgemeine Theorie der Schalen. *In* "The Theory of Thin Elastic Shells," Proc. I.U.T.A.M. Sympos., Delft, 1959 (W. T. Koiter, ed.), pp. 34-42. North-Holland Publ., Amsterdam. [B]

Chapter II

The Generic Equations of 3-Dimensional Continuum Mechanics

Each of Chapters III-V, which are devoted, respectively, to birods, beam-shells, and axishells, begins with a derivation of the equations of motion by a descent from the equations of balance of linear and rotational momentum of a 3-dimensional material continuum. These equations are written in integral-impulse form, which allows us, in a simple way, to account for discontinuities in internal and external variables. Later, in each of these chapters, appropriate forms of the Second Law of Thermodynamics are derived by descent from the 3-dimensional, integral form of the Clausius-Duhem inequality.

The present chapter summarizes those laws of 3-dimensional continuum mechanics relevant to our approach to shell theory. For a detailed discussion of these laws, we refer the reader to books and articles by Truesdell & Toupin (1960), Truesdell & Noll (1965), Chadwick (1976), Truesdell (1977), Spencer (1980), and Gurtin (1981), among others.

A. The Integral Equations of Motion

A *body* $\mathcal{B}$ is represented by a set of points called *particles* that move through 3-dimensional Euclidean space. This set of points at time t, $\mathcal{S}_t$, is called the *shape* (or *image*) of $\mathcal{B}$ at t; $\mathcal{S}_0 \equiv \mathcal{S}$ is called the *reference shape* and is assumed to be a connected region.

In classical mechanics, a particular particle p_k in a set of n may be identified (i.e., indexed) by a positive integer $k = 1, 2, \cdots, n$. In continuum mechanics a particle in $\mathcal{B}$ may be identified by its *position* $\mathbf{x}$ in $\mathcal{S}$ relative to a given Cartesian frame $Oxyz$ represented by the standard orthonormal triad of basis vectors $\{\mathbf{e}_x, \mathbf{e}_y, \mathbf{e}_z\}$. Thus,

$$\mathbf{x} \equiv x\mathbf{e}_x + y\mathbf{e}_y + z\mathbf{e}_z . \tag{A.1}$$

If $\bar{\mathbf{x}}$ denotes the position (or *deformed image*) of the particle at time t, then its *motion* is a relation of the form

$$\bar{\mathbf{x}} = \bar{\mathbf{x}}(\mathbf{x}, t) \ , \ -\infty < t < \infty \ , \tag{A.2}$$

where $\bar{\mathbf{x}}(\mathbf{x}, 0) = \mathbf{x}$. Insofar as practical, we shall denote deformed images by an overbar. In writing (A.2), we have assumed that no particle fragments. Further, we shall assume that no two particles coalesce, i.e., we shall assume that $\bar{\mathbf{x}}(\mathbf{x}, t)$ is spatially 1:1. This implies that (A.2) has an inverse, though we shall not make explicit use of this fact in this book. Finally, we shall assume, except along boundaries or wavefronts where certain derivatives may be discontinuous or fail to exist, that $\bar{\mathbf{x}}(\mathbf{x}, t)$ is sufficiently smooth for all of the differentiations that follow to make sense. In particular, the velocity and acceleration of a particle are denoted and defined by

$$\dot{\bar{\mathbf{x}}} \equiv \partial\bar{\mathbf{x}}(\mathbf{x}, t)/\partial t \ , \ \ddot{\bar{\mathbf{x}}} \equiv \partial^2\bar{\mathbf{x}}(\mathbf{x}, t)/\partial t^2 \ . \tag{A.3}$$

The equations of motion state that in an *inertial frame*, henceforth identified with *Oxyz*, the change in the linear {rotational} momentum of a *fixed* set of particles over any time interval (t_1, t_2) is equal to the net linear {rotational} impulse acting on these particles. To be precise, let $\rho(\mathbf{x})$ denote the mass density of $\mathcal{B}$ in $\mathcal{S}$ and, for simplicity, assume that the rotational impulses arise from body and contact forces only. Further, let V denote any subregion of $\mathcal{S}$ having a piecewise smooth boundary ∂V with outward unit normal $\mathbf{n}$, and let dV and dA denote differential elements of volume and area. See Fig. 2.1.

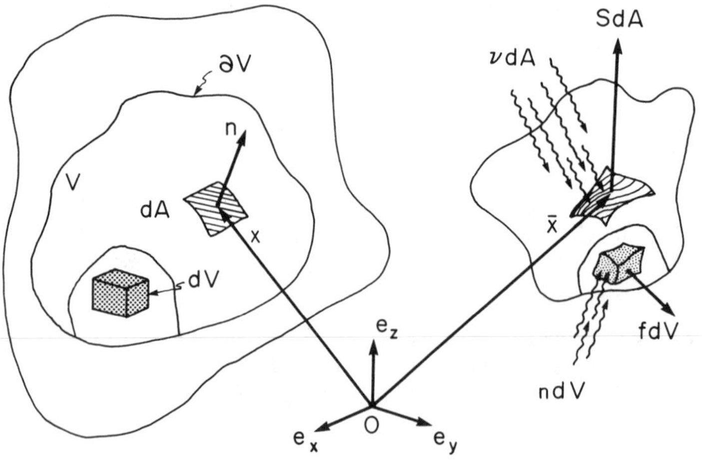

Fig. 2.1. The geometry of 3-dimensional continuum mechanics.

If, at time t, $\mathbf{f}dV$ denotes the body force on the particles initially in dV and $\mathbf{S}dA$ denotes the contact force on the particles initially in dA exerted by the particles initially on the outward side of dA, then the equations of motion take the form

$$\int_{t_1}^{t_2}\Big(\int_{\partial V}\mathbf{S}dA + \int_V \mathbf{f}dV\Big)dt = \int_V \rho\dot{\bar{\mathbf{x}}}dV\Big|_{t_1}^{t_2} \tag{A.4}$$

$$\int_{t_1}^{t_2}\Big(\int_{\partial V}\bar{\mathbf{x}}\times\mathbf{S}dA + \int_V \bar{\mathbf{x}}\times\mathbf{f}dV\Big)dt = \int_V \bar{\mathbf{x}}\times\rho\dot{\bar{\mathbf{x}}}dV\Big|_{t_1}^{t_2}. \tag{A.5}$$

B. Stress Vectors

Consider an arbitrary point P in $\mathcal{S}$ with position $\mathbf{x}$. As in Fig. 2.2, take V in (A.4) to be a tetrahedron T_δ within $\mathcal{S}$ having 3 faces parallel to the 3 Cartesian coordinate planes. Denote the area of the yz-face, which has an outward unit normal $-\mathbf{e}_x$, by $A_\delta(-\mathbf{e}_x)$, etc. The 4th face has an outward unit normal $\mathbf{n}$, area $A_\delta(\mathbf{n})$, and lies a distance δ from P. The volume of T_δ is equal to $(1/3)\delta A_\delta(\mathbf{n})$. If we set

$$\mathbf{n} = n_x\mathbf{e}_x + n_y\mathbf{e}_y + n_z\mathbf{e}_z\,, \tag{B.1}$$

then $A_\delta(-\mathbf{e}_x) = A_\delta(\mathbf{n})n_x$, etc.

A useful device for establishing the equality of two vectors is to note that $\mathbf{u}=\mathbf{v}$ if and only if $\mathbf{u}\cdot\mathbf{c}=\mathbf{v}\cdot\mathbf{c}$ for all vectors $\mathbf{c}$. Thus, applying (A.4) to T_δ, taking the dot product of both sides with an arbitrary constant vector $\mathbf{c}$, assuming that $\mathbf{S}(\mathbf{x}, t, \mathbf{n})$ is a continuous function of $\mathbf{x}$ and t, and applying the first mean-value theorem for integrals, we obtain an expression that we can rearrange in the form

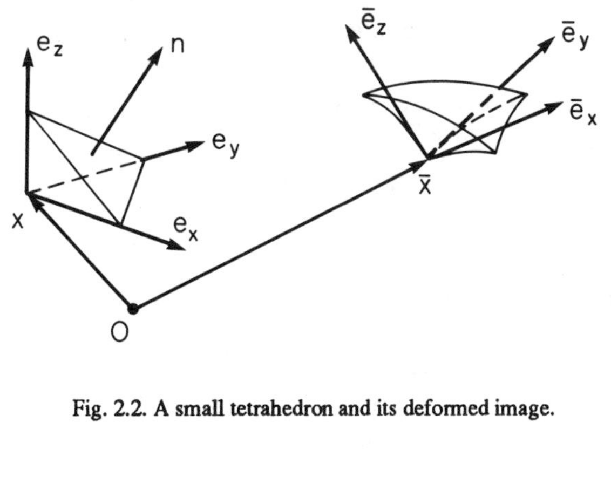

Fig. 2.2. A small tetrahedron and its deformed image.

$$\mathbf{c}\cdot\int_{t_1}^{t_2}[\mathbf{S}(\mathbf{x}_x,t,-\mathbf{e}_x)n_x+\mathbf{S}(\mathbf{x}_y,t,-\mathbf{e}_y)n_y+\mathbf{S}(\mathbf{x}_z,t,-\mathbf{e}_z)n_z$$

$$+\mathbf{S}(\mathbf{x}_n,t,\mathbf{n})]dtA_\delta(\mathbf{n})=\mathbf{c}\cdot(\int_{T_\delta}\rho\dot{\bar{\mathbf{x}}}dV\,\Big|_{t_1}^{t_2}-\int_{t_1}^{t_2}\int_{T_\delta}\mathbf{f}dVdt)\,, \tag{B.2}$$

where $\mathbf{x}_x$ is a vector from O to some unknown point on the yz-face of T_δ, etc., and $\mathbf{x}_n$ is a vector from O to some unknown point on the fourth face of T_δ. Let us assume further that $|\rho\dot{\bar{\mathbf{x}}}|\le K_1(t)$ and $|\mathbf{f}|\le K_2(t)$, where K_1 is bounded and K_2 is integrable. Dividing both sides of (B.2) by $A_\delta(\mathbf{n})$, taking the limit as $\delta\to 0$, and noting that the interval (t_1,t_2) and the vector $\mathbf{c}$ are arbitrary, we have

$$\mathbf{S}(\mathbf{x},t,-\mathbf{e}_x)n_x+\mathbf{S}(\mathbf{x},t,-\mathbf{e}_y)n_y+\mathbf{S}(\mathbf{x},t,-\mathbf{e}_z)n_z+\mathbf{S}(\mathbf{x},t,\mathbf{n})=\mathbf{0}\,. \tag{B.3}$$

Assuming that $\mathbf{S}$ is also a continuous function of $\mathbf{n}$, we find upon letting $\mathbf{n}\to\mathbf{e}_x$ in (B.3) that

$$\mathbf{S}(\mathbf{x},t,\mathbf{e}_x)=-\mathbf{S}(\mathbf{x},t,-\mathbf{e}_x)\equiv\mathbf{S}_x(\mathbf{x},t)\,. \tag{B.4}$$

We call $\mathbf{S}_x$ the *x-stress vector*. Returning to (B.3) and letting $\mathbf{n}$ approach $\mathbf{e}_y$ or $\mathbf{e}_z$ and defining stress vectors $\mathbf{S}_y$ and $\mathbf{S}_z$ analogous to $\mathbf{S}_x$, we can rewrite (B.3) in the form

$$\mathbf{S}=\mathbf{S}_x n_x+\mathbf{S}_y n_y+\mathbf{S}_z n_z\,. \tag{B.5}$$

If $\mathbf{x}$ is given as a function of a set of curvilinear coordinates σ^i, $i=1,2,3$, the associated *covariant base vectors* are denoted and defined by

$$\mathbf{g}_i\equiv\partial\mathbf{x}/\partial\sigma^i\,. \tag{B.6}$$

More explicitly, if $x=x(\sigma^1,\sigma^2,\sigma^3)$, etc.,

$$\mathbf{g}_1=\frac{\partial x}{\partial\sigma^1}\mathbf{e}_x+\frac{\partial y}{\partial\sigma^1}\mathbf{e}_y+\frac{\partial z}{\partial\sigma^1}\mathbf{e}_z\,,\ \text{etc.} \tag{B.7}$$

In the literature, $(\sigma^1,\sigma^2,\sigma^3)$ are referred to as Lagrangian, material, convected, or referential coordinates.

If the curvilinear coordinates are taken to be the Cartesian coordinates (x,y,z) themselves, i.e., if $\sigma^1=x$, $\sigma^2=y$, $\sigma^3=z$, then $\mathbf{g}_1=\mathbf{e}_x$, *etc.*. It follows immediately that the equation that corresponds to (B.5) in curvilinear coordinates is

$$\mathbf{S}=\mathbf{S}^i n_i\,, \tag{B.8}$$

where we sum over a repeated index from 1 to 3 and $n_i=\mathbf{n}\cdot\mathbf{g}_i$. We call the $\mathbf{S}^i$ the *σ^i–stress vectors* or, as explained in Section E, the *contravariant vector components* of the first Piola-Kirchhoff stress tensor **P**.

If the integrands in (A.4) and (A.5) are sufficiently smooth, it follows that we may replace $\mathbf{S}$ by $\mathbf{S}^i n_i$ and apply the divergence theorem. Further, if the resulting integrands are continuous for all volumes V within some fixed volume and all intervals (t_1,t_2) within some fixed interval, then the following *local* equations of balance of linear and rotational momentum must hold:

$$\frac{1}{\sqrt{g}}\frac{\partial(\sqrt{g}\,\mathbf{S}^i)}{\partial\sigma^i}+\mathbf{f}=\rho\ddot{\mathbf{x}}\ ,\ \bar{\mathbf{g}}_i\times\mathbf{S}^i=\mathbf{0}\ , \tag{B.9}$$

where $g \equiv \det(\mathbf{g}_i \cdot \mathbf{g}_j)$.

C. Heat

Consider, again, the same arbitrary subregion V of $\mathcal{S}$ and let νdA be the heat *influx* across the image of dA and ndV the rate of heat production[1] by the particles that initially composed dV. The rate of heating of the material initially within V is denoted and defined by

$$\mathcal{Q}\equiv\int_{\partial V}\nu dA+\int_V ndV\ . \tag{C.1}$$

By arguments similar to those that established (B.5), it may be shown that

$$\nu=\mathbf{c}\cdot\mathbf{n}\ , \tag{C.2}$$

where $\mathbf{c}$ is the *cold flux* vector.[2]

D. The Clausius-Duhem (-Truesdell-Toupin) Inequality

The generally accepted form of the Second Law of Thermodynamics is

$$\mathcal{H}\Big|_{t_1}^{t_2}\geq\int_{t_1}^{t_2}\mathcal{J}dt\ , \tag{D.1}$$

where

$$\mathcal{H}\equiv\int_V\eta\, dV\Big|_{t_1}^{t_2}\ \text{and}\ \mathcal{J}\equiv\int_{\partial V}\frac{\nu}{\theta}dA+\int_V\frac{n}{\theta}dV\ . \qquad \text{(D.2)}^3$$

Here, ηdV is the *entropy* of the particles that initially composed dV and $\theta > 0$ is the *absolute temperature*.

The reader needs no further results to proceed to any of Chapters III-V where 1- or 2-dimensional equations of motion and thermodynamics are developed by descent from (A.4), (A.5), (C.1), and (D.1).

---

[1]We have used the symbol n rather than the more conventional symbol r to avoid confusion with the radial coordinate r used in Chapter V on axishells.

[2]We work with $\mathbf{c}$ instead of the *heat flux* vector $\mathbf{q}$ to avoid a plethora of minus signs later on.

[3] Truesdell & Toupin (1960, Section 255) added the volume integral on the right side. See also Truesdell (1984, p. 42).

E. The First Piola-Kirchhoff Stress Tensor

Equations (B.5) and (B.8) show that $\mathbf{S}(\mathbf{x}, t, \mathbf{n})$ is a linear, homogeneous function of the components of $\mathbf{n}$. The right sides of these equations may therefore be viewed as representing the action of a *linear operator* P, depending on $\mathbf{x}$ and t, that sends $\mathbf{n}$ into $\mathbf{S}$. Such an operator is called a *2nd order tensor*. In particular, P is called the *first Piola-Kirchhoff stress tensor* (Truesdell & Toupin, 1965) and (B.5) or (B.8) can be written symbolically as

$$\mathbf{S} = \mathsf{P}\cdot\mathbf{n} \,. \tag{E.1}$$

The direct product of two vectors $\mathbf{u}$ and $\mathbf{v}$, denoted by $\mathbf{uv}$, is a special type of 2nd order tensor called a *dyad.* By definition, if $\mathbf{w}$ is a third vector, then $\mathbf{w}\cdot\mathbf{uv} \equiv (\mathbf{w}\cdot\mathbf{u})\mathbf{v}$ and $\mathbf{uv}\cdot\mathbf{w} \equiv \mathbf{u}(\mathbf{v}\cdot\mathbf{w})$. Using dyadic notation and the fact that $(\mathbf{t}+\mathbf{u})\cdot\mathbf{vw} = (\mathbf{t}\cdot\mathbf{v}+\mathbf{u}\cdot\mathbf{v})\mathbf{w}$, we can represent P in Cartesian coordinates as

$$\mathsf{P} = \mathbf{S}_x\mathbf{e}_x + \mathbf{S}_y\mathbf{e}_y + \mathbf{S}_z\mathbf{e}_z \tag{E.2}$$

or, in general coordinates, as

$$\mathsf{P} = \mathbf{S}^i\mathbf{g}_i \,. \tag{E.3}$$

In view of (E.3), the $\mathbf{S}^i$ may be called the contravariant *vector* components of P in the basis $\{\mathbf{g}_i\}$.

Since $\mathbf{S}dA$ is the force acting on the *image* of dA in the *deformed* body, it is natural to resolve $\mathbf{S}^i$ into components along the *deformed covariant base vectors* $\overline{\mathbf{g}}_i$. Thus, with

$$\mathbf{S}^i = S^{ij}\overline{\mathbf{g}}_j \,, \tag{E.4}$$

(B.8) and (E.3) take the form

$$\mathbf{S} = S^{ij}n_i\overline{\mathbf{g}}_j \tag{E.5}$$

$$\mathsf{P} = S^{ij}\overline{\mathbf{g}}_j\mathbf{g}_i \,. \tag{E.6}$$

Even in Cartesian coordinates, the deformed images $\{\overline{\mathbf{e}}_x, \overline{\mathbf{e}}_y, \overline{\mathbf{e}}_z\}$ of the base vectors $\{\mathbf{e}_x, \mathbf{e}_y, \mathbf{e}_z\}$, as suggested in Fig. 2.2, are, in general, neither orthogonal nor of unit length.

Although, in general, $\mathsf{P} \neq \mathsf{P}^T$, we see, by inserting (E.4) into $(B.9)_2$ that $S^{ij} = S^{ji}$. The symmetry of the components S^{ij} is a consequence of representing P in the *mixed* dyadic basis $\{\overline{\mathbf{g}}_i\mathbf{g}_j\}$.

*F. Gross Equations of Motion

The purpose of this section is to show that the equations of balance of linear and rotational momentum, when applied to all of $\mathcal{B}$, can be cast into a form that is identical to those for a *rigid* body. These results, of interest in their own right, are particularly useful in suggesting the proper way to define the *spin* in the shell theories of Chapters IV-VI and the boundary conditions for rods and shells in Chapters III-VI.

Let

$$M \equiv \int \rho dV \tag{F.1}$$

denote the mass of $\mathcal{B}$, where the integral is taken over the shape $\mathcal{S}$ of $\mathcal{B}$ at $t=0$. The position $\overline{\mathbf{X}}$ of the center of gravity of $\mathcal{B}$ at any time t is defined by

$$M\overline{\mathbf{X}} \equiv \int \rho \overline{\mathbf{x}} dV \,. \tag{F.2}$$

With

$$\mathbf{F} \equiv \int_{\partial \mathcal{S}} \mathbf{S} dA + \int \mathbf{f} dV \tag{F.3}$$

denoting the net external force acting on $\mathcal{B}$, (A.4) applied to $\mathcal{S}$ takes the form

$$\int_{t_1}^{t_2} \mathbf{F} dt = \mathbf{P} \,\Big|_{t_1}^{t_2} , \tag{F.4}$$

where

$$\mathbf{P} \equiv M\dot{\overline{\mathbf{X}}} \tag{F.5}$$

is the *linear momentum of* $\mathcal{B}$. Thus, as with a point-mass, the net impulse on $\mathcal{B}$ in the interval (t_1, t_2) is equal to the change in its linear momentum.

To obtain an analogous equation for the gross balance of rotational momentum, we set

$$\overline{\mathbf{x}} = \overline{\mathbf{X}} + \overline{\mathbf{z}} \tag{F.6}$$

and note by (F.2) that

$$\int \rho \overline{\mathbf{z}} dV = \mathbf{0} \,. \tag{F.7}$$

Applying the right side of (A.5) to $\mathcal{S}$, inserting (F.6), and noting (F.7), we get

$$\int \overline{\mathbf{x}} \times \rho \dot{\overline{\mathbf{x}}} dV = \overline{\mathbf{X}} \times \mathbf{P} + \mathbf{R} \,, \tag{F.8}$$

where

$$\mathbf{R} \equiv \int \rho \overline{\mathbf{z}} \times \dot{\overline{\mathbf{z}}} dV \tag{F.9}$$

is the *rotational momentum of* $\mathcal{B}$ *with respect to its center of gravity.*

Turning to the left side of (A.5), we replace $\overline{\mathbf{x}}$ by $\overline{\mathbf{X}}+\overline{\mathbf{z}}$ and make use of (F.3) and (F.8) to conclude that

$$\int_{t_1}^{t_2}(\overline{\mathbf{X}}\times\mathbf{F}+\mathbf{T})dt=(\overline{\mathbf{X}}\times\mathbf{P}+\mathbf{R})_{t_1}^{t_2}, \tag{F.10}$$

where

$$\mathbf{T}\equiv\int\overline{\mathbf{z}}\times\mathbf{S}dA+\int\overline{\mathbf{z}}\times\mathbf{f}dV \tag{F.11}$$

is the *net external torque with respect to the center of gravity of $\mathcal{B}$* and the first integral is taken over the boundary of $\mathcal{S}$.

In analogy with the kinematics of a rigid body, we define (implicitly) a *gross spin* Ω by setting

$$\mathbf{R}\equiv\int_{\mathcal{S}}\rho\overline{\mathbf{z}}\times(\Omega\times\overline{\mathbf{z}})dV\equiv\overline{\mathbf{I}}\cdot\Omega, \tag{F.12}$$

where

$$\overline{\mathbf{I}}\equiv\int\rho[(\overline{\mathbf{z}}\cdot\overline{\mathbf{z}})\mathbf{1}-\overline{\mathbf{z}}\,\overline{\mathbf{z}}]dV \tag{F.13}$$

is the *moment of inertia tensor of $\mathcal{B}$ at time t with respect to its center of gravity.* In (F.13), $\mathbf{1}$ is the identity tensor ($\mathbf{1}\cdot\mathbf{v}=\mathbf{v}$ for all vectors $\mathbf{v}$) and $\overline{\mathbf{z}}\,\overline{\mathbf{z}}$ is a dyadic tensor [$\overline{\mathbf{z}}\,\overline{\mathbf{z}}\cdot\mathbf{v}\equiv\overline{\mathbf{z}}(\overline{\mathbf{z}}\cdot\mathbf{v})$ for all vectors $\mathbf{v}$].

Problem: Express

$$K\equiv\tfrac{1}{2}\int_{\mathcal{S}}\rho\dot{\overline{\mathbf{x}}}\cdot\dot{\overline{\mathbf{x}}}dV, \tag{F.14}$$

the kinetic energy of $\mathcal{B}$, in terms of M, $\overline{\mathbf{X}}$, $\overline{\mathbf{I}}$, and Ω.

Solution (by Mrs. Fisher): Let $\{\mathbf{E}_x,\mathbf{E}_y,\mathbf{E}_z\}$ be a frame that rotates relative to the inertial frame $\{\mathbf{e}_x,\mathbf{e}_y,\mathbf{e}_z\}$ with spin Ω. Then a basic kinematic result is that

$$\dot{\overline{\mathbf{z}}}=\overline{\mathbf{z}}^{*}+\Omega\times\overline{\mathbf{z}}, \tag{F.15}$$

where * denotes the time rate of change relative to $\{\mathbf{E}_x,\mathbf{E}_y,\mathbf{E}_z\}$. By (F.6), (F.7), and (F.15),

$$K=\tfrac{1}{2}\int_{\mathcal{S}}\rho\dot{\overline{\mathbf{X}}}\cdot\dot{\overline{\mathbf{X}}}dV+\tfrac{1}{2}\int_{\mathcal{S}}\rho\overline{\mathbf{z}}^{*}\cdot\overline{\mathbf{z}}^{*}dV+\Omega\cdot\int_{\mathcal{S}}\rho\overline{\mathbf{z}}\times\overline{\mathbf{z}}^{*}dV+\tfrac{1}{2}\int_{\mathcal{S}}\rho(\Omega\times\overline{\mathbf{z}})\cdot(\Omega\times\overline{\mathbf{z}})dV. \tag{F.16}$$

From (F.9) and (F.15),

$$\mathbf{R}=\int_{\mathcal{S}}\rho\overline{\mathbf{z}}\times\overline{\mathbf{z}}^{*}dV+\int_{\mathcal{S}}\rho\overline{\mathbf{z}}\times(\Omega\times\overline{\mathbf{z}})dV \tag{F.17}$$

which, by (F.12), implies that

$$\int_{\mathcal{S}}\rho\overline{\mathbf{z}}\times\overline{\mathbf{z}}^{*}dV=\mathbf{0}. \tag{F.18}$$

Again, by (F.12), the last term on the right of (F.16) is equal to $\tfrac{1}{2}\Omega\cdot\overline{\mathbf{I}}\cdot\Omega$.

Hence, by (F.1),

$$K = \tfrac{1}{2}M\dot{\overline{\mathbf{X}}}\cdot\dot{\overline{\mathbf{X}}} + \tfrac{1}{2}\Omega\cdot\overline{\mathsf{I}}\cdot\Omega + K_E\,, \tag{F.19}$$

where

$$K_E \equiv \tfrac{1}{2}\int_S \rho\overline{\mathbf{z}}^*\cdot\overline{\mathbf{z}}^* dV \tag{F.20}$$

is the *kinetic energy of $\mathcal{B}$ relative to the frame* $\{\mathbf{E}_x, \mathbf{E}_y, \mathbf{E}_z\}$*;* K_E vanishes if $\mathcal{B}$ is rigid □.

It is reasonable to ask if the definition (F.12) of Ω is optimal in any sense. The answer, supplied by Mrs. Fisher, begins by assuming that Ω is not defined by (F.12). Then (F.19) takes the form

$$K = \tfrac{1}{2}M\dot{\overline{\mathbf{X}}}\cdot\dot{\overline{\mathbf{X}}} + \tfrac{1}{2}\Omega\cdot\mathsf{I}\cdot\Omega + K_E + \Omega\cdot\int\rho\overline{\mathbf{z}}\times\overline{\mathbf{z}}^* dV\,, \tag{F.21}$$

which, in view of (F.17), may be rewritten as

$$K_E = K - \tfrac{1}{2}M\dot{\overline{\mathbf{X}}}\cdot\dot{\overline{\mathbf{X}}} + (\tfrac{1}{2}\Omega\cdot\mathsf{I}\cdot\Omega - \mathbf{R}\cdot\Omega)\,. \tag{F.22}$$

But $\mathbf{R}$ and I, as defined by (F.9) and (F.13), are independent of Ω, as are the first two terms in (F.22). Thus, the spin that minimizes K_E is the spin that minimizes the quadratic form in parenthesis; that is, the value of Ω defined by (F.12): what seems a reasonable definition from rotational momentum considerations proves to be a reasonable definition from an energetic view as well.

Problem: If $\mathcal{B}$ *were* rigid we should have

$$\dot{K} = \mathbf{F}\cdot\dot{\overline{\mathbf{X}}} + \mathbf{T}\cdot\Omega\,, \tag{F.23}$$

since for a rigid body $\dot{\overline{\mathsf{I}}}\cdot\Omega = (\Omega\times\mathsf{I})\cdot\Omega = \mathbf{0}$. Determine how the right side of this equation is modified if $\mathcal{B}$ is deformable. Assume that $\mathbf{F}, \dot{\mathbf{P}}, \mathbf{T}$, and $\dot{\mathbf{R}}$ are continuous so that (F.4) and (F.10) may be replaced by

$$\mathbf{F} = \dot{\mathbf{P}} = M\ddot{\overline{\mathbf{X}}}\;,\;\; \mathbf{T} = \dot{\mathbf{R}} = \overline{\mathsf{I}}\cdot\dot{\Omega} + \dot{\overline{\mathsf{I}}}\cdot\Omega\,. \tag{F.24}$$

Solution: Taking the time derivative of (F.19) and using (F.24), we have

$$\begin{aligned}\dot{K} &= M\dot{\overline{\mathbf{X}}}\cdot\ddot{\overline{\mathbf{X}}} + \Omega\cdot\overline{\mathsf{I}}\cdot\dot{\Omega} + \tfrac{1}{2}\Omega\cdot\dot{\overline{\mathsf{I}}}\cdot\Omega + \dot{K}_E \\ &= \mathbf{F}\cdot\dot{\overline{\mathbf{X}}} + \mathbf{T}\cdot\Omega - \tfrac{1}{2}\Omega\cdot\dot{\overline{\mathsf{I}}}^*\cdot\Omega + \dot{K}_E\,\square\,.\end{aligned} \tag{F.25}$$

If, for a rigid body, there exists a scalar P such that the right side of (F.23) is equal to $-\dot{P}$, then we have *the first integral of motion*

$$K + P = \text{Constant}\,, \tag{F.26}$$

and we say that $\mathbf{F}$ and $\mathbf{T}$ are *potential (conservative)*. Conditions on $\mathbf{F}$ and $\mathbf{T}$ under which P exists may be developed along the lines given by Antman (1972) and Simmonds (1984) for the case $\mathbf{F} = \mathbf{0}$.

We can discuss a first integral of motion for deformable bodies by defining what we shall call the *reduced kinetic energy*,

$$\mathcal{K} \equiv \tfrac{1}{2} M \dot{\bar{\mathbf{X}}} \cdot \dot{\bar{\mathbf{X}}} + \tfrac{1}{2} \boldsymbol{\omega} \cdot \mathbf{I} \cdot \boldsymbol{\omega} \,, \tag{F.27}$$

where $\mathbf{I}$ is the moment of inertia of $\mathcal{B}$ in its reference shape (i.e., $\bar{\mathbf{I}}$ at $t=0$) and $\boldsymbol{\omega}$ is the *effective spin of* $\mathcal{B}$, defined implicitly by

$$\mathbf{R} \equiv \mathbf{I} \cdot \boldsymbol{\omega} \,. \tag{F.28}$$

The adjective "reduced" implies that, in general, $\mathcal{K} \leq K$. For example, if $\mathcal{B}$ undergoes a pure radial expansion, $K = K_E > 0$ but $\mathcal{K} = 0$.

Since $\mathbf{I}$ is time-independent, $\dot{\mathbf{R}} = \mathbf{I} \cdot \dot{\boldsymbol{\omega}}$, and it follows from (F.24) and (F.27) that

$$\dot{\mathcal{K}} = \mathbf{F} \cdot \dot{\bar{\mathbf{X}}} + \mathbf{T} \cdot \boldsymbol{\omega} \,. \tag{F.29}$$

For a deformable body, we define $\mathbf{F}$ and $\mathbf{T}$ to be potential if there exists a scalar $\mathcal{P}$ such that the right side of (F.29) is equal to $-\dot{\mathcal{P}}$. In this case, we have as a first integral of motion of $\mathcal{B}$,

$$\mathcal{K} + \mathcal{P} = \text{Constant} \,. \tag{F.30}$$

References

[A letters in brackets indicates the section where the reference appears.]

Antman, S. S. (1972). Solution of problem 71-24*, angular velocity and moment potential for a rigid body. *SIAM Rev.* **14,** 649-652. [F]

Chadwick, P. (1976). "Continuum Mechanics." Wiley, New York. [Intro]

Gurtin, M. E. (1981). "An Introduction to Continuum Mechanics." Academic Press, New York, etc. [Intro]

Simmonds, J. G. (1984). Moment potentials. *Am. J. Phys.* **52,** 851-852. [F]

Spencer, A. J. M. (1980). "Continuum Mechanics." Longman, London and New York. [Intro]

Truesdell, C. A. (1977). "A First Course in Rational Continuum Mechanics," vol. I. Academic Press, New York. [Intro]

Truesdell, C. A. (1984). "Rational Thermodynamics," 2nd ed. Springer-Verlag, New York, etc. [D]

Truesdell, C. A., and Noll, W. (1965). The non-linear field theories of mechanics. *In* "Encyclopedia of Physics" (S. Flügge, ed.), vol. III/3. Springer-Verlag, Berlin, etc. [Intro]

Truesdell, C. A., and Toupin, R. A. (1960). The classical field theories. *In* "Encyclopedia of Physics" (S. Flügge, ed.), vol. III/1. Springer-Verlag, Berlin and New York. [Intro, D, E]

Chapter III

Longitudinal Motion of Straight Rods with Bi-Symmetric Cross Sections (Birods)

This chapter illustrates, in as simple a way as possible, our approach to deriving complete field equations for 1- or 2-dimensional continua from the integral equations of motion and thermodynamics of a 3-dimensional continuum. In it are such basic concepts as descent from 3 dimensions by weighted averages, shock waves, classical and weak solutions, mechanical boundary conditions, the Principle of Virtual Work (for initial/boundary-value problems), mechanical versus thermomechanical theories of elastic behavior, displacement versus intrinsic equations of motion, and variational principles. Missing, of course, is a presentation of those concepts associated with curvature effects—an essential feature of shell theory. Furthermore, although we mention *material* stability, there is no discussion of *buckling* as we do not permit our rod to bend.

A. Geometry of the Undeformed Rod

A straight, bi-symmetric rod is a body whose reference shape is the set of all points with a position vector of the form

$$\mathbf{x}(x, y, z) = x\mathbf{e}_x + y\mathbf{e}_y + z\mathbf{e}_z \ , \ (x, y) \in C_z \ , \ z \in [0, L] . \tag{A.1}$$

As in Fig. 3.1, C_z, the undeformed cross section of the rod, is a closed, connected, planar set that may depend on z and is symmetric with respect to the x- and y-axes. The left and right *ends* of the rod are, respectively, C_0 and C_L. The *side* of the rod is the set of all points such that $(x, y) \in \partial C_z, z \in [0, L]$, where ∂C_z is the boundary of C_z.

B. Integral Equation of Motion

To derive an integral equation of motion, we take the arbitrary volume V of Chapter II to be any slice of the rod between, say, C_a and C_b, $0 \leq a < b \leq L$. As the reference shape of the rod is described in Cartesian coordinates, $dV = dxdydz$. We remind the reader that in what follows and throughout the

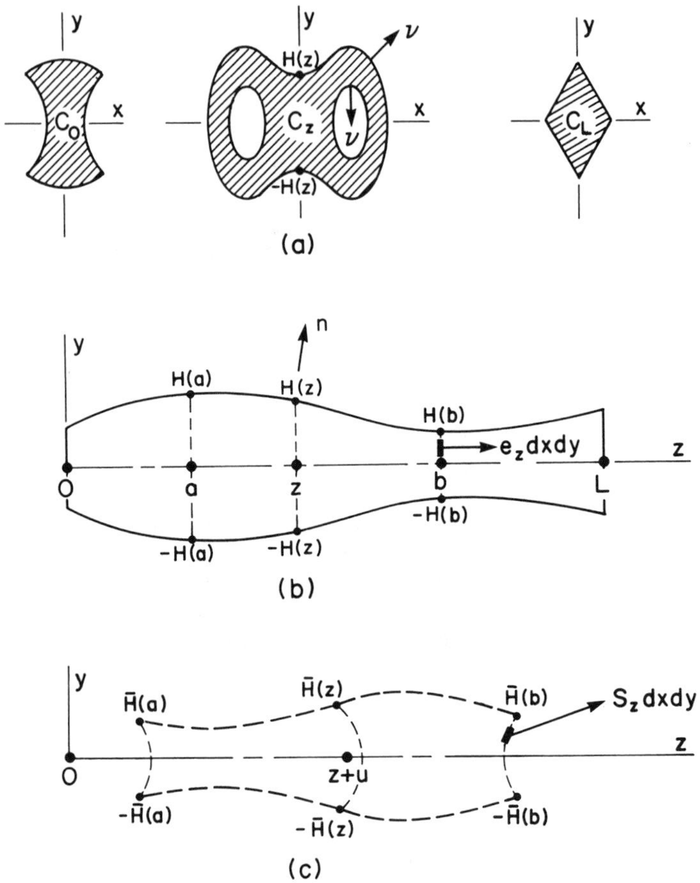

Fig. 3.1. Deformation of a birod of variable cross-section: (a) typical cross-sections; (b) side view of undeformed birod; (c) deformed slice of birod.

book, we adopt a *Lagrangian viewpoint* (actually Euler's) in which the motion of a body is referred to its reference shape. That is, the deformed position $\bar{\mathbf{x}}$ of a particle is taken as a function of time and its initial position $\mathbf{x}$.

On the ends of the slice

$$\mathbf{n}dA = \pm \mathbf{e}_z dxdy \;, \tag{B.1}$$

where, as in Chapter II, $\mathbf{n}$ is an *outward* unit normal to the reference surface of the body.

To compute $\mathbf{n}dA$ on the side of the slice, we assume that ∂C_z has the parametric representation

$$\partial C_z:\ x=\hat{x}(s,z)\ ,\ y=\hat{y}(s,z)\ ,\ 0\leq s\leq p(z)\ , \tag{B.2}$$

where s is arc length along ∂C_z and $p(z)$ is the length of the perimeter of ∂C_z. With $\hat{\mathbf{x}}\equiv\hat{x}(s,z)\mathbf{e}_x+\hat{y}(s,z)\mathbf{e}_y+z\mathbf{e}_z$ denoting the position of a point on ∂C_z, we have

$$\mathbf{n}dA=\hat{\mathbf{x}},_s\times\hat{\mathbf{x}},_z dzds=(\mathbf{\nu}+J\mathbf{e}_z)dzds\ , \tag{B.3}$$

where $\mathbf{\nu}\equiv\hat{y},_s\mathbf{e}_x-\hat{x},_s\mathbf{e}_y$ is an outward normal vector to ∂C_z and $J\equiv\hat{x},_s\hat{y},_z-\hat{x},_z\hat{y},_s$. See Fig. 3.1. In Figs. 3.1a and 3.1b, $\pm H(z)$ denotes the symmetrically located points of intersection of ∂C_z with the y-axis. In Fig. 3.1c, $\pm\overline{H}(z)$ denotes the deformed images of these two points.

From (B.1), (B.3), and (II.B.5), it follows that the contact force acting on the ends of the slice is

$$\mathbf{S}dA=\pm\mathbf{S}_z dxdy\ , \tag{B.4}$$

while on the side of the slice the contact force is

$$\mathbf{S}dA=(\hat{y},_s\mathbf{S}_x-\hat{x},_s\mathbf{S}_y+J\mathbf{S}_z)dzds\ \equiv\hat{\mathbf{S}}dzds\ . \tag{B.5}$$

As the ends and side of the slice make up ∂V, the net contact force on the slice at any instant, by (B.4) and (B.5), may be expressed as

$$\int_{\partial V}\mathbf{S}dA=\int_{C_z}\mathbf{S}_z dxdy\,\Big|_a^b+\int_a^b\int_{\partial C_z}\hat{\mathbf{S}}dsdz\ . \tag{B.6}$$

By (IIA.4), the slice is subjected to the net body force

$$\int_V\mathbf{f}dV=\int_a^b\Big(\int_{C_z}\mathbf{f}dxdy\Big)dz \tag{B.7}$$

and has linear momentum

$$\int_V\rho\dot{\mathbf{x}}dV=\int_a^b\Big(\int_{C_z}\rho\bar{\mathbf{x}}dxdy\Big)^{\bullet}dz\ . \tag{B.8}$$

Substituting the right sides of (B.6)-(B.8) into (IIA.4), we may write the resulting equation of balance of linear momentum in the form

$$\mathbf{L}\equiv\int_{t_1}^{t_2}[\mathbf{N}_z\,\Big|_a^b+\int_a^b\mathbf{p}dz\,]dt-\int_a^b m\dot{\bar{\mathbf{y}}}dz\,\Big|_{t_1}^{t_2}=\mathbf{0}\ , \tag{B.9}$$

where

$$\mathbf{N}_z\equiv\int_{C_z}\mathbf{S}_z dxdy \tag{B.10}$$

is the *stress resultant* (with dimensions of FORCE),

$$\mathbf{p}\equiv\int_{\partial C_z}\hat{\mathbf{S}}ds+\int_{C_z}\mathbf{f}dxdy \tag{B.11}$$

is the *external force* per unit length of the z-axis (the undeformed centerline of the rod),

$$m \equiv \int_{C_*} \rho dxdy \tag{B.12}$$

is the *mass per unit length* of the z-axis, and

$$\bar{\mathbf{y}} \equiv \int_{C_*} \rho \bar{\mathbf{x}} dxdy / m \tag{B.13}$$

is the *weighted average position* of the deformed rod. Henceforth, we assume that our inertial frame $Oxyz$ has been chosen so that $\bar{\mathbf{y}}(z, 0) = z\mathbf{e}_z$. Because $\bar{\mathbf{y}}$ is a weighted average, $\bar{\mathbf{y}}$ is not, in general, the deformed position of the particle with initial position $z\mathbf{e}_z$.

We now assume that the external disturbances (mechanical or thermal) on the slice produce gross longitudinal motion only and that the balance of rotational momentum is satisfied identically. For conciseness, we shall henceforth refer to rods with bi-symmetric cross sections that undergo such motion as *birods*. More precisely, these motions are such that

$$(\mathbf{N}_z, \mathbf{p}, \bar{\mathbf{y}}) \equiv (N, p, z + u)\mathbf{e}_z \,, \tag{B.14}$$

where u is the *weighted average longitudinal displacement*. Under (B.14), the equation of balance of linear momentum, (B.9), takes the scalar form

$$L \equiv \int_{t_1}^{t_2} (N \big|_a^b + \int_a^b p dz) dt - \int_a^b mv \big|_{t_1}^{t_2} dz = 0 \,, \tag{B.15}$$

where

$$v \equiv \dot{u} = \partial u(z, t)/\partial t \,. \tag{B.16}$$

Three-dimensional considerations are now behind us; henceforth, we may view a birod as a 1-dimensional continuum endowed with pointwise properties of mass and stiffness.

Suppose that the initial velocity of the birod is a prescribed function $\bar{v}(z)$, $0 < z < L$. Then if we set $t_1 = 0$ and let $t_2 \to 0$ through positive values, we obtain from the integral equation of motion, (B.15), Antman's (1980) form of the *natural initial condition*,

$$\int_a^b (\overset{0}{v} - \bar{v}) dz = 0 \;, \quad 0 < a < b < L \,, \tag{B.17}$$

where $\overset{0}{v}$ is the limit of $v(z, t)$ as $t \to 0$ through positive values. If the integrand is continuous, then, because the interval (a, b) is arbitrary, $\overset{0}{v} = \bar{v}(z)$. Likewise, suppose that $N_0(t)$ and $N_L(t)$, $0 < t$, denote prescribed values of the axial force on the ends $z = 0$ and $z = L$ of the birod. Then in (B.15) if we set $a = 0$ and let $b \to 0$ from the right or set $b = L$ and let $a \to L$ from the left, we obtain Antman's form of the *natural boundary conditions*,

$$\int_{t_1}^{t_2} (N_0 - \hat{N}_0)dt = \int_{t_1}^{t_2} (\hat{N}_L - N_L)dt = 0 \ , \ \forall\, 0 < t_1 < t_2 \ , \tag{B.18}$$

where N_0 is the limit of $N(z, t)$ as $z \to 0$ from the right and N_L is the limit of $N(z, t)$ as $z \to L$ from the left. The name "natural" comes from the Principle of Virtual Work (Section H) where the vanishing of a certain functional for a class of test functions implies (B.17) and (B.18). In contrast, the *essential initial condition*

$$u(z, 0) = \bar{u}(z) \ , \ 0 < z < L \ , \tag{B.19}$$

and any *essential kinematic boundary constraints*, as discussed in Section G, must be imposed a priori on the fields entering the Principle of Virtual Work.

C. Differential Equation of Motion

If N' and $\dot{v}$ are, respectively, integrable functions of z and t, where a prime now denotes differentiation with respect to z, then, by the Fundamental Theorem of Calculus, we can rewrite (B.15) in the form

$$\int_{t_1}^{t_2}\int_a^b (N' + p - m\dot{v})dzdt = 0 \ . \tag{C.1}$$

Furthermore, if the integrand is continuous on $(0, L)\times(0, \infty)$, then, as a, b, t_1, t_2 are arbitrary, save only that $0 \le a < b \le L$ and $0 \le t_1 < t_2 < \infty$, it follows that (C.1) implies the differential equation

$$N' + p - m\dot{v} = 0 \ , \ 0 < z < L \ , \ 0 < t \ . \tag{C.2}$$

The argument in going from the global statement (C.1) to the local statement (C.2) is standard: suppose that at some point z^* and time t^* the right side of (C.2) were not zero. Then, by the assumed continuity of $N' + p - m\dot{v}$, there exists an open rectangle centered at (z^*, t^*) and contained in $(0, L)\times(0,\infty)$ such that $N' + p - m\dot{v}$ is of one sign in this rectangle. Choosing the region of integration in (C.1) to coincide with this rectangle, we get a non-zero integral—a contradiction.

Note that if $p = 0$, (C.2) states that the divergence in zt-coordinates of the *force-momentum vector* $(N, -mv)$ is zero. Equations that can be so written are said to be of *conservation form.*

*D. Jump Condition and Propagation of Singularities

In the linear theory of elastic, homogeneous rods with uniform cross sections of area A, the axial force is given by $N = EAu'$, where E is Young's modulus. Thus, with $v = \dot{u}$, (C.2) reduces to

$$EAu'' + p = m\ddot{u} \ . \tag{D.1}$$

If $p = 0$, (D.1) has the general solution

$$u = F(z - ct) + G(z + ct) \ , \ c = \sqrt{EA/m} \ , \tag{D.2}$$

where F and G are arbitrary functions. However, unless F and G are twice differentiable, (D.2) is merely a *formal* solution of (D.1). Yet such solutions are too useful to be ignored.

For example, consider a rod that is fixed at its left end, initially unstressed, and at rest. At time zero a tensile force $\hat{N}_L$ is applied and maintained at the right end. (This might represent a cable supporting an elevator that is suddenly loaded). The formal solution to this problem, found by choosing F and G in (D.2) to satisfy the boundary conditions

$$u(0, t) = 0 \ , \ u'(L, t) = \hat{N}_L/EA \ , \tag{D.3}$$

is displayed in Fig. 3.2a which shows various regions of the solution strip $(0, L)\times(0, \infty)$ separated by wave fronts across which u' and v have jumps. As shown in Fig. 3.2b, this wave front moves back and forth along the z-axis with velocity c. Note, however, that on either side of the wave front the solution is

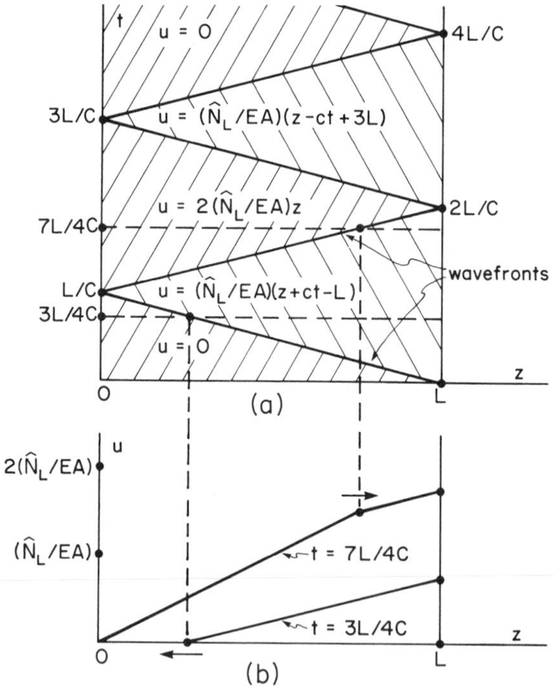

Fig. 3.2. (a) Formal solution of $c^2u'' = \ddot{u}$ in various regions of the zt-plane; (b) graph of solution at various times (arrows show direction of motion of wavefront).

smooth enough to satisfy (D.1). (If one imagines the graph of the solution in an orthogonal *ztu*-coordinate system, then Fig. 3.2b represents the intersection of this graph with various planes $t=\text{constant}$.)

The above solution may be legitimized by showing that it satisfies the integral form of the equation of motion (B.15) which admits rougher solutions than does the differential equation (D.1). However, rather than pursuing this very special problem, let us return to Section B (that precedes any consideration of boundary conditions or constitutive relations) and ask: Can there exist a piecewise smooth curve (or shock wave)

$$\mathcal{S}:\ z=s(t)\ ,\ 0<t \tag{D.4}$$

in the *zt*-plane across which the unknowns N and v suffer jumps? In answering this we shall assume that the *shock velocity* $\dot{s}(t)\neq 0$. [At $z=0, L$, $s(t)$ has corners and $\dot{s}(t)$ does not exist.]

Referring to Fig. 3.3a, we note first that (B.15) can be cast in the form

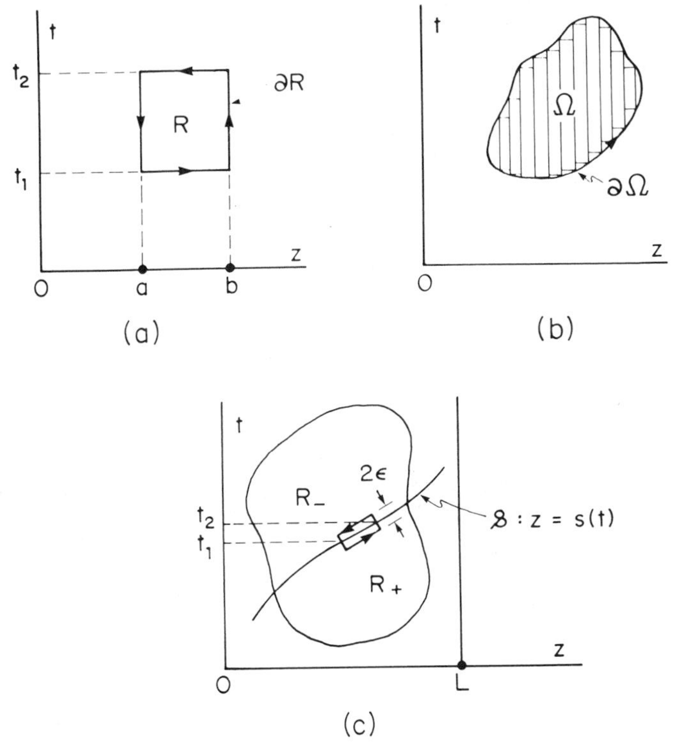

Fig. 3.3. Regions in the *zt*-plane and shock fronts.

$$\oint_{\partial R} (Ndt + mvdz) + \int_R pdzdt = 0\,, \tag{D.5}$$

where R denotes the rectangle $(a, b)\times(t_1, t_2)$ in the zt-plane with boundary ∂R and the first term on the left is a line integral taken in a counter-clockwise sense as the notation indicates. But *any* region Ω in the zt-plane may be regarded as the limit of the union of narrow vertical strips as indicated in Fig. 3.3b. Thus, as (D.5) holds for all such strips, we have the more general relation

$$\oint_{\partial\Omega} (Ndt + mvdz) + \int_\Omega pdzdt = 0\,. \tag{D.6}$$

Notice that (D.6) follows *formally* upon writing the equation of motion (C.2) in the form $N_z + (-mv)_t + p = 0$, where $N' \equiv N_z \equiv \partial N/\partial z$ and $\dot{v} \equiv v_t \equiv \partial v/\partial t$, and applying the divergence theorem.

As in Fig. 3.3c, we now take the region Ω in (D.6) to be a strip of width 2ε with centerline along the curve $\mathcal{S}$ and ends that intersect $\mathcal{S}$ at the points $P_1(s(t_1), t_1)$ and $P_2(s(t_2), t_2)$. Let the strip lie within an open, simply connected region $\mathcal{R}$, cut by $\mathcal{S}$ into two open, disjoint sets R_+ and R_-. Further, assume that $\mathcal{S}$ is smooth within $\mathcal{R}$.

Suppose that $f(z, t)$ is a function continuous on R_- and R_+ with unique limits f_- or f_+ as a point $P(s(t), t)$ on $\mathcal{S}$ is approached from R_- or R_+ respectively. Then the jump in f at P from R_- to R_+ is denoted and defined by

$$[f] \equiv f_+ - f_- = \lim_{\eta\to 0+} [f(s(t)+\eta, t) - f(s(t)-\eta, t)]\,. \tag{D.7}$$

If ε, the half width of the strip, approaches zero, then (D.6) reduces to

$$\int_{t_2}^{t_1} ([N] + m\dot{s}[v])dt = 0\,. \tag{D.8}$$

We assume that all fields are smooth in R_- and R_+. Thus, the integrand is continuous, and as (D.8) holds for all $t_2 > t_1$, we have

$$[N] = -m\dot{s}[v]\,. \tag{D.9}$$

To relate $[v]$ to a jump in the strain, we assume that u_z and v in R_- and R_+ approach unique limits, $(u_z)_-, (u_z)_+, \cdots$, as we approach the point P on $\mathcal{S}$. The (scaled) directional derivatives of u in R_- and R_+ parallel to $\mathcal{S}$ as we approach P are, respectively,

$$\left(\frac{du}{dt}\right)_- = (u_z)_-\dot{s} + v_-\,, \quad \left(\frac{du}{dt}\right)_+ = (u_z)_+\dot{s} + v_+\,. \tag{D.10}$$

But the rod must not break, i.e., $[u](t) = 0$. Hence,

$$\frac{d}{dt}[u] = \left(\frac{du}{dt}\right)_+ - \left(\frac{du}{dt}\right)_- = \dot{s}[u_z] + [v] = 0\,. \tag{D.11}$$

Thus, (D.9) takes the form

$$[N] = mc^2[u_z] \, , \tag{D.12}$$

where $c(t) \equiv |\dot{s}(t)|$ is the *speed of the shock.* This equation is the 1-dimensional analog of equation (205.4) of Truesdell & Toupin (1960), who give an exhaustive discussion of (and distinguish between) surfaces of discontinuity and shocks. A careful treatment of the propagation of finite disturbances in rubber-like rods may be found in Nowinski (1965).

For a linearly elastic rod, $[N]=EA[u_z]$, and (D.12) yields the well-known result found earlier: $c=(EA/m)^{1/2}$. For a nonlinearly elastic rod, (D.12) not only can lead to bounds on the shock speed without the need to solve the equation of motion, it also shows that the material composing the rod is *stable* only so long as the dimensionless stress at each point, N/EA, is an *increasing* function of the strain $e=u_z$. This second point is discussed in Section J.

E. The Weak Form of the Equation of Motion

Classical solutions of (C.2) are pairs of functions (N, v) such that each term in (C.2), as a function of z and t, is continuous on $(0,L)\times(0,T)$, where $T>0$ is arbitrary. *Weak solutions* of (C.2) are extensions of classical solutions, introduced to encompass certain rough functions which may arise in practice, such as those discussed in the preceding section. This is accomplished by replacing the *pointwise differential equation* (C.2) by a *global integral equation* whose solutions, if sufficiently smooth, are also classical solutions. We leave the consideration of boundary and initial conditions to the next section. A brief history of weak solutions may be found in a book review by Dieudonné (1984).

Let $V(z, t)$ be a *test function* which, for the moment, we shall characterize merely as being sufficiently smooth for the following operations to make sense. Assume that N and v are classical solutions. Multiply (C.2) by V and integrate over any subregion $(a, b)\times(t_1, t_2)$ of $(0,L)\times(0,T)$. The idea now is to take derivatives off N and v (so that we can weaken the smoothness conditions on a classical solution) and put them on the relatively smooth function V. Thus, integrating once by parts in space and time, we arrive at the *weak form of the equation of motion*:

$$\int_{t_1}^{t_2} NV \,\Big|_a^b dt - \int_a^b mvV \,\Big|_{t_1}^{t_2} dz - \int_{t_1}^{t_2}\int_a^b (NV' - mv\dot{V} - pV)dzdt = 0 \, , \; \forall \, V \, . \tag{E.1}$$

We emphasize that V is *not* required to satisfy any boundary conditions.

As it stands (that is, ignoring the assumptions under which it was derived), (E.1) makes sense whenever its integrals make sense. This implies that N, v, m, and p can be rough if V is sufficiently smooth. Physically, however, there is an obvious restriction on N, v, and p, namely, the integrals in (B.15), the integral-impulse form of the equation of motion, must make sense. In particular, as (B.15) implies the jump conditions (D.9) and (D.12), N, v, and u_z may suffer jumps but not u. Thus, we shall define the *class of test functions* $\mathcal{T}$ to comprise all functions V such that the integrals in (E.1) make sense whenever those in (B.15) do.

In (E.1), were we to set $v = \dot{u}$ and

$$-\int_a^b mvV\,\Big|_{t_1}^{t_2} dz + \int_{t_1}^{t_2}\int_a^b mv\dot{V}dzdt = \int_a^b m(u\dot{V} - vV)\Big|_{t_1}^{t_2} dz - \int_{t_1}^{t_2}\int_a^b mu\ddot{V}dzdt \,, \tag{E.2}$$

we would obtain a weak form of the equation of motion that admitted solutions for which $[u] \neq 0$. Although the definition of a weak solution often involves taking *all* derivatives off the unknown and putting them on a test function (e.g., Stackgold, 1979, p. 175, or Zauderer, 1983, pp. 288-290), such weak solutions for a birod could be too rough to make sense physically.

It is easy to see that (B.15) follows from (E.1) by taking V to be an arbitrary constant. The results of Antman (1980) (which are far from obvious) imply that we can go the other way: the weak form of the equations of motion can be obtained directly from (B.15)! Thus, weak solutions are seen to have more physical significance than classical solutions, for they are exactly those functions that satisfy the more fundamental integral-impulse form of the equation of motion.

F. The Mechanical Work Identity

In (E.1) let $V = v$, the actual velocity of the birod. Then we have the *Mechanical Work Identity*

$$\int_{t_1}^{t_2} \mathcal{W}dt = \mathcal{K}\,\Big|_{t_1}^{t_2} + \int_{t_1}^{t_2} \mathcal{D}dt \,. \tag{F.1}$$

Here

$$\mathcal{W} \equiv Nv\,\Big|_a^b + \int_a^b pvdz \tag{F.2}$$

is the *apparent external mechanical power*,

$$\mathcal{K} \equiv \tfrac{1}{2}\int_a^b mv^2 dz \tag{F.3}$$

is the *reduced kinetic energy*, and

$$\mathcal{D} \equiv \int_a^b N\dot{e}dz \tag{F.4}$$

is the *deformation power*, where

$$e \equiv u' \tag{F.5}$$

is the *strain*, sometimes called the *engineering strain* to distinguish it from the *Lagrangian strain*, $\varepsilon \equiv u' + \tfrac{1}{2}u'^2$.

The reason we use the adjectives "apparent" and "reduced" to characterize $\mathcal{W}$ and $\mathcal{K}$ is the following. Set

$$\bar{\mathbf{x}} = \bar{\mathbf{y}} + \bar{\mathbf{z}}\,. \tag{F.6}$$

Then, by (B.13) and (B.14),

$$\int_{C_z} \rho\bar{\mathbf{z}}dxdy = \mathbf{0} \ \ , \ \ \dot{\bar{\mathbf{x}}} = v\mathbf{e}_z + \dot{\bar{\mathbf{z}}}\,. \tag{F.7}$$

From (B.4), (B.5), and the above two expressions, we conclude that, from a 3-dimensional viewpoint, the actual rate-of-work, at any time t, of the external loads on the ends, side, and throughout a slice of a birod is

$$\begin{aligned} &\int_{C_z} \mathbf{S}_z\cdot\dot{\bar{\mathbf{x}}}dxdy \,\Big|_a^b + \int_a^b \int_{\partial C_z} \hat{\mathbf{S}}\cdot\dot{\bar{\mathbf{x}}}dsdz + \int_a^b \int_{C_z} \mathbf{f}\cdot\dot{\bar{\mathbf{x}}}dxdydz \\ &\qquad = \mathcal{W} + \int_{C_z} \mathbf{S}_z\cdot\dot{\bar{\mathbf{z}}}dxdy \,\Big|_a^b + \int_a^b \int_{\partial C_z} \hat{\mathbf{S}}\cdot\dot{\bar{\mathbf{z}}}dsdz + \int_a^b \int_{C_z} \mathbf{f}\cdot\dot{\bar{\mathbf{z}}}dxdydz\,. \end{aligned} \tag{F.8}$$

Thus, $\mathcal{W}$ represents only a part of the actual external mechanical power. Likewise, the actual kinetic energy of a slice of the birod is

$$\tfrac{1}{2}\int_a^b \int_{C_z} \rho\dot{\bar{\mathbf{x}}}\cdot\dot{\bar{\mathbf{x}}}dxdydz = \mathcal{K} + \tfrac{1}{2}\int_a^b \int_{C_z} \rho\dot{\bar{\mathbf{z}}}\cdot\dot{\bar{\mathbf{z}}}dxdydz\,, \tag{F.9}$$

which is always greater than or equal to $\mathcal{K}$. The last term in (F.9) represents, as it were, the "heat" of 3-dimensionality.

G. Mechanical Boundary Conditions

From (F.2) we infer that the (apparent) rate of work of the end forces acting on a birod is $Nv\,\big|_0^L$. The aim of this section is to wring from this term the form of the *mechanical boundary conditions*; "classical" boundary conditions fall out as a special case. In what follows, the reader must not be perturbed if it appears that we have introduced too elaborate a notation or unnecessarily invoked results from linear algebra for the analysis of a mere birod: save for working with larger matrices, virtually every result concerning boundary conditions can be applied, unaltered, to beamshells, axishells, or general shells.

To handle conditions at both ends of the birod efficiently, we introduce the notation

$$\mathcal{U} \equiv [u_0, u_L]^T \ \ , \ \ \mathcal{U}^\bullet \equiv [v_0, v_L]^T\,, \tag{G.1}$$

where $\mathcal{U}$ and $\mathcal{U}^\bullet$ are 2×1 matrices (i.e., column vectors) and $u_0 \equiv u(0,t)$, $u_L \equiv u(L,t)$.

Suppose that there are $m \leq 2$ *independent, nonholonomic constraints* of the form

$$A\mathcal{U}^{\bullet} - \mathcal{B} = 0\,, \tag{G.2}$$

where A and $\mathcal{B}$ are, respectively, $m \times 2$ and $m \times 1$ matrices that may depend on $\mathcal{U}$ and t but not $\mathcal{U}^{\bullet}$. (We use the symbol "m" because it is customary to refer to $m \times n$ matrices. There should be no confusion with the mass/length of the birod that we have also denoted by m). Often, when we are given geometrical constraints, we are given no information as to the physical mechanism that enforces the constraints—Is it an infinite rigid body? Is it a massless rigid body?—nor should a proper mathematical formulation of a complete problem require it. Nevertheless, it is useful occasionally to invent such mechanisms, as we shall see presently in an example problem. We can treat nonholonomic constraints as one does in classical mechanics (e. g., Goldstein, 1980).

By m independent constraints we mean that $\operatorname{rank} A = m$. Hence, from linear algebra, the solution of $A\mathcal{U}^{\bullet} = \mathcal{B}$ has the well-known form

$$\mathcal{U}^{\bullet} = \sum_{1}^{2-m} q_k \mathcal{M}_k(\mathcal{U}, t) + \mathcal{H}(\mathcal{U}, t)\,. \tag{G.3}$$

Here, the q_k's are unknown *generalized edge velocities*, the $\mathcal{M}_k$'s are any linearly independent vectors such that $A\mathcal{M}_k = 0$ (i.e., the $\mathcal{M}_k$'s span the null space of A), and $\mathcal{H}$ is uniquely determined by the conditions $A\mathcal{H} = \mathcal{B}$ and $\mathcal{H}^T\mathcal{M}_k = 0$, where the superscript T denotes "transpose" (i.e., $\mathcal{H}$ lies in the range of A^T). The beauty of (G.3) is that $\mathcal{U}^{\bullet}$ is represented as the sum of an unknown vector plus an orthogonal, known vector, computable in terms of the elements A and $\mathcal{B}$ of the given constraint (G.2).

Clearly, (G.2) is not the most general constraint that we could imagine. An amusing and not unimportant example is a thermally conducting rod, fixed at its left end. Initially, the right end of the rod is stress-free and lies a small distance from a rigid, thermally conducting wall. At time zero a uniform source of heat is placed at the left end of the rod. As the rod heats, it expands, and its right end eventually touches the wall, thus suddenly changing the mechanical and thermal boundary conditions at $z = L$. But as heat flows out of the right end of the rod into the wall—we assume it to be cooler that the end of the rod—the rod contracts and the original boundary conditions at $z = L$ are restored. This changing end condition is more general than (G.2) and involves, during moments of contact, integrals of the unknown solution over $[0, L]$.[1]

Now let

$$\mathcal{N} \equiv [-N_0, N_L]^T\,. \tag{G.4}$$

Then from (G.1), (G.3), and (G.4),

$$Nv \Big|_0^L = \mathcal{N}^T \mathcal{U}^{\bullet} = \sum_{1}^{2-m} \mathcal{N}^T \mathcal{M}_k q_k + \mathcal{N}^T \mathcal{H}\,. \tag{G.5}$$

Adopting the terminology of classical mechanics, we call the scalars $\mathcal{N}^T\mathcal{M}_k$ the

[1]A general discussion of mechanical-thermal boundary conditions is given by Batra (1972).

generalized edge forces and $\mathcal{N}^T\mathcal{H}$ the *power of the constraints.*

In classical mechanics, to have enough information to solve Lagrange's equations of motion, the generalized forces must be prescribed. In continuum mechanics, each differential element of material acts, in a sense, like a rigid body. In the interior of a continuum, the contact forces that enter the equations of motion are themselves determined by other field equations, e. g. the constitutive relations. But at a boundary of a continuum only a part of the loads acting on a material element come from the interior. To obtain enough information to determine the motion of boundary elements, we must, as with a rigid body, prescribe the generalized boundary forces. This suggests what we shall call the *Principle of Mechanical Boundary Conditions*, namely, there must be prescribed

- m constraints of the form $A\,\mathcal{U}^\bullet = \mathcal{B}$;
- $2-m$ generalized edge forces of the form $\mathcal{N}^T\mathcal{M}_k$.

Here is an application of the proposed Principle.

Problem: The ends of a birod are maintained at a prescribed distance $f(t)$ apart. Determine the boundary conditions.

Solution: There is one (holonomic) constraint, $u_L - u_0 = f(t) - L$, or, in the form (G.2),

$$v_L - v_0 = \dot{f}(t)\,. \tag{G.6}$$

This equation has the general solution

$$\mathcal{U}^\bullet = [v_0, v_L]^T = q_1[1, 1]^T + \tfrac{1}{2}\dot{f}[-1, 1]^T\,, \tag{G.7}$$

where $q_1 = \tfrac{1}{2}(v_0 + v_L)$. According to our Principle of Mechanical Boundary Conditions, we must prescribe in addition to (G.6) the generalized edge force

$$\mathcal{N}^T\mathcal{M}_1 = [-N_0, N_L][1, 1]^T = N_L - N_0\,. \tag{G.8}$$

Thus, because of the constraint, only the *net* axial force on the birod can be prescribed.

The mechanical meaning of (G.8) becomes apparent if we imagine the constraint $u_L - u_0 = f(t) - L$ maintained by an idealized, massless framework consisting of two end-plates attached to a movable piston running through an axial hole in the birod. If $N_L > 0$, the piston carries a compressive force N_L so that if $N_0 > 0$, the net force acting to the right on the left end-plate is $N_0 - N_L$. Were the left end-plate attached to a linear dashpot with a damping coefficient, c, some given function of time, we should have

$$N_0 - N_L = c(t)v_0 \tag{G.9}$$

and *not* $N_L = 0, N_0 = c(t)v_0$. □

The most common boundary conditions discussed in the literature are those we shall call *classical:* at each end of the birod, u or N is a prescribed function of time (often a constant). In this case our Principle reduces to one of four cases.

(1) If both u_0 and u_L are prescribed, $m=2$ and there are no generalized forces to be prescribed.

(2) If u_0 only is prescribed, $m=1$, $\mathcal{M}_1=[0,\,1]^T$ and there is one generalized force, $\mathcal{N}^T\mathcal{M}_1=N_L$.

(3) If u_L only is prescribed, a result analogous to (2) obtains.

(4) If neither u_0 nor u_L is prescribed, $m=0$ and we may take $\mathcal{M}_1=[1,0]^T$ and $\mathcal{M}_2=[0,\,1]^T$. The generalized forces are $\mathcal{N}^T\mathcal{M}_1=-N_0$ and $\mathcal{N}^T\mathcal{M}_2=N_L$.

H. The Principle of Virtual Work

The *Principle of Virtual Work* is the statement that if $\{N,\,u\}$ is a pair of sufficiently smooth functions and u satisfies the essential initial condition $u(z,\,0)=\bar{u}(z)$ plus any boundary kinematic constraints, then the vanishing of a certain functional of N and u for all test functions of a certain class implies that N and u are weak solutions of the integral equation of motion (B.15) and satisfy the natural conditions (B.17) and (B.18). An advantage of this Principle is that it leads immediately to methods for approximate solution, such as finite elements. We derive the Principle of Virtual Work by specializing the weak form of the equation of motion as follows.

In (E.1) we take $a=0, b=L, t_1=0$, and $t_2=T>0$, where T is positive, and introduce the *class of test functions* $\mathcal{T}_H$ whose elements satisfy the *final* condition

$$V(z,\,T)=0\,. \tag{H.1}$$

Suppose, for the moment, that there are no kinematic constraints on the ends of the birod. If we incorporate the natural initial and boundary conditions, (B.17) and (B.18), the left side of (E.1) takes the form

$$J[N,\,v;V]\equiv\int_0^T\hat{\mathcal{N}}^T\mathcal{V}dt+\int_0^L m\bar{v}\,\overset{0}{V}dz-\int_0^T\int_0^L(NV'-mv\dot{V}-pV)dzdt\;, \tag{H.2}$$

where

$$\mathcal{V}\equiv[V_0,\,V_L]^T\,, \tag{H.3}$$

$\overset{0}{V}\equiv V(z,\,0+)$, and $\hat{\mathcal{N}}$ is the *prescribed* load vector.[2]

The Principle of Virtual Work states that if $J=0$ for all test functions such that $V(z,\,T)=0$ and all displacements u such that $u(z,\,0)=\bar{u}(z)$, then N and u are weak solutions which, if sufficiently smooth, are also classical solutions of the differential equation of motion (C.2) and satisfy the pointwise form of the natural conditions (B.17) and (B.18).

---

[2]We shall use a "hat" (^) to denote *prescribed* or *specified* quantities throughout the book.

Suppose, however, that there are constraints of the form $A\mathcal{U}^{\bullet} - \mathcal{B} = 0$. To accommodate these, we multiply both sides of this expression on the left by the transpose of an arbitrary $m \times 1$ column vector $\mathcal{L}$—a *free* Lagrange multiplier vector—and add the resulting expression to J. To account for the associated *forces of constraint*, we must, at the same time, further restrict our test functions V so that they satisfy the *homogeneous* constraints

$$A\mathcal{V} = 0\,. \tag{H.4}$$

As A may depend on $\mathcal{U}$ and t, we denote this class of test functions by $\mathcal{T}_H(\mathcal{U}, t)$. We may enforce (H.4) either implicitly or explicitly: each has its advantages.

To satisfy $A\mathcal{V} = 0$ implicitly, we multiply this expression on the left by the transpose of a second Lagrange multiplier vector $\tilde{\mathcal{L}}$ having m *unknown* components and add the resulting expression to J. This and the above-mentioned addition to J yield the new functional

$$\begin{aligned} J^*[N, v, \tilde{\mathcal{L}}; V, \mathcal{L}] \equiv \int_0^T [(\hat{\mathcal{N}}^T + \tilde{\mathcal{L}}^T A)\mathcal{V} + \mathcal{L}^T(A\mathcal{U}^{\bullet} - \mathcal{B})]dt + \int_0^L m\bar{v}\overset{0}{V}dz \\ - \int_0^T\int_0^L (NV' - mv\dot{V} - pV)dzdt\,. \end{aligned} \tag{H.5}$$

Suppose that $J^* = 0$ for all V in $\mathcal{T}_H(\mathcal{U}, t)$ and all $\mathcal{L}$, and that the term NV' may be integrated by parts. Then in (H.5) there arises the boundary term

$$(\hat{\mathcal{N}}^T + \tilde{\mathcal{L}}^T A - \mathcal{N}^T)\mathcal{V}. \tag{H.6}$$

The constraint $A\mathcal{V} = 0$ implies that the first m components of $\mathcal{V}$ may be expressed in terms of its last $2-m$ components, which may be chosen arbitrarily. Thus, the last $2-m$ components of the vector in parentheses in (H.6) must vanish:

$$(\hat{\mathcal{N}}^T + \tilde{\mathcal{L}}^T A - \mathcal{N}^T)_k = 0\ ,\ k = m+1, \cdots 2\,. \tag{H.7}$$

Moreover, the m unknown components that constitute $\tilde{\mathcal{L}}$ may be chosen so that the first m components of this same vector vanish:

$$(\hat{\mathcal{N}}^T + \tilde{\mathcal{L}}^T A - \mathcal{N}^T)_k = 0\ ,\ k = 1, \cdots m\,. \tag{H.8}$$

Thus,

$$\hat{\mathcal{N}}^T + \tilde{\mathcal{L}}^T A - \mathcal{N}^T = 0\,. \tag{H.9}$$

But (H.9) is precisely the same conclusion we would have reached had we ignored the constraint $A\mathcal{V} = 0$ and simply taken V in $\mathcal{T}_H$: the price we pay for not enforcing $A\mathcal{V} = 0$ is the appearance of the m unknowns components of $\tilde{\mathcal{L}}$.

The disadvantage here is that, unless the nature of the mechanism enforcing the constraint $A\mathcal{U}^{\bullet} - \mathcal{B} = 0$ is known, it may not be at all obvious how to prescribe all of the components of $\hat{\mathcal{N}}$. Fortunately, the unknown vector $\tilde{\mathcal{L}}$ is easily eliminated from (H.9) by multiplying both sides of this expression from the right by the linearly independent vectors $\mathcal{M}_k$ that span the null space of A. [Recall that the $\mathcal{M}_k$'s were introduced in (G.3)]. This yields the boundary conditions

$$\mathcal{N}^T \mathcal{M}_k = \hat{\mathcal{N}}^T \mathcal{M}_k \ , \ k = 1, \cdots 2-m \tag{H.10}$$

which recapitulate the Principle of Mechanical Boundary Conditions: if there are kinematic constraints, only the generalized forces need be prescribed.

If we wish to recover (H.10) directly from the Principle of Virtual Work then we must satisfy the constraint $A\mathcal{V}=0$ explicitly. This is done, in analogy with (G.3), by noting that the constraint on $\mathcal{V}$ has the solution

$$\mathcal{V} = \sum_{1}^{2-m} r_k \mathcal{M}_k(\mathcal{U}, t) \ , \tag{H.11}$$

where the r_k's are arbitrary. Inserting (H.11) into (H.5), we obtain the new virtual work functional

$$\begin{aligned} \bar{J}[N, v; V, \mathcal{L}] \equiv \int_0^T [\sum_{1}^{2-m} \hat{\mathcal{N}}^T \mathcal{M}_k r_k + \mathcal{L}^T(A\mathcal{U}^{\bullet} - \mathcal{B})]dt + \int_0^L mg\overset{0}{V}dz \\ - \int_0^T \int_0^L (NV' - mv\dot{V} - pV)dzdt \ . \end{aligned} \tag{H.12}$$

The Principle of Virtual Work now states that if $\bar{J}=0$ for all u satisfying $u(z, 0) = \bar{u}(z)$, for all $\mathcal{L}$, and for all V in $\mathcal{T}_H(\mathcal{U})$ [which implies, in particular, that on the ends of the birod, $\mathcal{V} = \sum_{1}^{2-m} r_k \mathcal{M}_k$], then N and u are weak solutions which, if sufficiently smooth, are also classical solutions satisfying the differential equation of motion (C.2), the initial conditions $\dot{u}(z, 0) = \bar{v}(z)$, the kinematic constraint $A\mathcal{U}^{\bullet} = \mathcal{B}$, and the generalized force boundary conditions (H.10). The disadvantage here is, of course, that the constraint $A\mathcal{V}=0$ must be satisfied explicitly, which could be cumbersome in practice.

I. The Mechanical Theory of Birods

By definition, *the Mechanical Theory of Birods* ignores thermal effects and assumes that there exists an *internal energy* $\mathcal{E}$ such that

$$\int_{t_1}^{t_2} \mathcal{D}dt = \mathcal{E}\Big|_{t_1}^{t_2} . \tag{I.1}$$

In general, $\mathcal{E}$ is a functional of the strain e which means that $\mathcal{E}$ might depend on the strain history, as in a viscoelastic bar. If we substitute (I.1) into the Mechanical Work Identity (F.1), we get the *Principle of Conservation of (Reduced) Mechanical Energy*:

$$\int_{t_1}^{t_2} \mathcal{W}dt = (\mathcal{K} + \mathcal{E})_{t_1}^{t_2} . \tag{I.2}$$

When extended to include thermal effects (Section L), this becomes the First Law of Thermodynamics.

J. The Mechanical Theory of Elastic Birods

The Mechanical Theory of Elastic Rods assumes that N is a function of e and z and that there exists a *strain-energy density* Φ, reckoned per unit length of the birod's undeformed centerline and depending on e and z only, such that

$$N\dot{e} = \dot{\Phi} = \Phi,_e \dot{e} \; . \tag{J.1}$$

Because $\Phi,_e$ is independent of $\dot{e}$ and because we may always construct a displacement field such that $\dot{e} = v'$ takes on any prescribed value at any z and t, (J.1) implies the *stress-strain relation*

$$N = \Phi,_e \; . \tag{J.2}$$

As Φ has dimensions of FORCE and as the only inhomogeneity, if the rod is materially uniform, comes from the variation of the cross-sectional area A, it is natural to introduce a dimensionless strain-energy density ϕ by setting

$$\Phi \equiv EA\phi(e) \; , \tag{J.3}$$

where it is understood that E (Young's modulus), A, and the dimensionless strain-energy density ϕ may depend on z. Furthermore, in a manner that we shall not attempt to make explicit, ϕ may also depend on the variable geometry of the cross section.

1. Restrictions on the strain-energy density

There are four restrictions that we may place on the functional form of $\phi(e)$[3]. First, we may set $\phi(0)=0$ because it is only $\phi,_e$ that affects the stress in the rod. Second, if there are no initial stresses, $\phi,_e(0)=0$. Third, $\phi(e)$ must be a continuous function of e (so that small changes in e produce only small changes in N). Fourth, and most important, if $\phi(e)$ is to represent a *stable material* in the sense of Rivlin, then *the birod must transmit infinitesimal, longitudinal waves at every level of strain.* (D.12) shows that this implies that if $[e] > 0$, then $[N] > 0$, i.e., for each z, N must be a *strictly monotonically increasing* function of e. See Hayes & Rivlin (1961), Rivlin (1980, 1981), and Sawyers & Rivlin (1984) for further references on this concept in 3-dimensions.

2. The equation of motion in displacement and intrinsic forms

Substitution of $N=\Phi,_e$ into the differential equation of longitudinal motion, (C.2), yields the widely-used *displacement form*

$$(\Phi,_e)' + p = m\ddot{u} \; , \tag{J.4}$$

where the strain-energy density $\Phi = EA\phi(u')$, via (F.5) and (J.3). If $\phi = \tfrac{1}{2}u'^2$, then (J.4) reduces to the well-known linear equation of rod dynamics

---

[3]It is to be understood that, in general, ϕ depends on the material coordinate z as well.

$$(EAu')' + p = m\ddot{u} . \tag{J.5}$$

Typical boundary conditions for (J.4) or (J.5) require that at the ends of the rod, u or N be specified functions of time; these include the case where u is specified at one end and N at the other. The most general boundary conditions in a mechanical theory are relations of the form $B_0(u_0, u_L, N_0, N_L, t)=0$, $B_L(u_0, u_L, N_0, N_L, t)=0$, where $u_0 = u(0, t)$, etc. in the notation of Section G.

Initial conditions require the location and velocity of each point of the centerline at $t=0$, and hence are of the form

$$u(z, 0) = \overline{u}(z) \ , \ \dot{u}(z, 0) = \overline{v}(z) . \tag{J.6}$$

If the rod is initially unstressed, then $\overline{u}(z) = \text{constant}$.

In anticipation of its importance in the formulation of plate and shell equations, we present now the *intrinsic form* of the equation of motion of a birod. This form is obtained by dividing (J.4) by m, differentiating the resulting equation with respect to z and setting $u' = e$, with the result

$$[(\Phi,_e)'/m]' + (p/m)' = \ddot{e} . \tag{J.7}$$

Now e is the field variable instead of u. It is necessary, of course, that $(p/m)'$ be, at most, a function of e and z, i.e., it must not depend on u itself. If $\Phi = ½EAe^2$, (J.7) becomes

$$[(EAe)'/m]' + (p/m)' = \ddot{e} . \tag{J.8}$$

A true intrinsic formulation should also avoid the appearance of u in the initial and boundary conditions. To obtain initial conditions on e, we merely differentiate (J.6) with respect to z to obtain

$$e(z, 0) = \overline{u}'(z) \ , \ \dot{e}(z, 0) = \overline{v}'(z) . \tag{J.9}$$

Geometric boundary conditions for e follow upon differentiating boundary conditions on u with respect to time and substituting the result into the equation of motion (J.4), evaluated at $z = 0, L$. Thus, if $u_0 = h_0(t)$ and $u_L = h_L(t)$, the boundary conditions on e take the form

$$(\Phi,_e)' = m\ddot{h}_0 - p \ \text{at} \ z = 0 \ , \ (\Phi,_e)' = m\ddot{h}_L - p \ \text{at} \ z = L . \tag{J.10}$$

Boundary conditions on N do not change.

The derivation of the intrinsic initial/boundary value problem for the longitudinal motion of a birod required space and time differentiations. While this introduces smoothness restrictions on the functions involved, it offers an important advantage: uniform distributed loads (such as those resulting from taking the equation of motion in a frame in constant acceleration) drop out of the equation of motion, reappearing only in kinematic boundary conditions of the form (J.10). Also, the intrinsic initial/boundary value problem is unaffected by a superimposed uniform longitudinal velocity. This may be of value in dynamical problems in which small strains are superimposed on large, almost uniform motions.

K. Variational Principles

The Calculus of Variations generalizes the ordinary Calculus of One Variable. In the latter, we deal with numerical-valued functions defined on sets of real numbers; in the former, we deal with function-valued *operators* defined on sets of *admissible functions*. In calculus, the concept and computation of the differential of a function f, $df=f'(x)dx$, is basic. Here, $f'(x)$ is the derivative of f at x and dx is any real number. Thus, df is a function of *two* variables, x and dx, and is *linear* in the second. The smaller dx the better df approximates the change in f when x is replaced by $x+dx$.

In the Calculus of Variations, the generalization of the differential of a function f is the *variation* of an operator $\mathcal{F}$, denoted variously by

$$\delta\mathcal{F} = \delta\mathcal{F}[u, \delta u] = \mathcal{F}'(u)\delta u \ . \tag{K.1}$$

Here, u belongs to some set $\mathcal{A}$ of admissible functions and δu, called the *variation of u*, belongs to a related linear vector space $\mathcal{A}_\delta$, called the *set of admissible variations of u*, such that if $u \in \mathcal{A}$ then $u+\delta u \in \mathcal{A}$. The middle expression in (K.1) is to be read: the variation of $\mathcal{F}$ at u in the "direction" of δu; the last expression in (K.1) emphasizes the analogy with $df=f'(x)dx$ and the fact that $\delta\mathcal{F}$ is linear in δu; $\mathcal{F}'(u)$ is called the *Fréchet derivative of $\mathcal{F}$ at u*.

Of particular importance in what follows are numerical-valued operators or *functionals*. The smaller the magnitude of δu, measured in some appropriate norm, the better the variation of a functional approximates the change in the functional when u is replaced by $u+\delta u$. We refer the reader to books by Gelfand & Fomin (1963), Vainberg (1964), and Oden & Reddy (1983) where precise definitions are given and relevant theorems established.

The aim of the various discussions of Variational Principles, in this and other chapters, boils down to determining when an expression of the form $\mathcal{G}(u)\delta u$ can be written as the variation of a functional $\mathcal{H}(u)$. There seem to be at least two advantages over and above the Principle of Virtual Work in formulating a mechanics problem as $\delta\mathcal{F}=0$. First, if the solution of $\delta\mathcal{F}=0$ represents a minimum or maximum of $\mathcal{F}$ and not merely a stationary point, then valuable upper or lower bounds on $\mathcal{F}$ may be constructed by approximate methods. Second, a variational principle may sometimes be used to prove the existence of a solution. See, for example, the book by Berger (1977).

We shall assume that the mechanics of computing $\delta\mathcal{F}$, given $\mathcal{F}$, is familiar to the reader and we shall adopt, as needed and without elaboration, results of the Calculus of Variations. Thus, the variations of the space and time partial derivative operators are given by

$$\delta(u') = (\delta u)' \ , \ \delta(v) = (\delta u)^{\cdot} \ . \tag{K.2}$$

The starting point for our derivation of variational principles is the weak form of the equation of motion, (E.1). Setting $a=0$ and $b=L$, we obtain

$$\int_{t_1}^{t_2} NV\,|_0^L dt - \int_0^L mvV\,|_{t_1}^{t_2} dz - \int_{t_1}^{t_2}\int_0^L (NV' - mv\dot{V} - pV)dzdt = 0\,. \quad \text{(K.3)}$$

For a birod, we shall take $\mathcal{A}$ to consist of all functions u that have piecewise continuous partial derivatives u' and $\dot{u}=v$ on the open rectangle $(0,L)\times(t_1,t_2)$ and that, for the present, satisfy classical boundary conditions (as defined at the end of Section G). Further, in (K.3), we set $V=\delta u$ and take $\mathcal{A}_\delta$ to consist of all functions with similar smoothness conditions on $(0,L)\times(t_1,t_2)$, but which satisfy *homogeneous forms* of the classical boundary conditions. Finally, we denote by ∂U an end where displacement or velocity is specified and by ∂N an end where an axial force $N=\hat{N}$ is specified. Obviously, $\delta u=0$ on ∂U so that (K.3) may be written

$$\int_{t_1}^{t_2} \hat{N}\delta u|_{\partial N} dt - \int_0^L mv\delta u\,|_{t_1}^{t_2} dz - \int_{t_1}^{t_2}\int_0^L (N\delta u' - mv\delta v - p\delta u)dzdt = 0\,. \quad \text{(K.4)}$$

We shall proceed further in two stages.

1. Hamilton's Principle

We assume that the shell is elastic in the sense of Section J. Then, with $\mathcal{K}$ denoting the reduced kinetic energy as in (F.3), and with

$$\mathcal{E} = \int_0^L \Phi dz \quad \text{(K.5)}$$

denoting the internal energy, we can write (K.4) in the form

$$\delta\int_{t_1}^{t_2} (\mathcal{K} - \mathcal{E})dt = -\int_{t_1}^{t_2} (\hat{N}\,\delta u\,|_{\partial N} + \int_0^L p\delta u dz)dt + \int_0^L mv\delta u dz\,|_{t_1}^{t_2}\,. \quad \text{(K.6)}$$

If the distributed load p is a function of u, z, and t only and if the given edge forces are functions of u and t only, then we can introduce a *load potential*

$$\mathcal{P} \equiv -\int_0^u \hat{N}(w,t)\,|_{\partial N} dw - \int_0^L [\int_0^u p(w,z,t)dw]dz \quad \text{(K.7)}$$

so that (K.6) reduces to

$$\delta\int_{t_1}^{t_2} (\mathcal{K} - \mathcal{E} - \mathcal{P})dt = \int_0^L mv\delta u dz\,|_{t_1}^{t_2}\,. \quad \text{(K.8)}$$

If conditions are such that $v\delta u$ vanishes at both t_1 and t_2, then the right side of this equation vanishes and we have *Hamilton's Principle* for a birod:

$$\delta\int_{t_1}^{t_2} (\mathcal{K} - \mathcal{E} - \mathcal{P})dt = 0\,. \quad \text{(K.9)}$$

The integral in (K.9) is sometimes referred to as the *action integral* and its integrand, the *action*. Note that the steps leading from (K.4) to Hamilton's Principle may be reversed.

In summary, then, any function u that: (1), has piecewise continuous derivatives on $(0,L)\times(t_1,t_2)$; (2), satisfies classical kinematic boundary conditions; (3), is such that either v vanishes or u is specified at t_1 and t_2; and (4), renders the action integral stationary, also satisfies (K.4) and hence is a weak solution. Conversely, if u is a weak solution, δu an admissible variation, the birod elastic, and the external loads derivable from a potential, then u renders the action integral stationary.

The requirement that $v\delta u$ vanish at t_2 is inconvenient because we are usually given u and v at t_1 only, i.e., we usually deal with *initial value problems* where nothing is known a priori at t_2. This restricts the *direct use* of the stationary properties of the action integral to cases where the physics of the problem (e.g., periodicity) yields a priori information at t_2 or where a sequence of approximations has a fixed but unknown value at t_2.

The development of variational principles for *linear* initial/boundary value problems was initiated by Gurtin (1964), the key idea being to replace time integration from t_1 to t_2 of the product of two functions by their *convolution* over this interval. See Chapter III of Gurtin's 1972 article in the *Encyclopedia of Physics* and the paper by Herrera & Bielak (1974) for further development of these ideas. Whether they will prove to be fruitful in nonlinear shell problems remains an open question.

Hamilton's Principle (K.9) has been used widely as a starting point for the development of approximate methods of solution, such as the popular finite element method, or for deriving restricted forms of the equation of motion. However, when the details of many of these approximations are examined, it is seen that, in reality, the more fundamental Principle of Virtual Work or the weak form of the equations have been used. The "test functions" V in (E.1) have not necessarily been variations of u and $v\delta u$ has not necessarily been set to zero at t_2. For further (somewhat controversial) discussion on using Hamilton's Principle as a tool for the approximate solution of time-dependent problems, see papers by Baruch & Riff (1982), Bailey (1983), and the ensuing discussions by Leipholz (1983a,b), Bailey (1984), Baruch & Riff (1984), and Smith (1984a,b), where additional references will be found.

2. *Variational principles for elastostatics*

(a) Principle of stationary total potential energy. If the birod is in static equilibrium so that $v = \mathcal{K} = 0$ in (K.8), then, because the time interval of integration is arbitrary, we have

$$\delta(\mathcal{E} + \mathcal{P}) = 0\,. \tag{K.10}$$

The functional $\mathcal{E} + \mathcal{P}$ is the *total potential energy* and (K.10) expresses the well-known *Principle of Stationary Total Potential Energy* for a birod: among all admissible displacements u, i.e., among all sufficiently smooth functions that satisfy the kinematic constraints (if any) at $z = 0$ and $z = L$, only those that satisfy the equilibrium equation $(\Phi,_e)' + p = 0$ and the prescribed boundary conditions on ∂N make the functional $\mathcal{E} + \mathcal{P}$ stationary. Many regard this to be the premier variational principle in elastostatics. In *linear* elastostatics, where $\mathcal{E}$ and $\mathcal{P}$ are, respectively, quadratic and linear functionals of u, it can be shown that the solution of the equilibrium equations renders the functional $\mathcal{E}+\mathcal{P}$ an absolute minimum. This need not be so in nonlinear elastostatics.

(b) Extended variational principles. The total potential energy functional $\mathcal{E}+\mathcal{P}$ depends on admissible displacements only. The strain e and the force N are *derived* quantities which do not appear explicitly in the functional. For several theoretical and practical reasons related to ill-conditioning and overly restrictive smoothness requirements (which reveal their true significance only in 2-dimensional shell theories), it is useful to try to replace this principle by alternative, *extended* principles. The following paths may be explored.

(i) The set of admissible displacements may be enlarged by relaxing smoothness or constraint conditions.

(ii) More of the field variables may be allowed to vary independently.

(iii) One may attempt to make N (instead of u) the only quantity to be varied. (Complementary Energy Principle).

The key idea in fashioning extended variational principles from the Principle of Stationary Total Potential Energy is to turn relations that are satisfied a priori in the latter, such as stress-strain and strain-displacement equations or displacement boundary conditions, into stationary conditions in the former. This is achieved with Lagrange multipliers. An enlightening discussion of these ideas in a purely mathematical setting may be found in Courant & Hilbert (1953, vol. 1, Chapt. IV, Sect. 9).

One of the better-known principles is that of Hu (1955) and Washizu (1955, 1982) which states that

$$\delta\Pi_1 \equiv \delta[\mathcal{E}+\mathcal{P}-\int_0^L (e-u')N dz - (u-\hat{u})\tilde{N}\,|_{\partial U}] = 0\,. \tag{K.11}$$

Here, the set of admissible functions are displacements, strains, and axial forces, *plus* the variable $\tilde{N}$ which is taken to be independent of N on ∂U. The fields u, e, and N are subject to *no* constraints save that they be smooth enough for the integrals that appear explicitly and implicitly in (K.11) to exist; a hat (ˆ) denotes a prescribed value. Note that N and N play the role of Lagrange multipliers in (K.11).

To show that all the relations assumed a priori in the Principle of Stationary Total Potential Energy are stationary conditions in the Hu-Washizu Principle, we carry out the variation in (K.11) and integrate by parts to obtain

$$\int_0^L [(\Phi,_e - N)\delta e - (N' + p)\delta u - (e - u')\delta N]dz$$
$$+ (N-\tilde{N})\delta u\,|_{\partial U} - (u-\hat{u})\delta\tilde{N}\,|_{\partial U} + (N-\hat{N})\delta u\,|_{\partial N} = 0\,. \tag{K.12}$$

Since the variations of e, u, N, and $\tilde{N}$ are independent of one another, the vanishing of all terms in parentheses is a necessary condition for stationarity, *if it exists*. It is possible, of course, to retrace the steps to obtain (K.11) from (K.12) so that the converse of the theorem can also be inferred: if the expressions in parentheses vanish, then (K.11) holds.

A modified, slightly less general form of the Hu-Washizu functional Π_1 is obtained by taking $\tilde{N}=N|_{\partial U}$, which leads to the principle

$$\delta\Pi_2 \equiv \delta[\mathcal{E}+\mathcal{P}-\int_0^L (e-u')Ndz-(u-\hat{u})N|_{\partial U}]=0\,. \tag{K.13}$$

Another variational principle, associated with the names of Hellinger (1914) and E. Reissner (1950, 1970), can be obtained from (K.12) if we set $\Phi,_e=N$, $e=u'$, and assume that the former equation can be solved for e as a function of N. We may then introduce a *stress-energy density* Ψ by the Legendre transformation

$$\Phi - Ne = \Psi(N) \tag{K.14}$$

from which it follows that $\Psi,_N=-e$ and hence

$$\delta\Psi = -e\delta N\,. \tag{K.15}$$

These results, substituted into (K.11), yield

$$\delta\Pi_3 \equiv \delta[\int_0^L (Nu'+\Psi)dz+\mathcal{P}-(u-\hat{u})\tilde{N}|_{\partial U}]=0\,. \tag{K.16}$$

The admissible functions in the Hellinger-Reissner Principle are u, N, and $\tilde{N}$, subject to no constraints save that u and N be sufficiently smooth for all integrals in (K.16) to exist. Carrying-out the variation in (K.16) and integrating the u'-term by parts, we obtain

$$\begin{aligned}\int_0^L [(u'+\Psi,_N)\delta N-(N'+p)\delta u]dz+(N-\hat{N})\delta u\,|_{\partial N}\\ +[(N-\tilde{N})\delta u-(u-\hat{u})\delta\tilde{N}]_{\partial U}=0\,.\end{aligned} \tag{K.17}$$

The stationary conditions are that the expressions in parentheses vanish. The use of Ψ as an energy density in shell theory has several advantages, as we shall discuss and illustrate in Sections N and O of the next chapter on beamshells.

The possibility of deriving a true *Complementary Energy Principle* hinges on completely eliminating displacements from (K.16). The analogous step for shells may prove to be difficult if the strain-energy density is not quadratic. [For a discussion of the problems encountered in 3-dimensional bodies, see de Veubeke (1972), Koiter (1973), Ogden (1977), Washizu (1982), and Wempner (1980), where additional references may be found]. However, for birods, the elimination of u is easy.

We assume, as before, that $p=p(u)$ and $\hat{N}=\hat{N}(u)$, but now require that these functions have unique inverses on $[0, L]$ and ∂N respectively. Under these conditions we can introduce, via a Legendre transformation, a *complementary load potential*

$$\mathcal{P}(p,\hat{N}) \equiv \int_0^L pu(p)dz + \hat{N}u(\hat{N})|_{\partial N} + \mathcal{P}(p,\hat{N}) \,, \tag{K.18}$$

where the load potential $\mathcal{P}$ comes from (K.7). The variation of $\mathcal{P}$ is given by

$$\delta\mathcal{P} = \int_0^L u\delta p dz + u\delta\hat{N}|_{\partial N} \,, \tag{K.19}$$

so that for constant p and $\hat{N}$, $\delta\mathcal{P}=0$. Introducing $\mathcal{P}$ into (K.16), integrating the term Nu' by parts, and identifying $\tilde{N}$ with N on ∂U, we arrive at

$$\delta\{-\int_0^L [-\Psi + (N' + p)u]dz + \hat{u}N|_{\partial U} + (N - \hat{N})u|_{\partial N} + \mathcal{P}\} = 0 \,. \tag{K.20}$$

We now restrict the admissible axial forces to satisfy $N' + p = 0$ on $0 < z < L$ and $N = \hat{N}$ on ∂N. Under these conditions all the u-terms in (K.20) (except $\hat{u}$) drop out, $\mathcal{P}$ becomes a function of N only, and we obtain the *Complementary Energy Principle*

$$\delta\Pi_4 \equiv \delta[\hat{u}N|_{\partial U} + \mathcal{P}(N) + \int_0^L \Psi dz] = 0 \,, \tag{K.21}$$

N being the only function to be varied. The stationary conditions for this principle can be obtained by performing the indicated variation in (K.21) while taking into account (K.19) and the admissibility requirements on N. The result is

$$\int_0^L (u' + \Psi_{,N})\delta N dz - (u - \hat{u})\delta N|_{\partial U} = 0 \,, \tag{K.22}$$

so that the strain-displacement and edge compatibility relations must be satisfied for stationarity.

In anticipation of subsequent chapters and to preserve uniformity of presentation, we construct a new functional from Π_4 by first introducing the *stretch* $\lambda \equiv 1 + e$ as a new argument of the strain-energy density Φ (a common procedure in strongly nonlinear problems). We then introduce the *modified stress-energy density*

$$\Omega \equiv \Psi - N = \Phi(\lambda) - \lambda N \,, \tag{K.23}$$

where $\lambda = \lambda(N)$. Substituting (K.23) into (K.21) and using the relation

$$\int_0^L (N - pz)dz = (zN)_0^L \,, \tag{K.24}$$

we obtain, finally,

$$\delta\Pi_5 \equiv \delta[(\hat{\bar{z}N})_{\partial U} + \mathcal{P}_1 + \int_0^L \Omega dz] = 0 \,, \tag{K.25}$$

where $\bar{z} \equiv z + u$ and

$$\mathcal{P}_1(p,\hat{N}) \equiv \int_0^L p\bar{z}(p)dz + \hat{N}\bar{z}(\hat{N})|_{\partial N} + \mathcal{P}(p,\hat{N}) \,. \tag{K.26}$$

It can be easily shown for the important case of "dead" loads, where p and $\hat{N}$ do not depend on u (or $\bar{z}$), that $\mathcal{P} = \mathcal{P}_1 = 0$.

The main outcome of the above modifications is that the *displacement u* has been replaced by the *deformed position* as the argument in the variational principle. The functional Π_3 can be modified in a similar way.

This concludes our review of some of the better-known variational principles as applied to birods. The particularly simple and transparent form these principles have taken should help in their application to beamshells, axishells, and general shells. The reader must be forewarned, however, that some of the most important and complicated aspects of 2-dimensional shell theory, such as curvature, large rotations, strain compatibility, and conditions along curved boundaries, do not have counterparts in the theory of 1-dimensional birods so that challenges and disappointments await.

L. Thermal Equations

We now apply (II.C.1) and (II.D.1) to a slice of a birod between $z=a$ and $z=b$, using the geometrical relations developed in Section B.

With the cold flux vector of Section II.C written in the component form

$$\mathbf{c} = c_x\mathbf{e}_x + c_y\mathbf{e}_y + c_z\mathbf{e}_z \ , \tag{L.1}$$

it follows from (B.1), (B.3), and (II.C.1) that the heating of the slice of the birod can be expressed as

$$\mathcal{Q} = q\,\Big|_a^b + \int_a^b s\,dz \ , \tag{L.2}$$

where

$$q \equiv \int_{C_z} c_z\,dxdy \tag{L.3}$$

is the *heat influx* from right to left across a section and

$$s \equiv \int_{\partial C_z} \hat{\mathrm{v}}\,ds + \int_{C_z} n\,dxdy \tag{L.4}$$

is the external *centerline heating* resulting from heat flux through the side of the birod plus heat generated in the birod (say from radiation or chemical reactions). From (II.C.2) and (B.3),

$$\hat{\mathrm{v}} \equiv (c_x\hat{y}_{,s} - c_y\hat{x}_{,s} + c_z J) \ . \tag{L.5}$$

Turning to the Clausius-Duhem inequality and proceeding as above, we find that (II.D.2) reduces to

$$\mathcal{H} = \int_a^b \iota\,dz\,\Big|_{t_1}^{t_2} \ \text{and}\ \mathcal{J} = \int_{t_1}^{t_2} (j\,\Big|_a^b + \int_a^b \nu\,dz)\,dt \ . \tag{L.6}$$

Here

$$\iota \equiv \int_{C_*} \eta\, dxdy \tag{L.7}$$

is the *entropy resultant*,

$$j \equiv \int_{C_*} \frac{c_z}{\theta} dxdy \tag{L.8}$$

is the *entropy flux* from right to left through the section C_z, and

$$\nu \equiv \int_{\partial C_*} \frac{\hat{\nu}}{\theta} ds + \int_{C_*} \frac{n}{\theta} dxdy \tag{L.9}$$

is the *entropy supply*.

In our descent from 3-dimensions, 4 *internal* thermal variables have emerged: q, ι, j, and ν. The 1-dimensional heat influx s, like its mechanical analog, the 1-dimensional external load p, is to be regarded as prescribed. As we shall see in Section N, a 1-dimensional temperature field $T(z, t)$ appears automatically as the coefficient of ι in a reduced energy equation, just as the 1-dimensional strain e appeared automatically as the coefficient of the axial force N in the Mechanical Work Identity of Section F.

M. The First Law of Thermodynamics

The longitudinal equation of motion (B.15), the Mechanical Work Identity (F.1), the heat equation (L.2), and the Clausius-Duhem inequality (II.D.1) with (L.6) are *exact* (though incomplete) consequences of the 3-dimensional equations of continuum mechanics. To couple the mechanical and thermal variables in these equations, we *postulate* (rather than attempt to derive from 3-dimensions) the following *First Law of Thermodynamics* (Conservation of Energy): There exists an *internal energy* I such that

$$\int_{t_1}^{t_2} (\mathcal{W} + \mathcal{Q}) dt = (\mathcal{K} + I)_{t_1}^{t_2} . \tag{M.1}$$

Here $\mathcal{W}$ is defined by (F.2), $\mathcal{Q}$ by (L.2), and $\mathcal{K}$ by (F.3). In view of the Mechanical Work Identity (F.1), (M.1) is equivalent to the *reduced equation of energy balance*,

$$\int_{t_1}^{t_2} (\mathcal{D} + \mathcal{Q}) dt = I \Big|_{t_1}^{t_2} . \tag{M.2}$$

N. Thermoelastic Birods

The definition of a *thermoelastic birod* is based on several assumptions. The first is that there exists an *internal energy density* $i(e, \iota, z)$ such that

$$I = \int_a^b i dz \, . \tag{N.1}$$

Inserting (F.4), (L.2), and (N.1) into (M.2) and assuming that all fields are sufficiently smooth, we conclude, in view of the arbitrariness of the intervals (a, b) and (t_1, t_2), that (M.2) implies the *local equation of reduced energy balance*

$$N\dot{e} + q' + s = \dot{i} \, . \tag{N.2}$$

Likewise, (II.D.1) and (L.6) imply the *local Clausius-Duhem inequality,*

$$\dot{\iota} \geq j' + \nu \, . \tag{N.3}$$

To obtain a more useful form of (N.2), we introduce the so-called *free-energy density* by setting

$$\Phi \equiv i - T\iota \, , \tag{N.4}$$

where

$$T \equiv -i_{,\iota} \tag{N.5}$$

is the *apparent absolute temperature* of the birod. We shall assume that $T > 0$. Using (N.4), we may express (N.2) as

$$N\dot{e} - \iota\dot{T} - \dot{\Phi} + q' = T\dot{\iota} - s \, . \tag{N.6}$$

With a temperature field now defined, it is useful to rewrite (N.3) so that it resembles as closely as possible the local form of the 3-dimensional Clausius-Duhem inequality which, from (II.D.1), states that the time-rate-of-change of the entropy density is greater than or equal to the divergence of a quantity (the heat influx vector divided by the absolute temperature) plus the local heating divided by the absolute temperature. To this end we set

$$\alpha \equiv Tj \tag{N.7}$$

$$\beta \equiv \int_{\partial C_\cdot} \hat{\nu} \left[\frac{1}{\theta} - \frac{1}{T} \right] ds + \int_{C_\cdot} n \left[\frac{1}{\theta} - \frac{1}{T} \right] dA \tag{N.8}$$

so that (N.3) reads

$$\dot{\iota} \geq (\alpha/T)' + s/T + \beta \, . \tag{N.9}$$

Inserting (N.9) into (N.6), we have

$$N\dot{e} - \iota\dot{T} - \dot{\Phi} + \delta' + \alpha T'/T \geq T\beta \, , \tag{N.10}$$

where $\delta \equiv q - \alpha$.

Our definition of a thermoelastic birod assumes a constitutive equation for the free-energy density of the form

$$\Phi = \hat{\Phi}(\Lambda, z), \tag{N.11}$$

where

$$\Lambda \equiv (e, T, \beta, T') \tag{N.12}$$

is an *argument list.* The *Principle of Equipresence* states that "A variable present as an independent variable in one constitutive equation should be present in all." (Truesdell & Toupin, 1960, p. 704). Thus, to be consistent with (N.11), we assume that

$$N = N(\Lambda, z) \ , \ \iota = \iota(\Lambda, z) \ , \ q = q(\Lambda, z) \ , \ \alpha = \alpha(\Lambda, z) . \tag{N.13}$$

We now wring several important consequences from (N.10).

Using (N.11), (N.13), and the chain rule to express $\dot{\Phi}$ and δ' in (N.10) in terms of time and space derivatives of e, T, β, and T', we obtain an expression that we may rewrite as follows:

$$\begin{aligned} &[N - \Phi,_e]\dot{e} - [\iota + \Phi,_T]\dot{T} - [\Phi,_\beta]\dot{\beta} - [\Phi,_{T'}]\dot{T}' \\ &+ [\delta,_e]e' + [\delta,_\beta]\beta' + [\delta,_{T'}]T'' \\ &+ [(\delta,_T + \alpha/T)T'] \geq [T\beta] . \end{aligned} \tag{N.14}$$

Now consider the fields

$$e = e_* + e'_*(z - z_*) + \dot{e}_*(t - t_*) \tag{N.15}$$

$$T = T_* + T_*'(z - z_*) + \dot{T}_*(t - t_*) + \tfrac{1}{2}T''_*(z - z_*)^2 + \dot{T}'_*(z - z_*)(t - t_*) \tag{N.16}$$

$$\beta = \beta_* + \beta'_*(z - z_*) + \dot{\beta}_*(t - t_*) , \tag{N.17}$$

where quantities with an asterisk are constants that we assume can be assigned at will. Associated values of $(\Phi, N, \iota, q, \alpha)$ follow from (N.7) and (N.13) and we choose p and s so that the equation of motion (C.2) and the reduced energy equation (N.6) are satisfied.

Turning to (N.14), we note that the bracketed terms become functions of Λ_* only when evaluated at (z_*, t_*). Thus, such terms *in the first two lines of* (N.14) must be zero, for if any were not, we could always choose their coefficients, which are elements of Λ'_* and $\dot{\Lambda}_*$, so that the inequality is violated. That is, (N.11)-(N.14) imply that

$$\Phi = \hat{\Phi}(e, T) \tag{N.18}$$

$$N = \Phi,_e \ , \ \iota = -\Phi,_T , \tag{N.19}$$

and $\delta = \hat{\delta}(T)$. But if z is replaced by $-z$ (a permissible change of variable), δ changes sign while T does not. Thus, $\delta \equiv 0$, i.e., $\alpha = q$, and (N.14) at (z_*, t_*) reduces to

$$q_* T'_* \geq T_*^2 \beta_* . \tag{N.20}$$

Note by (N.18) and (N.19) that the reduced energy equation (N.6) becomes

$$q' = T\dot{\iota} - s\,. \tag{N.21}$$

Now unless $q_* \to \infty$ as $T'_* \to 0$, which is physically absurd, it follows from (N.20) that

$$\beta_* \leq 0\,, \tag{N.22}$$

since β_* is assumed to be independent of T'_*.

Finally, because $q = \hat{q}(e, T, \beta, T')$ must change sign if z is replaced by $-z$, q must have the functional form

$$q = T'\hat{K}(e, T, \beta, T'^2)\,, \tag{N.23}$$

where K is the *thermal conductivity* of the birod.

As the role of β appears to be minor, it seems simplest in the end to assume that K is independent of β and then to note, in view of (N.22), that a *sufficient* condition for (N.20) to hold is that K be non-negative, i.e.,

$$K = \hat{K}(e, T, T'^2) \geq 0\,. \tag{N.24}$$

The theory we have outlined represents the simplest possible extension of the Mechanical Theory of Section I.

But is $\beta \leq 0$ a reasonable requirement? Consider the definition of β, (N.8). This is a peculiar expression because it involves both prescribed and unknown variables. Suppose that $n = 0$. In a "slow" process in which $\hat{v} > 0$ everywhere, heat flows into the birod through ∂C_z. Thus, because T, in some sense, measures the average of θ over C_z, θ on ∂C_z can be expected to be greater than T so that the integrand $\hat{v}(\theta^{-1} - T^{-1})$ in (N.8) should be negative. Conversely, if $\hat{v} < 0$ everywhere, then heat flows out across ∂C_z and θ should be less than T. Again, $\hat{v}(\theta^{-1} - T^{-1})$ will be negative. Of course, if $\hat{v}$ is positive on some parts of ∂C_z and negative on others or if $n \neq 0$, the picture is not so clear. Moreover, it would seem that β could be made positive in a "fast" thermal process, for if $\beta < 0$, then one could instantaneously change the sign of $\hat{v}$ and n which, because of the lagging temperature fields, would make β positive, at least briefly. We have no doubt that a 3-dimensional analysis—that could be linear yet convincing—would reveal that if an oscillating heat flux (or source) were applied, then any period of oscillation that produced positive values of β would be small compared to any typical time scale in conventional, 1-dimensional rod theory. We leave this as an open problem.

To the reader uneasy with (N.22), we offer two alternatives.

(a) Ignore the $T\beta$-term in (N.10) as one of the assumptions of the theory in thermoelastic birods. This is not as *ad hoc* as it might appear if one observes that the 1-dimensional energy balance equation (N.21) is already approximate: it follows from its 3-dimensional counterpart (by integrating over C_z) only if one neglects the difference

$$T\dot{e} - \int_{C_*} \theta\dot{\eta}\, dA \, . \tag{N.25}$$

[See the analogous discussion in Section F regarding the difference between the actual (3-dimensional) kinetic energy of a birod and the reduced (1-dimensional) kinetic energy].

(b) Make the more radical assumption—as we must do with shells—that β and its derivatives at (z_*, t_*) cannot be assigned independently of the elements of Λ_* and their derivatives, but rather that there exists a differential equation of the form

$$\Phi,_{\beta}\dot{\beta} = T\beta + C \, , \tag{N.26}$$

where $C = \hat{C}(\Lambda, z)$ is a constitutive function to be determined by experiment or by descent from 3-dimensions; C must vanish if $\beta = 0$. A reconsideration of the analysis stemming from (N.14) and (N.15) shows that it is still true that $\alpha = q$ while (N.18) and (N.24) need only be modified by adding β to the list of arguments.

Again, we leave it as on open problem to justify these two alternatives.

1. Linearized constitutive equations for N, h, and ι

Let the birod be stress- and entropy-free in its reference shape and at a constant temperature T_0. Then, with $T = T_0 + \tau$, it follows from (N.18) and (N.19) that, if the constitutive equations for N and ι are to be linear, the free-energy density must have the quadratic form

$$\Phi = \tfrac{1}{2}Ee^2 - \tilde{\alpha}e\tau - \tfrac{1}{2}(cm/T_0)\tau^2 \, , \tag{N.27}$$

where $\tilde{\alpha}$ is the *coefficient of thermal expansion* and c is the *specific heat at constant volume* (Dawson, 1976). Of course, E, $\tilde{\alpha}$, and c may vary with z. The constitutive equation for q is simply

$$q = K\tau' \, , \tag{N.28}$$

where $K = K(z) > 0$.

The equation of motion (C.2) and the equation of reduced energy balance (N.21) take the form

$$(Eu' - \tilde{\alpha}\tau)' + p = m\ddot{u} \tag{N.29}$$

$$(K\tau')' - (T_0 + \tau)(\tilde{\alpha}u' + cm/T_0\tau)^{\bullet} + s = 0 \, . \tag{N.30}$$

These two equations form the basis for a coupled thermoelasticity theory of birods. Lack of space precludes any development in this book. However, except for the nonhomogeneous terms, (N.29) and (N.30) are formally identical to the two equations that are studied extensively in a monograph by Day (1985).

References

[A letter in brackets indicates the section where the reference appears.]

Antman, S. S. (1980). The equations for large vibrations of strings. *Am. Math. Monthly,* **87,** 359-370. [B]

Bailey, C. D. (1983). Comments on 'Finite elements for initial value problems in dynamics.' *AIAA J.* **21,** 159-160. [K]

Bailey, C. D. (1984). Reply to remarks by C. V. Smith, Jr. *AIAA J.* **22,** 1179-1181. [K]

Baruch, M., and Riff, R. (1982). Hamilton's principle, Hamilton's law—6^n correct formulations. *AIAA J.* **20,** 687-692. [K]

Baruch, M., and Riff, R. (1984). Reply by authors to C. V. Smith, Jr. *AIAA J.* **22,** 1182-1183. [K]

Batra, R. C. (1972). On non-classical boundary conditions. *Arch. Rat. Mech. Anal.* **48,** 163-191. [G]

Berger, M. S. (1977). "Nonlinearity and Functional Analysis." Academic Press, New York, etc. [K]

Courant, R., and Hilbert, D. (1953). "Methods of Mathematical Physics," vol 1. Interscience, New York. [K]

Dawson, T. H. (1976). "Theory and Practice of Solid Mechanics." Plenum, New York and London. [N]

Day, W. A. (1985). "Heat Conduction within Linear Thermoelasticity." Springer-Verlag, New York, etc. [N]

Dieudonné, J. A. (1984). Review of the book "The Prehistory of the Theory of Distributions," by J. Lützen. *Am. Math. Monthly* **91,** 374-379. [E]

de Veubeke (1972). A new variational principle for finite elastic displacements. *Int. J. Engr. Sci.* **10,** 745-763. [K]

Gelfand, I. M., and Fomin, S. V. (1963). "Calculus of Variations" (Transl. from the Russian by R. Silverman). Prentice-Hall, Englewood Cliffs, New Jersey. [K]

Goldstein, H. (1980). "Classical Mechanics," 2nd ed. Addison-Wesley, Reading, Massachusetts, etc. [G]

Gurtin, M. E. (1964). Variational principles for linear elastodynamics. *Arch. Rat. Mech. Anal.* **16,** 34-50. [K]

Gurtin, M. E. (1972). The linear theory of elasticity. *In* "Encyclopedia of Physics" (C. Truesdell, ed.), vol. VIa/2. Springer-Verlag, Berlin, Heidelberg. [K]

Hayes, M., and Rivlin, R. S. (1961). Propagation of a plane wave in an isotropic elastic material subjected to pure homogeneous deformation. *Arch. Rat. Mech. Anal.* **8,** 15-22. [J]

Hellinger, E. (1914). Der allgemeine Ansätz der Mechanik der Kontinua. *Encylopadie der Mathematischen Wissenschaften* **4,** part 4, 602-694. [K]

Herrera, I., and Bielak, J. (1974). A simplified form of Gurtin's variational principle. *Arch. Rat. Mech. Anal.* **53,** 131-149. [K]

Hu, H. -C. (1955). On some variational principles in the theory of elasticity and the theory of plasticity. *Scientia Sinica* **4,** 33-54. [K]

Koiter, W. T. (1973). On the principle of stationary complementary energy in the nonlinear theory of elasticity. *SIAM J. Appl. Math.* **25,** 424-434. [K]

Leipholz, H. H. E. (1983a). The Galerkin formulation and the Hamilton-Ritz formulation: a critical review. *Acta Mech.* **47,** 283-290. [K]

Leipholz, H. H. E. (1983b). On direct methods in the calculus of variations. *Computer Methods Appl. Mech. Engr.* **37,** 57-78. [K]

Nowinski, J. L. (1965). On the propagation of finite disturbances in bars of rubberlike materials. *J. Engr. Industry* **87,** 523-529. [D]

Oden, J. T., and Reddy, J. N. (1983). "Variational Methods in Theoretical Mechanics," 2nd ed. Springer-Verlag, Berlin, etc. [K]

Ogden, R. W. (1977). Inequalities associated with the inversion of elastic stress-deformation relations and their implications. *Proc. Cambridge Phil. Soc.* **81,** 313-324. [K]

Reissner, E. (1950). On a variational theorem in elasticity. *J. Math. & Phys.* **29,** 90-95. [K]

Reissner, E. (1970). Variational methods and boundary conditions in shell theory. *Studies in Optimization* **1,** 78-94. [K]

Rivlin, R. (1980). Some reflections on material stability. *Mechanics Today* **5,** 409-425. [J]

Rivlin, R. (1981). Some thoughts on material stability. *In* "Proc. I.U.T.A.M. Sympos. Finite Elasticity" (D. E. Carlson & R. T. Shield, eds.), pp. 105-122. Martinus Nijhoff, the Hague, etc. [J]

Sawyers, K. N., and Rivlin, R. S. (1984). The incremental shear modulus in a compressible isotropic elastic material. *ZAMP* **35,** 1-12. [J]

Smith, C. V., Jr. (1984a). Remark on a comment by C. D. Bailey. *AIAA J.* **22,** 1178-1179. [K]

Smith, C. V., Jr. (1984b). Comment on 'Hamilton's principle, Hamilton's law—6^n correct formulations.' *AIAA J.* **22,** 1181-1182. [K]

Stackgold, I. (1979). "Green's Functions and Boundary Value Problems." Wiley, New York, etc. [E]

Truesdell, C. A., and Toupin, R. A. (1960). The classical field theories. *In* "Encyclopedia of Physics" (S. Flügge, ed.), vol. III/1. Springer-Verlag, Berlin and New York. [D, N]

Vainberg, M. M. (1964). "Variational Methods for the Study of Nonlinear Operators" (Transl. from the Russian by A. Feinstein). Holden-Day, Inc. San Francisco, etc. [K]

Washizu, K. (1955). On the variational principles of elasticity and plasticity. Aeroelastic and Structures Research Laboratory, M.I.T., T.N. 25-18. [K]

Washizu, K. (1982). "Variational Methods in Elasticity and Plasticity," 3rd ed. Pergamon, Elmsford, New York, etc. [K]

Wempner, G. A. (1980). Complementary theorems of solid mechanics. *In* "Variational Methods in the Mechanics of Solids" (S. Nemat-Nasser, ed.), pp. 127-135. Pergamon, Oxford and New York. [K]

Zauderer, E. (1983). "Partial Differential Equations of Applied Mathematics." Wiley, New York, etc. [E]

Chapter IV

Cylindrical Motion of Infinite Cylindrical Shells (Beamshells)

In this chapter we encounter our first shells—shells that act like beams. The single greatest difference from the chapter on birods is that curvature effects now infest, at least implicitly, nearly every equation we set down. Consequent new features that we shall discuss at some length are: (1) necessary and sufficient conditions for a *load potential* $\mathcal{P}$ to exist if there are distributed as well as edge loads; (2) definitions of different *extensional, shear*, and *bending strains*; (3) restrictions on the strain-energy density imposed by invariance requirements and implied by thinness; and (4) structural stability (buckling) including thermodynamic considerations.

A. Geometry of the Undeformed Shell and Planar Motion

An infinite, right cylindrical shell is a body whose initial shape is generated by translating a plane region normal to itself. See Fig. 4.1a. Mathematically, the plane region is the image of a rectangular region; the images of the top and bottom of the rectangular region generate the *faces* of the shell while the images of the ends of the rectangular region generate the *edges* of the shell. While our intuitive picture is that of a thin-walled panel of, possibly, variable thickness, the notion of thinness shall not enter explicitly until we come to constitutive relations in Section N.

For a quantitative description, consider a fixed, right-handed Cartesian reference frame $Oxyz$. Let $\{\mathbf{e}_x, \mathbf{e}_y, \mathbf{e}_z\}$ denote the associated set of orthonormal base vectors with $\mathbf{e}_z$ perpendicular to the plane region that generates the shell. The initial shape may then be given the vector-parametric form

$$\mathbf{x}(\sigma, \zeta, z) = \mathbf{r}(\sigma, \zeta) + z\mathbf{e}_z \;\;,\;\; 0 \le \sigma \le L \;\;,\;\; -H \le \zeta \le H \;\;,\;\; -\infty < z < \infty, \quad \text{(A.1)}$$

where $\mathbf{r}\cdot\mathbf{e}_z = 0$ and, for convenience, L and H have the dimensions of LENGTH (though σ and ζ need not measure arc length). The faces of the shell are given by $\mathbf{x}(\sigma, \pm H, z)$ and its edges by $\mathbf{x}(0, \zeta, z)$ and $\mathbf{x}(L, \zeta, z)$. We call $\mathbf{r}(\sigma, 0)$ the parametric equation of the *base curve* $\mathcal{B}$. (See Fig. 4.1b.)

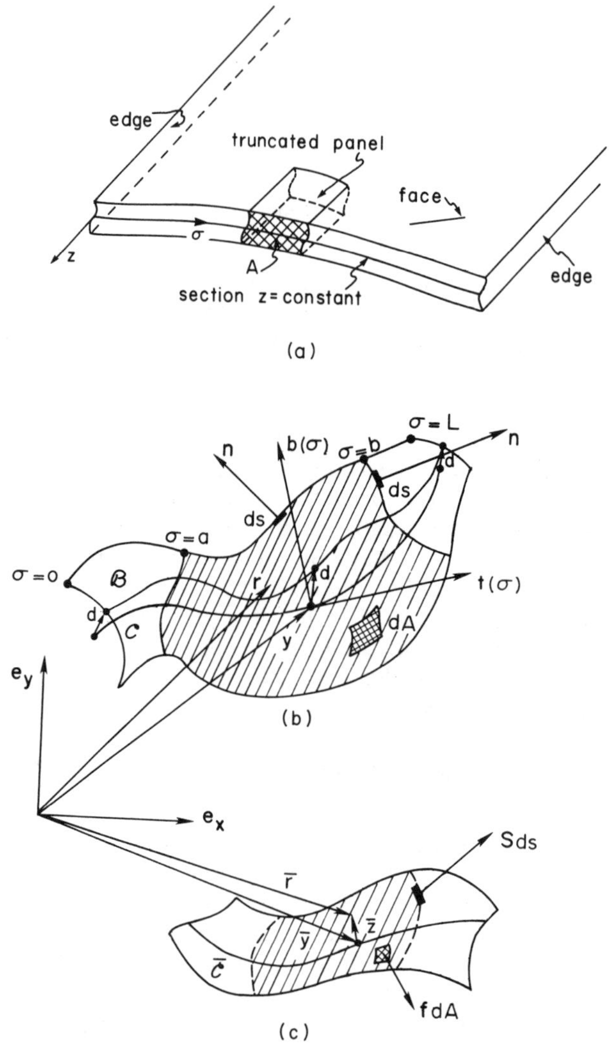

Fig. 4.1. Geometry of a beamshell: (a) perspective view; (b) undeformed and (c) deformed section z = constant.

The function $\mathbf{r}(\sigma, \zeta)$ is usually dictated by the construction of a shell, and often has a simple form. Thus, if the shell is materially isotropic, homogeneous, and of constant thickness h, it is natural to take

$$\mathbf{r}(\sigma, \zeta) = \mathbf{r}(\sigma, 0) + \zeta \mathbf{e}_z \times \mathbf{r}_{,\sigma}(\sigma, 0) \ , \ 0 \leq \sigma \leq L \ , \ -h/2 \leq \zeta \leq h/2 \, , \qquad \text{(A.2)}$$

where, now, σ is arc length along $\mathcal{B}$ and ζ is normal distance from $\mathcal{B}$. In this case, σ and ζ are called *shell coordinates*. More generally, if the shell is made of layers of different (possibly inhomogeneous) materials, each of variable thickness, then it is natural to choose $\mathbf{r}(\sigma, \zeta)$ so that ζ=constant along interfaces. Another example is offered by the flow of a fluid over a beamshell where the initial interface between the two media might be a natural choice for $\mathcal{B}$.

Until we reach Section N on strain-energy densities, we shall conform with Dawn Fisher's (1986) suggestion that leaving more latitude in the specification of $\mathbf{r}(\sigma, \zeta)$ not only simplifies the descent from 3-dimensions, but also prevents us, in some instances, from making a premature—and what can prove to be an unnatural—choice of the (initial or undeformed) reference curve, especially if the shell is nonhomogeneous or of variable thickness.

We now assume that all material properties, all external disturbances, and all boundary conditions are independent of z and such that every cross section undergoes the same motion while remaining in its initial plane. We call this *plane* or *cylindrical motion.*

B. Integral Equations of Cylindrical Motion

These are obtained by specializing (II.A.4) and (II.A.5), the integral equations of motion for a general body derived in Chapter II. There V denotes an arbitrary volume in the reference shape. The boundary of V, ∂V, is piecewise smooth with outward unit normal $\mathbf{n}$. All vectors in (II.A.4) and (II.A.5) lie in a fixed inertial frame.

To get shell equations, we take V, as shown in Fig. 4.1a, to be a truncated panel of the infinite cylindrical shell specified by (A.1). The panel has unit width in the z-direction and cross section

$$A \equiv \{\mathbf{r}(\sigma, \zeta) \mid 0 \le a \le \sigma \le b \le L \ , \ -H \le \zeta \le H\}\ , \tag{B.1}$$

as indicated by the shaded region in Fig. 4.1b. For cylindrical motion, (II.A.4) and (II.A.5) reduce to

$$\mathbf{L} \equiv \int_{t_1}^{t_2}\Big(\int_{\partial A} \mathbf{S}ds + \int_A \mathbf{f}dA\Big)dt - \int_A \rho\dot{\bar{\mathbf{r}}}dA \,\Big|_{t_1}^{t_2} = \mathbf{0} \tag{B.2}$$

$$R\mathbf{e}_z \equiv \int_{t_1}^{t_2}\Big(\int_{\partial A} \bar{\mathbf{r}}\times\mathbf{S}ds + \int_A \bar{\mathbf{r}}\times\mathbf{f}dA\Big)dt - \int_A \bar{\mathbf{r}}\times\rho\dot{\bar{\mathbf{r}}}dA \,\Big|_{t_1}^{t_2} = \mathbf{0}\ . \tag{B.3}$$

Here ∂A, the boundary of A, is assumed to be piecewise smooth with arc length s and outward unit normal $\mathbf{n}$. At time t, $\mathbf{S}ds$ is the contact force exerted on the deformed image of ds by the material outside of A, $\mathbf{f}dA$ is the body force exerted on the particles initially composing dA, and $\bar{\mathbf{r}}$ is the position of a particle with initial position $\mathbf{r}$, as shown in Fig. 4.1c. We note that the stress vector $\mathbf{S}_z$ acting over one of the ends z = constant of the panel will be equal and opposite to the stress vector acting on the opposite end. Hence, the effects of such stresses cancel out of (B.2) and (B.3).

From (A.1) and (B.1),

$$dA = |\mathbf{r},_\sigma \times \mathbf{r},_\zeta| d\sigma d\zeta \equiv \mu(\sigma, \zeta) d\sigma d\zeta \ , \ \mathbf{r},_\sigma \equiv \partial \mathbf{r}/\partial\sigma \text{ ,etc.} \tag{B.4}$$

Two parts make up ∂A: the *edges*, $\sigma = a, b$, on which $\mathbf{n}ds = \pm[\mathbf{r},_\zeta(^b_a, \zeta) \times \mathbf{e}_z]d\zeta$; and the *faces*, $\zeta = \pm H$, on which $\mathbf{n}ds = \pm[\mathbf{e}_z \times \mathbf{r},_\sigma(\sigma, \pm H)]d\sigma$. If we set $\mathbf{P} = \mathbf{S}^\sigma \mathbf{r},_\sigma + \mathbf{S}^\zeta \mathbf{r},_\zeta$, where $\mathbf{P}, \mathbf{S}^\sigma$, and $\mathbf{S}^\zeta$ are, respectively, the Piola-Kirchhoff tensor and its contravariant vector components, introduced in Section II.E, then, on the edges, $\mathbf{S}ds = \pm\mathbf{S}^\sigma \mu d\zeta$, and on the faces, $\mathbf{S}ds = \pm\mathbf{S}^\zeta \mu d\sigma$. Thus, (B.2) takes the form

$$\mathbf{L} \equiv \int_{t_1}^{t_2} [(\int_{-H}^{H} \mathbf{S}^\sigma \mu d\zeta)_a^b + \int_a^b (\int_{-H}^{H} \mathbf{f} \mu d\zeta + \mathbf{S}^\zeta \mu \mid_{-H}^{H}) d\sigma] dt$$
$$- \int_a^b (\int_{-H}^{H} \rho \dot{\bar{\mathbf{r}}} \mu d\zeta)_{t_2}^{t_1} d\sigma . \tag{B.5}$$

To give (B.5) the form of a shell equation, we first let

$$m \equiv \int_{-H}^{H} \rho\mu d\zeta = m(\sigma) \tag{B.6}$$

denote a mass (per unit width) per unit length and

$$\bar{\mathbf{y}} \equiv \int_{-H}^{H} \rho \bar{\mathbf{r}} \mu d\zeta / m = \bar{\mathbf{y}}(\sigma, t) \tag{B.7}$$

denote the *deformed, weighted position*. We call

$$\mathbf{y} \equiv \bar{\mathbf{y}}(\sigma, 0) \tag{B.8}$$

the parametric equation of *the initial (or undeformed) reference curve* C, and call (B.7) the parametric equation of its deformed image, $\bar{C}$. Arc length along C is given by

$$\tilde{\sigma} \equiv \int_0^\sigma |\mathbf{y},_\sigma(u)| du \equiv \psi(\sigma) . \tag{B.9}$$

The length of C is $\tilde{L} \equiv \psi(L)$.

The reader should note that, for fixed σ, (B.7) locates the *center of mass* of a differential panel of thickness $d\sigma$ and unit width in the z-direction, lying between the faces of the beamshell. Our method of defining C and $\bar{C}$ is thus equivalent to defining a *line of centers of mass* of the undeformed and deformed beamshell. See subsection N.1 for an example.

In linear beam theory, Prescott (1942) was apparently the first to suggest using a weighted vertical deflection. Stephen (1981), building on the work of Cowper (1966), has concluded that use of a weighted average displacement gives the best correlation with the predictions of linear, 3-dimensional elasticity of the low frequency vibrations of straight beams.

We emphasize that $\bar{\mathbf{y}}$ is an *average* and *not* the first term in an expansion of $\bar{\mathbf{r}}$ in powers of ζ. We also emphasize that, in general, $\bar{\mathbf{y}}$ *is not the deformed position of some particle in the shell.* Finally, we remind the reader that (B.7) comes from the inertial terms in the 3-dimensional equations of motion. Therefore, in statics (where t may be regarded as a parameter measuring the magnitude of the loads) other reasonable relations between $\bar{\mathbf{y}}$ and $\bar{\mathbf{r}}$ may be defined, e.g., $\bar{\mathbf{y}}=\bar{\mathbf{r}}(\sigma, 0, t)$ (or, what comes to the same thing, we may, in a static problem, choose ρ at our convenience). This point must be recognized if one attempts to derive constitutive relations for a shell by postulating a specific form for $\bar{\mathbf{r}}$ and then averaging in some way through the thickness. For further details see Section N.

In general, σ is not arc length along the reference curve C. However, there are theoretical and practical advantages in taking arc length along C as the independent spatial variable. To do so, we note that the arc length function $\psi(\sigma)$ defined in (B.9) is monotonically increasing and differentiable; hence so is its inverse, ψ^{-1}. Thus, if we introduce the new variable $\sigma=\psi^{-1}(\tilde{\sigma})$, then, in (A.1)-(B.8), we merely replace σ by $\tilde{\sigma}$, L by $\tilde{L}$, and μ by $\tilde{\mu}$. We may then drop the tildes.

Returning to our reduction of (B.5), we let

$$\mathbf{F} = \int_{-H}^{H} \mathbf{S}^{\sigma}\mu d\zeta \tag{B.10}$$

denote the *internal force (per unit width of the beamshell)*, let

$$\mathbf{p} = \int_{-H}^{H} \mathbf{f}\mu d\zeta + \mathbf{S}^{\zeta}\mu \,\Big|_{-H}^{H} \tag{B.11}$$

denote the *external force (per unit width) per unit length of* C, and set

$$\mathbf{v} \equiv \dot{\bar{\mathbf{y}}} \,. \tag{B.12}$$

Then (B.5) takes the form

$$\mathbf{L} \equiv \int_{t_1}^{t_2}(\mathbf{F}\,\Big|_a^b + \int_a^b \mathbf{p}d\sigma)dt - \int_a^b m\mathbf{v}d\sigma\,\Big|_{t_1}^{t_2} = \mathbf{0} \,. \tag{B.13}$$

To obtain an equation for the balance of rotational momentum in our shell, we set

$$\bar{\mathbf{r}} = \bar{\mathbf{y}} + \bar{\mathbf{z}} \tag{B.14}$$

and note from (B.6) and (B.7) that

$$\int_{-H}^{H} \rho\bar{\mathbf{z}}\mu d\zeta = \mathbf{0} \,. \tag{B.15}$$

In view of the remarks in the paragraph following (B.7) about the implications of that equation, we call (B.15) *the dynamic consistency condition.*

The discrepancy between the base curve $\mathcal{B}$ and the reference curve C may be measured by the *base-reference deviation*

$$\mathbf{d}(\sigma) \equiv \bar{\mathbf{z}}(\sigma, 0, 0)\,. \tag{B.16}$$

If $\mathbf{r}(\sigma, \zeta)$ has the form (A.2), then it follows from (B.6) and (B.7) that $\mathbf{d} = d\mathbf{e}_z \times \mathbf{r},_\sigma(\sigma, 0)$, where $d = \int_{-h/2}^{h/2} \rho\mu\zeta d\zeta/m$, $\mu = 1 - \zeta b(\sigma)$, and b is the curvature of the base curve. If the beamshell is also homogeneous and of constant thickness, h, then $d = h^2 b(\sigma)/12$, and, as we shall see, d/h can then be set to zero within the error in shell theory coming from the constitutive relations. In such a case, the midcurve, the base curve, and the undeformed reference curve can be regarded as virtually coincident and the extensive literature based on using the midcurve as the undeformed reference curve is applicable.

Turning to the equation of balance of rotational momentum, (B.3), we obtain, with the aid of (B.6), (B.10)-(B.12), (B.14), and (B.15),

$$\begin{aligned} R \equiv \int_{t_1}^{t_2} \{ [\mathbf{e}_z \cdot (\bar{\mathbf{y}} \times \mathbf{F}) + M]_a^b + \int_a^b [\mathbf{e}_z \cdot (\bar{\mathbf{y}} \times \mathbf{p}) + l] d\sigma \} dt \\ - \int_a^b [\mathbf{e}_z \cdot (\bar{\mathbf{y}} \times m\mathbf{v}) + I\omega] d\sigma \Big|_{t_1}^{t_2} = 0\,. \end{aligned} \tag{B.17}$$

Here

$$M = \mathbf{e}_z \cdot \int_{-H}^{H} \bar{\mathbf{z}} \times \mathbf{S}^\sigma \mu d\zeta \tag{B.18}$$

is the *moment* (per unit width),

$$l = \mathbf{e}_z \cdot \Big(\int_{-H}^{H} \bar{\mathbf{z}} \times \mathbf{f} \mu d\zeta + \bar{\mathbf{z}} \times \mathbf{S}^\zeta \mu \Big|_{-H}^{H} \Big) \tag{B.19}$$

is the *external torque per unit length of* C (per unit width),

$$I = \int_{-H}^{H} \rho \mathbf{z} \cdot \mathbf{z} \mu d\zeta \tag{B.20}$$

is the *moment of inertia per unit length of* C (per unit width), and

$$\omega = \mathbf{e}_z \cdot \int_{-H}^{H} \rho \bar{\mathbf{z}} \times \dot{\bar{\mathbf{z}}} \mu d\zeta / I \tag{B.21}$$

is the *spin*. Note that it is only the *combination* $I\omega$ that appears in (B.17), so that our definition of I (and hence that of ω) is, to some extent, arbitrary. However, (B.20) is certainly reasonable: given the reference shape, I can be computed *a priori* and, by extracting the time independent factor I from the rotational momentum density, we can write the time rate of change of the rotational kinetic energy density as $\tfrac{1}{2}(I\omega^2)^{\bullet}$. (See the related discussion in Section II.F).[1] Upon the further assumption that $\bar{\mathbf{z}}$ is a single, straight, rotating, *material* line of constant length, our definitions of I and ω reduce to those of conventional shear-deformable beamshell theory.

---

[1]The alternative, time-dependent definition, $I = \int_{-H}^{H} \rho \bar{\mathbf{z}} \cdot \bar{\mathbf{z}} \mu d\zeta$, has the drawback that $I\omega\dot{\omega}$ is then not the time derivative of $\tfrac{1}{2} I\omega^2$.

In (B.13) and (B.17) we have exact integral equations of motion that, formally, are identical to those for the planar motion of a beam. For conciseness, we shall henceforth refer to a cylindrical shell undergoing cylindrical motion as a *beamshell.*

C. Initial and Spin Bases

Assume that the vector parametric equation of the initial curve C has the form

$$\mathbf{y} = x(\sigma)\mathbf{e}_x + y(\sigma)\mathbf{e}_y \, . \tag{C.1}$$

Then the unit tangent to C is given by

$$\mathbf{t} \equiv \mathbf{y}'(\sigma) = x'(\sigma)\mathbf{e}_x + y'(\sigma)\mathbf{e}_y \equiv \cos\alpha(\sigma)\mathbf{e}_x + \sin\alpha(\sigma)\mathbf{e}_y \, , \tag{C.2}$$

where α is the angle the tangent to C makes with the x-axis. A unit normal to C is given by

$$\mathbf{b} \equiv \mathbf{e}_z \times \mathbf{t} = -\sin\alpha\,\mathbf{e}_x + \cos\alpha\,\mathbf{e}_y \, . \tag{C.3}$$

We call $\{\mathbf{t}, \mathbf{b}\}$ the *initial basis.* The *curvature* of C is denoted and defined by

$$\alpha' \equiv \mathbf{y}''\cdot\mathbf{b} \, . \tag{C.4}$$

Let $\{\mathbf{T}, \mathbf{B}\}$ denote a rigid, orthonormal basis that rotates in the xy-plane with spin ω with respect to $\{\mathbf{t}, \mathbf{b}\}$ and coincides with this latter basis at $t=0$. Here, ω is defined by (B.21). With

$$\beta \equiv \int_0^t \omega dt \tag{C.5}$$

denoting the *rotation* of the beamshell, we have

$$\mathbf{T} = \cos\beta\,\mathbf{t} + \sin\beta\,\mathbf{b} = \cos(\alpha+\beta)\mathbf{e}_x + \sin(\alpha+\beta)\mathbf{e}_y \tag{C.6}$$

$$\mathbf{B} = -\sin\beta\,\mathbf{t} + \cos\beta\,\mathbf{b} = -\sin(\alpha+\beta)\mathbf{e}_x + \cos(\alpha+\beta)\mathbf{e}_y \, . \tag{C.7}$$

We call $\{\mathbf{T}, \mathbf{B}\}$ the *spin basis.* See Fig. 4.2.

*D. Jump Conditions and Propagation of Singularities

Consider the rectangle $\Omega \equiv (a, b)\times(t_1, t_2)$ in the σt-plane. Along the sides parallel to the σ-axis, $dt=0$, and along the sides parallel to the t-axis, $d\sigma=0$. With this observation we can write (B.13) and (B.17) in the form

$$\mathbf{L} = \oint_{\partial\Omega} (\mathbf{F}dt + m\mathbf{v}d\sigma) + \int_{\Omega} \mathbf{p}d\sigma dt = \mathbf{0} \tag{D.1}$$

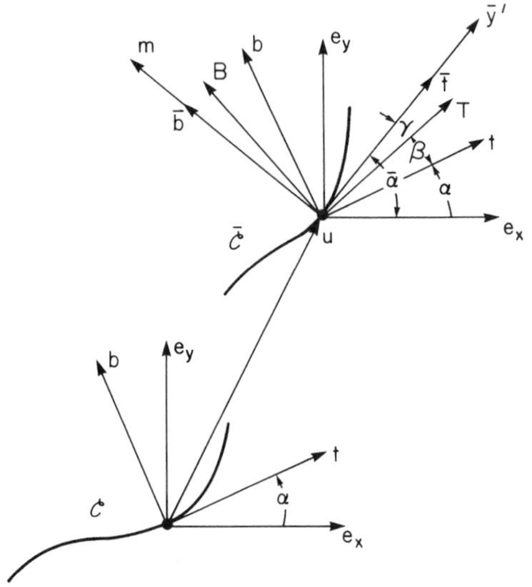

Fig. 4.2. Vectors and angles associated with the initial curve $\mathcal{C}$ and its deformed image $\bar{\mathcal{C}}$.

$$R = \oint_{\partial\Omega} \{[\mathbf{e}_z \cdot (\bar{\mathbf{y}} \times \mathbf{F}) + M]dt + [\mathbf{e}_z \cdot (\bar{\mathbf{y}} \times m\mathbf{v}) + I\omega]d\sigma\}$$
$$+ \int_{\Omega} [\mathbf{e}_z \cdot (\bar{\mathbf{y}} \times \mathbf{p}) + l]d\sigma dt = 0 \,. \tag{D.2}$$

Arguments identical to those in Section III.D show that (D.1) and (D.2) hold in *any region* Ω that can be represented as the limit of the union of vertical strips.

To extract shock conditions from (D.1) and (D.2), let $\mathcal{S}$: $\sigma = s(t)$ denote a smooth *shock line* in the σt-plane. Applying (D.1) and (D.2) to an arbitrarily small, strip-like region centered on $\mathcal{S}$ of the type shown in Fig. 3.3c, we obtain the jump conditions

$$[\mathbf{F}] + \dot{s}m[\mathbf{v}] = \mathbf{0} \tag{D.3}$$

$$[\mathbf{e}_z \cdot (\bar{\mathbf{y}} \times \mathbf{F}) + M] + \dot{s}[\mathbf{e}_z \cdot (\bar{\mathbf{y}} \times m\mathbf{v}) + I\omega] = 0 \,, \tag{D.4}$$

where, if $\mathcal{F}(\sigma, t)$ is any scalar or vector field, the jump in $\mathcal{F}$ as we move from left to right across $\mathcal{S}$ is defined by

$$[\mathcal{F}] \equiv \lim_{\eta \to 0} [\mathcal{F}(s(t) + \eta, t) - \mathcal{F}(s(t) - \eta, t)] \,. \tag{D.5}$$

As the beamshell must not pull apart, $[\bar{\mathbf{y}}] = 0$. Hence, $[\mathbf{e}_z \cdot (\bar{\mathbf{y}} \times \mathbf{F})] = \mathbf{e}_z \cdot (\bar{\mathbf{y}} \times [\mathbf{F}])$, etc., so that, in view of (D.3), (D.4) reduces to

$$[M] + \dot{s}I[\omega] = 0 \,. \tag{D.6}$$

To express jumps in the kinematic variables in terms of jumps in spatial derivatives, we assume that $\bar{\mathbf{y}}', \dot{\bar{\mathbf{y}}} = \mathbf{v}, \beta'$, and $\dot{\beta} = \omega$ are continuous in some neighborhood $\mathcal{N}_-$ to the left of $\mathcal{S}$ and on $\mathcal{S}$ itself, with a similar condition holding in some neighborhood $\mathcal{N}_+$ to the right of $\mathcal{S}$, where, here and henceforth, a prime shall denote partial or total differentiation with respect to σ. Let $\underline{\tan} \equiv (\dot{s}, t)_P$ denote a tangent vector at some point P on the shock line $\mathcal{S}$. At points P_+ and P_- in $\mathcal{N}_+$ and $\mathcal{N}_-$, we compute the directional derivatives of $\bar{\mathbf{y}}$ and β parallel to $\underline{\tan}$. Letting P_+ and P_- approach P, we find, in analogy with (III.D.11) for the birod, that

$$\frac{d}{dt}[\bar{\mathbf{y}}] = \dot{s}[\bar{\mathbf{y}}'] + [\dot{\bar{\mathbf{y}}}] \ , \ \frac{d}{dt}[\beta] = \dot{s}[\beta'] + [\dot{\beta}] \ . \tag{D.7}$$

Because the beamshell must not pull apart nor kink, $[\bar{\mathbf{y}}] = \mathbf{0}$ and $[\beta] = 0$. Hence, with (D.7), (D.3) and (D.6) may be written

$$[\mathbf{F}] = c^2 m[\bar{\mathbf{y}}'] \ , \ [M] = c^2 I[\beta'] \ , \tag{D.8}$$

where $c \equiv |\dot{s}(t)|$ is the *speed of the shock* (if a shock exists).

E. Differential Equations of Cylindrical Motion

If the various fields in the integral equations of motion, (B.13) and (B.17), are sufficiently smooth, then differential equations of motion may be obtained as follows. For any function f whose derivative is integrable, set $f|_\alpha^\beta = \int_\alpha^\beta f' dx$ and rewrite (B.13) and (B.17) in the form $\int_{t_1}^{t_2}\int_a^b g\, d\sigma dt = 0$, where g stands for the integrand in either equation. As this relation must hold for all $t_1 < t_2$ and for all $0 \le a < b \le L$, it follows that if g is a continuous function of σ and t, then $g = 0$. Thus, from (B.13),

$$\mathbf{F}' + \mathbf{p} - m\dot{\mathbf{v}} = \mathbf{0} \ , \tag{E.1}$$

and from (B.17), when (E.1) is used to simplify the resulting equation,

$$M' + \mathbf{m}\cdot\mathbf{F} + l - I\dot{\omega} = 0 \ . \tag{E.2}$$

In (E.2)

$$\mathbf{m} \equiv \mathbf{e}_z \times \bar{\mathbf{y}}' \tag{E.3}$$

is the *rational* normal to the deformed initial curve $\mathcal{C}$, so-called because if we set

$$\bar{\mathbf{y}} \equiv \bar{x}\mathbf{e}_x + \bar{y}\mathbf{e}_y \ , \tag{E.4}$$

then $\mathbf{m}$ is rational (in fact linear) in $\bar{x}'$ and $\bar{y}'$.

F. The Weak Form of the Equations of Motion

Suppose that $\mathbf{F}, \mathbf{v}, M$, and ω are classical solutions of the differential equations of motion (E.1) and (E.2) for $0 < \sigma < L$ and $0 < t < T$. Let $\mathbf{V}(\sigma, t)$ and

$\Omega(\sigma, t)$ be test functions which, for the present, are unspecified save that they be sufficiently smooth for the following operations to make sense. Take the dot product of (E.1) with $\mathbf{V}$ and multiply (E.2) by Ω. Add the resulting equations and integrate over any subregion $(a, b)\times(t_1, t_2)$ of $(0, L)\times(0, T)$. Finally, integrate by parts to remove spatial derivatives on $\mathbf{F}$ and M and time derivatives on $\mathbf{v}$ and ω to obtain

$$\int_{t_1}^{t_2}(\mathbf{F}\cdot\mathbf{V}+M\Omega)_a^b dt-\int_a^b(m\mathbf{v}\cdot\mathbf{V}+I\omega\Omega)_{t_1}^{t_2}d\sigma$$
$$-\int_{t_1}^{t_2}\int_a^b[\mathbf{F}\cdot(\mathbf{V}'-\mathbf{m}\Omega)+M\Omega'-m\mathbf{v}\cdot\dot{\mathbf{V}}-I\omega\dot{\Omega}-\mathbf{p}\cdot\mathbf{V}-l\Omega]d\sigma dt\equiv 0\,,\ \forall\ \mathbf{V},\Omega\,. \tag{F.1}$$

Because integration is a smoothing process, (F.1) may hold for more functions than those which satisfy (E.1) and (E.2). Thus, (F.1) is called the *weak form of the equations of motion.* Note that if $\mathbf{A}$ and B are arbitrary vector and scalar constants and if we take $\mathbf{V}=\mathbf{A}+B\mathbf{e}_z\times\bar{\mathbf{y}}$ and $\Omega=B$, then (F.1) reduces to $\mathbf{A}\cdot\mathbf{L}+BR\equiv 0$, which implies the integral equations of motion, (B.13) and (B.17).

G. The Mechanical Work Identity

If we take $\mathbf{V}=\mathbf{v}$ and $\Omega=\omega$ in (F.1), then the weak form of the equations of motion can be rewritten as

$$\int_{t_1}^{t_2}\mathcal{W}dt=\mathcal{K}\Big|_{t_1}^{t_2}+\int_{t_1}^{t_2}\mathcal{D}dt\,, \tag{G.1}$$

where

$$\mathcal{W}\equiv(\mathbf{F}\cdot\mathbf{v}+M\omega)_a^b+\int_a^b(\mathbf{p}\cdot\mathbf{v}+l\omega)d\sigma \tag{G.2}$$

is the *apparent external mechanical power,*

$$\mathcal{K}\equiv \tfrac{1}{2}\int_a^b(m\mathbf{v}\cdot\mathbf{v}+I\omega^2)d\sigma \tag{G.3}$$

is the *kinetic energy*, and

$$\mathcal{D}\equiv\int_a^b[\mathbf{F}\cdot(\mathbf{v}'-\omega\mathbf{m})+M\omega']d\sigma \tag{G.4}$$

is the *deformation power.* We call (G.1) the *Mechanical Work Identity.*

H. Mechanical Boundary Conditions

The edges of a beamshell are frequently subject to kinematical constraints. The simplest conditions are those of total fixity: $\bar{\mathbf{y}}_0=\mathbf{y}_0$, $\bar{\mathbf{y}}_L=\mathbf{y}_L$, $\beta_0=\beta_L=0$, where $\bar{\mathbf{y}}_0$ is short for $\bar{\mathbf{y}}(0,t)$, etc. Here, there are 6 scalar constraints. If one or the other edge is simply-supported or partially clamped, there will be fewer constraints. If the motion of one edge of a beamshell is prescribed or, more generally, if one edge is required to lie on a moving curve, we have a time-dependent constraint.

1. General boundary conditions for nonholonomic constraints

Let

$$y \equiv [\bar{\mathbf{y}}_0, \beta_0, \bar{\mathbf{y}}_L, \beta_L]^T \tag{H.1}$$

denote the 6-component *edge kinematic vector*, where it is understood that the the vector $\bar{\mathbf{y}}$ is to be represented by 2 scalar components. We shall assume that the kinematic boundary conditions consist of $m \leq 6$ *nonholonomic constraints* of the form

$$A\dot{y} - \mathcal{B} = 0\,, \tag{H.2}$$

where A and $\mathcal{B}$ are $m \times 6$ and $m \times 1$ matrices that depend on y and t only. By differentiation with respect to t, *holonomic constraints* of the form $\mathcal{R}(y, t) = 0$ can be put into the form (H.2). *Unilateral constraints* cannot be put into this form.

To say that we have m independent constraints means that $\operatorname{rank} A = m$. Thus, from linear algebra, it follows that the edge kinematic vector has the form

$$\dot{y} = \sum_{1}^{6-m} q_k \mathcal{M}_k(y, t) + \mathcal{H}(y, t)\,, \tag{H.3}$$

where the q_k's are unknown scalars (the generalized edge velocities), the $\mathcal{M}_k$'s are $6-m$ linearly independent vectors that span the null space of A, and $\mathcal{H}$ is uniquely determined by the conditions $A\mathcal{H} = \mathcal{B}$, $\mathcal{H}^T\mathcal{M}_k = 0$, $k = 1, \cdots, 6-m$.

With

$$\mathcal{N} \equiv [-\mathbf{F}_0, -M_0, \mathbf{F}_L, M_L]^T \tag{H.4}$$

denoting the *edge load vector*, we have for the rate of work of the edge loads [see (G.2)],

$$(\mathbf{F}\cdot\mathbf{v} + M\omega)_0^L = \mathcal{N}^T\dot{y} = \sum_{1}^{6-m} \mathcal{N}^T\mathcal{M}_k q_k + \mathcal{N}^T\mathcal{H}. \tag{H.5}$$

We call the $6-m$ scalars $\mathcal{N}^T\mathcal{M}_k$ the *generalized edge loads* (associated with the generalized edge velocities q_k) and the scalar $\mathcal{N}^T\mathcal{H}$ the *power of the forces of constraint*.

In classical mechanics, an explicit solution of the initial value problem for Lagrange's equations of motion of a rigid body requires that the generalized forces be specified. For beamshells, we shall call the analogous requirement *the Principle of Mechanical Boundary Conditions*. That is, on the edges of the beamshell there must be prescribed

- $m \leq 6$ kinematic constraints
- $6-m$ generalized edge loads.

The generalized edge velocities and the power of the forces of constraint can be found only after we have a solution for $\bar{\mathbf{y}}$ and β. Let us illustrate the Principle with an example.

Problem: As in Fig. 4.3, the left edge of a beamshell moves to the right with a prescribed horizontal velocity $v(t)$. The right edge of the beamshell slides along a rough, moving curve given by the equation $F(x, y, t)=0$. The coefficient of friction between the beamshell and the moving curve is μ. (Such conditions are typical for sheets of paper moving through copying machines. See Benson, 1981). Determine the mechanical boundary conditions.

Solution: There are 3 constraints which, in scalar form, read

$$\dot{\bar{x}}_0 = v\ ,\ \dot{\bar{y}}_0 = 0\ ,\ F_{,x}(\bar{x}_L, \bar{y}_L, t)\dot{\bar{x}}_L + F_{,y}(\bar{x}_L, \bar{y}_L, t)\dot{\bar{y}}_L + F_{,t}(\bar{x}_L, \bar{y}_L, t) = 0\,. \quad \text{(H.6)}$$

For convenience, let

$$\mathbf{n} \equiv \frac{F_{,x}\mathbf{e}_x + F_{,y}\mathbf{e}_y}{(F_{,x}^2 + F_{,y}^2)^{1/2}} \equiv n_x\mathbf{e}_x + n_y\mathbf{e}_y \ \text{ at } x = \bar{x}_L\ ,\ y = \bar{y}_L \qquad \text{(H.7)}$$

be a unit normal to the moving curve. After writing (H.6) in the form $A\dot{y} = \mathcal{B}$, we find that

$$\begin{aligned} \mathcal{M}_1 &= [0, 0, 1, 0, 0, 0]^T \\ \mathcal{M}_2 &= [0, 0, 0, n_y, -n_x, 0]^T \\ \mathcal{M}_3 &= [0, 0, 0, 0, 0, 1]^T \end{aligned} \qquad \text{(H.8)}$$

$$\mathcal{H} = [v, 0, 0, 0, 0, 0]^T + (F_{,x}^2 + F_{,y}^2)^{-1/2} F_{,t}[0, 0, 0, n_x, n_y, 0]^T\,. \qquad \text{(H.9)}$$

Hence, with $q_1 = \dot{\beta}_0$, $q_2 = \dot{\bar{x}}_L n_y - \dot{\bar{y}}_L n_x$, and $q_3 = \dot{\beta}_L$, we have from the above and (H.5),

$$(\mathbf{F}\cdot\mathbf{v} + M\omega)_0^L = \mathcal{N}^T \mathcal{M}_2 q_2 + M_L\dot{\beta}_L - M_0\dot{\beta}_0 + \mathcal{N}^T\mathcal{H}\,. \qquad \text{(H.10)}$$

By the Principle of Mechanical Boundary Conditions, we must prescribe M_0, M_L, and $\mathcal{N}^T\mathcal{M}_2 = (\mathbf{e}_z \times \mathbf{n})\cdot\mathbf{F}$, the force acting on the right edge of the beamshell tangent to the moving curve. We now assume that the left edge is hinged and that the curve, though rough, exerts no moment on the right edge of the beamshell, i.e., $M_0 = M_L = 0$. Finally, as the normal *compressive* force between

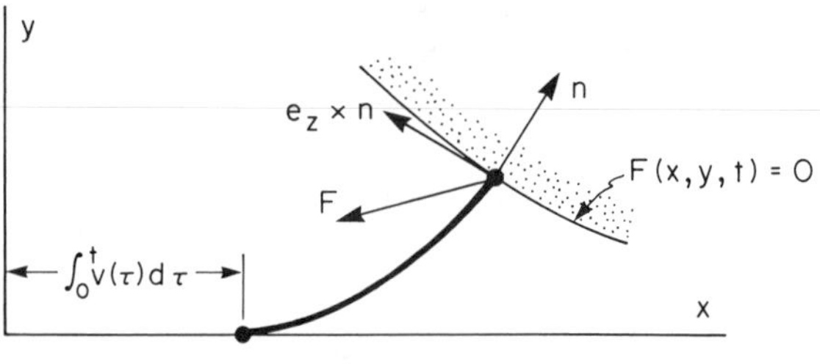

Fig. 4.3. A beamshell with moving edge constraints.

the right edge of the beamshell and the beamshell is $\mathbf{F}\cdot\mathbf{n}$,

$$(\mathbf{e}_z \times \mathbf{n})\cdot\mathbf{F} = \mu\mathbf{F}\cdot\mathbf{n} . \tag{H.11}$$

Our analysis has assumed that the right edge of the beamshell stays in contact with the moving curve. A more elaborate analysis might consider a unilateral constraint in which the third condition in (H.6) is replaced by a load-free condition if $\mathbf{F}\cdot\mathbf{n} \geq 0$. □

2. Classical boundary conditions

If we prescribe m (≤ 6) components of the *edge velocity vector*

$$\dot{y} \equiv v = [\mathbf{v}_0, \omega_0, \mathbf{v}_L, \omega_L]^T \tag{H.12}$$

plus the $6-m$ *conjugate* components of the edge load vector $\mathcal{N} = [-\mathbf{F}_0, -M_0, \mathbf{F}_L, M_L]^T$, i.e., the components of $\mathcal{N}$ corresponding to the unknown components of v, then we have *classical boundary conditions.* In dealing with classical boundary conditions, the following decomposition can be useful:

Let $v = \hat{v} + v_*$, where $\hat{v}$ is equal to v with the *unknown* components of v set to zero and v_* is equal to v with the *known* components of v set to zero. The edge load vector, obviously, can be given the analogous decomposition $\mathcal{N} = \hat{\mathcal{N}} + \mathcal{N}_*$. Thus, (H.5) may be given the more explicit form

$$(\mathbf{F}\cdot\mathbf{v} + M\omega)_0^L = \mathcal{N}^T v = \hat{\mathcal{N}}^T v_* + \mathcal{N}_*^T \hat{v} \equiv (\mathbf{F}\cdot\mathbf{v} + M\omega)_{\partial N} + (\mathbf{F}\cdot\mathbf{v} + M\omega)_{\partial U} , \tag{H.13}$$

where ∂N indicates those components of $\mathcal{N}$ which are prescribed and ∂U indicates those components of v which are prescribed. On each edge of the beamshell there may be contributions to ∂N, or ∂U, or *both.* We may refer to such edges, respectively, as ∂N-edges or ∂U-edges, or both.

For example, suppose that $\mathbf{F} = H\mathbf{e}_x + V\mathbf{e}_y$ and $\mathbf{v} = v\mathbf{e}_x + w\mathbf{e}_y$, and that we are given $\hat{w}_0, \hat{\omega}_L, \hat{H}_0, \hat{M}_0, \hat{H}_L$, and $\hat{V}_L$. Then

$$v = [v_0, \hat{w}_0, \omega_0, v_L, w_L, \hat{\omega}_L]^T \tag{H.14}$$

$$\hat{v} = [0, \hat{w}_0, 0, 0, 0, \hat{\omega}_L]^T \tag{H.15}$$

$$v_* = [v_0, 0, \omega_0, v_L, w_L, 0]^T \tag{H.16}$$

$$\mathcal{N} = [-\hat{H}_0, -V_0, -\hat{M}_0, \hat{H}_L, \hat{V}_L, M_L]^T \tag{H.17}$$

$$\hat{\mathcal{N}} = [-\hat{H}_0, 0, -\hat{M}_0, \hat{H}_L, \hat{V}_L, 0]^T \tag{H.18}$$

$$\mathcal{N}_* = [0, -V_0, 0, 0, 0, M_L]^T , \tag{H.19}$$

and hence

$$(\mathbf{F}\cdot\mathbf{v} + M\omega)_{\partial N} = \hat{H}_L v_L + \hat{V}_L w_L - \hat{H}_0 v_0 - \hat{M}_0 \omega_0 \qquad \text{(H.20)}$$

$$(\mathbf{F}\cdot\mathbf{v} + M\omega)_{\partial U} = M_L \hat{\omega}_L - V_0 \hat{w}_0 \, . \qquad \text{(H.21)}$$

Here, $\sigma=0$ as well as $\sigma=L$ is both a ∂N- and a ∂U-edge, because at least one load component and one velocity component is prescribed on each. We call such classical boundary conditions *mixed.* In an appendix to this section, Dawn Fisher has cast our discussion of general and classical boundary conditions into the framework of vector spaces.

In the boundary conditions of elastostatics, the displacement $\mathbf{u}=\bar{\mathbf{y}}-\mathbf{y}$ appears in place of the velocity $\mathbf{v}$ and the rotation β appears in place of the spin ω.

3. Typical boundary conditions

The edges of a beamshells are either free or attached to adjacent bodies. Conditions at these edges come from kinematic continuity, force, and moment equilibrium, and may be strongly nonlinear, multidirectional, or require prior knowledge of the response of the adjacent bodies to the loads transmitted by the beamshell. To simplify edge conditions, the relative stiffnesses of some parts of the adjoining bodies (with respect to the beamshell or with respect to each other) are taken to be zero or infinity.

Some typical boundary conditions at $\sigma=L$ are listed below and indicated schematically in Fig. 4.4, where a part of the undeformed and deformed beamshell near this edge are shown. In discussing these boundary conditions, it is convenient to take $\mathbf{t}_L$ and $\mathbf{b}_L$, the unit tangent and normal vectors to the initial curve C at the right edge, to coincide with $\mathbf{e}_x$ and $\mathbf{e}_y$. We warn the reader that most of these boundary conditions, which have evolved over years—nay, centuries—have sprung from the linearized Euler-Bernoulli theory of initially straight beams; their extrapolation to the nonlinear theory of initially curved beamshells may, in some cases, be nonunique or of questionable practical importance.

(a) Fixed edge (Fig. 4.4a). This is a ∂U-edge, attached to a body that is "rigid" (relative to the beamshell) so that

$$\mathbf{u}_L = \mathbf{0} \ , \ \beta_L = 0 \, . \qquad \text{(H.22)}$$

(b) Free edge (Fig. 4.4b). This is a ∂N-edge, free of loads, i.e.,

$$\mathbf{F}_L = \mathbf{0} \ , \ M_L = 0 \, . \qquad \text{(H.23)}$$

(c) Simply-supported edge (Fig. 4.4c). This is both a ∂U-edge and a ∂N-edge. In linearized (small deflection) beam theory, the moment and transverse displacement are zero at a simply-supported edge; the axial displacement may also be zero (a *hinged* support or so-called SS4 condition) or the corresponding axial force may be zero (a *diaphragm* or *roller* support or SS3 condition). The SS labels are used widely in shell theory.

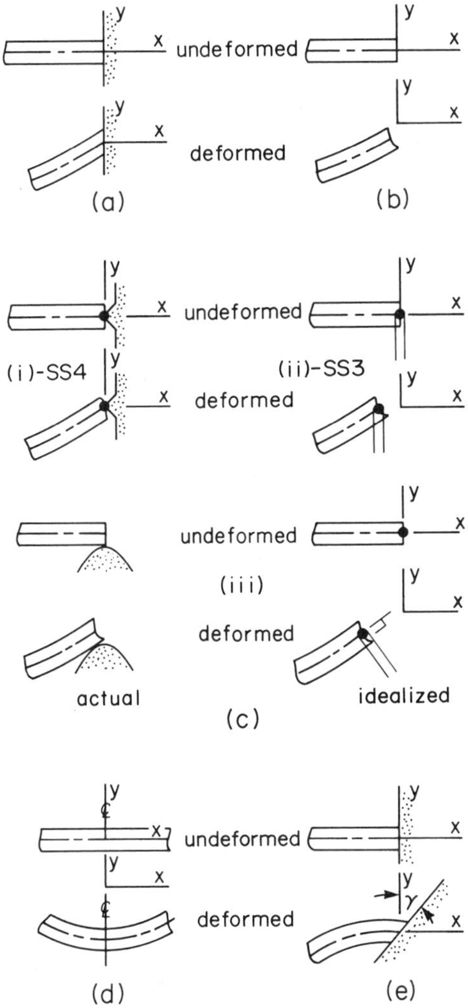

Fig. 4.4. Some typical boundary conditions: (a) fixed edge; (b) free edge; (c) simply-supported edge: (i) hinged; (ii) extrinsic diaphragm; (iii) intrinsic diaphragm; (d) symmetry; (e) clamped edge.

The extension of simple-support conditions to nonlinear theories of beamshells is not unique because the "transverse" and "normal" directions may be considered either to rotate with the edge, yielding *intrinsic* conditions, or else to remain fixed in the frame $\{\mathbf{e}_x, \mathbf{e}_y\}$, yielding *extrinsic* conditions. We list some possibilities.

(i) **Hinged edge (SS4)** [Fig. 4.4c(i)].

$$M_L = 0 \ , \ \mathbf{u}_L = \mathbf{0} \ . \tag{H.24}$$

(ii) **Extrinsic diaphragm support (SS3)** [Fig. 4.4c(ii)].

$$M_L = 0 \ , \ (\mathbf{u}\cdot\mathbf{b})_L = 0 \ , \ (\mathbf{F}\cdot\mathbf{t})_L = 0 \, . \tag{H.25}$$

(iii) **Intrinsic diaphragm support** [Fig. 4.4c(iii)]. The associated constraint, which is nonholonomic, is expressed conventionally in terms of the beamshell velocity **v**. If the supporting diaphragm rotates with the frame $\{\bar{\mathbf{t}}_L, \bar{\mathbf{b}}_L\}$, then

$$M_L = (\mathbf{v}\cdot\mathbf{m})_L = 0 \ , \ (\mathbf{F}\cdot\bar{\mathbf{y}}')_L = 0 \, , \tag{H.26}$$

or, in terms of the fixed directions $\mathbf{t}_L$ and $\mathbf{b}_L$,

$$M_L = (\mathbf{v}\cdot\mathbf{b})_L + \mathbf{e}_z\cdot(\mathbf{u}'\times\mathbf{v})_L = (\mathbf{F}\cdot\mathbf{t})_L + (\mathbf{F}\cdot\mathbf{u}')_L = 0 \, ; \tag{H.27}$$

if the supporting diaphragm rotates with the frame $\{\mathbf{T}, \mathbf{B}\}$, then

$$M_L = (\mathbf{v}\cdot\mathbf{B})_L = N_L = 0 \, . \tag{H.28}$$

(d) Symmetry conditions (Fig.4.4d). Symmetry and antisymmetry are very helpful in reducing the labor of solving nonlinear beamshell problems. A function $F(\sigma)$ is *symmetric* or *even* with respect to σ_0 if $F(\sigma_0+\Delta\sigma)=F(\sigma_0-\Delta\sigma)$ for all $\Delta\sigma$ and is *antisymmetric* or *odd* with respect to σ_0 if $F(\sigma_0+\Delta\sigma)=-F(\sigma_0-\Delta\sigma)$ for all $\Delta\sigma$. The latter condition implies that $F(\sigma_0)=0$ while the former implies that $F'(\sigma_0)=0$ if F is differentiable at σ_0.

(i) **Symmetry.** If a beamshell deforms symmetrically with respect to $\sigma=L$, we have

$$Q_L = (\mathbf{u}\cdot\mathbf{t})_L = \beta_L = 0 \, . \tag{H.29}$$

Other types of symmetry may also be discussed.

(ii) **Antisymmetry.** The 3 types of simple-support conditions listed in subsection H.3.c are precisely the 3 possible types of antisymmetric conditions; the edge is therefore a ∂U- and a ∂N-edge. The most common antisymmetric edge conditions are provided by the extrinsic diaphragm support, represented by (H.25).

(e) Clamped edge (Fig. 4.4e). The edge is held so that the slopes of $\bar{C}$ and C are the same at $\sigma=L$. Thus, noting that $\bar{\alpha}=\alpha+\beta+\gamma$, we have from the figure

$$\mathbf{u}_L = \mathbf{0} \ , \ \beta_L = -\gamma_L \, . \tag{H.30}$$

In contrast to the other edge conditions, this one is not classical because an expression for β itself can be found only by introducing a constitutive relation for γ, say $\gamma=-\Psi,_Q(N, Q, k)$. For further discussion on the distinction between fixed and clamped edges in a theory of beamshells that admits shear strains, see Antman (1972, p. 675) and Timoshenko & Goodier (1970, pp. 44-46). In a Kirchhoff motion, the boundary conditions for fixed and clamped edges are identical.

The reader should realize that the above list of boundary conditions is not exhaustive, but is rather a summary of typical, though idealized, cases. These can be expected to produce relatively simple boundary value problems.

*Appendix: The Principle of Mechanical Boundary Conditions

by Dawn Fisher

The aim of this appendix is to show how the elementary theory of vector spaces provides an ideal framework for the presentation of classical boundary conditions and the Principle of Mechanical Boundary Conditions.

For fixed y and t, the set of all possible edge velocity vectors v forms a 6-dimensional vector space. In this space, let ∂N be the null space of the constraint matrix A, and let ∂U be the orthogonal complement of ∂N. The map Av is 1:1 on ∂U and each vector in ∂U corresponds to exactly one choice of the nonhomogeneous term $\mathcal{B}$. If $\mathcal{B}$ is fixed, every vector v which satisfies the constraint $Av=\mathcal{B}$ has the unique decomposition $v=v_{\partial N}+v_{\partial U}=v_{\partial N}+\mathcal{H}$, where $A\,v_{\partial N}=0$ and $A\mathcal{H}=\mathcal{B}$.

The space of all possible edge-load vectors $\mathcal{N}$ is also 6-dimensional and can, for simplicity, be identified with the space of edge velocity vectors. Thus, each edge-load vector also has the unique decomposition $\mathcal{N}=\mathcal{N}_{\partial N}+\mathcal{N}_{\partial U}$, and its rate of work is given by

$$(\mathbf{F}\cdot\mathbf{v}+M\omega)_0^L=\mathcal{N}^T v=\mathcal{N}_{\partial N}^T v_{\partial N}+\mathcal{N}_{\partial U}^T v_{\partial U}\,. \tag{H.31}$$

In the classical case, every element of A is 0 except for a 1 in each row, with all the 1's in different columns. Now, v is in ∂N if v has 0's in the columns where A has 1's and v is in ∂U if v has 0's in the columns where A doesn't have 1's. (In the notation of this Section, $v_{\partial N}=\tilde{v}$, $v_{\partial U}=\hat{v}$, $\mathcal{N}_{\partial N}=\hat{\mathcal{N}}$ and $\mathcal{N}_{\partial U}=\check{\mathcal{N}}$.) Suppose now that y is a component of y for which the corresponding component v of v is constrained by the equation $v=f(y,t)$. If the value of y is given at $t=t_0$, all future values of y are determined by the differential equation $dy/dt=f(y,t)$, independent of any other information we might have. The component of generalized force corresponding to y is irrelevant for predicting the future behavior of the system. On the other hand, if y is a component of y for which the corresponding v is not constrained, then even if we know the values of y and v at t_0, we usually cannot predict the future values of y unless we also know the current and future values of the corresponding generalized force. The generalized forces which correspond to unconstrained generalized velocities must be specified to predict the future behavior of the system.

The Principle of Mechanical Boundary Conditions generalizes this observation. As $v_{\partial N}$ is the unconstrained portion of v, $\mathcal{N}_{\partial N}$ must be specified to predict the future behavior of the system. If an orthonormal basis $\mathcal{M}_1,\cdots,\mathcal{M}_n$ is chosen for ∂N, we need to know $\mathcal{N}^T\mathcal{M}_1,\cdots,\mathcal{N}^T\mathcal{M}_n$.

If $\partial N=0$, then $A\,v=\mathcal{B}$ is a solvable system of 6 quasilinear equations in 6 unknowns and the present value of y determines all future values, without reference to $\mathcal{N}$. If $\partial U=0$, then $A\equiv 0$ is not a constraint matrix.

I. The Principle of Virtual Work

The Principle of Virtual Work for a beamshell is the statement that the vanishing of the integral in (F.1), for certain test functions $\mathbf{V}$ and Ω, implies that $\mathbf{F}, M, \bar{\mathbf{y}}$, and β satisfy initial and boundary conditions and are solutions of the integral-impulse equations of motion, (B.13) and (B.17). We derive various forms of the Principle of Virtual Work from the weak form of the equations of motion following the line of reasoning set forth in Section H of the preceding chapter on birods.

First, we set $a=0, b=L, t_1=0$, and $t_2=T$, where T is positive and arbitrary. Next, we assume initial conditions of the form

$$\mathbf{v}(\sigma,0)=\overset{0}{\mathbf{v}}\ ,\quad \omega(\sigma,0)=\overset{0}{\omega}\ , \tag{I.1}$$

force boundary conditions of the form $\mathcal{N}=\hat{\mathcal{N}}$ and restrict the test functions in (F.1) to satisfy the final conditions

$$\mathbf{V}(\sigma,T)=\mathbf{0}\ ,\quad \Omega(\sigma,T)=0\,. \tag{I.2}$$

We denote all such pairs $(\mathbf{V},\Omega)$ of test functions by $\mathcal{T}_H$. Thus, with $\mathcal{N}=[-\mathbf{F}_0,-M_0,\mathbf{F}_L,M_L]^T$ and

$$\mathcal{V}\equiv[\mathbf{V}_0,\Omega_0,\mathbf{V}_L,\Omega_L]^T\ , \tag{I.3}$$

(F.1) takes the special form

$$J[\mathbf{F},M,\mathbf{v},\omega;\mathbf{V},\Omega]\equiv\int_0^T\hat{\mathcal{N}}^T\mathcal{V}dt+J_R=0\ ,\quad \forall\,(\mathbf{V},\Omega)\in\mathcal{T}_H\ , \tag{I.4}$$

where

$$\begin{aligned} J_R\equiv&\int_0^L(m\overset{0}{\mathbf{v}}\cdot\overset{0}{\mathbf{V}}+I\overset{0}{\omega}\overset{0}{\Omega})d\sigma\\ &-\int_0^T\int_0^L[\mathbf{F}\cdot(\mathbf{V}'-\mathbf{m}\Omega)+M\Omega'-m\mathbf{v}\cdot\dot{\mathbf{V}}-I\omega\dot{\Omega}-\mathbf{p}\cdot\mathbf{V}-l\Omega]d\sigma dt\ , \end{aligned} \tag{I.5}$$

$\overset{0}{\mathbf{V}}\equiv\mathbf{V}(\sigma,0)$, and $\overset{0}{\Omega}\equiv\Omega(\sigma,0)$.

Suppose that there is a kinematic constraint at the edge of the beamshell of the form $A\,v-\mathcal{B}=0$. To recover from a virtual work principle this constraint as well as the boundary conditions for the associated generalized edge loads, as discussed in Section H, we may, as we did with birods, either introduce free and unknown Lagrange multiplier vectors, $\mathcal{L}$ and $\tilde{\mathcal{L}}$, or else further restrict the test functions to satisfy the homogeneous constraint $A\mathcal{V}=0$. The former method leads from (I.4) to the following principle which is the analogue of (III.H.5):

$$J^*[\mathbf{F}, M, \mathbf{v}, \omega, \tilde{\mathcal{L}}; \mathbf{V}, \Omega, \mathcal{L}] \equiv \int_0^T [(\hat{\mathcal{N}}^T + \tilde{\mathcal{L}}^T A)\mathcal{V} + \mathcal{L}^T(A v - \mathcal{B})]dt + J_R = 0 , \quad (I.6)$$

for all $(\mathbf{V}, \Omega) \in \mathcal{T}_H$ and for all $\mathcal{L}$.

Alternatively, to account for the constraint $A\mathcal{V}=0$ explicitly, we set

$$\mathcal{V} = \sum_1^{6-m} r_k \mathcal{M}_k(y, t) \quad (I.7)$$

and denote by $\mathcal{T}_H(y)$ the subset of $\mathcal{T}_H$ whose elements have the form (I.7) on the edges of the beamshell. When (I.7) is introduced into (I.6) we obtain the principle

$$\bar{J}[\mathbf{F}, M, \mathbf{v}, \omega, r_k; \mathbf{V}, \Omega, \mathcal{L}] \equiv \int_0^T [\sum_1^{6-m} \hat{\mathcal{N}}^T \mathcal{M}_k r_k + \mathcal{L}^T(A v - \mathcal{B})]dt + J_R = 0 , \quad (I.8)$$

for all $(\mathbf{V}, \beta) \in \mathcal{T}_H(y)$ and all $\mathcal{L}$. This equation is the analogue of (III.H.12) for a birod.

See (Q.5)-(Q.8) for a variational form of the Principle of Virtual Work.

J. Potential (Conservative) Loads

We define the external loads to be *potential* or *conservative* if there exists a *functional* $\mathcal{P}[\bar{\mathbf{y}}, \beta]$ such that

$$\mathcal{W} = -\dot{\mathcal{P}} . \quad (J.1)$$

In this case, (G.1) reduces to

$$(\mathcal{K} + \mathcal{P})\Big|_{t_1}^{t_2} + \int_{t_1}^{t_2} \mathcal{D} dt = 0 . \quad (J.2)$$

As (J.1) is unaffected by adding a constant to $\mathcal{P}$, it is sometimes useful to define $\mathcal{P}$ so that $\mathcal{P}[\mathbf{y}, 0] = 0$.

One of the main advantages of a load potential, if one exists, is that it simplifies the analysis of the stability of an equilibrium configuration. Another advantage is that, if the beamshell is elastic, then energy is conserved. This fact is useful in vibration problems, in checking approximate methods, and in establishing variational principles. A major problem of both practical and theoretical interest is therefore to find necessary and sufficient conditions for a load potential to exist. Before addressing this, let us determine $\mathcal{P}$ for 3 common surface loads.[2]

1. Dead loading (and a torsional spring)

Here $\mathbf{p} = \mathbf{d}(\sigma)$, a vector function independent of deformation and time.[3] If

[2]Some of these examples are given by Antman (1976).

[3]If the loading is due to gravity in the negative y-direction, then $\mathbf{p} = -mg\mathbf{e}_y$, where g is the local gravitational constant.

the rate of work of the edge loads is zero, then, from (G.2),

$$\mathcal{W} = \int_0^L \mathbf{d}\cdot\dot{\bar{\mathbf{y}}}d\sigma = \left(\int_0^L \mathbf{d}\cdot\bar{\mathbf{y}}d\sigma\right)^{\bullet} \equiv -\dot{\mathcal{P}}\,. \tag{J.3}$$

If the left edge of the beamshell is built in and the right edge hinged to a nonlinear torsional spring such that $\hat{M}_L = k_T f(\beta_L)$, where k_T is a constant, then we have, in place of the the potential in (J.3),

$$\mathcal{P} = -\left[\int_0^L \mathbf{d}\cdot\bar{\mathbf{y}}d\sigma + k_T\int_0^{\beta_L} f(\beta)d\beta\right]. \tag{J.4}$$

2. Centrifugal loading

Let the beamshell rotate with constant angular velocity Ω about the z-axis. Then

$$\mathbf{p} = m\bar{\mathbf{y}}\Omega^2\,, \tag{J.5}$$

and, if the rate of work of the edge loads is zero,

$$\mathcal{W} = m\Omega^2\int_0^L \bar{\mathbf{y}}\cdot\dot{\bar{\mathbf{y}}}d\sigma = \left[\tfrac{1}{2}m\Omega^2\int_0^L \bar{\mathbf{y}}\cdot\bar{\mathbf{y}}d\sigma\right]^{\bullet} \equiv -\dot{\mathcal{P}}\,. \tag{J.6}$$

3. Pressure loading (constant or hydrostatic)[4]

Let the beamshell be subject to a force p *per unit length of* $\bar{\mathcal{C}}$ directed along $-\bar{\mathbf{b}}$. From (E.3),

$$\mathbf{p} = -p\mathbf{m} = p\bar{\mathbf{y}}'\times\mathbf{e}_z\,. \tag{J.7}$$

Thus, with $\bar{\mathbf{y}} = \bar{x}\mathbf{e}_x + \bar{y}\mathbf{e}_y$, we have

$$\int_0^L \mathbf{p}\cdot\dot{\bar{\mathbf{y}}}d\sigma = \int_0^L p\,(\bar{y}'\dot{\bar{x}} - \bar{x}'\dot{\bar{y}})d\sigma\,. \tag{J.8}$$

If $p = p_0$, a constant, then, after an integration by parts,

$$\mathcal{W} = (\mathbf{F}\cdot\mathbf{v} + M\omega + p_0\bar{y}\dot{\bar{x}})_0^L - p_0\dot{\bar{A}}\,, \tag{J.9}$$

where

$$\bar{A} \equiv \int_0^L \bar{y}\bar{x}'d\sigma \tag{J.10}$$

is the area between $\bar{\mathcal{C}}$ and the x-axis. If the boundary conditions are such that the integrated terms in (J.9) are the time derivative of an *edge* potential $-\Gamma$, then

$$\mathcal{P} = \Gamma + p_0\bar{A}\,. \tag{J.11}$$

If $p = -\rho g\bar{y}$, i.e., if there is a hydrostatic pressure produced by a fluid of density ρ whose free surface coincides with the x-axis, then

---

[4]Some authors, e.g., Sills & Budiansky (1978), take "hydrostatic" pressure to mean constant pressure.

$$\int_0^L \mathbf{p}\cdot\dot{\mathbf{y}}d\sigma = \rho g\int_0^L \bar{y}(\bar{y}'\dot{\bar{x}} - \bar{x}'\dot{\bar{y}})d\sigma = \rho g\,(\tfrac{1}{2}\bar{y}^2\dot{\bar{x}}\,|_0^L - \dot{\bar{B}})\,, \tag{J.12}$$

where

$$\bar{B} \equiv \tfrac{1}{2}\int_0^L \bar{y}^2\bar{x}'d\sigma\,. \tag{J.13}$$

Geometrically, $\bar{B}/\bar{A}$ is the centroid of the area between $\mathcal{C}$ and the x-axis (if $\mathcal{C}$ lies below the x-axis, $\bar{A}$ is negative). If there exists an edge potential such that

$$(\mathbf{F}\cdot\mathbf{v} + M\omega + \tfrac{1}{2}\rho g\bar{y}^2\dot{\bar{x}})_0^L = -\dot{\Gamma}\,, \tag{J.14}$$

then

$$\mathcal{P} = \Gamma + \rho g\bar{B}\,. \tag{J.15}$$

*4. *General discussion and examples*

Necessary and sufficient conditions for the existence of a load potential in 3-dimensional continuum mechanics have been discussed at length by Sewell (1967), Romano (1972), Buffler (1984), and Fisher (1987), among others, and we refer the reader to these papers for references to earlier and related work. The requisite mathematical machinery from functional analysis (and variational theory in particular) may be found in Gelfand & Fomin (1963), Vainberg (1964), Stackgold (1979), and Oden & Reddy (1983). Here, we shall give an informal presentation of the basic ideas, culminating in a test for the potentiality of $\mathcal{W}$ which, if passed, leads to a formula for $\mathcal{P}$.

Let us start with a careful analysis of the analogous but simpler problem of finding necessary and sufficient conditions for the potentiality of a 2-dimensional force field **F** defined on some connected domain $\mathcal{D}$ of the xy-plane. (**F** is not to be confused with the force **F** in a beamshell). In analogy with (J.1), we could define **F** to be potential (or conservative) if there exists a scalar function $\mathcal{P}(\mathbf{x})$ such that

$$-\mathbf{F} = \nabla\mathcal{P}(\mathbf{x}) \equiv \mathcal{P}'(\mathbf{x})\,. \tag{J.16}$$

An alternative that emphasizes the underlying mechanical idea is to define **F** to be potential on $\mathcal{D}$ if the work done by **F** along any piecewise smooth path γ_{PQ} in $\mathcal{D}$ joining any 2 points P and Q depends on P and Q only. Consequently, if P is fixed and Q is free with position **x**, then

$$\mathcal{P}(\mathbf{x}) \equiv -\int_{\gamma_{PQ}} \mathbf{F}(\mathbf{y})\cdot d\mathbf{y} \tag{J.17}$$

is a scalar function of **x** such that (J.16) holds, i.e., $\mathcal{P}$ is a potential for **F**. If $\mathcal{D}$ is convex (and therefore simply-connected) and contains the origin O, then, with $P = O$ and γ_{PQ} a straight line from O to Q, (J.17) reduces to

$$\mathcal{P}(\mathbf{x}) = -\int_0^1 \mathbf{F}(\tau\mathbf{x})\cdot\mathbf{x}d\tau\,, \tag{J.18}$$

where τ is a normalized distance parameter along $\mathbf{x}$.

The above definition of a potential force involves a *global* criterion on $\mathbf{F}$. To derive an easy-to-apply *local* test, let $\mathbf{a}$ and $\mathbf{b}$—$\mathbf{b}$ having nothing to do with the unit normal defined in (C.3)—be any constant vectors and assume that $\mathcal{D}$ is open so that if the point $Q(\mathbf{x})$ is in $\mathcal{D}$, so is $S(\mathbf{x}+h\mathbf{a}+k\mathbf{b})$, provided h and k are sufficiently small non-zero constants.

Consider the parallelogram $QRST$ in Fig. 4.5. If $\mathbf{F}$ is potential, then, by definition,

$$\int_{QRS} \mathbf{F}(\mathbf{y})\cdot d\mathbf{y} = \int_{QTS} \mathbf{F}(\mathbf{y})\cdot d\mathbf{y}\,, \tag{J.19}$$

where QRS is the broken straight line from Q to R to S, etc. We rewrite (J.19) as follows:

$$\int_{RS} \mathbf{F}(\mathbf{y})\cdot d\mathbf{y} - \int_{QT} \mathbf{F}(\mathbf{y})\cdot d\mathbf{y} = \int_{TS} \mathbf{F}(\mathbf{y})\cdot d\mathbf{y} - \int_{QR} \mathbf{F}(\mathbf{y})\cdot d\mathbf{y}\,. \tag{J.20}$$

On RS, $\mathbf{y}=\mathbf{x}+h\mathbf{a}+\tau k\mathbf{b}$ and on QT, $\mathbf{y}=\mathbf{x}+\tau k\mathbf{b}$, $0\le\tau\le 1$. Hence, the left side of (J.20) may be written

$$k\int_0^1 \mathbf{b}\cdot[\mathbf{F}(\mathbf{x}+h\mathbf{a}+\tau k\mathbf{b}) - \mathbf{F}(\mathbf{x}+\tau k\mathbf{b})]d\tau \equiv k\int_0^1 f(\tau)d\tau\,. \tag{J.21}$$

We now assume that $\mathbf{F}$ is continuous on $\mathcal{D}$ so that the scalar f is a continuous function of τ. This implies that there exists some number $\bar{\tau}$ between 0 and 1 such that $\int_0^1 f(\tau)d\tau = f(\bar{\tau})$. This result plus a similar one coming from the right side of (J.20) involving a number $\tilde{\tau}$ between 0 and 1 allows us to rewrite (J.20) in the form

$$k\mathbf{b}\cdot[\mathbf{F}(\mathbf{x}+h\mathbf{a}+\bar{\tau}k\mathbf{b}) - \mathbf{F}(\mathbf{x}+\bar{\tau}k\mathbf{b})] = h\mathbf{a}\cdot[\mathbf{F}(\mathbf{x}+\tilde{\tau}h\mathbf{a}+k\mathbf{b}) - \mathbf{F}(\mathbf{x}+\tilde{\tau}h\mathbf{a})]\,. \tag{J.22}$$

Finally, we assume that the scalar $\mathbf{c}\cdot\mathbf{F}$ has a gradient $(\mathbf{c}\cdot\mathbf{F})'(\mathbf{x})$ on $\mathcal{D}$. By the Mean Value Theorem there exists numbers $\hat{\tau}, \tau^* \in (0, 1)$ such that (J.22) may be replaced by

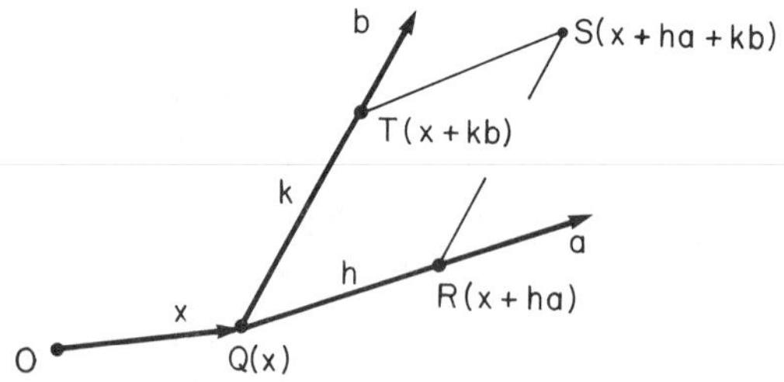

Fig. 4.5. Diagram for establishing whether a force field is potential.

$$kh\mathbf{a}\cdot(\mathbf{b}\cdot\mathbf{F})'(\mathbf{x}+\hat{\tau}h\mathbf{a}+\bar{\tau}k\mathbf{b}) = hk\mathbf{b}\cdot(\mathbf{a}\cdot\mathbf{F})'(\mathbf{x}+\tau^* h\mathbf{a}+\tilde{\tau}k\mathbf{b}) . \quad (J.23)$$

Dividing both sides by hk and letting $h^2+k^2 \to 0$, we obtain

$$\mathbf{a}\cdot(\mathbf{b}\cdot\mathbf{F})'(\mathbf{x}) = \mathbf{b}\cdot(\mathbf{a}\cdot\mathbf{F})'(\mathbf{x}) , \quad (J.24)$$

for all constant vectors **a** and **b**. This is the *necessary condition* for potentiality that we sought. In Cartesian coordinates, with $\mathbf{F}=F_x\mathbf{e}_x+F_y\mathbf{e}_y$, (J.24) implies the familiar condition $F_{x},_{y}=F_{y},_{x}$. By arguments found in texts on advanced calculus, e.g., Buck (1978), it may be shown that if $\int \mathbf{F}\cdot d\mathbf{y}$ vanishes around every hole in $\mathcal{D}$ (should it be multiply-connected), then (J.24) is also a *sufficient condition* for a potential to exist.

The importance of the steps in above analysis, especially condition (J.24), is that they carry over by analogy to the study of the potentiality of the external loads on a beamshell.

In place of the force field **F**, which may be regarded as an operator that sends each element **x** in its domain into a new vector **F**(**x**), we have, with a beamshell, a load operator $\mathcal{F}$ that sends any element

$$\overline{w} \equiv (\overline{\mathbf{y}}, \beta, \overline{\mathbf{y}}_0, \beta_0, \overline{\mathbf{y}}_L, \beta_L) \quad (J.25)$$

in a certain domain $\mathcal{A}$ (the class of admissible functions) into a new class of functions

$$\mathcal{F}(\overline{w}) \equiv (\mathbf{p}, l, \mathbf{F}_0, M_0, \mathbf{F}_L, M_L) . \quad (J.26)$$

Here, $\overline{\mathbf{y}}_0 \equiv \overline{\mathbf{y}}(0, t)$, etc.

Just as we introduced the dot (or inner) product $\mathbf{F}(\mathbf{x})\cdot d\mathbf{x}=\mathbf{F}(\mathbf{x})\cdot\dot{\mathbf{x}}dt$, so may we define an inner product involving the load operator:

$$\begin{aligned} <\mathcal{F}(\overline{w}), \delta\overline{w}> &\equiv \int_0^L (\mathbf{p}\cdot\delta\overline{\mathbf{y}} + l\delta\beta)d\sigma + (\mathbf{F}\cdot\delta\overline{\mathbf{y}} + M\delta\beta)_0^L \\ &= [\int_0^L (\mathbf{p}\cdot\dot{\overline{\mathbf{y}}} + l\dot{\beta})d\sigma + (\mathbf{F}\cdot\dot{\overline{\mathbf{y}}} + M\dot{\beta})_0^L]dt \\ &= <\mathcal{F}(\overline{w}), \dot{\overline{w}}>dt . \end{aligned} \quad (J.27)$$

If $\gamma: \mathbf{x}=\mathbf{x}(t)$, $t_1 \le t \le t_2$ denotes a path in the xy-plane, then $\int_\gamma \mathbf{F}(\mathbf{x})\cdot d\mathbf{x}= \int_{t_1}^{t_2}\mathbf{F}(\mathbf{x})\cdot\dot{\mathbf{x}}dt$ is the work done by **F** in the interval (t_1, t_2). Likewise, if $\gamma: \overline{w}=\overline{w}(t)$, $t_1 \le t \le t_2$ is a path in the class of admissible functions $\mathcal{A}$, then by (G.2) and (J.27),

$$\int_{t_1}^{t_2} \mathcal{W}dt = \int_\gamma <\mathcal{F}(\overline{w}), \delta\overline{w}> = \int_{t_1}^{t_2} <\mathcal{F}(\overline{w}), \dot{\overline{w}}>dt \quad (J.28)$$

is the work done on the beamshell during the interval (t_1, t_2). If the left side of (J.28) depends on $\overline{w}(t_1)$ and $\overline{w}(t_2)$ only, then $\mathcal{F}$ is *potential* or *conservative*. If $\mathcal{A}$ is convex, i.e., if the deformation field $(1-\tau)\overline{w}(t_1)+\tau\overline{w}(t_2)$, $0 \le \tau \le 1$, is always admissible, as we shall assume, then a potential for $\mathcal{W}$ is given by

$$\mathcal{P}[\overline{w}] \equiv -\int_0^1 <\mathcal{F}(w+\tau(\overline{w}-w)), \overline{w}-w>d\tau\,, \tag{J.29}$$

where $w=(\mathbf{y}, 0, \mathbf{y}_0, 0, \mathbf{y}_L, 0)$ is the *initial* kinematic field. The above equation is the analogue of (J.19).

Let $\mathcal{A}_\delta$ denote the set of all functions such that if v and w are in $\mathcal{A}$, then $w-v$ is in $\mathcal{A}_\delta$. [The elements of $\mathcal{A}_\delta$ may be denoted by $(\delta\overline{\mathbf{y}}, \delta\beta, \cdots, \delta\beta_L)$.] We shall restrict attention henceforth to *classical boundary conditions* which means that each component of either $\mathcal{N}=[-\mathbf{F}_0, M_0, \mathbf{F}_L, M_L]^T$ or $y=[\overline{\mathbf{y}}_0, \beta_0, \overline{\mathbf{y}}_L, \beta_L]^T$ (but *not* both) is specified, where it is understood that $\mathbf{F}$ and $\overline{\mathbf{y}}$ are to be represented in component form. Thus, $\mathcal{A}_\delta$ is a *linear vector space.*

The analog of the gradient of $\mathbf{F}\cdot\mathbf{c}$ evaluated at $\mathbf{x}$ is the *Fréchet derivative* of the scalar $<\mathcal{F}, v>$ evaluated at $\overline{w}$, which may be denoted by $\mathcal{F}'(\overline{w})v$. The analog of the necessary condition, (J.24), that $\mathbf{F}$ be potential is

$$<\mathcal{F}'(\overline{w})u, v> = <\mathcal{F}'(\overline{w})v, u>\ ,\ \ \forall\, u,v \in \mathcal{A}_\delta\,. \tag{J.30}$$

In the parlance of functional analysis, a *necessary condition* that $\mathcal{W}$ be potential is that $\mathcal{F}'(\overline{w})$ be *symmetric*. As we have assumed that $\mathcal{A}$ is convex, it may be shown, e.g., Vainberg (1964), that (J.30) is also a *sufficient condition* for a potential to exist.

In practice we compute each side of (J.30) using the Calculus of Variations where, in perhaps more familiar notation, $\mathcal{F}'(\overline{w})u = \delta\mathcal{F}[\overline{w}, u]$, to be read "the variation of $\mathcal{F}$ at $\overline{w}$ in the direction of u." As an application of (J.30), we consider 2 examples.

Problem: A beamshell is free of distributed loads, built in at $\sigma=0$ and subject, at $\sigma=L$, to a force $\mathbf{F}_L$ and moment M_L that each depend on $\overline{\mathbf{y}}_L$ and β_L. Determine when this loading is potential.

Solution: Here

$$<\mathcal{F}(\overline{w}), \delta\hat{w}> = \mathbf{F}_L\cdot\delta\hat{\mathbf{y}}_L + M_L\delta\hat{\beta}_L \tag{J.31}$$

so that

$$<\mathcal{F}'(\overline{w})\delta\overline{w}, \delta\hat{w}> = \delta\mathbf{F}_L\cdot\delta\hat{\mathbf{y}}_L + \delta M_L\delta\hat{\beta}_L\,. \tag{J.32}$$

If $\mathbf{F}\equiv H\mathbf{e}_x + V\mathbf{e}_y$, $\delta\overline{\mathbf{y}}_L = \delta\overline{x}_L\mathbf{e}_x + \delta\overline{y}_L\mathbf{e}_y$, etc., and $H,_1 \equiv \partial H/\partial\overline{x}$ at $(\overline{\mathbf{y}}_L, \beta_L)$, etc., then the right side of (J.32) takes the expanded form

$$\begin{aligned}(H,_1\delta\overline{x}_L + H,_2\delta\overline{y}_L + H,_3\delta\beta_L)\delta\hat{x}_L + (V,_1\delta\overline{x}_L + \cdots)\delta\hat{y}_L \\ + (M,_1\delta\overline{x}_L + \cdots)\delta\hat{\beta}_L\,.\end{aligned} \tag{J.33}$$

Hence, the edge loads are potential if and only if

$$H,_2 = V,_1\ ,\ \ H,_3 = M,_1\ ,\ \ V,_3 = M,_2\,, \tag{J.34}$$

a well-known condition from vector calculus. In particular, the *follower load*

$$\mathbf{F}_L = -P\,[\cos(\alpha+\beta)\mathbf{e}_x + \sin(\alpha+\beta)\mathbf{e}_y] \ , \ M_L = 0 \tag{J.35}$$

is *not* potential because $H,_3 - M,_1 = P\sin(\alpha+\beta) \neq 0$. □

Problem: For simplicity assume that $\bar{\mathbf{y}}$ and β are specified at $\sigma = 0, L$. Determine the form (J.30) takes if the elements of the 3×1 load matrix

$$\mathcal{L} = [\mathbf{p}, l]^T = [p_x, p_y, l]^T \tag{J.36}$$

are functions of

$$\bar{w} \equiv [\bar{\mathbf{y}}, \beta]^T = [\bar{x}, \bar{y}, \beta]^T \tag{J.37}$$

and $\bar{w}'$, its derivative with respect to σ. In particular, show that *if the distributed couple l is zero, then a normal pressure p is potential if and only if p depends on $\bar{\mathbf{y}}$ only* and determine the potential $\mathcal{P}$ in this case.

Solution: Let

$$A \equiv \mathcal{L},_{\bar{w}} \equiv \begin{bmatrix} p_x,_{\bar{x}} & p_x,_{\bar{y}} & p_x,_{\beta} \\ p_y,_{\bar{x}} & p_y,_{\bar{y}} & p_y,_{\beta} \\ l,_{\bar{x}} & l,_{\bar{y}} & l,_{\beta} \end{bmatrix} \tag{J.38}$$

with a similar definition of a matrix $B \equiv \mathcal{L},_{\bar{w}'}$, in which differentiation with respect to $\bar{x}$ is replaced by differentiation with respect to $\bar{x}'$ etc. Thus,

$$<\mathcal{F}(\bar{w}), \delta\bar{w}> = \int_0^L \mathbf{p}\cdot\delta\bar{\mathbf{y}}\,d\sigma = \int_0^L \mathcal{L}^T\delta\bar{w}\,d\sigma \tag{J.39}$$

and the condition for potentiality, (J.30), takes the form

$$\begin{aligned} &<\mathcal{F}'(\bar{w})\delta\bar{w}, \delta\hat{w}> - <\mathcal{F}'(\bar{w})\delta\hat{w}, \delta\bar{w}'> \\ &\quad = \int_0^L [\delta\hat{w}^T(A - A^T)\delta\bar{w} + \delta\hat{w}^T B\,\delta\bar{w}' - \delta\bar{w}^T B\,\delta\hat{w}']d\sigma = 0\,. \end{aligned} \tag{J.40}$$

Integrating the $\delta\bar{w}'$-term by parts, we have

$$\int_0^L [\delta\hat{w}^T(A - A^T - B') - \delta\hat{w}'^T(B + B^T)]\delta\bar{w}\,d\sigma = 0\,. \tag{J.41}$$

We now claim, because $\delta\bar{w}$ and $\delta\hat{w}$ are essentially arbitrary, that the matrices in parentheses in (J.41), if continuous, must vanish. That is,

$$B = -B^T \ , \ B' = A - A^T\,. \tag{J.42}$$

These are the necessary and sufficient conditions for $\int_0^L (\mathbf{p}\cdot\dot{\bar{\mathbf{y}}} + l\dot{\beta})d\sigma$ to be potential. The proof of (J.42) has 3 parts.

First, let v denote the row vector in brackets in (J.41). Because $\delta\bar{w}$ is an arbitrary differentiable function that need vanish only at $\sigma = 0, L$, we may choose $\delta\bar{w} = \sigma(L-\sigma)v^T(\sigma)$. This yields $\int_0^L \sigma(L-\sigma)vv^T(\sigma)d\sigma = 0$, which, if v is continuous, implies that $v(\sigma) \equiv 0$ on $[0, L]$.

Second, note that if there were a point $\sigma_* \in [0, L]$ where $B+B^T \neq 0$, then we could always choose $\delta\hat{w}$ so that $\delta\hat{w}(\sigma_*)=0$, $\delta\hat{w}'(\sigma_*)\equiv u$, an arbitrary constant column vector, to obtain at $\sigma=\sigma_*$, $u^T(B+B^T)\neq 0$ for all u. Clearly, this is a contradiction.

Finally, with $B+B^T=0$, it follows that the factor $A-A^T-B'$ in the remaining term in the row vector v must vanish everywhere. For if it failed to at, say, $\tilde{\sigma} \in (0, L)$ then we would have $\delta\hat{w}^T(A-A^T-B')\neq 0$ at $\sigma=\tilde{\sigma}$ which, since $\delta\hat{w}(\tilde{\sigma})$ may be chosen arbitrarily, is a contradiction. By continuity, if $(A-A^T-B')$ vanishes on $(0, L)$ it vanishes on $[0, L]$.

If $\mathbf{p}=-p\mathbf{m}=p(\bar{y}'\mathbf{e}_x-\bar{x}'\mathbf{e}_y)$ and $l=0$, then

$$B=\begin{bmatrix} (p\bar{y}')_{,\bar{x}'} & (p\bar{y}')_{,\bar{y}'} & (p\bar{y}')_{,\beta'} \\ -(p\bar{x}')_{,\bar{x}'} & -(p\bar{x}')_{,\bar{y}'} & -(p\bar{x}')_{,\beta'} \\ 0 & 0 & 0 \end{bmatrix}. \tag{J.43}$$

Hence, $B=-B^T$ implies that

$$p_{,\bar{x}'}=p_{,\bar{y}'}=p_{,\beta'}=0\,, \tag{J.44}$$

i.e., $p=p(\bar{x}, \bar{y}, \beta, \sigma)$. Furthermore, $B'=A-A^T$ reduces to the equation

$$\begin{bmatrix} 0 & p' & 0 \\ -p' & 0 & 0 \\ 0 & 0 & 0 \end{bmatrix}=\begin{bmatrix} 0 & (p\bar{x}')_{,\bar{x}}+(p\bar{y}')_{,\bar{y}} & (p\bar{y}')_{,\beta} \\ -(p\bar{x}')_{,\bar{x}}-(p\bar{y}')_{,\bar{y}} & 0 & -(p\bar{x}')_{,\beta} \\ -(p\bar{y}')_{,\beta} & (p\bar{x}')_{,\beta} & 0 \end{bmatrix}, \tag{J.45}$$

which implies that $p_{,\sigma}=p_{,\beta}=0$. That is, *a normal pressure is potential if and only if* $p=p(\bar{x}, \bar{y})$, as asserted earlier.

To compute the potential itself associated with p, let

$$q\equiv\int_0^{\bar{y}} p(\bar{x}, \tau)d\tau\,. \tag{J.46}$$

Then

$$<\mathcal{F}(\bar{w}), \delta\bar{w}>=\int_0^L p(\bar{y}'\delta\bar{x}-\bar{x}'\delta\bar{y})d\sigma=\int_0^L(q'\delta\bar{x}-\bar{x}'\delta q)d\sigma\,. \tag{J.47}$$

Integrating by parts and recalling that, for simplicity, we have assumed $\bar{y}$ and β to be prescribed on the edges of the beamshell, we have

$$<\mathcal{F}(\bar{w}), \delta\bar{w}>=-\delta\int_0^L\bar{x}'q\,d\sigma\equiv-\delta\mathcal{P}. \tag{J.48}$$

It is easy to verify that if $p=p_0$ or $p=\rho g\bar{y}$, then the expression for $\mathcal{P}$ given by (J.48) reduces to (J.11) or (J.15) (with $\Gamma=0$). □

K. Strains

In our approach to shell theory, strains are not defined *a priori*, but rather fall out naturally from the deformation power. Here we follow Reissner (1962).

If $\bar{\mathbf{y}}' \neq \mathbf{0}$, as we shall henceforth assume, then there exist unique scalars e and g, called the *extensional and shearing strains*, respectively, such that

$$\bar{\mathbf{y}}' = (1+e)\mathbf{T} + g\mathbf{B} . \tag{K.1}$$

From (C.5)-(C.7),

$$\dot{\mathbf{T}} = \omega\mathbf{B} \ , \ \dot{\mathbf{B}} = -\omega\mathbf{T} . \tag{K.2}$$

Hence the time derivative of (K.1) can be expressed as

$$\mathbf{v}' = \dot{\bar{\mathbf{y}}}' = \dot{e}\mathbf{T} + (1+e)\omega\mathbf{B} + \dot{g}\mathbf{B} - \omega g\mathbf{T} = \dot{e}\mathbf{T} + \dot{g}\mathbf{B} + \omega\,\mathbf{m} . \tag{K.3}$$

Substituting (K.3) into (G.4), we obtain for the deformation power,

$$\mathcal{D} = \int_a^b \tau{:}\dot{\varepsilon}\, d\sigma , \tag{K.4}$$

where

$$\tau{:}\dot{\varepsilon} = N\dot{e} + Q\dot{g} + M\dot{k} \tag{K.5}$$

is the *deformation power density*,

$$k = \beta' \tag{K.6}$$

is the *bending strain*, and

$$N = \mathbf{F}\cdot\mathbf{T} \ , \ Q = \mathbf{F}\cdot\mathbf{B} . \tag{K.7}$$

In a *reduced* Kirchhoff motion $g = 0$[5] and N and Q—the *normal* and *shear* stress resultants—are the components of $\mathbf{F}$ along and normal to the deformed initial curve. Note that in (K.6), β is differentiated with respect to the arc length of the *initial* curve which geometrically uncouples bending and stretching. For example, if a circular cylindrical shell undergoes a pure radial deflection, $k = 0$.

From (K.1), (K.6), and (K.7) and the fact that $[\beta] = 0$, the jump conditions (D.8) reduce to

$$[N] = c^2 m[e] \ , \ [Q] = c^2 m[g] \ , \ [M] = c^2 I[k] . \tag{K.8}$$

Reissner (1972) and Antman (1974a) have developed beam theories based, essentially, on (K.1) and (K.6). The analogous theory for shells was initiated, independently, by Wempner (1969) and Simmonds & Danielson (1970,

---

[5] We speak of a "reduced Kirchhoff motion" because g is a 1-dimensional transverse shearing strain, whereas "a Kirchhoff motion" (or "the Kirchhoff Hypothesis") usually means that the transverse shearing and normal strains in a shell-like, *3-dimensional body* are zero. See Novozhilov (1953, p. 189).

1972), and has been elaborated upon by Pietraszkiewicz (1979).

If e, g, and β are known, it follows immediately from (K.1) that the *displacement* is given by

$$\mathbf{u} \equiv \bar{\mathbf{y}} - \mathbf{y} = \int [(1+e)\mathbf{T} + g\mathbf{B}]d\sigma - \mathbf{y} . \tag{K.9}$$

At this point, we have a scalar and a 2-dimensional vector equation of motion, 3 static variables, N, Q, and M, and 3 kinematic quantities, e, g, and β. Constitutive relations will be discussed in Section N. Initial conditions on $\mathbf{u}$ are standard:

$$\text{at } t = 0 \text{ prescribe } \mathbf{u} \text{ and } \dot{\mathbf{u}} . \tag{K.10}$$

Constraints on the displacement $\mathbf{u}$ become constraints on e, g, and β via (K.9). In dynamics, the linear acceleration $\dot{\mathbf{v}} = \ddot{\mathbf{u}}$ that appears in the differential equation of motion (E.1) can be expressed in terms of time derivatives of e, g, and β by differentiating (E.1) with respect to σ, as discussed in Section P.

L. Alternative Strains and Stresses

This section is for readers who may prefer to use the displacement $\mathbf{u}$ as a basic kinematic variable. To avoid imposing the reduced Kirchhoff hypothesis, we introduce the *shear strain* γ defined in Fig. 4.2 as an additional kinematic variable. As the expressions for e and k in terms of $\mathbf{u}$ and γ are awkward, we shall introduce new extensional and bending strains ε and K that are *quadratic polynomials* in $\mathbf{u}$ and its derivatives; K also depends linearly on γ'. It turns out that the pair (ε, γ) is simple to visualize and in some cases (as we shall show in Section N on strain-energy densities and in section O on elastostatics) is more convenient to use than the pair (e, g). Associated with ε and K will be a new conjugate force and moment $\bar{N}$ and M^*.

Taking note of (C.2) and (K.9), we set

$$\bar{\mathbf{y}}' = \mathbf{t} + \mathbf{u}' \equiv \lambda[\cos(\beta+\gamma)\mathbf{t} + \sin(\beta+\gamma)\mathbf{b}] \equiv \lambda(\cos\bar{\alpha}\,\mathbf{e}_x + \sin\bar{\alpha}\,\mathbf{e}_y) \equiv \lambda\bar{\mathbf{t}} , \tag{L.1}$$

where $\bar{\mathbf{t}}$ is a unit tangent to the deformed reference curve $\bar{\mathcal{C}}$,

$$\bar{\alpha} = \alpha + \beta + \gamma \tag{L.2}$$

is the slope of $\bar{\mathcal{C}}$ (see Fig. 4.2), and λ is the *axial stretch*. Thus

$$\lambda^2 = \bar{\mathbf{y}}' \cdot \bar{\mathbf{y}}' \equiv 1 + 2\varepsilon , \tag{L.3}$$

where

$$\varepsilon = \mathbf{u}' \cdot \mathbf{t} + \frac{1}{2}\mathbf{u}' \cdot \mathbf{u}' \tag{L.4}$$

is the *Lagrangian strain*. Noting by (E.3) and (L.1) that

$$\mathbf{m} = \lambda[-\sin(\beta+\gamma)\,\mathbf{t} + \cos(\beta+\gamma)\,\mathbf{b}] = \lambda(-\sin\bar{\alpha}\,\mathbf{e}_x + \cos\bar{\alpha}\,\mathbf{e}_y) \equiv \lambda\bar{\mathbf{b}}\,, \quad \text{(L.5)}$$

we have from (L.1),

$$\dot{\bar{\mathbf{y}}}' - \dot{\beta}\mathbf{m} = \mathbf{v}' - \omega\mathbf{m} = (\dot{\lambda}/\lambda)\bar{\mathbf{y}}' + \dot{\gamma}\mathbf{m}\,. \quad \text{(L.6)}$$

If we now set

$$\mathbf{F} = \tilde{N}\bar{\mathbf{y}}' + \lambda^{-2}Q^*\mathbf{m} \quad \text{(L.7)}$$

and note that $\lambda\dot{\lambda} = \dot{\varepsilon}$, then the deformation power density, from (G.4) and (K.4), takes the form

$$\tau{:}\dot{\varepsilon} = \tilde{N}\dot{\varepsilon} + Q^*\dot{\gamma} + M\dot{k}\,. \quad \text{(L.8)}$$

Comparing (K.1) to (L.1) and noting (C.6) and (C.7)—see also Fig. 4.2—we have,

$$1 + e = \lambda\cos\gamma\ , \quad g = \lambda\sin\gamma\,. \quad \text{(L.9)}$$

When (L.9) is substituted into (K.5) and the resulting expression compared with (L.8) (noting that $\lambda\dot{\lambda} = \dot{\varepsilon}$), we obtain the following relation between the two sets of stress measures:

$$\lambda^2\tilde{N} = (1+e)N + gQ\ , \quad Q^* = (1+e)Q - gN\,. \quad \text{(L.10)}$$

To introduce the new bending strain K, we first note that the curvatures of $\mathcal{C}$ and $\bar{\mathcal{C}}$ are, respectively, α' and $\bar{\alpha}'/\lambda$. From (C.2), (L.1), and $\mathbf{u} = \bar{\mathbf{y}} - \mathbf{y}$ follows

$$\begin{aligned} \lambda^2\bar{\alpha}' - \alpha' &= \mathbf{e}_z\cdot(\bar{\mathbf{y}}'\times\bar{\mathbf{y}}'') - \mathbf{e}_z\cdot(\mathbf{y}'\times\mathbf{y}'') \\ &= \mathbf{e}_z\cdot(\mathbf{u}'\times\mathbf{y}'' + \mathbf{y}'\times\mathbf{u}'' + \mathbf{u}'\times\mathbf{u}'')\,. \end{aligned} \quad \text{(L.11)}$$

Hence, $\lambda^2\bar{\alpha}' - \alpha'$, which is clearly a measure of bending, is *quadratic* in $\mathbf{u}$ and its derivatives.

To relate $k = \beta'$ to $\lambda^2\bar{\alpha}' - \alpha'$, recall that $\bar{\alpha} = \alpha + \beta + \gamma$ and $\lambda^2 = 1 + 2\varepsilon$. Thus

$$\lambda^2 k = \lambda^2\beta' = \lambda^2(\bar{\alpha} - \gamma)' - \lambda^2\alpha' = K - 2\varepsilon\alpha'\,, \quad \text{(L.12)}$$

where

$$K \equiv \lambda^2\bar{\alpha}' - \alpha' - \lambda^2\gamma'\,. \quad \text{(L.13)}$$

This bending strain is the analog of the meridional bending strain introduced by Simmonds (1985a) to simplify the displacement form of the equations of motion of axishells. (See Section V.P).

Since $(\lambda^2)^{\cdot} = 2\dot{\varepsilon}$, it follows from (L.12) that

$$\lambda^2\dot{k} = \dot{K} - 2(\alpha' + K)\lambda^{-2}\dot{\varepsilon}\,. \quad \text{(L.14)}$$

Substituting this expression into (L.8), we have

$$\tau{:}\dot{\varepsilon} = \bar{N}\dot{\varepsilon} + Q^*\dot{\gamma} + M^*\dot{K}\,, \quad \text{(L.15)}$$

where we have introduced the *modified* moment and force[6]

$$M^* = \lambda^{-2}M \ , \ \bar{N} = \tilde{N} - 2(\alpha' + K)\lambda^{-2}M^* \ . \tag{L.16}$$

It is of interest to compare K with the bending strain of Budiansky (1968)—call it κ_*—derived assuming $\gamma=0$[7]; κ_* is polynomial in $\mathbf{u}$ and defined so that, when linearized, it reduces to that of the "best" linear theory (Budiansky & Sanders, 1963) of Sanders (1959) and Koiter (1960). This latter theory, which enjoys a particularly simple static-geometric duality, may be characterized as one in which the bending strains are expressed in terms of rotations only. Thus, because k is expressible in terms of rotations only ($k=\beta'$), it follows from (L.12) and (L.13) that

$$\kappa_* = K - 2\alpha'\varepsilon + \lambda^2\gamma' \ . \tag{L.17}$$

Still another bending strain,

$$\kappa \equiv \lambda^2 k = \lambda^2(\bar{\alpha} - \alpha - \gamma)' = \mathbf{e}_z\cdot(\bar{\mathbf{y}}' \times \bar{\mathbf{y}}'' - \lambda^2\mathbf{y}' \times \mathbf{y}'') - \lambda^2\gamma' \ , \tag{L.18}$$

was introduced by Libai & Simmonds (1983a) to extend Budiansky's definition to non-Kirchhoffian motions. This introduces a new modified axial force

$$N^* \equiv \tilde{N} - 2\kappa\lambda^{-2}M^* \ , \tag{L.19}$$

so that

$$\tau{:}\dot{\varepsilon} = N^*\dot{\varepsilon} + Q^*\dot{\gamma} + M^*\dot{\kappa} \ . \tag{L.20}$$

If one choses to work with horizontal and vertical components of displacement, then the bending strain K seems simplest. Thus, with

$$\mathbf{u} = u_x\mathbf{e}_x + u_y\mathbf{e}_y \ , \tag{L.21}$$

we get from (L.11) and (L.13)

$$K = (\cos\alpha + u_x')u_y'' - (\sin\alpha + u_y')u_x'' + (u_x'\cos\alpha + u_y'\sin\alpha)\alpha' - \lambda^2\gamma' \ . \tag{L.22}$$

On the other hand, if one resolves the displacement into components tangential and normal to the initial curve C, then the bending strain κ seems simplest. Thus, setting

$$\mathbf{u} = u_\sigma\mathbf{t} + u_\zeta\mathbf{b} \ , \tag{L.23}$$

and noting from (C.2) and (C.3) that $\mathbf{t}' = \alpha'\mathbf{b}$ and $\mathbf{b}' = -\alpha'\mathbf{t}$, we have

$$\bar{\mathbf{y}}' = \mathbf{y}' + \mathbf{u}' = (1 + u_\sigma' - \alpha' u_\zeta)\mathbf{t} + (u_\zeta' + \alpha' u_\sigma)\mathbf{b} \equiv (1 + \bar{e})\mathbf{t} + \bar{\beta}\mathbf{b} \tag{L.24}$$

[6]Modified stress resultants and couples are a feature of many general shell theories.

[7]An erroneous factor of 1/2 in the the third term of equation (85) of Budiansky's paper must be replaced by 1/4.

$$\bar{\mathbf{y}}'' = \bar{e}'\mathbf{t} + \bar{\beta}'\mathbf{b} + \alpha'\mathbf{e}_z \times \bar{\mathbf{y}}' , \tag{L.25}$$

where $\bar{e}$ and $\bar{\beta}$ are a linearized strain and rotation. For future use we note that

$$\mathbf{m} = \mathbf{e}_z \times \bar{\mathbf{y}}' = -\bar{\beta}\mathbf{t} + (1 + \bar{e})\mathbf{b} . \tag{L.26}$$

Substituting (L.24) and (L.25) into (L.18), we obtain

$$\kappa = (1 + \bar{e})\bar{\beta}' - \bar{\beta}\bar{e}' - \lambda^2\gamma' . \tag{L.27}$$

Obviously, other bending strains could be defined. For example, $K - \alpha'\varepsilon$, when linearized, would have the simplest expression in terms of u_x and u_y. This plethora of possible bending strains is in contrast to what we found when β was one of the kinematic variables where a simple bending strain, $k = \beta'$, emerged.

M. The Mechanical Theory of Beamshells

So far, all of our equations have been exact consequences of the equations of motion of 3-dimensional continuum mechanics. In our approach, the approximate nature of shell theory appears when we postulate a form for *The Principle of Conservation of Energy.*

In practice, thermal effects are often negligible. This leads us to characterize the *Mechanical Theory of Beamshells* by the assumption that there exists an *internal energy* $\mathcal{E}$ which, in general, is a *functional* of (e, g, k)[8] such that

$$\int_{t_1}^{t_2} \mathcal{D}dt = \mathcal{E} \Big|_{t_1}^{t_2} . \tag{M.1}$$

Thus, from (G.1),

$$\int_{t_1}^{t_2} \mathcal{W}dt = (\mathcal{K} + \mathcal{E})_{t_1}^{t_2} . \tag{M.2}$$

We may refer to (M.2) as *The Principle of Conservation of Mechanical Energy.* When we come to treat thermal effects in Section T, we shall refer to the generalization of (M.2) as *The First Law of Thermodynamics.* Note that if the external loads are potential, then (J.2) and (M.1) imply that

$$\mathcal{K} + \mathcal{P} + \mathcal{E} = \text{constant}. \tag{M.3}$$

[8] This means, for example, that $\mathcal{E}$ could depend on the history of the strains. Dependence of $\mathcal{E}$ on the spatial derivatives of the strains is excluded because such terms do not appear in the deformation power density $\tau\dot{:}\varepsilon$. Put another way, the Mechanical Work Identity creates a strain for each stress. Strain measures beyond e, g, and k would appear in our approach only if we were to consider higher moments of the 3-dimensional equations of motion. In our opinion a major defect of the higher order shell theory proposed by Ericksen & Truesdell (1958) is that there are more strains than stresses. Our insistence that $\mathcal{E}$ not depend on derivatives of e, g, or k rests on more than mathematical consistency—it is borne out by observation and experiments. Without this correlation with physical reality, the postulates of beamshell theory would be vacuous.

N. Elastic Beamshells and Strain-Energy Densities

We define a beamshell to be *elastic* if the internal energy has the form

$$\mathcal{E} = \int_a^b \Phi(e, g, k; \sigma) d\sigma . \tag{N.1}$$

Without loss of generality we may set

$$\Phi(0, 0, 0; \sigma) = 0 . \tag{N.2}$$

We call Φ the *strain energy per unit length of C*, (or the *strain-energy density* for short).

We now assume that Φ is a differentiable function of e, g, and k. Then (K.4), (K.5), (M.1), and (N.1) imply that

$$\int_a^b [(N - \Phi,_e)\dot{e} + (Q - \Phi,_g)\dot{g} + (M - \Phi,_k)\dot{k}] d\sigma = 0 . \tag{N.3}$$

We assume that, like Φ, N, Q, and M depend on e, g, k, and σ only and that the terms in parentheses in (N.3) are continuous functions of σ. Because (N.3) must hold for all conceivable rotation and displacement fields, we may choose $\dot{e}, \dot{g}$, and $\dot{k}$ arbitrarily. It now follows by a standard argument that (N.3) implies the following *constitutive relations:*

$$N = \Phi,_e \ , \ Q = \Phi,_g \ , \ M = \Phi,_k \ , \ \forall\,(e, g, k) \ , \ 0 < \sigma < L . \tag{N.4}$$

In some problems it turns out that (ε, γ, k) is a more convenient set of strains to use. Then, in the above reasoning, we merely replace (K.5) by (L.8) to obtain, instead of (N.4),

$$\tilde{N} = \Phi,_\varepsilon \ , \ Q^* = \Phi,_\gamma \ , \ M = \Phi,_k \ , \ \forall\,(\varepsilon, \gamma, k) \ , \ 0 < \sigma < L . \tag{N.5}$$

In many beamshell problems, the material and loads are apt to produce nearly unshearable and nearly inextensional deformations. In such cases, it is best to take N and Q as the unknown field variables rather than e and g. Otherwise the field equations may be ill-conditioned. To effect this change, we assume that $(N.4)_{1,2}$ can be solved uniquely for e and g. If we then introduce, via a Legendre transformation, the *mixed elastic-energy density*

$$\Psi = \Phi - Ne - Qg = \Psi(N, Q, k) , \tag{N.6}$$

it follows that

$$e = -\Psi,_N \ , \ g = -\Psi,_Q \ \text{or} \ \varepsilon = -\Psi,_{\tilde{N}} \ , \ \gamma = -\Psi,_{Q^*} \ \text{and} \ M = \Psi,_k . \tag{N.7}$$

Special beamshell theories emerge when one or more arguments of Ψ are assumed to be absent: in a *classical* beamshell, $\Psi = \Psi(k,N)$; in an *elastica,* $\Psi = \Psi(k)$; in a *string or membrane,* $\Psi = \Psi(N)$; and in an *idealized chain,* $\Psi = 0$.

In the analysis to follow, the strains are taken to be e, g, and k. However, the same arguments apply to any consistent triplet of extensional, shear, and bending strains. As Φ has dimensions of MASS$\times$(TIME)$^{-2}$, it is convenient to set Φ equal to a nominal "Young's Modulus" times a quantity with dimensions of LENGTH times a dimensionless strain-energy density ϕ. Four or more dimensionalizing lengths come to mind: L, the length of the reference curve C; $1/\alpha'(\sigma)$, the radius of curvature of C; $d(\sigma)$, the base-reference deviation—see Fig. 4.1b—defined in Section B; and $h_1(\sigma), \cdots, h_m(\sigma)$, the thicknesses of the m layers of which the shell might be constructed.[9] However, the total thickness $h(\sigma) \equiv h_1(\sigma) + \cdots + h_m(\sigma)$ seems the obvious choice, first, because Φ must vanish with h; second, because L is non-local; and third, because α', d or $h_2, \cdots, h_m$ could be zero.

We also want to make the arguments of ϕ dimensionless. The extensional and shear strains, e and g, are so, but the bending strain k has dimensions of (LENGTH)$^{-1}$. For the same reasons as above, h is again a natural dimensionalizing factor. Moreover, if the shear strain is small—it vanishes under the Kirchhoff hypothesis—then hk is a measure of the 3-dimensional extensional strains at the faces of the beamshell which is a convenient quantity for qualitative reasoning. Thus, we shall set

$$\Phi = Eh\phi(e, g, hk; \sigma), \tag{N.8}$$

where E is Young's Modulus and h is the (total) beamshell thickness.

1. Quadratic strain-energy densities

Many materials remain elastic under loads only if the strains are sufficiently small. Thus, most of the literature on the nonlinear deformation of elastic beamshells has been concerned with *geometric nonlinearity* in the presence of linear stress-strain relations. If the reference shape is stress free, the associated dimensionless strain-energy density ϕ must then be a homogeneous quadratic function of e, g, and k. Moreover, simple arguments based on dimensional analysis and geometric invariance (that we shall spell out in the next subsection) show that it is consistent with the error made in assuming ϕ to be quadratic to further assume that it has the form

$$\phi = \tfrac{1}{2}(C_e e^2 + C_s g^2 + C_c ehk + C_b h^2 k^2), \tag{N.9}$$

where C_e, C_s, C_c, and C_b are elastic-geometric parameters that depend on the $\sigma\zeta$-coordinate system, the ratios of various elastic moduli, d/h, and, if there are m layers, $h_1/h, \cdots, h_m/h$. Moreover, because k changes sign if the normal to C is reversed, so must C_c.

As an example, consider a beamshell made of two, initially flat sheets, glued together (Fig. 4.6). The sheets are elastically isotropic and homogeneous, obey Hooke's Law, and have thicknesses, densities, Young's Moduli, and Poisson ratios of h_+, ρ_+, E_+, ν_+ and h_-, ρ_-, E_-, ν_-, respectively. We take the

---

[9]The thickness of a layer may be defined as the length of a ζ-coordinate curve, from one face to the other. In shell coordinates—see (A.2)—thickness is measured normal to the reference curve C.

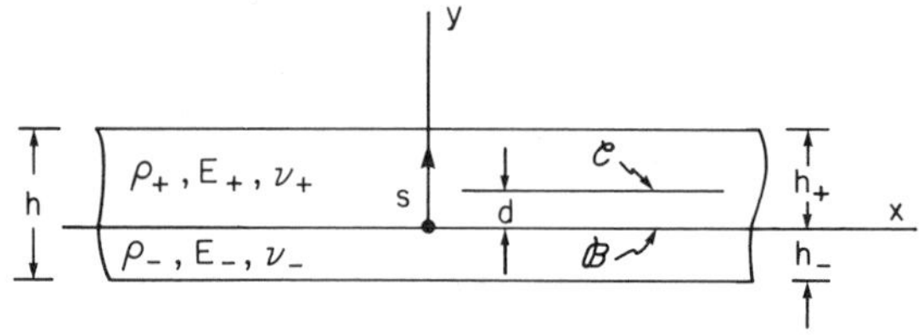

Fig. 4.6. Geometry of an initially flat, two-layered beamshell.

interface between the sheets to coincide with the xz-plane so that, in (A.1),

$$\mathbf{r}(\sigma,\zeta)=\sigma\mathbf{e}_x+\frac{h_+(|\zeta|+\zeta)-h_-(|\zeta|-\zeta)}{2H}\mathbf{e}_y\,,\ 0\le\sigma\le L\,,\ -H\le\zeta\le H\,. \quad \text{(N.10)}$$

Using (B.4), we find that

$$\mu=\frac{h_\pm}{H}\ \text{ if }\ \zeta\gtrless 0\,. \quad \text{(N.11)}$$

From (B.6)-(B.8), it follows that

$$\mathbf{y}=\sigma\mathbf{e}_x+\frac{1}{2}\left[\frac{\rho_+h_+^2-\rho_-h_-^2}{\rho_+h_++\rho_-h_-}\right]\mathbf{e}_y=\mathbf{r}(\sigma,0)+d\mathbf{e}_y\,, \quad \text{(N.12)}$$

which lies at the *center of mass* in the thickness direction; the same is true for more than two layers or for a continuously varying mass density $\rho(\zeta)$.

By hypothesis, there is no strain in the z-direction. Further, assume that the normal stress in the y-direction as well as the shear strains are negligible. Then, with ε_x denoting the component of the 3-dimensional strain in the x-direction, we have for the 3-dimensional strain-energy densities of each sheet,

$$W_\pm=\frac{E_\pm\varepsilon_x^2}{2(1-\nu_\pm^2)}\,. \quad \text{(N.13)}$$

Finally, assume that we may approximate ε_x by $e+(s-d)k$, where s is directed distance along $\mathbf{e}_y$ from $\mathcal{B}$. Thus[10],

$$\Phi=\frac{1}{2}\left\{\frac{E_-}{1-\nu_-^2}\int_{-h_-}^{0}[e+(s-d)k]^2ds+\frac{E_+}{1-\nu_+^2}\int_0^{h_+}[e+(s-d)k]^2ds\right\} \quad \text{(N.14)}$$

$$=\tfrac{1}{2}Eh(C_ee^2+C_cehk+C_bh^2k^2)\,,$$

[10]In this example, we use a method of descent from 3 dimensions in which we integrate a postulated, 3-dimensional form of W through the thickness. (See subsection 3 to follow.)

where $E \equiv \frac{1}{2}(E_- + E_+)$, $h \equiv h_- + h_+$, and

$$C_e = \frac{E_-/E}{1-\nu_-^2}\frac{h_-}{h} + \frac{E_+/E}{1-\nu_+^2}\frac{h_+}{h}$$

$$C_c = \frac{E_+/E}{1-\nu_+^2}\left[\left(\frac{h_+}{h}\right)^2 - 2\frac{dh_+}{h^2}\right] - \frac{E_-/E}{1-\nu_-^2}\left[\left(\frac{h_-}{h}\right)^2 + 2\frac{dh_-}{h^2}\right] \tag{N.15}$$

$$C_b = \frac{E_-/E}{1-\nu_-^2}\left[\frac{1}{3}\left(\frac{h_-}{h}\right)^3 + \left(\frac{h_-}{h}\right)^2\frac{d}{h} + \frac{h_-}{h}\left(\frac{d}{h}\right)^2\right] + \frac{E_+/E}{1-\nu_+^2}\left[\frac{1}{3}\left(\frac{h_+}{h}\right)^3 - \left(\frac{h_+}{h}\right)^2\frac{d}{h} + \frac{h_+}{h}\left(\frac{d}{h}\right)^2\right].$$

Note that under an inversion of the normal to C, we are to replace d by $-d$ and interchange the + and - signs so that C_c changes sign but C_b does not.

We said in Section B that other choices of **y** can be made in statics. Particularly useful is one for which $C_c = 0$, in which case, from $(N.15)_2$,

$$d = \frac{1}{2C_e Eh}\left(\frac{E_+}{1-\nu_+^2}h_+^2 - \frac{E_-}{1-\nu_-^2}h_-^2\right). \tag{N.16}$$

Just as (N.12) shows that choosing **y** to satisfy the dynamic consistency condition (B.15) at $t = 0$ makes d the centroid of a *density-weighted* thickness, so the above expression shows that another choice of **y** makes d the centroid of a *modulus-weighted* thickness. This latter interpretation holds for more than two layers and for a continuously varying modulus. (In beam theory, d locates the *centroidal axis*.) Conversely, if we choose d, we can, if we wish, use (N.12) to determine the corresponding density ratio ρ_+/ρ_-. (As noted before, in static problems ρ can be chosen arbitrarily.)

If the beamshell consists of a single layer, homogeneous in the thickness direction, then, adopting the results of elementary beamshell theory, we can take $C_c = 0$ in (N.9) and

$$C_e = \frac{1}{1-\nu^2}, \quad C_s = \frac{5}{12+11\nu}, \quad C_b = \frac{1}{12(1-\nu^2)}, \tag{N.17}$$

where ν is Poisson's ratio and the value of C_s is that computed by Cowper (1966) for a beam of rectangular cross section which, as he notes, is independent of the aspect ratio of the rectangle. (Slightly different values of C_s computed by other authors are also listed by Cowper.) In *classical beamshell theory* the shearing strain term $C_s g^2$ in (N.9) is set to zero. [In the theory of *beams*, where one assumes a state of plane stress rather than plane strain in the *xy*-plane, Poisson's ratio is set to zero in $(N.17)_1$ and $(N.17)_3$].

From (N.6) and (N.7) with $EC_b = E_b$, etc., the associated mixed-energy density is

$$\Psi = \frac{1}{2}\left[E_b h^3 k^2 - \frac{N^2}{E_e h} - \frac{Q^2}{E_s h}\right] \equiv Eh\psi(hk, n, q; \sigma)\,, \qquad \text{(N.18)}$$

where n and q are dimensionless stress resultants. To the degree of approximation inherent in the assumption that the strain-energy density is quadratic, we can also write

$$\Psi = \frac{1}{2}\left[E_b h^3 k^2 - \frac{\tilde{N}^2}{E_e h} - \frac{Q^{*2}}{E_s h}\right] \equiv Eh\psi(hk, \tilde{n}, q^*; \sigma)\,, \qquad \text{(N.19)}$$

where $\tilde{n}$ and q^* are dimensionless forms of the modified stress resultants defined by (L.7). In *classical beamshell theory* the Q-terms in (N.18) and (N.19) are set to zero.

Taking ϕ to be quadratic in the strains raises two related questions: what error do we make in adopting (N.9) and, more generally, what can be said about the form of ϕ when the 3-dimensional strains in a beamshell are large, say in a rubber-like body? The next two subsections address these questions.

*2. *General strain-energy densities*

The aim here is to see what simplifications in the *functional form* of ϕ can be deduced from locality, geometric inhomogeneity, geometric invariance, material stability, thinness, smoothness, and small-strain quadricity. These terms are defined and discussed below. For simplicity, we shall assume, henceforth, that the beamshell consists of a single, transversely homogeneous, elastic layer, and we shall use the shell coordinates in (A.2). However, we shall not require that $\mathcal{C}$ coincide with the midline and d will measure the deviation.

By *locality* we mean the assumption that ϕ at σ does not depend on the derivatives of h at σ, though ϕ may depend on $h(\sigma)$ itself. Locality seems plausible in beamshells in which the thickness varies slowly with σ and can be substantiated by experiments or additional arguments if desired; see Niordson (1971). (We are aware of, but shall not discuss, recent studies in which shells of rapidly varying thickness are replaced by equivalent shells of constant or slowly varying thickness.) Thus, we assume that the dimensionless strain-energy density has the functional form

$$\phi = \phi(e, g, hk; d/h, \mathcal{G})\,, \qquad \text{(N.20)}$$

where $\mathcal{G}$ represents the geometry of $\mathcal{C}$. In (N.20) it is understood that ϕ may depend on σ (in some dimensionless fashion) through h, d, $\mathcal{G}$, and through *longitudinal* variations of material properties.

Consider next the effects of $\mathcal{G}$ on ϕ. Two beamshells that differ only in their reference shapes can be expected to have different strain-energy densities. However, a theory that did not predict the influence of the reference shape on ϕ, i.e., a theory that left us the task of determining ϕ experimentally for each different shape, would have diminished power. To avoid this we adopt a principle we call *geometric inhomogeneity*, namely, that ϕ depends on the geometry of $\mathcal{C}$ only through its curvature α'. Thus,

$$\phi = \phi(e,\, g,\, hk\,;d/h,\, h\alpha')\,, \tag{N.21}$$

where we have multiplied α' by h to keep the arguments of ϕ dimensionless. (Essentially, locality and geometric inhomogeneity imply that there are no action-at-a-distance effects.)

Next, we examine the implications of *geometric invariance* and show that the dependence of ϕ on α' must be quite specific. From (K.1),

$$g = \overline{\mathbf{y}}'\cdot\mathbf{B}\,. \tag{N.22}$$

Hence, if σ is replaced by $-\sigma$, or $\mathbf{B}$ by $-\mathbf{B}$ (but not both simultaneously), g changes sign. Moreover, from (C.2)-(C.7), and (K.6),

$$\alpha' = \mathbf{t}'\cdot\mathbf{b}\ \ ,\ \ k = \mathbf{T}'\cdot\mathbf{B} - \alpha'\,. \tag{N.23}$$

Replacing σ by $-\sigma$ leaves α' and k unchanged because $\mathbf{t}'$ and $\mathbf{T}'$ remain unchanged (the latter because $\mathbf{t}$ and $\mathbf{T}$ *do* change sign with σ). However, replacing $\mathbf{b}$ by $-\mathbf{b}$ and $\mathbf{B}$ by $-\mathbf{B}$ does change the sign of α' and k. The situation is summarized by the first 5 columns in Table 4.2 of Section T. Furthermore, as we have noted before, the directed distance d from the base curve to the reference curve changes sign with $\mathbf{b}$. As ϕ must remain invariant when distances and normals are reversed—such changes merely represent a different mathematical description—it follows that ϕ must have the functional form

$$\phi = \phi(e,\, g^2,\, d^2/h^2;\, dk,\, h^2k^2;\, h^2\alpha' k,\, d\alpha',\, h^2\alpha'^2) \equiv \phi(\mathcal{X};\mathcal{Y};\mathcal{Z})\,. \tag{N.24}$$

The implications of *material stability* follow from equations (K.8) which, in view of (N.4), (N.8), and (N.24), take the form

$$\begin{gathered}[Eh\phi_{(1)}] = c^2 m[e]\ \ ,\ \ [2Ehg\phi_{(2)}] = c^2 m[g] \\ [Eh^3(2k\phi_{(5)} + \alpha'\phi_{(6)}) + Ehd\phi_{(4)}] = c^2 I[k]\,.\end{gathered} \tag{N.25}$$

Here, $\phi_{(i)}$ denotes the partial derivative of ϕ with respect to its ith argument. According to the interpretation of material stability of Rivlin and his co-workers (see subsection III.J.1 for references and a short discussion), ϕ must be such that (N.25) gives identical, positive values of c^2—there may be more than one—for arbitrary, infinitesimal jumps in e, g, and k.

By *thinness* we mean that $|h\alpha', d\alpha', hk, dk| \ll 1$. That is, at every point of $\mathcal{C}$, the ratio of the thickness or the base-reference deviation (which is of the order of the thickness) to the radius of curvature of $\mathcal{C}$ or $\overline{\mathcal{C}}$ is small. Of course, one can easily conceive of deformations in which there are points where $|hk|$ or $|dk|$ is large, even though $|h\alpha'|$ or $|d\alpha'|$ is small; for example, the two points of application on an initially circular beamshell of equal and opposite outward radial line loads. However, if there are only a finite number of such points, we have every reason to expect—we offer no proofs, only experience—

that beamshell theory will predict the overall deformation quite well.

To exploit thinness, we make our first *smoothness* assumption, namely that the mean value theorem holds with respect to the last three arguments of ϕ [denoted collectively by $\mathcal{Z}$ in (N.24)]. That is,

$$\phi = \phi(\mathcal{X}; \mathcal{Y}; 0) + [\phi_{(6)}(\mathcal{X}; \mathcal{Y}; \tau\mathcal{Z})h^2k + \phi_{(7)}(\mathcal{X}; \mathcal{Y}; \tau\mathcal{Z})d + \phi_{(8)}(\mathcal{X}; \mathcal{Y}; \tau\mathcal{Z})h^2\alpha']\alpha' , \tag{N.26}$$

where τ is some constant between 0 and 1.

To proceed further, we assume that ϕ approaches a non-degenerate, homogeneous quadratic form in e, g, and hk as $\| e \| \to 0$, where

$$\| e \|^2 \equiv e^2 + g^2 + h^2k^2 . \tag{N.27}$$

In other words, we assume that there are constants $\underline{C}$ and $\bar{C}$ such that

$$\underline{C} \| e \|^2 \leq \phi \leq \bar{C} \| e \|^2 , \quad \| e \| \leq e_0 , \tag{N.28}$$

where e_0 is some third constant.[11] This is what *small-strain quadricity* means. Thus, the reference shape is stress-free and the stress-strain relations, for sufficiently small strains, are linear, homogeneous, and non-degenerate. Applied to (N.26), small-strain quadricity implies that $\phi_{(6)}$ and $(\phi_{(7)}, \phi_{(8)})$ must vanish to first and second order in $\sqrt{e^2+g^2}$, respectively. Hence, we shall assume that there exist constants C_1 and C_2 such that

$$\begin{aligned} |\phi_{(6)}(\mathcal{X}; \mathcal{Y}; \mathcal{Z})| &\leq C_1\sqrt{e^2+g^2} \\ |\phi_{(7)}(\mathcal{X}; \mathcal{Y}; \mathcal{Z}), \phi_{(8)}(\mathcal{X}; \mathcal{Y}; \mathcal{Z})| &\leq C_2(e^2+g^2) , \end{aligned} \tag{N.29}$$

uniformly in $\mathcal{X}$, $\mathcal{Y}$, and $\mathcal{Z}$.

Let

$$\phi(\mathcal{X}; \mathcal{Y}; 0) \equiv \tilde{\phi}(\mathcal{X}; \mathcal{Y}) = \tilde{\phi}(e, g^2, d^2/h^2; dk, h^2k^2) . \tag{N.30}$$

Then, from (N.26)-(N.30) and Schwarz' inequality, we have

$$\begin{aligned} |\phi(\mathcal{X}; \mathcal{Y}; \mathcal{Z}) - \tilde{\phi}(\mathcal{X}; \mathcal{Y})| &\leq [C_1h^2|k|\sqrt{e^2+g^2} + C_2(|d| + h^2|\alpha'|)(e^2+g^2)]|\alpha'| \\ &\leq [C_1h + C_2(|d| + h^2|\alpha'|)]|\alpha'| \, \| e \|^2 \\ &= \mathrm{O}(h\alpha'\phi) \text{ if } \| e \| \leq e_0 . \end{aligned} \tag{N.31}$$

Thus, if we neglect curvature effects, that is, if we replace the strain-energy density by that of a flat beamshell, we make a relative error of $\mathrm{O}(h\alpha')$.

To exploit the smallness of dk and h^2k^2, we further assume that $\tilde{\phi}$ is sufficiently smooth in these arguments to have a finite Taylor expansion of the form

[11] We expect $e_0 = \mathrm{O}(1)$, although (N.28) could conceivably hold for all $\| e \|$.

$$\tilde{\phi} = \tilde{\phi}^0 + \tilde{\phi}^0_{(4)} dk + \tilde{\phi}^0_{(5)} h^2k^2 + \frac{1}{2}(\tilde{\phi}^0_{(4,4)} d^2k^2 + 2\tilde{\phi}^0_{(4,5)} dh^2k^3 + \underline{\tilde{\phi}^0_{(5,5)} h^4k^4}) + \frac{1}{3!}(\tilde{\phi}^0_{(4,4,4)} d^3k^3 + \underline{3\tilde{\phi}^0_{(4,4,5)} d^2h^2k^4 + \cdots}) + \frac{1}{4!}(\underline{\tilde{\phi}^*_{(4,4,4,4)} d^4k^4 + \cdots}) . \tag{N.32}$$

Here, $\tilde{\phi}_{(i,j)}$ denotes the 2nd partial derivative of $\tilde{\phi}$ with respect to its ith and jth arguments, etc.. Further, a superscript 0 denotes evaluation of a function at $(\mathcal{X}; 0)$ while an asterisk denotes evaluation of a function at $(\mathcal{X}; \tau\mathcal{Y})$, where τ is some number between 0 and 1.

Let

$$\bar{\phi} \equiv \tilde{\phi}^0 + \tilde{\phi}^0_{(4)} dk + (\tilde{\phi}^0_{(5)} + \tfrac{1}{2}\tilde{\phi}^0_{(4,4)} d^2/h^2)h^2k^2 + (\tilde{\phi}^0_{(4,5)} + \frac{1}{3!}\tilde{\phi}^0_{(4,4,4)} d^2/h^2)dh^2k^3$$
$$\equiv \bar{A}(X) + \bar{B}(X)dk + \bar{C}(X)h^2k^2 + \bar{D}(X)dh^2k^3 \ , \ X = e, g^2, d^2/h^2 \tag{N.33}$$

denote the expression we obtain by neglecting the underlined terms on the right side of (N.32). [Note that the condition of small strain quadricity, (N.28), implies that $\bar{B}/\sqrt{e^2+g^2}$ must be bounded as $\| e \| \to 0$.] If we assume that the neglected coefficients

$$\tilde{\phi}^0_{(5,5)}, \tilde{\phi}^0_{(4,4,5)}, \cdots, \tilde{\phi}^*_{(4,4,4,4)}, \cdots$$

in (N.32) are uniformly bounded functions of $\mathcal{X}$ and $\mathcal{Y}$, then, so long as $\| e \| \le e_0$, we make a relative error of $O(h^2k^2)$ when we approximate $\tilde{\phi}$ by $\bar{\phi}$ because $h^4k^4 = O(h^2k^2 \| e \|^2)$. This result coupled with (N.31) implies that

$$\phi = \bar{\phi}[1 + O(h\alpha' + h^2k^2)] \ , \ \| e \| \le e_0 . \tag{N.34}$$

We emphasize that we have neither assumed nor shown that the shear strain g is small, nor need it be. We also point out a fact that might, at first, seem contradictory. Consider the term $\phi_{(6)}(\mathcal{X}; \mathcal{Y}; \tau\mathcal{Z})h^2\alpha' k$ in (N.26), which we neglected, and the term $\tilde{\phi}^0_{(5)} h^2k^2$ in (N.33), which we kept. We do not claim that the former is always small compared to the latter. Rather, all we claim is that the former is always small compared to ϕ; when the two terms are of the same order of magnitude, *both* are small compared to ϕ which, of course, must then be dominated by the contributions of the extensional and shearing strain. However, if we are studying small perturbations superimposed upon large known deformations, then the a priori neglect of small terms in the strain-energy density becomes a delicate matter that, in specific cases, needs to be examined with more care than we have done here.

We close this subsection with several comments related to (N.34). First, the advantage of having error estimates is that when we come to introduce the constitutive equations into parts of the equations of motion, we are often able to discard certain terms because they are of the same order of magnitude as errors in $\overline{\phi}$.

Second, to obtain an approximation to $\overline{\phi}$ for small shear strains, we apply Taylor's Theorem (with remainder) to (N.33) to get

$$\begin{aligned}\overline{\phi} &= \hat{\phi} + \tfrac{1}{2}\overline{A}_{(2,2)}(X^*)g^4 \\ &\quad + [\overline{B}_{(2)}(X^{**})dk + \overline{C}_{(2)}(X^{**})h^2k^2 + \overline{D}_{(2)}(X^{**})dh^2k^3]g^2 \,.\end{aligned} \tag{N.35}$$

Here, $X^* = e, \tau_1 g^2, d^2/h^2$ and $X^{**} = e,\ \tau_2 g^2, d^2/h^2$, where $0 < \tau_1, \tau_2 < 1$, and

$$\begin{aligned}\hat{\phi} &= \overline{A}(e, 0, d^2/h^2) + \overline{A}_{(2)}(e, 0, d^2/h^2)g^2 \\ &\quad + \overline{B}(e, 0, d^2/h^2)dk + \overline{C}(e, 0, d^2/h^2)h^2k^2 + \overline{D}(e, 0, d^2/h^2)dh^2k^3 \\ &\equiv \hat{A}(e, d^2/h^2) + \hat{A}_s(e, d^2/h^2)g^2 + \hat{B}(e, d^2/h^2)dk \\ &\quad + \hat{C}(e, d^2/h^2)h^2k^2 + \hat{D}(e, d^2/h^2)dh^2k^3 \,.\end{aligned} \tag{N.36}$$

Using Schwarz' inequality and assuming that the coefficient functions $\overline{A}_{(2,2)}, \cdots, \overline{D}_{(2)}$ in (N.35) are uniformly bounded in X for $\| e \| < e_0$, we arrive, via (N.34), at an expression of the form

$$\phi = \hat{\phi}[1 + O(h\alpha' + h^2k^2 + g^2)] + O(g^4) \ , \quad \| e \| < e_0 \,. \tag{N.37}$$

If we approximate ϕ by $\hat{\phi}$, then, by using a straight, homogeneous beamshell, we could find $\hat{A}(e, 0)$ from a simple tension test and $\hat{A}_s(e, 0)$ and $\hat{C}(e, 0)$ by measuring the speed of propagation of small sinusoidal shear and transverse (flexural) waves at various levels of extensional strain. Realistically, we expect the term $\hat{A}_s(e, 0)g^2$ to be important only in shells that are highly nonhomogeneous in the thickness direction or orthotropic, with a very low transverse shear modulus. A typical case is a sandwich shell with a core very weak in shear.

**3. Strain-energy density by descent from 3-dimensions*

In a sense, it is circular to ask how to derive the 2-dimensional strain-energy density Φ from a 3-dimensional strain-energy density W because 3-dimensional constitutive relations are usually inferred from tests on thin sheets! Nevertheless, if for no other reason than theoretical simplicity, one sometimes postulates special forms for W. The classic example is a rubberlike body in which one assumes that

$$W = Ew(I_1, I_2) \ , \ I_3 = 1 \,. \tag{N.38}$$

Here, E is some nominal elastic constant with the same dimensions as W and I_1, I_2, and I_3 are the standard 3-dimensional strain invariants given, for example, on p. 57 of Green & Zerna (1968). $I_3 = 1$ means that the body is incompressible.

The aim of this subsection is to deduce from (N.38) an expression for $\hat{\phi}$ for a rubber-like beamshell by making a *plausible* assumption on the variation in the thickness direction of the 3-dimensional deformed position $\overline{\mathbf{x}}$, *there being no way, in general, to infer the actual 3-dimensional deformation from the 1-dimensional strain fields of extension, shear, and bending delivered by beamshell theory.*

As we showed in the preceding subsection, the form of $\hat{\phi}$ for a curved beamshell is the same as that for a flat beamshell. Therefore, in deriving $\hat{\phi}$ from w, we may confine attention to 3-dimensional plates. This simplifies the geometry considerably. Furthermore, as $\hat{\phi}$ is explicitly independent of time, we need consider only static deformations. Our approach will produce an explicit formula for $\hat{\phi}$ in terms of ε and k plus error terms.

Consider an infinite plate with reference shape

$$0 \le x \le L \ , \ -\tfrac{1}{2}h \le y \le \tfrac{1}{2}h \ , \ -\infty < z < \infty . \tag{N.39}$$

For simplicity we take the material density ρ to be constant.

The 3-dimensional metric tensor $\mathbf{G}$ of the deformed plate is defined implicitly by

$$d\bar{\mathbf{x}} \cdot d\bar{\mathbf{x}} \equiv d\mathbf{x} \cdot \mathbf{G} \cdot d\mathbf{x} . \tag{N.40}$$

By (A.1) and (B.14),

$$\bar{\mathbf{x}} = \bar{\mathbf{y}}(\sigma) + \bar{\mathbf{z}}(\sigma, \zeta) + z\mathbf{e}_z , \tag{N.41}$$

where we have set $x = \sigma$ and $y = \zeta$. Hence, the matrix of *physical components* of $\mathbf{G}$ may be expressed as

$$\begin{bmatrix} G_\sigma & 2\Gamma & 0 \\ 2\Gamma & G_\zeta & 0 \\ 0 & 0 & 1 \end{bmatrix} \equiv \begin{bmatrix} (\lambda\bar{\mathbf{t}} + \bar{\mathbf{z}},_\sigma)\cdot(\lambda\bar{\mathbf{t}} + \bar{\mathbf{z}},_\sigma) & (\lambda\bar{\mathbf{t}} + \bar{\mathbf{z}},_\sigma)\cdot\bar{\mathbf{z}},_\zeta & 0 \\ (\lambda\bar{\mathbf{t}} + \bar{\mathbf{z}},_\sigma)\cdot\bar{\mathbf{z}},_\zeta & \bar{\mathbf{z}},_\zeta\cdot\bar{\mathbf{z}},_\zeta & 0 \\ 0 & 0 & 1 \end{bmatrix} \tag{N.42}$$

and thus

$$I_1 = \mathrm{tr}\,\mathbf{G} = G_\sigma + G_\zeta + 1 \tag{N.43}$$

$$I_2 = \tfrac{1}{2}(\mathrm{tr}\,\mathbf{G})^2 - \tfrac{1}{2}\mathrm{tr}\,\mathbf{G}^2 = G_\sigma + G_\zeta + G_\sigma G_\zeta - 4\Gamma^2 \tag{N.44}$$

$$I_3 = \det \mathbf{G} = G_\sigma G_\zeta - 4\Gamma^2 , \tag{N.45}$$

where tr denotes trace and det denotes determinant.

The incompressibility condition $I_3 = 1$ implies that

$$G_\zeta = \frac{1 + 4\Gamma^2}{G_\sigma} . \tag{N.46}$$

Hence,

$$J \equiv I_1 = I_2 = 1 + G_\sigma + \frac{1}{G_\sigma} + \frac{4\Gamma^2}{G_\sigma} . \tag{N.47}$$

In the nonlinear theory of shells undergoing *small* strains, John (1965a, 1965b, 1969, 1971) and Berger (1973) have established rigorously the earlier observations by Hildebrand, Reissner & Thomas (1949), Koiter (1960), and others that a shell is in an approximate state of plane stress. We may see this formally by letting L^* denote some relevant length scale along the deformed reference curve $\mathcal{C}$ of our beamshell ("the smallest 'wavelength' of the deformation pattern" in Koiter's words). If τ denotes the maximum stress in the beamshell (viewed as a 3-dimensional body), the 3-dimensional equilibrium equations, written in directions parallel and perpendicular to $\mathcal{C}$, imply that the transverse shearing stresses are $O(h\tau/L^*)$ while the transverse normal stress is $O(h^2\tau/L^{*2} + h\tau/R)$, where R is a characteristic length scale of $\mathcal{C}$. If W is an isotropic, quadratic function of the 3-dimensional strains, then the above results imply that if the normal strain is $O(e)$, where e is the maximum strain, then the transverse shearing strains are only $O(he/L^*)$. We shall assume that these order-of-magnitude estimates continue to hold in homogeneous, isotropic, rubber-like materials if $e = O(1)$.

To account for shear strain through quadratic terms, we assume that w in (N.38) is a sufficiently smooth function of its arguments for the following Taylor expansion to hold:

$$w(J, J) \equiv v(J) = v(K) + v'(K)(4\Gamma^2/G_\sigma) + \tfrac{1}{2}v''(K^*)(4\Gamma^2/G_\sigma)^2 , \quad \text{(N.48)}$$

where

$$K = 1 + G_\sigma + G_\sigma^{-1} \quad \text{(N.49)}$$

and $K^* = K + 4\tau\Gamma^2/G_\sigma$, $0 < \tau < 1$. Assuming $v''(K^*)/G_\sigma^2$ to be bounded [at least for $e = O(1)$], we have

$$\phi \equiv h^{-1}\int_{-h/2}^{h/2} w\,d\zeta = h^{-1}\int_{-h/2}^{h/2} [v(K) + v'(K)(4\Gamma^2/G_\sigma)]d\zeta + O(\overline{\Gamma}^4) , \quad \text{(N.50)}$$

where

$$\overline{\Gamma}^4 \equiv h^{-1}\int_{-h/2}^{h/2} \Gamma^4 d\zeta . \quad \text{(N.51)}$$

We now set

$$\bar{\mathbf{z}} = p(\sigma, \zeta)\bar{\mathbf{t}} + q(\sigma, \zeta)\bar{\mathbf{b}} \quad \text{(N.52)}$$

and note that the dynamic consistency condition (B.15) implies that[12]

$$\int_{-h/2}^{h/2} p\,d\zeta = \int_{-h/2}^{h/2} q\,d\zeta = 0 . \quad \text{(N.53)}$$

To account for thinness, we introduce the change of variables and parameter

[12]We remind the reader that in a static problem where the inertial terms are zero, the density in (B.15) may be assumed to have any convenient form.

$$(\sigma, \bar{x}, \bar{y}, d\sigma/d\bar{\alpha}) = L^*(\sigma^*, \cdots, \bar{\kappa}^{-1}) \ , \ (\zeta, p, q) = \tfrac{1}{2}h(\zeta^*, \delta p^*, q^*)$$
$$\delta = \tfrac{1}{2}h/L^* \ , \ \partial/\partial\sigma^* = {}' \ , \ \partial/\partial\zeta^* = \dot{} \ , \tag{N.54}$$

where $\bar{\alpha} = \beta + \gamma$. (The reader will have no trouble in distinguishing the above δ, which appears in this subsection only, from the δ used in other sections to indicate a variation.) We then have by (L.1), (L.5), and (N.52),

$$\bar{\mathbf{z}},_{\sigma} = \delta[(-\bar{\kappa}q + \delta p')\bar{\mathbf{t}} + (q' + \delta\bar{\kappa}p)\bar{\mathbf{b}}] \ , \ \bar{\mathbf{z}},_{\zeta} = \dot{q}\bar{\mathbf{b}} + \delta\dot{p}\bar{\mathbf{t}} \tag{N.55}$$

On the right side of (N.55) and henceforth, we drop the asterisk to avoid a cluttered notation. We now have from (N.42),

$$G_{\sigma} = \lambda^2 - 2\lambda\bar{\kappa}q\delta + (2\lambda p' + q'^2 + \bar{\kappa}^{2} q^2)\delta^2 + 2\bar{\kappa}(pq' - qp')\delta^3 + (p'^2 + \bar{\kappa}^{2} p^2)\delta^4$$
$$\equiv G_{\sigma}^{(0)} + G_{\sigma}^{(1)}\delta + G_{\sigma}^{(2)}\delta^2 + O(\delta^3) \tag{N.56}$$

$$G_{\zeta} = \dot{q}^2 + \dot{p}^2\delta^2 \equiv G_{\zeta}^{(0)} + G_{\zeta}^{(1)}\delta + O(\delta^2) \tag{N.57}$$

$$2\Gamma = [\lambda\dot{p} + q'\dot{q} + \bar{\kappa}(p\dot{q} - q\dot{p})\delta + p'\dot{p}\delta^2]\delta = O(\delta) \ . \tag{N.58}$$

Hence, from (N.49) and (N.56),

$$K = 1 + \lambda^2 + \frac{1}{\lambda^2} + \left[1 - \frac{1}{\lambda^4}\right] G_{\sigma}^{(1)}\delta + \left[\left[1 - \frac{1}{\lambda^4}\right] G_{\sigma}^{(2)} + \frac{G_{\sigma}^{(1)^2}}{\lambda^6}\right]\delta^2 + \cdots$$
$$\equiv K_0 + K_1\delta + K_2\delta^2 + O(\delta^3) \ . \tag{N.59}$$

Substituting this expansion into the integrand in (N.50), we have

$$v(K) + v'(K)(4\Gamma^2/G_{\sigma}) = v(K_0) + v'(K_0)K_1\delta + [v'(K_0)K_2 + \tfrac{1}{2}v''(K_0)K_1^2]\delta^2$$
$$+ [4v'(K_0)/\lambda^2]\Gamma^2 + O(\delta^3 + \delta\Gamma^2) \ . \tag{N.60}$$

But from (N.56), (N.59), and the dynamic consistency condition, $(N.53)_2$, $h^{-1}\int_{-h/2}^{h/2} G_{\sigma}^{(1)} d\zeta^* \equiv \int G_{\sigma}^{(1)} = 0$. Thus, as $K_0 = K_0(\sigma)$, (N.50) reduces to

$$\phi = v(K_0) + [v'(K_0)\int K_2 + \tfrac{1}{2}v''(K_0)\int K_1^2]\delta^2$$
$$+ [4k_s v'(K_0)/\lambda^2]\gamma^2 + O(\delta^3 + \delta\gamma^2) \ , \tag{N.61}$$

where *we have identified the square of the shear strain* γ *with some shear constant* k_s *times the average through the thickness of the square of the shear strain* Γ, i.e., we have set

$$\gamma^2 = k_s \int \Gamma^2 \ . \tag{N.62}$$

Note from $(N.53)_1$, (N.56), and (N.59) that

$$\int K_2 = \left[1 + \frac{3}{\lambda^4}\right] \bar{\kappa}^{2} \int q^2 + \left[1 - \frac{1}{\lambda^4}\right] \int q'^2 \tag{N.63}$$

$$\int K_1^2 = 4\lambda^2 \left[1 - \frac{1}{\lambda^4}\right]^2 \bar{\kappa}^{2} \int q^2 \tag{N.64}$$

and that the unknown function p in (N.52) does not affect the terms explicitly displayed in (N.61).

We now assume that q has an expansion of the form

$$q(\sigma, \zeta, \delta) = q_0(\sigma, \zeta) + O(\delta) . \tag{N.65}$$

Substituting (N.65) into (N.56)-(N.58) and the resulting expressions into the incompressibility condition (N.46), we find, to lowest order, that

$$\dot{q}_0 = \lambda^{-1} . \tag{N.66}$$

Furthermore, the dynamic consistency condition $(N.53)_2$ implies that $\int q_0 = 0$; hence,

$$q_0 = \lambda^{-1} \zeta . \tag{N.67}$$

Substituting this into (N.63) and (N.64) and the resulting expressions into (N.61), and replacing $\delta\bar{\kappa}$ by $\frac{1}{2} h d\bar{\alpha}/d\sigma$, we obtain for the dimensionless strain-energy density of a flat, rubberlike beamshell,

$$\phi = \nu(K_0) + \frac{1}{12}\left[\frac{1}{\lambda^2}\left[1 + \frac{3}{\lambda^4}\right]\nu'(K_0) + 2\left[1 - \frac{1}{\lambda^4}\right]^2 \nu''(K_0)\right] h^2 \left[\frac{d\bar{\alpha}}{d\sigma}\right]^2 + \left[\frac{4k_s \nu'(K_0)}{\lambda^2}\right]\gamma^2 + O(\int q_0'^2 \delta^2 + \delta^3 + \delta\gamma^2) . \tag{N.68}$$

For a neo-Hookean material,

$$w(J, J) = \nu(J) = \frac{1}{6}(J - 3) . \tag{N.69}$$

Hence, with $\lambda^2 = 1 + 2\varepsilon$, (N.68) without the error term, reduces to

$$\hat{\phi} = \frac{2}{3}\frac{\varepsilon^2}{1 + 2\varepsilon} + \frac{1}{24}\frac{1 + \varepsilon + \varepsilon^2}{(1 + 2\varepsilon)^3} h^2 \left[\frac{d\bar{\alpha}}{d\sigma}\right]^2 + \frac{2k_s\gamma^2}{3\lambda^2} . \tag{N.70}$$

This expression agrees with (N.9) in the limit of small strains if we take $\nu = 1/2$ in (N.17) and $k_s = 3/14$.

If (N.70) with $\gamma = 0$ is compared with equation (4.6) of Simmonds (1985b)—where $\lambda\kappa' = d\bar{\alpha}/d\sigma$—or with related equations in Libai & Simmonds (1981, 1983b), a 1 will be found in place of the numerator $1 + \varepsilon + \varepsilon^2$ in (N.70). The reason is that the dynamic consistency condition, (B.15), is not enforced in these papers. [Recall that (B.15) follows from the definition (B.7) of $\bar{y}$, which need not be used in static situations.] However, we can bring the different expressions for ϕ into agreement as follows.

In Section 4 of Simmonds (1985b) the results for an arbitrary flat plate are applied to a beamshell. Write the form assumed there for $\bar{\mathbf{x}}$ as

$$\bar{\mathbf{x}} = \tilde{\mathbf{y}} + \tilde{q}(\sigma, \zeta)\tilde{\mathbf{b}} + z\mathbf{e}_z \,. \tag{N.71}$$

Let us nondimensionalize the terms in (N.71) as in (N.54) and then drop the asterisks. From (B.15) follows

$$\bar{\mathbf{y}} = \tilde{\mathbf{y}} + (\tfrac{1}{2}\delta\int_{-1}^{1} \tilde{q}\, d\zeta)\tilde{\mathbf{b}} \,. \tag{N.72}$$

From equations (3.20) and (4.5) of Simmonds (1985b), we have

$$\tilde{q} = \tilde{\lambda}^{-1}\zeta + \tfrac{1}{2}\tilde{\lambda}^{-3}\bar{\kappa}\zeta^2\delta + \cdots \,, \tag{N.73}$$

where $\tilde{\lambda} = |\tilde{\mathbf{y}}'|$. Thus

$$\bar{\mathbf{y}} = \tilde{\mathbf{y}} + (\frac{1}{6}\tilde{\lambda}^{-3}\bar{\kappa}\delta + \cdots)\tilde{\mathbf{b}} \,. \tag{N.74}$$

Proceeding formally, we have from (N.74),

$$\lambda\bar{\mathbf{t}} = \bar{\mathbf{y}}' = (\tilde{\lambda} - \frac{1}{6}\tilde{\lambda}^{-3}\tilde{\alpha}'\bar{\kappa}\delta^2 + \cdots)\tilde{\mathbf{t}} + [\frac{1}{6}(\tilde{\lambda}^{-3}\bar{\kappa})'\delta^2 + \cdots]\tilde{\mathbf{b}} \,, \tag{N.75}$$

$$\lambda^2 = \tilde{\lambda}^2 - \frac{1}{3}\tilde{\lambda}^{-2}\tilde{\alpha}'\bar{\kappa}\delta^2 + \cdots \,, \tag{N.76}$$

and

$$\bar{\mathbf{b}} = \mathbf{e}_z \times \bar{\mathbf{t}} = (1 + \cdots)\tilde{\mathbf{b}} + (\cdots)\tilde{\mathbf{t}} \,. \tag{N.77}$$

From (N.77),

$$\bar{\kappa} = (1 + \cdots)\tilde{\alpha}' \,. \tag{N.78}$$

Substituting (N.78) into (N.76) and (N.76) into (N.68), we find that 6 times the resulting expression agrees with equation (4.6) of Simmonds (1985b) to within terms of $O(\delta^2)$. The factor of 6 arises because w for a Mooney material in the above reference was taken as $J - 3$ and not $(1/6)(J - 3)$ as in (N.69).

O. Elastostatics

The statics of beamshells is an old field and its literature is enormous, even for the special case of the *elastica*. [A review by Gorski (1976) on the nonlinear theory cites over 200 papers.] Therefore, we shall confine our attention to just a few topics that are either new, non-standard, or else give insight into the behavior of more general shells.

For applications, it is convenient to present the force equilibrium equations in component form either in the horizontal ($\mathbf{e}_x$) and vertical ($\mathbf{e}_y$) directions, in the directions of **T** and **B**, or in the directions of $\bar{\mathbf{t}}$ and $\bar{\mathbf{b}}$. As we saw in Section J on potential loads, the components of **p** are often simplest in the frames $\{\mathbf{e}_x, \mathbf{e}_y\}$ or $\{\bar{\mathbf{y}}', \mathbf{m}\}$. Thus, we represent **p** in the alternative forms

$$\begin{aligned}\mathbf{p} &= p_x\mathbf{e}_x + p_y\mathbf{e}_y \\ &= \bar{p}_\sigma\bar{\mathbf{y}}' - p\mathbf{m} \\ &= [(1+e)\bar{p}_\sigma + gp]\mathbf{T} + [\bar{p}_\sigma g - (1+e)p]\mathbf{B} \\ &\equiv p_T\mathbf{T} + p_B\mathbf{B}\,.\end{aligned} \tag{O.1}$$

Note that $\bar{p}_\sigma$ and p are reckoned per unit length of $\bar{\mathcal{C}}$ (per unit width of the beamshell) while p_x, p_y, p, p_T, and p_B are reckoned per unit length of $\mathcal{C}$.

With

$$\mathbf{F} \equiv H\mathbf{e}_x + V\mathbf{e}_y\,, \tag{O.2}$$

(L.5) and (L.7) yield

$$Q^* = \mathbf{F}\cdot\mathbf{m} = \lambda(V\cos\bar{\alpha} - H\sin\bar{\alpha})\,. \tag{O.3}$$

Hence, from (E.1) and (E.2),

$$H' + p_x = 0\ ,\ V' + p_y = 0 \tag{O.4}$$

$$M' + \lambda(V\cos\bar{\alpha} - H\sin\bar{\alpha}) + l = 0\,. \tag{O.5}$$

The advantage of this form of the equilibrium equations is obvious: indefinite integration reduces (O.4) to

$$H = -\int p_x d\sigma\ ,\ V = -\int p_y d\sigma\,, \tag{O.6}$$

so that (O.5) may be written

$$M' = \lambda[(\int p_y d\sigma)\cos\bar{\alpha} - (\int p_x d\sigma)\sin\bar{\alpha}] - l\,. \tag{O.7}$$

In the frame $\{\mathbf{T}, \mathbf{B}\}$,

$$\mathbf{F} = N\mathbf{T} + Q\mathbf{B}\,. \tag{O.8}$$

Noting from (C.6), (C.7), and (K.6) that

$$\mathbf{T}' = (\alpha' + k)\mathbf{B}\ ,\ \mathbf{B}' = -(\alpha' + k)\mathbf{T}\,, \tag{O.9}$$

we have from (E.1), (E.2), and (O.1)

$$\begin{aligned}&N' - (\alpha' + k)Q + (1+e)\bar{p}_\sigma + gp = 0 \\ &Q' + (\alpha' + k)N + g\bar{p}_\sigma - (1+e)p = 0\end{aligned} \tag{O.10}$$

$$M' + (1+e)Q - gN + l = 0\,. \tag{O.11}$$

Finally, relative to the frame $\{\bar{\mathbf{y}}', \mathbf{m}\}$, we have from (L.1), (L.5), and (L.7),

$$\mathbf{F} = \lambda\tilde{N}\bar{\mathbf{t}} + \lambda^{-1}Q^*\bar{\mathbf{b}} \tag{O.12}$$

and

$$\bar{\mathbf{t}}' = (\alpha' + k + \gamma')\bar{\mathbf{b}} \ , \ \bar{\mathbf{b}}' = -(\alpha' + k + \gamma')\bar{\mathbf{t}} . \tag{O.13}$$

Substituting these expressions into (E.1) and (E.2), we get

$$\begin{aligned} &(\lambda\tilde{N})' - (\alpha' + k + \gamma')\lambda^{-1}Q^* + \lambda\bar{p}_\sigma = 0 \\ &(\lambda^{-1}Q^*)' + (\alpha' + k + \gamma')\lambda\tilde{N} - \lambda p = 0 \end{aligned} \tag{O.14}$$

$$M' + Q^* + l = 0 . \tag{O.15}$$

To obtain a complete set of field equations, we must add constitutive equations. These follow from (N.4), (N.5), or (N.7), once a strain-energy density Φ or a mixed strain-energy density Ψ is given. A systematic discussion might consider the behavior of a beam shell with various boundary conditions and distributed loads for each of the 8 possible forms of the mixed energy-density: $\Psi = \Psi(\bar{N}, Q^*, k)$, $\Psi = \Psi(Q^*, k)$, $\cdots$, $\Psi = \Psi(\bar{N})$, $\Psi = 0$. Such an ambitious, pedantic catalog would contain too many cases of little practical interest and too many cases with nearly identical solutions. Instead, we shall give a brief survey of two important general cases: inextensional beamshells under different distributed loads and general beamshells under normal pressure.

1. Inextensional beamshells

Here, we consider one of the most widely used approximations in beamshell theory. The governing equations can be derived in two ways. The *physical approach* assumes that $\Psi = \Psi(Q^*, k)$ which, by (N.7), implies that $\varepsilon = 0$. ($e = 0$ does not seem to be what is usually meant by inextensionality). The alternative *mathematical approach* scales all variables in such a way that the governing equations of the full nonlinear theory of beamshells contain a small, dimensionless parameter. If this parameter is set to zero, the equations reduce to a dimensionless form of those of the physical approach. The advantages of the mathematical approach are that (a), it shows precisely under what conditions a beamshell behaves as if it were truly inextensional, and (b), it suggests that the exact solutions of the full equations can be expressed as perturbation series whose higher order terms reflect the effects of non-zero extensional strains; often these series are the sum of a rapidly decaying edge-zone contribution and a smoothly varying interior contribution. As we are interested only in pure inextensional deformation, we shall, for simplicity, use the physical approach.

By assuming successively simpler forms for Ψ, we obtain the following hierarchy of inextensional theories.

(a) Shearable, inextensional beamshells: $\Psi = \Psi(Q^*, k)$.

Because $\lambda = 1$, (O.14) reduces to

$$\tilde{N}' - (\alpha' + k + \gamma')Q^* + \bar{p}_\sigma = 0 \ , \ Q^{*\prime} + (\alpha' + k + \gamma')\tilde{N} - p = 0 \ . \quad \text{(O.16)}$$

In particular, if

$$\Psi = \frac{1}{2}E_b h^3 k^2 - \frac{1}{2}\frac{Q^{*2}}{E_s h} \ , \quad \text{(O.17)}$$

we have the nonlinear generalization of Rankine-Timoshenko[13] beam theory. Our equations for an initially straight beam ($\alpha = 0$) should be compared with those of Buchanan, Huang & Cheng (1970).

(b) Unshearable, inextensional beamshells: $\Psi = \Psi(k)$.

This is the nonlinear Euler-Bernoulli theory of beamshells. Here, because $\mathbf{T} = \bar{\mathbf{t}}$ and $\mathbf{B} = \bar{\mathbf{b}}$, (O.10) and (O.11) take the same form as (O.16) and (O.15), namely,

$$N' - (\alpha' + k)Q + \bar{p}_\sigma = 0 \ , \ Q' + (\alpha' + k)N - p = 0 \quad \text{(O.18)}$$

$$M' + Q + l = 0 \ . \quad \text{(O.19)}$$

If the thickness, h, is a constant, the deformed initial curve $\mathcal{C}$ is called an *elastica*. A useful collection of results for initially straight or curved beams subject to end loads or to concentrated, in-span forces may be found in the book by Frisch-Fay (1962).

Problem: Determine a dimensionless, differential equation for the angle of rotation β of a horizontally cantilevered, Euler-Bernoulli beamshell under its own weight. Also, determine the position of the tip of the beamshell as a function of a dimensionless weight parameter.

Solution: From (O.6), $H = 0$ and $V = -mg(L - \sigma)$. Since $\varepsilon = \alpha = \gamma = l = 0$, (O.5) reduces to

$$M' - mg(L - \sigma)\cos\beta = 0 \ . \quad \text{(O.20)}$$

Take $\Psi = \frac{1}{2}E_b h^3 k^2$ and introduce the dimensionless distance and parameter

$$s \equiv \frac{\sigma}{L} \ , \ \mu \equiv \frac{mgL^3}{E_b h^3} \ . \quad \text{(O.21)}$$

Substituting (O.21) into (O.20) and adjoining the boundary conditions that the angle of rotation be 0 at the built-in end and that the moment be 0 at the free end, we obtain the 2-point boundary value problem:

---

[13]The shear strain correction term to classical (Euler-Bernoulli) beam theory was proposed by Rankine (1858, pp. 342-344). Timoshenko (1921, 1922) pointed out that in the dynamic equations, the transverse shear strain effect was of the same order of magnitude as the rotary inertia term.

$$\ddot{\beta} - \mu(1-s)\cos\beta = 0 \ , \ \beta(0) = \dot{\beta}(1) = 0 \ , \ \dot{\beta} \equiv d\beta/ds \ . \qquad (O.22)$$

From (C.6), (C.7), (K.1), and (O.21), we have for the position of the tip of the deformed beamshell

$$\bar{y}(L) = L\{[\int_0^1 \cos\beta(s,\mu)ds]\mathbf{e}_x + [\int_0^1 \sin\beta(s,\mu)ds]\mathbf{e}_y\} \ . \qquad (O.23)$$

This problem was solved in terms of a power series by Hummel & Morton (1927), numerically by Bickley (1934), again, in terms of a powers series, by Rohde (1953), and, again, numerically by Holden (1972), who noted that a first integral of (O.22) offered by T. M. Wang (1969) is wrong because $d^2\theta/dsd\bar{x} = d^2\theta/d\bar{x}ds$ only if $\bar{x}$ is a linear function of s. Table 4.1, computed by Neil Fergusson, gives the angle of rotation and the dimensionless Cartesian coordinates of the tip of the beamshell. □

A related problem—that of the "stiff" catenary—is to determine the shape of a beamshell under its own weight if its 2 ends are built in to piers less than a distance L apart and not necessarily at the same height. An approximate solution in which the effect of bending stiffness is treated as a small, singular perturbation of the classic catenary problem may be found in van der Heijden (1973) or Simmonds & Mann (1986, Chapter 9).

More elaborate problems involving the shape of beams under a variety of loads and supports, many of which have application to the handling of paper, have been considered by C. Y. Wang (1981a, 1981b, 1981c, 1986a), Wang & Watson (1981a, 1983), Watson (1981), Watson & Wang (1982, 1983), and Wang & Shodja (1984). See C. Y. Wang (1986b) for a review of the literature on the heavy elastica including a discussion of contact with an elastic foundation.

(c) Membrane-inextensional beamshells: $\Psi \equiv 0$.

From the constitutive relations (N.7) we have $\varepsilon = \gamma = M = 0$. Assuming that the distributed external couple l is also 0, it follows from (O.19) that $Q = 0$ so that (O.18) reduces to

$$N' + \bar{p}_\sigma = 0 \ , \ \bar{\alpha}'N - p = 0 \ , \qquad (O.24)$$

where $\bar{\alpha}' = \alpha' + k$ is the curvature of the deformed initial curve $\bar{\mathcal{C}}$. Equations (O.24) show that $\bar{\mathcal{C}}$ *adapts itself* so that the external loads may be carried by the membrane force N alone.

Application of these equations to cables and inflatable dams may be found in the book by Irvine (1981). Wang & Watson (1981b) and Wang (1984a) use (O.24) to study long cylindrical containers that store fluids or grain and that may be in contact with the ground or suspended above it; some of their approximate methods have been improved upon by Namais (1985) who considers fluid-filled membranes resting on a rigid, horizontal surface and pressed upon from above by another rigid, horizontal surface.

Table 4.1
Tip Deflection and Rotation of a Heavy Elastica

μ	$\int_0^1 \cos\beta(s, \mu)ds$	$-\int_0^1 \sin\beta(s, \mu)ds$	$-\beta(1, \mu)$
0	1	0	0
.5	.9978	.0623	.0831
1.0	.9912	.1235	.1651
1.5	.9808	.1825	.2449
2.0	.9669	.2385	.3216
2.5	.9503	.2910	.3946
3.0	.9315	.3396	.4635
3.5	.9113	.3843	.5282
4.0	.8901	.4252	.5886
4.5	.8684	.4623	.6448
5.0	.8467	.4959	.6970
5.5	.8250	.5264	.7454
6.0	.8037	.5539	.7904
6.5	.7829	.5789	.8321
7.0	.7628	.6015	.8708
7.5	.7432	.6220	.9067
8.0	.7244	.6406	.9401
8.5	.7063	.6576	.9712
9.0	.6890	.6731	1.0002
9.5	.6723	.6872	1.0273
10.0	.6564	.7002	1.0526
10.5	.6411	.7121	1.0763
11.0	.6264	.7231	1.0985
11.5	.6124	.7332	1.1194
12.0	.5990	.7425	1.1389
12.5	.5862	.7511	1.1573
13.0	.5739	.7591	1.1747
13.5	.5621	.7666	1.1910
14.0	.5508	.7735	1.2065
14.5	.5400	.7800	1.2210
15.0	.5296	.7860	1.2348
∞	0	1	$\pi/2$

Problem: Express the shape of $\overline{\mathcal{C}}$ in terms of elliptic integrals if the only distributed load is the hydrostatic pressure

$$p = p_0 - \rho g \bar{y} . \tag{O.25}$$

Here, p_0 is a reference pressure, ρg is the specific weight of the fluid, and $(\bar{x}, \bar{y})$ are the Cartesian coordinates of a point on $\overline{\mathcal{C}}$. See Fig. 4.7a.

Solution: From $(O.24)_1$, $N = N_0$, a constant, so by (E.4), (L.1), $(O.24)_2$, and (O.25), we obtain the system of 3, 1st-order differential equations

$$\bar{x}' = \cos\bar{\alpha} \ , \ \bar{y}' = \sin\bar{\alpha} \ , \ \bar{\alpha}' = a - b\bar{y} \, , \tag{O.26}$$

where $a = p_0 / N_0$ and $b = \rho g / N_0$. Integration yields

$$a - b\bar{y} = (2b\cos\bar{\alpha} + C_1)^{1/2}$$

$$\sigma = \int \frac{d\bar{\alpha}}{(2b\cos\bar{\alpha} + C_1)^{1/2}} + C_2 \ , \ \bar{x} = \int \frac{\cos\bar{\alpha} d\bar{\alpha}}{(2b\cos\bar{\alpha} + C_1)^{1/2}} + C_3 \, , \tag{O.27}$$

where C_1, C_2, and C_3 are constants of integration that depend on the boundary conditions of the specific problem. The integrals in (O.27) are elliptic, of the 1st and 2nd kinds. □

Newman (1975) has pointed out 2 other interesting, unrelated problems that reduce to solving (O.26). The first is to find the shape of a 2-dimensional drop of fluid where, now, N_0 is the surface tension. See Fig. 4.7b. The second is to find the shape of an initially flat, inextensional beamshell subject to an end force P_0 and an end moment M_0 only. See Fig. 4.7c. Here

$$M = E_b h^3 \bar{\alpha}' = M_0 - P_0 \bar{y} \, . \tag{O.28}$$

Solutions for many different boundary conditions can be found in Frisch-Fay's book.

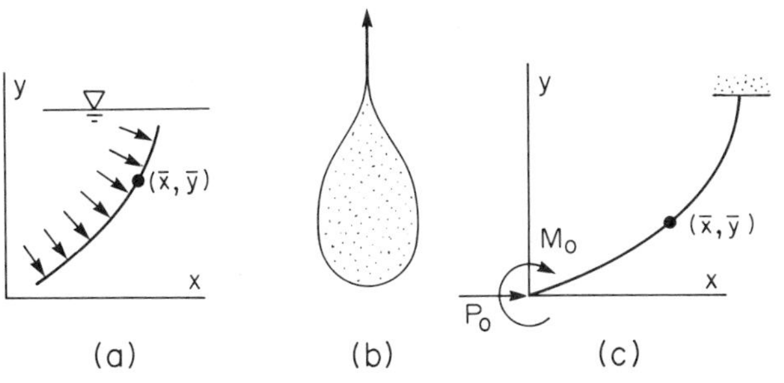

Fig. 4.7. Typical problems with a linear variation of the deformed curvature: (a) membrane containing a fluid; (b) 2-dimensional fluid drop; (c) elastica subject to moment and axial end force.

2. *Pressure loaded beamshells*

(a) Semi-qualitative results.

The following informal discussion, accompanied by Fig. 4.8, derives from our experimental and mathematical experience with very flexible, nearly unshearable beams. For simplicity, we shall assume that $\Psi = \Psi(N, k)$ in this subsection, i.e., we shall invoke the Kirchhoff hypothesis as a *constitutive assumption.*

First consider an initially straight beamshell held at both ends against translation by hinged supports. (Fig. 4.8a.) If the exact solution is developed as a series in some dimensionless displacement parameter, we find that the first approximation $\bar{\mathbf{y}}^{(0)}$ to the deformed position of $\mathcal{C}$ is valid if $|\mathbf{u}|/h \ll 1$ and can be obtained from linear, inextensional Euler-Bernoulli theory. However, an *inextensional model* has, in this case, a very limited range (Parikh, Schmidt & DaDeppo, 1972): if the load increases to where $|\mathbf{u}|/h = O(1)$, extensions can no

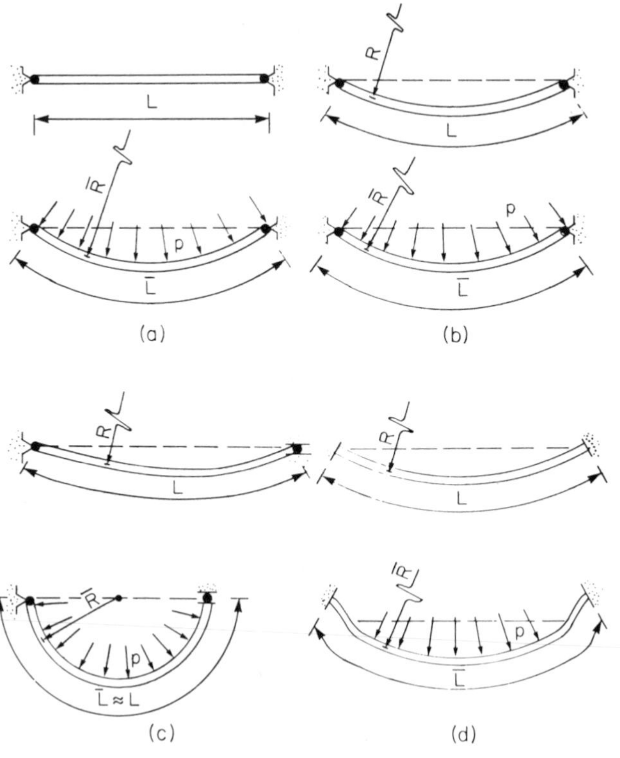

Fig. 4.8. Large deflections of elastic, pressure loaded beamshells subject to various boundary conditions: (a) initially flat and hinged; (b) initially circular and hinged; (c) initially flat, right end sliding; (d) initially circular (or flat) and clamped.

longer be ignored. The consideration of this and other geometric and material nonlinearities forms the basis of a more elaborate analysis of the nonlinear behavior of beamshells.

If the pressure is very high, the influence of bending stiffness is small and $\mathcal{C}$ approaches, asymptotically, a circular arc. The *momentless* or *membrane* beamshell model in which $\Psi = \Psi(N)$ or, equivalently, $\Phi = \Phi(e)$, yields the circular arc configuration directly, with e given implicitly by

$$\tfrac{1}{2}(pL/\Phi,_e) = \sin\left[\tfrac{1}{2}(1+e)(pL/\Phi,_e)\right]. \tag{O.29}$$

If the deformation is shallow, i.e., if $kL = (1 + epL/\Phi,_e) \ll 1$, then retention of the first 2 terms in the expansion of the sine function leads to the simpler expression

$$e = \frac{1}{6}\left[\frac{p^2L^2}{24\Phi,_e^2}\right], \tag{O.30}$$

which, if $|e| \ll 1$, yields, by (N.9)

$$e \approx \frac{1}{2}\left[\frac{pL}{3E_s h}\right]^{1/3}. \tag{O.31}$$

Consider next a hinged beamshell initially in the shape of a circular arc with curvature $\alpha' = 1/R$. (Fig. 4.8b.) The major change from the solution for the flat beamshell is that, due to *curvature effects*, the membrane model now represents a first-approximation to the exact solution for all pressures, unless the beamshell is shallow ($L/R \ll 1$). In this latter case the solution for low pressure is more complex, but the asymptotic behavior at high pressure is very similar to the first case.

Consider again a flat beamshell with the right support hinged but now free to move horizontally. (Fig. 4.8c.) The situation has changed drastically: it is now the *inextensional* model that provides an excellent approximation, regardless of the magnitude of the pressure. This is an elementary illustration that shells tend to exhibit inextensional deformation if the kinematic conditions at the boundaries permit it. At high pressure, the effects of bending stiffness decrease and $\mathcal{C}$ approaches a semicircle. In fact, we could construct a *membrane-inextensional* model in which equilibrium and kinematics alone determine the solution. In practice, very high pressure would induce some stretching, but the shape of the beamshell would remain semicircular with an axial force $N = (pL/\pi)(1 + \Psi,_N)$. The membrane-inextensional model is not common in beamshell theory, but does find more applications in other forms of shells.

Finally, consider the first or second case but with clamped supports. (Fig. 4.8d.) The situation at high pressure will be very similar to the corresponding hinged cases, except that there will be *edge-zones* or *boundary-layers* where the bending stresses may be of the same order or larger than the direct stresses. Within the edge-zones there exist solutions that decay rapidly into the interior. These supplement the smoothly varying membrane solutions and permit satisfaction of the clamping condition, $\beta = 0$. Edge-zone corrections exist in the other 3 cases, but are insignificant.

The examples discussed above are qualitatively simple, but they do exhibit some of the more important effects in shells of boundary conditions, curvature, and loading.

(b) Pressure loaded beamshells—quantitative discussion.

Most treatments of beamshells in the literature that deal with geometric and material nonlinearities are devoted to the formulation and numerical solution of the governing equations. In a few cases, such as the *inextensional elastica*, analytic solutions are available, as has been noted in subsection O.1. A discussion of the *linearly extensible elastica*, which may be considered to be a special case of the *small strain, finite rotation model* of thin shells, is given in the book by Pflüger (1975).

A more general model of a beamshell involving large strains and rotations as well as material nonlinearities has been studied extensively by Antman (1968, 1969, 1972, 1974b). For the important special case of straight or circular beamshells under constant normal pressure and end loads, Antman has obtained first integrals of the governing equations, thereby reducing their solution to quadratures. The details are as follows.

Set $\mathbf{p} = -p\mathbf{m}$, $l = 0$, take the dot product of (E.1) with $\mathbf{F}$, multiply (E.2) by p, and add to obtain

$$\mathbf{F} \cdot \mathbf{F}' + pM' = 0 . \tag{O.32}$$

For *constant* p, this equation may be integrated to yield the *equilibrium integral*

$$N^2 + Q^2 + 2pM = A , \tag{O.33}$$

where A is a constant of integration. Note that (O.33) holds regardless of the shape of the beamshell. Next, combine (N.4), (O.10), and (O.11), and take Φ to depend on σ only through e, g, and k. It follows that

$$\left[(1+e)\frac{\partial \Phi}{\partial e} + g\frac{\partial \Phi}{\partial g} + (\alpha' + k)\frac{\partial \Phi}{\partial k} - \Phi \right]' = \alpha'' M . \tag{O.34}$$

If $\alpha'' = 0$ (straight or circular beamshell), then, regardless of the loading, (O.34) implies the *energy integral*

$$(1+e)\frac{\partial \Phi}{\partial e} + g\frac{\partial \Phi}{\partial g} + (\alpha' + k)\frac{\partial \Phi}{\partial k} - \Phi = B , \tag{O.35}$$

where B is a constant of integration.

Equations (O.33), (O.35), and (N.4) or (N.7) are 5 algebraic equations in the 6 variables e, g, k, M, N, and Q. We assume that 5 of them can be expressed in terms of either M, N, or Q, chosen for convenience. Subsequent substitution into that equilibrium equation among (O.10) and (O.11) which contains the derivative of the chosen variable reduces the solution for this variable to quadrature. The rectangular coordinates of $\bar{y}$ follow from (K.1) and (E.4) as

$$\begin{aligned}\bar{x} &= \int [(1+e)\cos(\alpha+\beta) - g\sin(\alpha+\beta)]d\sigma \\ \bar{y} &= \int [(1+e)\sin(\alpha+\beta) + g\cos(\alpha+\beta)]d\sigma \,.\end{aligned} \tag{O.36}$$

The procedure outlined above leads to a complete solution, but is not devoid of numerical difficulties. Special techniques may have to be devised because the unknown constants, A and B, appear inside some of the integrals.

An alternative form of (O.35) is

$$N + (\alpha' + k)\Psi,_k - \Psi = B \,, \tag{O.37}$$

which comes from the Legendre transformation (N.6) expressing Φ in terms of Ψ. The explicit dependence of this form of the energy integral on N and Q makes (O.37) particularly useful in what follows.

As an application of the equations of this section, consider a ring beamshell of initial radius, R, subject to 2 equal and opposite outward radial forces of magnitude $2P$. (Fig. 4.9). Here, the kinematics permits *inextensional deformation* which will dominate until the ring becomes nearly flat. At this point the

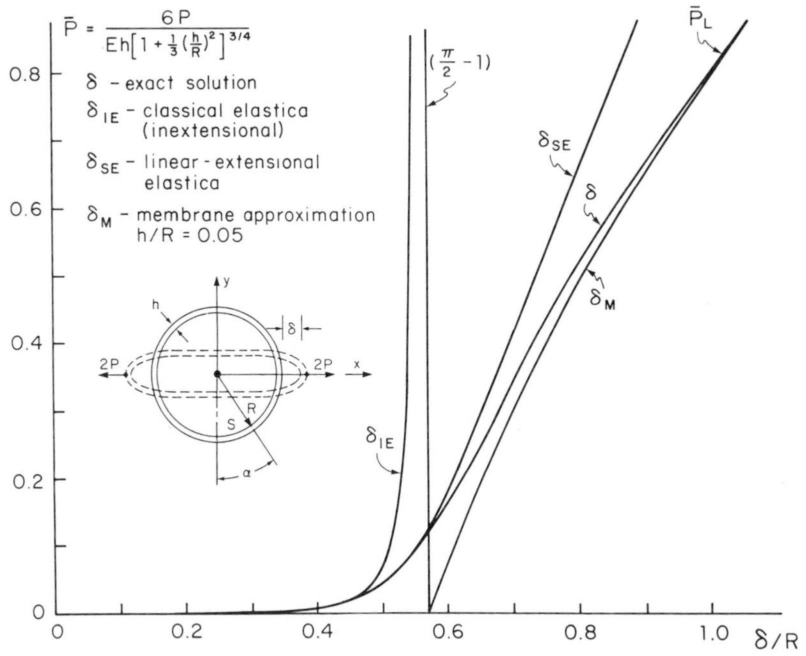

Fig. 4.9. Load-deflection curves for a line-loaded beamshell.

linearly extensional model, in which Ψ takes the quadratic form (N.18), becomes a good predictor of the deformation. As the loads increase further, the full nonlinear model must come into play. At very high loads, the solution approaches, asymptotically, the *membrane model* which lets the beamshell straighten without resistance and then stretch according to the material law, $\Psi = \Psi(N)$. An outline of the solution process is as follows:

Let the arc length, $\sigma = \alpha R$, be measured from a point equidistant from the loads. Then the equilibrium equations for the ring are simply

$$N = P\cos(\alpha + \beta) \ , \ Q = -P\sin(\alpha + \beta) \ . \tag{O.38}$$

Substitution of (O.38) into (O.37) facilitates the expression of the bending strain, $k = \beta'$, in terms of β itself. Hence, the implicit solution for $\beta(\sigma)$ becomes

$$\sigma = \int \frac{d\beta}{k(\beta)} \ . \tag{O.39}$$

The constant B in the energy integral (O.37) is determined by the symmetry condition, $\beta(\pi R/2) = 0$. Results for 2 material models are presented below. The reader is referred to Section 4.6 of Frisch-Fay's book where a solution for the classical elastica ($\Psi = \Phi = \frac{1}{2} E_b h^3 k^2$) is given.

(i) **Strongly nonlinear, incompressible, unshearable material.** The strain-energy density in this example is taken from Libai & Simmonds (1981):

$$\Phi = \frac{Eh}{3(1 + 2\varepsilon)} \left[\varepsilon^2 + \frac{h^2}{12}(k - 2\alpha'\varepsilon)^2 \right] \ . \tag{O.40}$$

The above expression for Φ, with $E \approx (3/2)E_e$, is a simplification of a strain-energy density that was developed with the *deformed initial material midsurface replacing* $\mathcal{C}$. For further discussion, see Section N.

Considering that all material models are approximate, we may, for the purposes of this section, simply take (O.40) as a *postulated* form of Φ. Fig. 4.10 reproduces results from Libai & Simmonds (1983b), giving the variation of the displacement, δ, at the point of application of a pulling force, P. Also shown in the figure are the predictions of various simplified theories from which their range of application can be inferred. It appears as if the *linear extensional model* gives the best overall agreement with the "exact" solution.

(ii) **Material with the quadratic mixed-energy density (N.18).** Here, the aim is get some idea of the effect of shear deformation on the behavior of the loaded ring. No restrictions are placed on the size of the deformations, but the results for large strains should be viewed cautiously, given that, for most materials, (N.18) is approximate. Substitution of (N.18) into the energy integral (O.37) yields

$$N + E_b h^3 k(\alpha' + \tfrac{1}{2}k) + \tfrac{1}{2}(N^2/E_e h + Q^2/E_s h) = B\ . \tag{O.41}$$

Expressing N and Q in terms of $(\alpha+\beta)$ and integrating, as in (O.39), we find that $\beta(\alpha)$ is given implicitly by

$$\alpha = \mu \int_0^{\alpha+\beta} [\mu^2(1 + k_0 R)^2 + (P/E_e h)^2 (E_e/E_s - 1)\sin^2\theta \\ - 2(P/E_e h)\sin\theta]^{-1/2} d\theta\ , \tag{O.42}$$

where $\mu^2 = E_b h^2/E_e R^2$ and k_0 is the bending strain at $\alpha=0$, to be determined from $\beta(\pi/2)=0$. Equation (O.42) can be expressed in terms of elliptic integrals of the first kind.

The variation of the length of the minor axis of the deformed ring with the loads for several values of E_e/E_s in given in Fig. 4.10, which is reproduced from Libai (1984) where more details are given and other cases studied.

P. Elastodynamics

In this section, we discuss 5 forms of the equations of motion for a beamshell. The first takes the displacement components and the shear strain, γ, as the basic unknowns. The resulting equations have 2 drawbacks: they are ill-conditioned for motion that is essentially inextensional and, for nonlinear problems, they are unduly complicated.

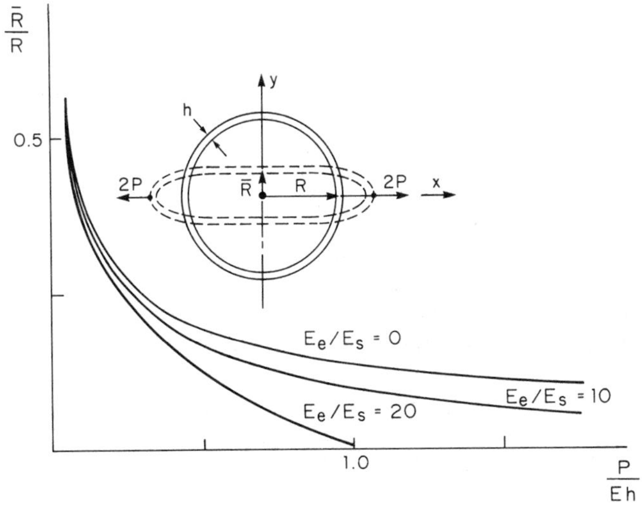

Fig. 4.10. Load-deflection curves for a beamshell with various degrees of shear deformability.

The 2nd form of the equations of motion involves β, N, and Q as the basic unknowns. Not only are the full nonlinear equations simpler than those of the first form, but inextensional motion causes no problems.

The 3rd form, useful for applying the method of characteristics, expresses the equations of motion as a system of 6, 1st-order, quasilinear partial differential equations in the unknowns N, Q, k, ω, v_T, and v_B, where the latter 2 are components of the beamshell velocity, **v**, along **T** and **B**.

The 4th form, which is restricted to classical flexural motion, uses an idea of Budiansky & Payton [& Carrier] (1961) for a circular beamshell to cast the equations of motion into a form involving a real-valued unknown, r, and a complex-valued unknown, z, of unit modulus that makes the role of nonlinearities transparent.

The last form, which is also restricted to classical flexural motion, is cast in the Cartesian frame $\{\mathbf{e}_x, \mathbf{e}_y\}$ and uses the deformation potentials introduced by Mansfield & Simmonds (1987) to integrate the equations of balance of linear momentum, in analogy with a well-known simplifying step in static problems.

1. Displacement-shear strain form

Given a specific strain-energy density, Φ, the resulting form of the differential equations of motion will depend on the basis we choose for the component representation of the force, **F**, and the displacement, **u**. Because we know of no compelling advantages of a displacement-shear strain form for the equations of motion, we shall be content with considering just 2 of several possible bases: $\{\mathbf{e}_x, \mathbf{e}_y\}$ and $\{\mathbf{t}, \mathbf{b}\}$.

As we discussed in Section L, it appears simplest to use the strains, ε, γ, and K, in the basis $\{\mathbf{e}_x, \mathbf{e}_y\}$. Because the right side of (L.15) is equal to $\Phi(\varepsilon, \gamma, K)$ if the beamshell is elastic, the stress resultants and moment conjugate to these strains are given by

$$\overline{N} = \Phi,_{\varepsilon} \ , \ Q^* = \Phi,_{\gamma} \ , \ M^* = \Phi,_K \, . \tag{P.1}$$

Furthermore, with

$$\overline{\mathbf{y}}' = (\cos\alpha + u_x{}')\mathbf{e}_x + (\sin\alpha + u_y{}')\mathbf{e}_y \tag{P.2}$$

and

$$\mathbf{m} = \mathbf{e}_z \times \overline{\mathbf{y}}' = -(\sin\alpha + u_y{}')\mathbf{e}_x + (\cos\alpha + u_x{}')\mathbf{e}_y \, , \tag{P.3}$$

it follows from (L.7), (L.16), (P.2), and (P.3) that

$$H \equiv \mathbf{F}\cdot\mathbf{e}_x = [\overline{N} + 2(\alpha' + K)\lambda^{-2}M^*](\cos\alpha + u_x{}') - \lambda^{-2}Q^*(\sin\alpha + u_y{}') \tag{P.4}$$

$$V \equiv \mathbf{F}\cdot\mathbf{e}_y = [\overline{N} + 2(\alpha' + K)\lambda^{-2}M^*](\sin\alpha + u_y{}') + \lambda^{-2}Q^*(\cos\alpha + u_x{}') \, , \tag{P.5}$$

where K is given by (L.22) in terms of u_x and u_y and

$$\lambda^2 = \overline{\mathbf{y}}'\cdot\overline{\mathbf{y}}' = 1 + 2(u_x'\cos\alpha + u_y'\sin\alpha) + u_x'^2 + u_y'^2 \,. \tag{P.6}$$

Thus, the equations of motion (E.1) and (E.2) take the component form

$$H' + p_x = m\ddot{u}_x \;,\; V' + p_y = m\ddot{u}_y \tag{P.7}$$

$$(\lambda^2 M^*)' + Q^* + l = I\ddot{\beta} \,. \tag{P.8}$$

It remains to express $\ddot{\beta}$ in terms of u_x, u_y, γ, and their derivatives. From (L.1), (L.2), and (P.2) follows

$$\begin{aligned}\ddot{\beta} &= \left[\tan^{-1}\left(\frac{\sin\alpha + u_y'}{\cos\alpha + u_x'}\right)\right]^{\bullet} - \ddot{\gamma} \\ &= \{\lambda^{-2}[(\cos\alpha + u_x')\dot{u}_y' - (\sin\alpha + u_y')\dot{u}_x']\}^{\bullet} - \ddot{\gamma} \,.\end{aligned} \tag{P.9}$$

If (P.1) is substituted into (P.4) and (P.5), and the resulting expressions substituted into (P.7) and (P.8), we get 3 equations of motion in terms of u_x, u_y, and γ of immense complexity. We shall defer any substantial discussion of simplification of the equations of motion until subsection P.2.

Turning to the basis $\{\mathbf{t}, \mathbf{b}\}$, we set

$$\mathbf{F} = n\mathbf{t} + q\mathbf{b} \tag{P.10}$$

and use (L.24), (L.26), and (O.1) to write

$$\begin{aligned}\mathbf{p} &= (p_x\cos\alpha + p_y\sin\alpha)\mathbf{t} + (-p_x\sin\alpha + p_y\cos\alpha)\mathbf{b} \\ &= [(1+\overline{e})\overline{p}_\sigma + \overline{\beta}p\,]\mathbf{t} + [\overline{\beta}\overline{p}_\sigma - (1+\overline{e})p\,]\mathbf{b} \\ &\equiv p_\sigma\mathbf{t} + p_\zeta\mathbf{b} \,.\end{aligned} \tag{P.11}$$

Thus, with $\mathbf{u} = u_\sigma\mathbf{t} + u_\zeta\mathbf{b}$, $\mathbf{t}' = \alpha'\mathbf{b}$, and $\mathbf{b}' = -\alpha'\mathbf{t}$, the vector equation, (E.1), of balance of linear momentum takes the scalar form

$$n' - \alpha' q + p_\sigma = m\ddot{u}_\sigma \tag{P.12}$$

$$q' + \alpha' n + p_\zeta = m\ddot{u}_\zeta \,. \tag{P.13}$$

The equation of balance of rotational momentum is identical to (P.8).

As κ rather than K is the preferred bending strain in the basis $\{\mathbf{t}, \mathbf{b}\}$, we use (L.7), (L.19), (L.24), (L.26), and (P.10) to get

$$n = (1+\overline{e})(N^* + 2\kappa\lambda^{-2}M^*) - \lambda^{-2}\overline{\beta}Q^* \tag{P.14}$$

$$q = \lambda^{-2}(1+\overline{e})Q^* + \overline{\mathbf{y}}(N^* + 2\kappa\lambda^{-2}M^*) \,, \tag{P.15}$$

where, in view of (L.20),

$$N^* = \Phi_{,\varepsilon} \ , \ Q^* = \Phi_{,\gamma} \ , \ M^* = \Phi_{,\kappa} \, . \tag{P.16}$$

The final step in arriving at 3 equations of motion in 3 unknowns is to express $\ddot{\beta}$ in (P.8) in terms of u_σ, u_ζ, and γ. From Fig. 4.2 and (L.24), we have

$$\bar{\mathbf{y}}' = (1+\bar{e})\mathbf{t} + \bar{\beta}\mathbf{b} = \lambda[\cos(\beta+\gamma)\mathbf{t} + \sin(\beta+\gamma)\mathbf{b}] \, . \tag{P.17}$$

Hence,

$$\tan(\beta+\gamma) = \frac{\bar{\beta}}{1+\bar{e}} \, , \tag{P.18}$$

from which follows

$$\ddot{\beta} = \{\lambda^{-2}[(1+\bar{e})\dot{\bar{\beta}} - \bar{\beta}\dot{\bar{e}}]\}^{\cdot} - \ddot{\gamma} \, , \tag{P.19}$$

where $\bar{e}$ and $\bar{\beta}$ are given in terms of u_σ and u_ζ in (L.24) and

$$\lambda^2 = (1+\bar{e})^2 + \bar{\beta}^2 \, . \tag{P.20}$$

As with our results in the basis $\{\mathbf{e}_x, \mathbf{e}_y\}$, the equations of motion expressed explicitly in terms of u_σ, u_ζ, and γ are frightfully complex and singularly unenlightening, so we do not write them out.

In the classical theory of beamshells, $\gamma = I = l = 0$. The rotational momentum equation, (P.8), together with the expressions for κ and λ^2 in terms of $\bar{e}$ and $\bar{\beta}$, (L.27) and (P.20), may then be used to reduce (P.14) and (P.15) to

$$n = (1+\bar{e})N^* + 2\bar{\beta}'M^* + \bar{\beta}M^{*\prime} \tag{P.21}$$

$$q = -(1+\bar{e})M^{*\prime} - 2\bar{e}'M^* + \bar{\beta}N^* \, . \tag{P.22}$$

Further, if ε and $h\kappa$ are small and the beamshell is homogeneous in the thickness direction, we may take $\Phi = [Eh/2(1-\nu^2)](\varepsilon^2 + h^2\kappa^2/12)$, in which case (P.16) yields

$$N^* = Eh[\bar{e} + \tfrac{1}{2}(\bar{e}^2 + \bar{y}^2)] \ , \ M^* = [Eh^3/12(1-\nu^2)][(1+\bar{e})\bar{y}' - \bar{y}\bar{e}'] \, . \tag{P.23}$$

Substituting (P.23) into (P.21) and (P.22) and these into (P.12) and (P.13), we obtain 2 coupled equations for u_σ and u_ζ.

Despite the simplifications of the classical theory, these equations are still quite complicated. Their major fault, however, is that they become ill-conditioned when the motion is nearly inextensional, i.e., when there is a near dependence of the form $\varepsilon = u_\sigma' - \alpha' u_\zeta + \tfrac{1}{2}[(u_\sigma' - \alpha' u_\zeta)^2 + (u_\zeta' \alpha' u_\sigma)^2] \approx 0$. Such a difficulty with a displacement formulation already manifests itself in the linear, static theory, as discussed in Section 2 of Simmonds (1966).

2. Stress resultant-rotation form

Much simpler and better conditioned equations of motion are obtained if we take β, N, and Q as the basic unknowns and use (N.7) to express e, g, and M in the form

$$e=-\Psi_{,N}=e(N,Q,k)\,,\; g=-\Psi_{,Q}=g(N,Q,k)\,,\; M=\Psi_{,k}=M(N,Q,k)\,, \quad (P.24)$$

where $k=\beta'$.

To express the inertial term in the linear momentum equation, (E.1), in terms of e, g, and β [and hence in terms of N, Q, and $\beta'=k$, via (P.24)], we differentiate (E.1) with respect to σ.[14] With $\mathbf{F}=N\mathbf{T}+Q\mathbf{B}$ and the aid of (K.2), (K.3), (O.1), and $\omega=\dot{\beta}$, we obtain the component form

$$\begin{aligned} &N'' - (\alpha+\beta)'^2 N - 2(\alpha+\beta)'Q' - (\alpha+\beta)''Q + [(1+e)\bar{p}_\sigma + gp]' \\ &\quad + (\alpha+\beta)'[g\bar{p}_\sigma - (1+e)p] = m[\ddot{e} - \dot{\beta}^2(1+e) - \ddot{\beta}g - 2\dot{\beta}\dot{g}] \end{aligned} \quad (P.25)$$

$$\begin{aligned} &Q'' - (\alpha+\beta)'^2 Q + 2(\alpha+\beta)'N' + (\alpha+\beta)''N + [(1+e)p - g\bar{p}_\sigma]' \\ &\quad + (\alpha+\beta)'[(1+e)\bar{p}_\sigma + gp] = m[\ddot{g} - \dot{\beta}^2 g + \ddot{\beta}(1+e) + 2\dot{\beta}\dot{e}]\,. \end{aligned} \quad (P.26)$$

The rotational momentum equation, (E.2), takes the form

$$M' + (1+e)Q - gN + l = I\ddot{\beta}\,. \quad (P.27)$$

Insertion of (P.24) into (P.25)-(P.27) yields 3 equations in the 3 unknowns N, Q, and β.

If the centerline of the undeformed, transversely homogeneous beamshell coincides with the initial curve, and if e, g, $hk \ll 1$, we may use (N.18) for $\Psi=\Psi(N,Q,k)$.

3. A system of 1st-order equations

Rather than work with 3 reduced equations of motion for N, Q, and k, as in the preceding subsection, it is sometimes more convenient for analytical (and, perhaps, for numerical) work to go the other way and to write the governing equations as a system of 1st-order partial differential equations. Thus, extending the work of Lee (1971) who confined himself to straight beams with linear stress-strain relations, we start by expressing the velocity in the form

$$\mathbf{v} = v_T\mathbf{T} + v_B\mathbf{B}\,. \quad (P.28)$$

Since $\dot{\mathbf{T}}=\omega\mathbf{B}$ and $\dot{\mathbf{B}}=-\omega\mathbf{T}$, the acceleration has the representation

$$\dot{\mathbf{v}} = (\dot{v}_T - \omega v_B)\mathbf{T} + (\dot{v}_B + \omega v_T)\mathbf{B}\,. \quad (P.29)$$

Substituting $\mathbf{F}=N\mathbf{T}+Q\mathbf{B}$ along with (O.1) and (P.29) into (E.1) and (E.2), noting that $\mathbf{T}'=(\alpha'+k)\mathbf{B}$ and $\mathbf{B}'=-(\alpha'+k)\mathbf{T}$, and making use of the constitutive relations (P.24), we obtain the 3 scalar equations

$$N' - (\alpha'+k)Q + p_T = m(\dot{v}_T - \omega v_B) \quad (P.30)$$

---

[14]Libai (1962) introduced this idea in general shell theory. Both he and, later, Keller & Ting (1966) used it to study the nonlinear motion of an initially straight beam.

$$Q' + (\alpha' + k)N + p_B = m(\dot{v}_B + \omega v_T) \tag{P.31}$$

$$\Psi_{,kN}N' + \Psi_{,kQ}Q' + \Psi_{,kk}k' + (1 - \Psi_{,N})Q + \Psi_{,Q}N + l = I\dot{\omega} \tag{P.32}$$

These are 3 equations for the 6 unknowns N, Q, k, ω, v_T, and v_B.

To obtain 3 additional equations, note first that $\omega = \dot{\beta}$ and $k = \beta'$ imply the compatibility condition

$$\omega' = \dot{k} . \tag{P.33}$$

Likewise, as $\mathbf{v}' = \dot{\bar{\mathbf{y}}}'$, it follows from (P.28) and $\bar{\mathbf{y}}' = (1+e)\mathbf{T} + g\mathbf{B}$ that

$$v_T' - kv_B = -\Psi_{,NN}\dot{N} - \Psi_{,NQ}\dot{Q} + \Psi_{,Nk}\dot{k} + \omega\Psi_{,Q} \tag{P.34}$$

$$v_B' + kv_T = -\Psi_{,QN}\dot{N} - \Psi_{,QQ}\dot{Q} - \Psi_{,Qk}\dot{k} + \omega(1 - \Psi_{,N}) . \tag{P.35}$$

Thus, (P.30)-(P.35) is a system of *quasilinear* partial differential equations of the form

$$A(\underline{u})\underline{u}' + B(\underline{u})\dot{\underline{u}} + C(\underline{u}) = 0 , \tag{P.36}$$

where A, B, and C are 6×6 matrices and $\underline{u} \equiv [N, Q, k, \omega, v_T, v_B]^T$.

If we use the quadratic expression, (N.18), for Ψ and set $\alpha = 0$, then our equations reduce to those of Lee (1971), save that he uses $1+e$ as an unknown whereas we use N, for reasons explained as the beginning of this section. Lee determined the characteristics for his equations and showed (in our notation) that if $E_e e + E_s > 0$, then the characteristics are real and his system is totally hyperbolic. It remains an interesting problem to study the characteristics of the system (P.36) to see if and how Lee's results generalize to curved, rubber-like beamshells.

4. *Classical flexural motion*

Flexural motion is characterized by the order of magnitude relation $\dot{\beta} = O[R^{-2}(E_b h^3/m)^{1/2}\beta]$, where R is some typical length associated with the initial curve. This suggests the nondimensionalization

$$(N, Q) = (E_b h^3/R^2)(\underline{N}, \underline{Q}) \ , \ M = (E_b h^3/R)\underline{M}$$
$$(p_T, p_B) = (E_b h^3/R^3)(\underline{p_T}, \underline{p_B}) \ , \ l = (E_b h^3/R^2)\underline{l} \ , \ I = mh^2 S \tag{P.37}$$
$$\sigma = R\underline{\sigma} \ , \ t = (m/E_b h^3)^{1/2} R^2 \underline{t} \ , \ \underline{\varepsilon} = h/R \ ,$$

where $S = O(1)$ is a shape factor. Calculating e, g, and M from (N.18) and (P.24), substituting the result into (P.25)-(P.27), and introducing the underscored variables defined above, we obtain 3 coupled, dimensionless equations for β, $\underline{N}$, and $\underline{Q}$. If we give the elastic constants their classical values, $[E_b = E/12, E_e = E/(1-\nu^2), E_s/E = \infty$, where E is Young's modulus], let $\varepsilon \to 0$, and use (P.27) to eliminate $\underline{Q}$, we obtain coupled equations for β and $\underline{N}$. Denoting $\partial/\partial\underline{\sigma}$ by a prime and *dropping the underbars*, we have

$$N'' - (\alpha+\beta)'^2 N + 2(\alpha+\beta)'\beta''' + (\alpha+\beta)''\beta'' + p_T' + (\alpha+\beta)' p_B = -\dot{\beta}^2 \quad (P.38)$$

$$-\beta'''' + (\alpha+\beta)'^2\beta'' + 2(\alpha+\beta)'N' + (\alpha+\beta)''N - p_B' + (\alpha+\beta)' p_T = \ddot{\beta}. \quad (P.39)$$

For a circular beamshell of radius R, $\alpha' = 1$, and these equations reduce to equations (35) and (36) of Simmonds (1979).

We close this section by reemphasizing an observation we made in the beginning of Section O on elastostatics, namely, that use of the Cartesian basis, $\{\mathbf{e}_x, \mathbf{e}_y\}$, often leads to remarkably simple equations. We illustrate this first with a simplified derivation of a single, complex-valued equation of classical flexural motion that generalizes an equation derived by Budiansky & Payton (1961) for a circular ring.[15] With

$$\theta \equiv \alpha + \beta \quad (P.40)$$

and $\mathbf{F} = N\mathbf{T} + Q\mathbf{B}$, we have, from (C.6) and (C.7),

$$\mathbf{F} = (N\cos\theta - Q\sin\theta)\mathbf{e}_x + (N\sin\theta + Q\cos\theta)\mathbf{e}_y\,. \quad (P.41)$$

In classical flexural motion the extensional and shear strains, e and g, are 0. Thus, from (K.1),

$$\bar{\mathbf{y}}' = \mathbf{T} = \cos\theta\mathbf{e}_x + \sin\theta\mathbf{e}_y\,. \quad (P.42)$$

Differentiating both sides of (E.1), the equation of balance of linear momentum, with respect to σ, noting that $\dot{\mathbf{v}}' = \ddot{\bar{\mathbf{y}}}'$, and inserting (P.41) and (P.42), we obtain the component equations

$$(N\cos\theta - Q\sin\theta)'' + p_x' = m(\cos\theta)^{\cdot\cdot} \quad (P.43)$$

$$(N\sin\theta + Q\cos\theta)'' + p_y' = m(\sin\theta)^{\cdot\cdot}. \quad (P.44)$$

A further assumption in classical flexural motion is that $l = I = 0$, $\Phi = \tfrac{1}{2}E_b h^3\beta'^2$, and, because extensional and shear strains are taken to be 0, $\mathbf{m} = \mathbf{e}_z \times \mathbf{T} = \mathbf{B}$. Thus, (E.2), the equation of balance of rotational momentum, reduces to

$$Q = E_b h^3(\alpha - \theta)''\,, \quad (P.45)$$

where we have set $\beta = \theta - \alpha$. If $\alpha = 0$, (P.43)-(P.45) reduce to the equations used by Keller & Ting (1966) and Verma (1972) in their studies of periodic vibrations of initially straight beams.

Inserting (P.45) into (P.43) and (P.44), introducing the dimensionless forms of N, σ, and t defined in (P.37) as well as dimensionless forms for p_x and p_y similar to those for p_T and p_B, and, finally, dropping underbars, we obtain the 2 reduced equations of motion

---

[15]This derivation, which makes use of a key observation by George Carrier, is reproduced in the Appendix of Simmonds (1979).

$$[N\cos\theta - (\alpha-\theta)''\sin\theta]'' + p_x{}' = (\cos\theta)^{\cdot\cdot} \tag{P.46}$$

$$[N\sin\theta + (\alpha-\theta)''\cos\theta]'' + p_y{}' = (\sin\theta)^{\cdot\cdot}. \tag{P.47}$$

Multiplying (P.47) by the imaginary unit i, adding it to (P.46), and introducing the complex quantities

$$z \equiv e^{i\theta} \ , \ q \equiv p_x + ip_y \ , \tag{P.48}$$

we have

$$[Nz + i(\alpha-\theta)'']'' + q' = \ddot{z} \ . \tag{P.49}$$

But, $z' = ie^{i\theta}\theta'$; hence, $z'' = iz\theta'' - z\theta'^2$. Thus, (P.49) may be cast into the form

$$\ddot{z} + (z'' - i\alpha'' + \underline{rz})'' = q' \ , \tag{P.50}$$

where

$$r \equiv \theta'^2 - N \tag{P.51}$$

is an unknown, *real-valued* function. *All* nonlinear effects are manifested in the single underlined term! The full implications of this deceptively simple equation remain to be explored.

Our 2nd illustration of the advantages of using the Cartesian frame $\{\mathbf{e}_x, \mathbf{e}_y\}$ in beamshell dynamics comes from a paper by Mansfield & Simmonds (1987) in which they study the motion of an initially flat sheet of paper in a gravity field. They start from (O.4) and (O.5) with inertia terms added and assume classical flexural motion. Generalizing their equations to beamshells of general shape under arbitrary distributed loads, we have, with $\overline{\mathbf{y}} = \overline{x}\mathbf{e}_x + \overline{y}\mathbf{e}_y$,

$$H' + p_x = m\ddot{\overline{x}} \ , \ V' + p_y = m\ddot{\overline{y}} \tag{P.52}$$

$$M' + V\cos\theta - H\sin\theta = 0 \ , \tag{P.53}$$

where $\theta = \alpha + \beta$, as in (P.40). These equations are supplemented by the moment-change-of-curvature relation

$$M = E_b h^3 \beta' = E_b h^3 (\theta - \alpha)' \tag{P.54}$$

plus the inextensionality conditions

$$\overline{x}' = \cos\theta \ , \ \overline{y}' = \sin\theta \ , \tag{P.55}$$

which follow from (L.1) and (L.2).

To solve (P.52), we introduce, following Mansfield & Simmonds, the *deformation potentials*

$$\Xi \equiv \int_L^\sigma \overline{x}(s,t)ds \ , \ \Omega \equiv \int_L^\sigma \overline{y}(s,t)ds \ . \tag{P.56}$$

Indefinite integration then reduces (P.52) to

$$H = m\ddot{\Xi} - \int p_x d\sigma \ , \ V = m\ddot{\Omega} - \int p_y d\sigma \, . \tag{P.57}$$

Substituting these relations along with (P.54) into (P.53), we have

$$[E_b h^3(\theta - \alpha)']' = (m\ddot{\Xi} - \int p_x d\sigma)\sin\theta - (m\ddot{\Omega} - \int p_y d\sigma)\cos\theta \, . \tag{P.58}$$

We obtain a complete system of field equations for the 3 unknowns β, Ξ, and Ω by adjoining to the above equation the inextensionality conditions that follow from (P.55), (P.56), and (P.57) as

$$\Xi'' = \cos\theta \ , \ \Omega'' = \sin\theta \, . \tag{P.59}$$

In static problems, these equations uncouple and (P.58) reduces to the well-known Euler-Bernoulli equation (subsection O.1.b). We refer the reader to the paper by Mansfield & Simmonds (1987) for a numerical solutions of these equations applied to a sheet of paper, issuing from a horizontal guide at constant velocity into a gravitational field.

Equations for the large, classical, flexural motion of straight, cantilevered beams of *variable* cross section have been presented by Pielorz & Nadolski (1986) and solved approximately by Galerkin's method for a linearly tapered beam, initially deformed by a constant end load and then released from rest.

Additional references to and examples from the elastodynamics of beamshells can be found in the references cited in this section.

Q. Variational Principles for Beamshells

1. Hamilton's Principle

As in Chapter III, the starting point for our discussion of variational principles for beamshells is the weak form of the equations of motion which, for the case at hand, is (F.1). Setting

$$\mathbf{V} = \delta\overline{\mathbf{y}} \ , \ \Omega = \delta\beta \ \text{and} \ \delta\mathbf{v} \equiv \delta\dot{\overline{\mathbf{y}}} \ , \ \delta\omega \equiv \delta\dot{\beta} \, , \tag{Q.1}$$

we have, after an integration by parts and some rearranging,

$$\begin{aligned} &\int_{t_1}^{t_2} \{(\mathbf{F}\cdot\delta\overline{\mathbf{y}} + M\delta\beta)_a^b + \int_a^b [\mathbf{p}\cdot\delta\overline{\mathbf{y}} + l\delta\beta - \mathbf{F}\cdot(\delta\overline{\mathbf{y}}' - \mathbf{m}\delta\beta) - M\delta\beta']d\sigma\}dt \\ &+ \int_a^b [-(m\mathbf{v}\cdot\delta\overline{\mathbf{y}} + I\omega\delta\beta)_{t_1}^{t_2} + \int_{t_1}^{t_2} (m\mathbf{v}\cdot\delta\mathbf{v} + I\omega\delta\omega)dt]d\sigma = 0 \, . \end{aligned} \tag{Q.2}$$

We recognize the last double integral to be the variation of the kinetic energy, $\mathcal{K}$, defined by (G.3). Replacing time differentiation by a variation, we obtain from (K.3),

$$\delta\overline{\mathbf{y}}' = \mathbf{T}\delta e + \mathbf{B}\delta g + \mathbf{m}\delta\beta \, . \tag{Q.3}$$

Likewise, (G.4), (K.4), (K.5), and (N.4) imply that

$$\mathbf{F}\cdot(\delta\overline{\mathbf{y}}' - \mathbf{m}\delta\beta) + M\delta\beta' = N\delta e + Q\delta g + M\delta k = \delta\Phi(e, g, k)\,, \tag{Q.4}$$

where the last equality holds if and only if the beamshell is elastic. With $\mathcal{E} = \int_a^b \Phi d\sigma$, (Q.2) reduces to

$$\int_{t_1}^{t_2} [(\mathbf{F}\cdot\delta\overline{\mathbf{y}} + M\delta\beta)_a^b + \int_a^b (\mathbf{p}\cdot\delta\overline{\mathbf{y}} + l\delta\beta)d\sigma]dt - \int_a^b (m\mathbf{v}\cdot\delta\overline{\mathbf{y}} + I\omega\delta\beta)_{t_1}^{t_2} d\sigma$$
$$= \int_{t_1}^{t_2} [\int_a^b (N\delta e + Q\delta g + M\delta k)d\sigma - \delta\mathcal{K}]dt = \int_{t_1}^{t_2} \delta(\mathcal{E} - \mathcal{K})dt\,. \tag{Q.5}$$

We now extend (Q.5) to the entire beamshell by setting $a = 0$ and $b = L$. Further, we assume that the edge kinematic vector, $\mathcal{y} = [\overline{\mathbf{y}}_0, \beta_0, \overline{\mathbf{y}}_L, \beta_L]^T$, of Section H satisfies m *holonomic* constraints of the form

$$\mathcal{R}(\mathcal{y}) = 0 \tag{Q.6}$$

and that the variation of $\mathcal{y}$, namely,

$$\delta\mathcal{y} \equiv [\delta\overline{\mathbf{y}}_0, \delta\beta_0, \delta\overline{\mathbf{y}}_L, \delta\beta_L]^T\,, \tag{Q.7}$$

satisfies the m homogeneous, linear conditions

$$\delta\mathcal{R}(\mathcal{y}) \equiv A(\mathcal{y})\delta\mathcal{y} = 0\,. \tag{Q.8}$$

Smoothness conditions on $0 < \sigma < L$ and $t_1 < t < t_2$ plus (Q.6) and (Q.8) define, respectively, the admissible kinematic fields and the admissible kinematic variations. In (Q.6)-(Q.8), we have the Principle of Virtual Work in a variational setting. (See also Section I).

We also assume that there exists a load potential, $\mathcal{P}$, as discussed in Section J. If $\mathcal{P}$ does not exist or if $\mathcal{y}$ does not satisfy constraints of the form $\mathcal{R}(\mathcal{y}) = 0$, then we cannot characterize an initial/boundary value problem as being equivalent to the vanishing of the variation of a functional. As $\mathcal{W} = -\dot{\mathcal{P}}$, it follows from (G.2) and from the same steps that led to (H.5) that

$$\sum_1^{m-6} \mathcal{N}^T \mathcal{M}_k \delta q_k + \int_0^L (\mathbf{p}\cdot\delta\overline{\mathbf{y}} + l\delta\beta)d\sigma = -\delta\mathcal{P}\,, \tag{Q.9}$$

so that (Q.5) takes the form

$$\delta\int_{t_1}^{t_2} (\mathcal{K} - \mathcal{E} - \mathcal{P})dt = \int_0^L (m\mathbf{v}\cdot\delta\overline{\mathbf{y}} + I\omega\delta\beta)_{t_1}^{t_2} d\sigma\,. \tag{Q.10}$$

Finally, if conditions are such that $\mathbf{v}\cdot\delta\overline{\mathbf{y}}$ and $\omega\delta\beta$ both vanish at t_1 and t_2, then the right side of (Q.10) vanishes and we have *Hamilton's Principle for a beamshell*:

$$\delta\int_{t_1}^{t_2} (\mathcal{K} - \mathcal{E} - \mathcal{P})dt = 0\,. \tag{Q.11}$$

This equation has the same formal form as (III.K.9), but the functionals $\mathcal{K}$, $\mathcal{E}$, and $\mathcal{P}$, defined by (G.3), (M.1), and (J.1), are those of a beamshell.

For classical boundary conditions, we may, in view of (H.13) and (Q.7), set

$$\sum_{1}^{6-m} \mathcal{N}^T \mathcal{M}_k \delta q_k = \hat{\mathcal{N}}^T \delta y = \hat{\mathcal{N}}^T \delta y_* \ , \qquad \text{(Q.12)}$$

where, in the notation of Section H, y_* is equal to y with the prescribed components of y set to 0, whereas $\hat{\mathcal{N}}^T$ is equal to the prescribed component of the edge-load vector, (H.4), with the unprescribed component set to 0. [See the example in (H.14)-(H.21) with δy replacing $\mathcal{V}$.]

2. Variational principles for elastostatics

We follow the steps described in Section K.2 in Chapter III on birods for obtaining variational principles for elastostatics, adapting the variables and field equations to beamshells.

(a) Principle of stationary potential energy

In static problems, $\mathcal{K}$, **v**, and ω are 0 and (Q.11) reduces to the *Principle of Stationary Total Potential Energy for beamshells*:

$$\delta(\mathcal{E} + \mathcal{P}) = 0 \ , \qquad \text{(Q.13)}$$

which states that among all admissible deformed positions, $\bar{\mathbf{y}}$ and rotations, β, only those that satisfy the equations of static equilibrium render the functional $\mathcal{E} + \mathcal{P}$ stationary. These "equilibrium equations" consist of the field equations of Section O *plus* any prescribed edge loads and are to be expressed in terms of $\bar{\mathbf{y}}$ and β through the constitutive relations of Section N and the kinematic relations of Section K. For example, if we work with the strains, e, g, and k, then the field equations are (O.10) and (O.11), into which the constitutive relations, (N.4), have been inserted, together with the kinematic relations

$$\begin{aligned} e &= \mathbf{T}\cdot\bar{\mathbf{y}}' - 1 = (\mathbf{t}\cos\beta + \mathbf{b}\sin\beta)\cdot\bar{\mathbf{y}}' - 1 \\ g &= \mathbf{B}\cdot\bar{\mathbf{y}}' = (-\mathbf{t}\sin\beta + \mathbf{b}\cos\beta)\cdot\bar{\mathbf{y}}' \ , \quad k = \beta' \ . \end{aligned} \qquad \text{(Q.14)}$$

For some applications it is more convenient, as discussed in Section L, to use the strains ε, γ, and κ rather than e, g, and k. In these cases it is $\bar{\mathbf{y}} = \mathbf{y} + u_\sigma \mathbf{t} + u_\zeta \mathbf{b}$ and γ that appear in the Principle of Stationary Total Potential Energy. The field equations of static equilibrium are then (P.12) and (P.13), without the inertia terms, the constitutive relations, (N.5), and the kinematic relations,

$$\varepsilon = \tfrac{1}{2}(\bar{\mathbf{y}}'\cdot\bar{\mathbf{y}}' - 1) \ , \quad \kappa = \mathbf{e}_z\cdot(\bar{\mathbf{y}}'\times\bar{\mathbf{y}}'') - (1 + 2\varepsilon)(\alpha + \gamma)' \ . \qquad \text{(Q.15)}$$

The real advantage of using ε, γ, and κ would come from using (Q.13) to obtain approximate solutions by the direct methods of the Calculus of Variations.

(b) Extended variational principles

Following the pattern set in Chapter III, we use as our starting point the Hu-Washizu Variational Principle, namely

$$\delta\Pi_1[\lambda_T, g, k, N, Q, M, \bar{\mathbf{y}}, \beta, \mathcal{L}] = 0 , \tag{Q.16}$$

where

$$\Pi_1 \equiv \mathcal{E} + \mathcal{P} - \int_0^L [\lambda_T N + gQ - \mathbf{F}\cdot\bar{\mathbf{y}}' + (k - \beta')M]d\sigma - \mathcal{L}^T \mathcal{R}(y) . \tag{Q.17}$$

Here, $\lambda_T \equiv 1 + e$ and $\mathcal{L}$ is a Lagrange multiplier vector with m columns. If classical kinematic boundary conditions are prescribed, we may set

$$\mathcal{L}^T \mathcal{R}(y) = [\tilde{\mathbf{F}}\cdot(\bar{\mathbf{y}} - \hat{\bar{\mathbf{y}}}) + \tilde{M}(\beta - \hat{\beta})]_{\partial U} , \tag{Q.18}$$

where $\tilde{\mathbf{F}}$ and $\tilde{M}$ are Lagrange multipliers and, as before, a hat denotes a prescribed quantity; ∂U is defined and discussed in subsection H.2. The reader should compare (Q.16), after insertion of (Q.18), with the analogous principle for birods, (III.K.11). As the notation in (Q.16) indicates, Π_1 is a functional of the strains, e, g,and k, the stress resultants and couple, N, Q, and M, the kinematic variables, $\bar{\mathbf{y}}$ and β, and the constant Lagrange multiplier vector, $\mathcal{L}$. The field variables are subject to *no* constraints, save for the necessary smoothness conditions for the integrals in (Q.17) to exist.

The Hu-Washizu Principle states that those and only those values of the variables that make Π_1 stationary also satisfy the field equations of beamshell theory. To prove this, we carry out the variation of Π_1, use the identity

$$\delta(\mathbf{F}\cdot\bar{\mathbf{y}}') = \mathbf{F}\cdot\delta\bar{\mathbf{y}}' - \mathbf{m}\cdot\mathbf{F}\delta\beta + \bar{\mathbf{y}}'\cdot(\mathbf{T}\delta N + \mathbf{B}\delta Q) , \tag{Q.19}$$

and integrate by parts, to obtain,

$$\begin{aligned} \delta\Pi_1 = \int_0^L [&(\Phi,_e - N)\delta e + (\Phi,_g - Q)\delta g + (\Phi,_k - M)\delta k \\ &- (\lambda_T - \mathbf{T}\cdot\bar{\mathbf{y}}')\delta N - (g - \mathbf{B}\cdot\bar{\mathbf{y}}')\delta Q - (k - \beta')\delta M \\ &- (\mathbf{F}' + \mathbf{p})\delta\bar{\mathbf{y}} - (M' + \mathbf{m}\cdot\mathbf{F} + l)\delta\beta]d\sigma \\ &+ [\mathcal{N}^T - \hat{\mathcal{N}}^T - \mathcal{L}^T A(y)]\delta y - \delta\mathcal{L}^T \mathcal{R}(y) = 0 , \end{aligned} \tag{Q.20}$$

where, as in Section H, $\mathcal{N} = [-\mathbf{F}_0, -M_0, \mathbf{F}_L, M_L]^T$. For classical boundary conditions,

$$\begin{aligned} &[\mathcal{N}^T - \hat{\mathcal{N}}^T - \mathcal{L}^T A(y)]\delta y - \delta\mathcal{L}^T \mathcal{R}(y) \\ &\quad = [(\mathbf{F} - \tilde{\mathbf{F}})\cdot\delta\bar{\mathbf{y}} + (M - \tilde{M})\delta\beta - (\bar{\mathbf{y}} - \hat{\bar{\mathbf{y}}})\cdot\delta\tilde{\mathbf{F}} - (\beta - \hat{\beta})\delta\tilde{M}]_{\partial U} \\ &\quad\quad + [(\mathbf{F} - \hat{\mathbf{F}})\cdot\delta\bar{\mathbf{y}} + (M - \hat{M})\delta\beta]_{\partial N} , \end{aligned} \tag{Q.21}$$

where ∂N is defined and discussed in Subsection H.2. As the variations in (Q.20) are all independent, the vanishing of all the expressions in parentheses in this equation is a necessary condition for Π_1 to be stationary. These expressions include constitutive relations, the strain-kinematic (i.e., strain-displacement and strain-rotation) relations, the equilibrium equations, the boundary conditions, and the equation $\mathcal{L}^T A = \mathcal{N}^T - \hat{\mathcal{N}}^T$. Because the calculations leading to (Q.20) can be reversed, the equivalence of the satisfaction of the above-mentioned conditions with a stationary point of Π_1 is assured.

The displacement $\mathbf{u}$ can be used instead of $\bar{\mathbf{y}}$ in Π_1 by setting $\bar{\mathbf{y}} = \mathbf{y} + \mathbf{u}, \bar{\mathbf{y}}' = \mathbf{t} + \mathbf{u}'$, and $\delta\bar{\mathbf{y}} = \delta\mathbf{u}$, but no particular advantage is gained at this stage by such a change.

We now derive variational principles in terms of the mixed-energy density, Ψ, the definition, uses, and merits of which are given and discussed in Section N. Setting $\Omega(N, Q, k) = \Phi - N\lambda_T - Qg = \Psi(N, Q, k) - N$ in (Q.20), we eliminate the stretch, λ_T, and the strain, g, as independent variables and express the result in the form

$$\delta\Pi_2[k, N, M, Q, \bar{\mathbf{y}}, \beta, \mathcal{L}] = 0\,, \tag{Q.22}$$

where

$$\Pi_2 = \mathcal{P} + \int_0^L [\Omega + \mathbf{F}\cdot\bar{\mathbf{y}}' + (\beta' - k)M]\,d\sigma - \mathcal{L}^T\mathcal{R}(y)\,. \tag{Q.23}$$

This principle is very similar to the Hu-Washizu Principle, (Q.16), save that $\Omega_{,N}$ and $\Omega_{,Q}$ replace $-\lambda_T$ and $-g$ in the stationary conditions. As in Π_1, M and k are independent fields in Π_2.

Carrying the development a step further, we take $k = \beta'$. Then, the last term in the integrand in (Q.23) vanishes and $\Omega = \Omega(N, Q, \beta')$, leaving us with the new principle

$$\delta\Pi_3[N, Q, \bar{\mathbf{y}}, \beta, \mathcal{L}] = 0\,, \tag{Q.24}$$

where

$$\Pi_3 \equiv \int_0^L (\Omega + \mathbf{F}\cdot\bar{\mathbf{y}}')\,d\sigma - \mathcal{L}^T\mathcal{R}(y)\,. \tag{Q.25}$$

Except for obvious smoothness requirements, the arguments of Π_3 are subject to no constraints.

To find the stationary conditions for Π_3, we compute its first variation, using the relation

$$\delta\Omega = \Omega_{,N}\delta N + \Omega_{,Q}\delta Q + \Omega_{,\beta'}\delta\beta' \tag{Q.26}$$

and *defining* the moment by

$$M \equiv \Omega_{,\beta'}\,. \tag{Q.27}$$

After some manipulations, we obtain

$$\delta\Pi_3 = \int_0^L [(\Omega,_N + \mathbf{T}\cdot\bar{\mathbf{y}}')\delta N + (\Omega,_Q + \mathbf{B}\cdot\bar{\mathbf{y}}')\delta Q$$
$$- (\mathbf{F}' + \mathbf{p})\cdot\delta\bar{\mathbf{y}} - (M' + \mathbf{m}\cdot\mathbf{F} + l)\delta\beta]d\sigma \quad \text{(Q.28)}$$
$$+ [\mathcal{N}^T - \hat{\mathcal{N}}^T - \mathcal{L}^T A(y)]\delta y - \delta\mathcal{L}^T \mathcal{R}(y) = 0 .$$

Hence, satisfaction of the strain-kinematic relations, the equilibrium equations, the boundary conditions, and $\mathcal{L}^T A = \mathcal{N}^T - \hat{\mathcal{N}}^T$ guarantee that Π_3 is stationary.

Our aim in the next variational principle is to eliminate the deformed position, $\bar{\mathbf{y}}$, as an independent field variable. When this is possible, we obtain a "complementary-like" principle, except that the rotation, β, remains as an independent variable. [Variational principles involving stress and rotation in nonlinear 3-dimensional elasticity and in nonlinear shell theory were first formulated, respectively, by Fraeys de Veubeke (1972) and Simmonds & Danielson (1972). For references to more recent work see Lee & Shield (1980), Fukuchi & Atluri (1981), and Atluri (1984)].

We now assume that the distributed load, $\mathbf{p}$, depends on $\bar{\mathbf{y}}$ only and can be inverted to yield $\bar{\mathbf{y}} = \bar{\mathbf{y}}(\mathbf{p})$. We also assume (for simplicity) that the boundary conditions are classical and that $\hat{\mathbf{F}}(\bar{\mathbf{y}})$ on ∂N can be solved for $\bar{\mathbf{y}}$ in terms of $\hat{\mathbf{F}}$. Under these conditions, we can define a *mixed load potential*

$$\mathcal{P}[\mathbf{p}, \hat{\mathbf{F}}, \beta] \equiv \mathcal{P} + \int_0^L \mathbf{p}\cdot\bar{\mathbf{y}}(\mathbf{p})d\sigma + \hat{\mathbf{F}}\cdot\bar{\mathbf{y}}(\hat{\mathbf{F}})|_{\partial N} \quad \text{(Q.29)}$$

such that

$$\delta\mathcal{P} = \int_0^L (\bar{\mathbf{y}}\cdot\delta\mathbf{p} - l\delta\beta)d\sigma + \bar{\mathbf{y}}\cdot\delta\hat{\mathbf{F}}|_{\partial N} - \hat{M}\delta\beta|_{\partial N} . \quad \text{(Q.30)[16]}$$

Starting with (Q.25), we integrate the term $\mathbf{F}\cdot\bar{\mathbf{y}}'$ by parts, identify $\mathbf{F}$ on ∂U with $\tilde{\mathbf{F}}$, and restrict $\mathbf{F}$ to satisfy

$$\mathbf{F}' + \mathbf{p} = \mathbf{0} \ , \ 0 < \sigma < L \ , \ \mathbf{F} = \hat{\mathbf{F}} \text{ on } \partial N . \quad \text{(Q.31)}$$

Rearranging the terms in the resulting expression, we obtain, finally,

$$\delta\Pi_4[N, Q, \beta, \tilde{M}] = 0 , \quad \text{(Q.32)}$$

where

$$\Pi_4 \equiv \mathcal{P} + \int_0^L \Omega d\sigma + [\,\mathbf{F}\cdot\hat{\bar{\mathbf{y}}} - \tilde{M}(\beta - \hat{\beta})]_{\partial U} . \quad \text{(Q.33)}$$

In this principle, $\mathbf{F}$ is constrained by (Q.31) but β is free and we have retained the Lagrange multiplier(s), $\tilde{M}$.

---

[16]In the important special case of "dead loads" in which $\mathbf{p}(\sigma)$ and $\hat{\mathbf{F}}$ are independent of $\bar{\mathbf{y}}$ and β, $\delta\mathbf{p}$ and $\delta\hat{\mathbf{F}}$ vanish and all the $\mathbf{p}$ and $\hat{\mathbf{F}}$ terms *drop out* of $\mathcal{P}$.

In the last principle to be derived, we restrict the field β to satisfy any boundary conditions imposed upon it. Thus, the last term in (Q.33) vanishes, leaving

$$\delta\Pi_5[N, Q, \beta] = 0 \,, \tag{Q.34}$$

where

$$\Pi_5 \equiv \mathcal{P} + \int_0^L \Omega d\sigma + \mathbf{F}\cdot\hat{\bar{\mathbf{y}}}|_{\partial U} \,. \tag{Q.35}$$

The fields, N and Q, in this principle must satisfy the equations of force equilibrium on $0 < \sigma < L$ and ∂N, while the rotation field, β, need only assume its prescribed boundary values (if any). The conditions for Π_5 to be stationary are the force-kinematic (or in this case, the force-displacement) relations, the equation of moment equilibrium, the moment boundary conditions on ∂N, and the kinematic conditions for $\bar{\mathbf{y}}$ on ∂U. Note that constitutive relations for e, g, and M are not needed to compute $\delta\Pi_5$, but serve as auxiliary relations to be used only if e, g, or M is needed.

As an application, we specialize the principle $\delta\Pi_5 = 0$ to the edge-loaded beamshell in Fig. 4.11, fixed at $\sigma = L$ to a rigid wall such that $\beta_L = 0$ and $\bar{\mathbf{y}}_L = \mathbf{y}_L$ and subject to a constant force, P, at its free end, $\sigma = 0$. For the remaining boundary condition at the free end we shall consider 2 possibilities: (a) $M_0 = \hat{M}$ or (b) $\beta_0 = \hat{\beta}$. Finally, we obtain the Euler-Lagrange equations for a circular beamshell.

As overall force equilibrium demands, it is obvious that, without loss of generality, we may orient the frame $\{\mathbf{e}_x, \mathbf{e}_y\}$ so that $\mathbf{F}(\sigma) = -P\mathbf{e}_y$, a constant vector. Thus, as $\mathbf{p} = \mathbf{0}$, the complementary potential, $\mathcal{P}$, defined by (Q.29), reduces to $\mathcal{P} = \hat{M}\beta_0$ in case (a) and to $\mathcal{P} = 0$ in case (b). Furthermore, as $\mathbf{F}\cdot\hat{\bar{\mathbf{y}}}|_{\partial U}$ is now a known constant, we may delete this term from Π_5 in (Q.35).

From Fig. 4.11, it is clear that the components of $\mathbf{F}$ are given by

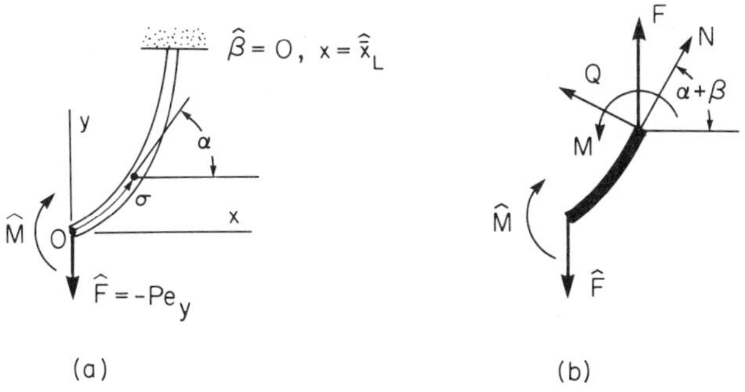

Fig. 4.11. Edge-loaded beamshell: (a) undeformed configuration; (b) deformed segment.

$$N = P\sin(\alpha+\beta)\ ,\ \ Q = P\cos(\alpha+\beta)\,. \tag{Q.36}$$

Putting (Q.36) and the previously discussed expressions for $\mathcal{P}$ into (Q.35), we have

$$\Pi_5\,[\beta,\beta'] = \int_0^L \Omega d\sigma + \hat{M}\beta_0\ , \tag{Q.37}$$

the last term being 0 in case (b). The variational principle associated with Π_5 states that of all sufficiently smooth fields, β, such that $\beta(L)=0$ [and, in case (b), $\beta(0)=\hat{\beta}_0$], only solutions[17] of the moment equilibrium equation (including the moment boundary conditions, if any) make Π_5 stationary. Direct methods of the Calculus of Variations can be applied at this point to obtain approximate solutions.

We now obtain the Euler-Lagrange equations implied by $\delta\Pi_5=0$ for the important special case of circular beamshells (or rings). (See Section O for a parallel, non-variational treatment). First, we rewrite our functional in the form

$$\Pi_5 = \int_0^L (\Omega - \hat{M}\beta')d\sigma \equiv \int_0^L \Lambda d\sigma\,. \tag{Q.38}$$

Because α' is a constant in circular beamshells, we introduce $\theta \equiv \alpha+\beta$ as a new field variable so that the integrand in (Q.38) is no longer an explicit function of σ. Thus, by a well-known result in the Calculus of Variations, the Euler-Lagrange equation implied by $\delta\Pi_5=0$ is

$$\Lambda - \theta'\,\frac{\partial\Lambda}{\partial\theta'} = C\ ,\ \text{a constant}\,. \tag{Q.39}$$

Substitution of the equations for Λ and θ yields

$$\Omega - \hat{M}\beta' - (\alpha+\beta)'(M-\hat{M}) = C\,. \tag{Q.40}$$

Since $\hat{M}\alpha'$ is a constant [equal to 0 in case (b)], we absorb it into C to get

$$(\alpha+\beta)'M - \Omega = B\ ,\ \text{another constant}\,. \tag{Q.41}$$

Comparing this expression with (O.37), we see that (Q.41) is precisely the energy integral of beamshell elastostatics. We may, therefore, interpret the energy integral of the edge-loaded, circular beamshell as the Euler-Lagrange equation of its force-rotation variational formulation.

Our approach in this section (and in others) has been to use the *deformed position*, $\bar{\mathbf{y}}$, rather than the *displacement*, $\mathbf{u}$, as a kinematic argument in the various variational principles. It is instructive at this point to recast Π_5 with the latter as the basic kinematic variable. In doing so, we note that $\bar{\mathbf{y}}$ appears explicitly in the boundary term $\mathbf{F}\cdot\hat{\bar{\mathbf{y}}}|_{\partial U}$ and in the mixed load potential, $\mathcal{P}$, through the

---

[17]Recall that we have said nothing about uniqueness.

terms[18] $\int_0^L \mathbf{p}\cdot\bar{\mathbf{y}}(\mathbf{p})d\sigma + \hat{\mathbf{F}}\cdot\bar{\mathbf{y}}(\hat{\mathbf{F}})|_{\partial N}$. We rewrite these terms as follows:

$$\begin{aligned} &\mathbf{F}\cdot\hat{\bar{\mathbf{y}}}|_{\partial U} + \hat{\mathbf{F}}\cdot\bar{\mathbf{y}}|_{\partial N} + \int_0^L \mathbf{p}\cdot\mathbf{y}d\sigma \\ &\quad = \mathbf{F}\cdot\hat{\mathbf{u}}|_{\partial U} + \hat{\mathbf{F}}\cdot\mathbf{u}(\hat{\mathbf{F}})|_{\partial N} + \int_0^L [\mathbf{p}\cdot\mathbf{u}(\mathbf{p}) + (\mathbf{F}\cdot\mathbf{y})' + \mathbf{p}\cdot\mathbf{y}]d\sigma \,, \end{aligned} \tag{Q.42}$$

where $\mathbf{y}$ is the *initial* position, such that $\bar{\mathbf{y}} = \mathbf{y} + \mathbf{u}$ and $\mathbf{y}' = \mathbf{t}$. Because $\mathbf{F}$ and $\mathbf{p}$ must satisfy the equilibrium condition, $\mathbf{F}' + \mathbf{p} = \mathbf{0}$, the last 2 terms in the integral on the right side reduce to $\mathbf{F}\cdot\mathbf{t}$, and Π_5 takes the form

$$\Pi_{5a} = \mathcal{P} + \int_0^L (\Omega + \mathbf{F}\cdot\mathbf{t})d\sigma + \mathbf{F}\cdot\hat{\mathbf{u}}|_{\partial U} \,, \tag{Q.43}$$

where

$$\mathcal{P} = \mathcal{P} + \int_0^L \mathbf{p}\cdot\mathbf{u}(\mathbf{p})d\sigma + \hat{\mathbf{F}}\cdot\mathbf{u}(\hat{\mathbf{F}})|_{\partial N} \tag{Q.44}$$

and

$$\mathbf{F}\cdot\mathbf{t} = N\cos\beta - Q\sin\beta \,. \tag{Q.45}$$

We observe that the change from position to displacement as an argument adds mixed, *force-rotation* terms to Ω. This, in itself, does not lead to any major difficulties since both N and Q as well as β are used as fields in Π_5. However, we do lose the "pure" form of Ω and, were we to construct a *full* complementary principle for nonlinear beamshells, then the appearance of mixed terms could be problematic. In this case, one possible approach is to switch to the total Lagrangian components $\mathbf{F}\cdot\mathbf{t}$ and $\mathbf{F}\cdot\mathbf{b}$ as basic stress variables. Another approach is to drop $\mathbf{u}$ in favor of $\bar{\mathbf{y}}$, as is done elsewhere in this book.

This concludes our discussion of some of the variational principles for beamshells. The reader will have realized by now that other principles may be constructed, including principles using the full *complementary stress-energy density*, $\Psi = \Phi - Ne - Qg - Mk$, such as the Hellinger-Reissner Principle. For a general exposition of variational methods in continuum mechanics see Oden & Reddy (1983); for survey papers dealing with nonlinear elasticity, including some discussions of finite element methods, see Nemat-Nasser (1974), Atluri (1984), and Reissner (1984).

R. The Mechanical Theory of Stability

Maintenance of stability is often vital in the design of thin-walled structures. The study of stability by experimental, analytical, and numerical methods has been a major topic in solid mechanics since the work of Euler (1744) on the buckling of columns. Today, there exist thousands of papers, review articles,

---

[18] $\bar{\mathbf{y}}$ also appears in $\mathcal{P}$ through the term $\int_0^L \mathbf{p}\cdot\delta\bar{\mathbf{y}}d\sigma$ which is a part of $\delta\mathcal{P}$. See (Q.5). However, $\delta\bar{\mathbf{y}} = \delta\mathbf{u}$.

and books on the subject and the flood continues. Our discussion in this section will therefore be necessarily limited and will concentrate on the *potential energy approach* and the *equations of neutral equilibrium* for elastic, shearable beamshells under conservative loads. Readers who are interested in a more exhaustive treatment of the stability problem are referred to Timoshenko & Gere (1961) for many classical results and to the 2 English translations (1967a, 1970) of Koiter's thesis (1945) for the pioneering work on the modern approach. Stability theory for discrete structural systems is presented by Thompson & Hunt (1973). The dynamic approach to structural stability is developed by Dym (1974) and Leipholz (1980). Nonconservative problems are studied by Bolotin (1963) and selected examples are presented by Leipholz (1980). Brush & Almroth (1975) emphasize engineering approaches to buckling and discuss numerical methods.

Of the many review articles on shell buckling, we mention those of Koiter (1968, 1976, 1981), Stein (1968), Budiansky (1974), Budiansky & Hutchinson (1979), Hutchinson & Koiter (1970), Tvergaard (1976), and Babcock (1983). Reviews of inelastic buckling analyses have been given by Hutchinson (1974) and Bushnell (1985).

The column, the ring, and the arch are the three types of beamshells most often studied and used in practice. The buckling and postbuckling of columns is extensively documented, so we shall not dwell on it. Chapters on the stability of rings and arches may be found, among other places, in Biezeno & Grammel (1939), Timoshenko & Gere (1961), Dym (1974), Brush & Almroth (1975), and Simitses (1976). Later in this section, we shall cite some of the many papers that have appeared in the literature.

As we pointed out in Section O, a problem in elastostatics may not have a unique solution unless a load history has been given. Thus, we distinguish between 2 types of static stability problems. In the first and more realistic one, we are given a load history characterized by a single, increasing dimensionless parameter, q, such that, at $q=0$, the shell is stress free and at rest.[19] As q is increased continuously from 0, we are asked to find the first (i.e., lowest) value, q_{cr}, where the beamshell becomes unstable. The solution of the beamshell equations for $0<q<q_{cr}$ is called *the fundamental equilibrium solution.*

In the 2nd type of static stability problem, we are given, not a load history, but an equilibrium configuration and asked to determine its stability. Here, all that is needed for an "infinitesimal" stability analysis is a knowledge of how the loads depend on the configuration of the beamshell in an arbitrarily small neigh-

[19]A shell can be made to pass from an undeformed configuration at rest to a deformed configuration at rest only by the application of loads that change with time. If we ignore inertia effects while the loads are changing, then we have a so-called *quasi-static process*. Thus, we may think of q as a dimensionless time in a loading that takes place so slowly that accelerations are negligible.

borhood of the equilibrium configuration.[20]

In studying the first type of stability problem, it is convenient to describe the response of the beamshell to increasing q in terms of a suitable *displacement norm*, δ. This could be a component of displacement evaluated at a specific point, say the vertical deflection under a point load, or some integrated measure of displacement, say a weighted displacement defined by a work expression. The graph of q versus δ is called *the fundamental equilibrium path*, but we should keep in mind that, in general, q is merely a convenient measure of the load level and δ is merely some measure of the distortion in an equilibrium configuration.

Physically, *loss of stability* means that the structure is unable to sustain a small increment in load by a small increment in shape: either some motion takes place and, by virtue of damping, the structure eventually assumes a new, relatively distant equilibrium configuration, or else there is violent motion leading to catastrophic failure. As linear elasticity predicts a unique, straight equilibrium path, nonlinearity must be introduced if loss of stability is to be predicted analytically (although the subsequent equations may sometimes be closely approximated by linear ones).

There is a third, rather common possibility: as the loading parameter, q, increases, δ becomes exceedingly large (though finite) and unacceptable from an engineering point of view. That is, the beamshell is *unstable for all practical purposes*[21], although it continues to be stable within the narrow, analytical definition. This special situation, which prevails in columns, has caused as many

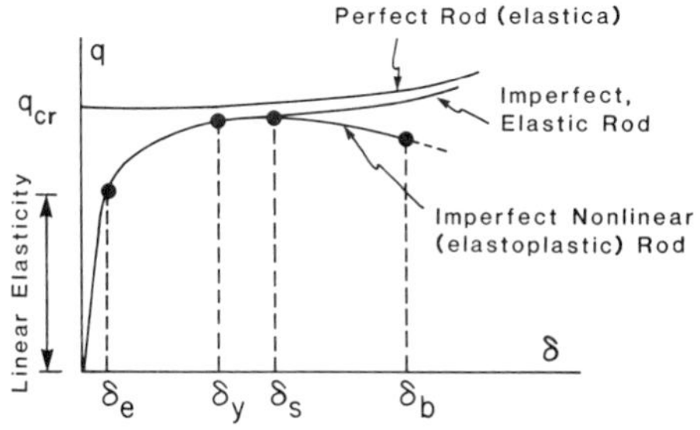

Fig. 4.12. The yielding, snapping, and breaking of an "unstable" rod.

[20]The 2 types of problems need not even yield the same result. A good example is the elastic perfectly-plastic rod: the first type yields the tangent-modulus theory while the second yields the reduced modulus theory. See Timoshenko & Gere (1961, Chapter 3).

[21]If the *material* of the beamshell is nonlinearly elastic or elasto-plastic, then the linear elastic material limit is exceeded if $\delta = \delta_y$ is sufficiently large. (See Fig.4.12.) Subsequently, the beamshell may snap nonlinearly-elastically or elasto-plastically ($\delta = \delta_s$) and then break, thus exceeding its *strength criterion* ($\delta = \delta_b$).

misconceptions of the basic concepts of stability as all other cases combined!

There are 2 types of loss of stability. The first is illustrated in Fig. 4.13a, where, at a so-called *limit point* on the the equilibrium path, q attains a local maximum, q_s, and neighboring equilibrium states with higher loads do not exist. The type of instability associated with such a limit point is called *snapping* and is observed in hinged, very shallow arches under uniform pressure. A horizontal tangent at a point on the q-δ curve implies strong, nonlinear interactions there and the full nonlinear equations [e.g., (K.1), (K.6), (N.4), (O.10), and (O.11)] must be solved to find q_s; except for special cases, such as shallow arches, numerical methods are required. Even then, difficulties may arise as q approaches the unknown value, q_s, and special techniques may have to be devised, as discussed by Weinitschke (1985).

The second type of instability occurs when another, *secondary equilibrium path* intersects the fundamental (or *prebuckling*) equilibrium path at a point, B, known as a *critical* or *bifurcation point.* For example, a straight column under a compressive axial load bifurcates by displacing transversely and a symmetric, non-shallow circular arch under fluid pressure bifurcates into an unsymmetric shape. Though there may be several such secondary paths, we shall assume in what follows that there is but one, as indicated in Figs. 4.13b,c,d. The load on a structure at a bifurcation point is called a *critical load*

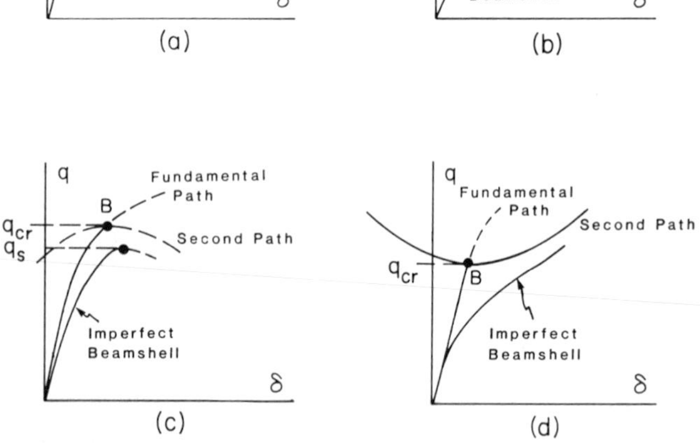

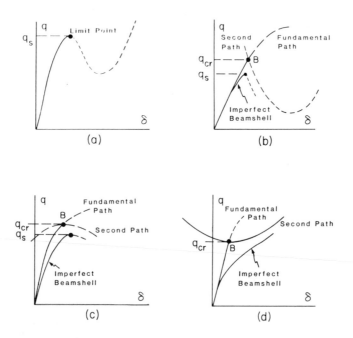

Fig. 4.13. Examples of equilibrium paths.

and the associated value of q is denoted by q_{cr}. The shape of the structure on the secondary equilibrium path at the bifurcation point is called the *buckling mode* and the study of how the buckling mode evolves along this path is called *postbuckling analysis*. The fate of the beamshell can usually be inferred from the shape of the secondary equilibrium path in a neighborhood of a bifurcation point. In Fig. 4.13d, q increases on the secondary equilibrium path no matter how we move away from B, which implies stable postbuckling behavior. In Figs. 4.13b,c, q always decreases on the secondary equilibrium path as we move away from B, which implies an unstable postbuckling behavior leading to structural collapse similar to snapping.

Bifurcation occurs only in perfect shapes: columns that are absolutely straight, rings that are absolutely circular, or beamshells that are absolutely symmetric. In reality, every structure deviates from its ideal form and will either snap at a value of $q < q_{cr}$ (Fig. 4.13b,c) or else exhibit stable, nonlinear behavior when q is close to q_{cr} (Fig. 4.13d). Still, bifurcation analysis is valuable because, with its aid, imperfect beamshells can be treated as perturbations of perfect ones.

In many important applications, bifurcation occurs at a sufficiently small load to permit a *linearized* prebuckling analysis, whereby the complications of a complete, nonlinear limit point calculation can be avoided. This is illustrated in Fig. 4.13b where B lies on a portion of the fundamental path that is nearly straight. Indeed, most of the studies in the literature on the buckling and postbuckling of planar beams are based on the neglect of prebuckling deformations, stress couples and shear stress resultants, i.e. only the effects of the axial force, N, are considered. Often this approximation is justified, but would not be for the situation illustrated in Fig. 4.13c. Also, it is usually not justified in shallow structures, in imperfection sensitive structures, in problems where there are concentrated loads, or in problems where there is secondary buckling arising from a bifurcation point on a secondary equilibrium path.

Another common approximation in buckling analyses, especially in columns, rings, and steep arches, is *inextensionality* [i.e., $e^{(1)} = 0$ in (R.11)]. A qualitative estimate of the partition of elastic energy between stretching and bending in non-shallow beamshells during buckling can be expected to support this simplification.

Structural *instability* is inherently a dynamic phenomenon and, strictly speaking, should be analyized using equations of motion. However, for elastic bodies under static, conservative loads, the *energy criterion*, first proposed by Lagrange (1788) for discrete systems, is almost universally accepted as a method for determining points of instability and the behavior of the structure in an immediate neighborhood of these points. For further discussion, see Koiter (1968, 1981), who discusses the thermodynamic foundations of the energy criterion which he defines in terms of an L_2 displacement norm, $\| \underline{u} \|$. Here, $\underline{u}$ represents a generalized displacement away from the state whose stability is to be studied. [A summary of Koiter's works, which rests on earlier ideas of Duhem (1911) and Ericksen (1966, 1967), may be found in Section U.] With beamshells, we shall use the notation $\mathbf{y} \equiv \mathbf{y}, \beta$ and $\underline{u} \equiv \mathbf{u}^{(1)}, \beta^{(1)}$. A lucid review of the problematics of the energy criterion has been presented by Ziegler (1967).

A heuristic, but physically meaningful statement of the energy criterion, is the following. Consider a point, $\mathbf{y}^{(0)}, q_0$, on an equilibrium path and an increment in potential energy, $\Delta(\mathcal{E}+\mathcal{P})$, defined for all generalized, incremental displacement fields $\underline{u}$ that are kinematically admissible. If $\Delta(\mathcal{E}+\mathcal{P})$ is positive definite for all $\|\underline{u}\|$ sufficiently small, that is, if $\Delta(\mathcal{E}+\mathcal{P})$ has a strict relative minimum at q_0, then the equilibrium state is stable. In simple words: we have to invest energy to displace a body from a stable state. Conversely, if $\Delta(\mathcal{E}+\mathcal{P})<0$ for *some* admissible $\underline{u}$, then the equilibrium state is unstable, i.e., disturbing an unstable equilibrium state releases stored energy that will be turned into motion. The equilibrium state is *critical* if there exists an admissible increment, $\underline{u}_{\rm cr}$, called a *buckling mode*, such that $\Delta(\mathcal{E}+\mathcal{P})>0$, with $\Delta(\mathcal{E}+\mathcal{P})=0$ at, at least, one point. The above definition implies that $\Delta(\mathcal{E}+\mathcal{P})[\underline{u}, q]$ is a minimum at $(\underline{u}_{\rm cr}, q_{\rm cr})$. Hence, the *Variational Criterion of Trefftz:*

$$\delta\Delta(\mathcal{E}+\mathcal{P})[\underline{u}_{\rm cr}, q_{\rm cr}]=0\,, \tag{R.1}$$

which provides, as stationary conditions, the field equations and boundary conditions for the determination of the critical load and buckling mode(s). Recalling the theorem of stationary total potential energy of Section Q, we see these stationary conditions will be equilibrium equations containing $q_{\rm cr}$ as a parameter. They, therefore, are called the *equations of neutral (or adjacent) equilibrium.* In view of the procedures involved in carrying out the variation in (R.1), these neutral equilibrium equations can also be obtained by perturbing the full nonlinear equilibrium equations; in fact they are usually obtained in this way.[22] It should be noted that an *actual* adjacent equilibrium state at the same q may not exist; in our analysis, $\underline{u}_{\rm cr}$ is merely an admissible field that describes a *possible configuration* of a secondary equilibrium path at $q_{\rm cr}$. The way q changes along this secondary path—see (R.4)—determines the *stability at the critical equilibrium state.*

If $\Delta(\mathcal{E}+\mathcal{P})$ is linear in q, say $\Delta(\mathcal{E}+\mathcal{P})=f[\underline{u}]-qg[\underline{u}]$, where f and g are functionals of the admissible class of deformations $\mathcal{A}$, then $q_{\rm cr}$ may be determined by minimizing the *Rayleigh quotient of f and g*, i.e.,

$$q_{\rm cr}=\min_{\underline{u}\in\mathcal{A}}\frac{f[\underline{u}]}{g[\underline{u}]}\,. \tag{R.2}$$

This method is convenient for approximate buckling analyses, but always yields upper bounds.

Suppose that $\Delta(\mathcal{E}+\mathcal{P})$ has a formal Taylor series in $\underline{u}$ of the form

$$\Delta(\mathcal{E}+\mathcal{P})=\Delta(\mathcal{E}+\mathcal{P})_2+\Delta(\mathcal{E}+\mathcal{P})_3+\Delta(\mathcal{E}+\mathcal{P})_4+\cdots\,, \tag{R.3}$$

---

[22]Recall that when we speak of equilibrium equations, we mean the field equations *plus* the load boundary conditions.

where a subscript, n, indicates a functional that depends on the nth power of $\underline{u}$ only. There is no linear term because $(\mathbf{y}^{(0)}, q_0)$ is an equilibrium point, i.e., by (Q.13) and (R.1), $\delta\Delta(\mathcal{E}+\mathcal{P})=\Delta(\mathcal{E}+\mathcal{P})_1=0$. As $\|\underline{u}\|$ can be made arbitrarily small, a necessary condition that $\Delta(\mathcal{E}+\mathcal{P})$ be positive definite is that $\Delta(\mathcal{E}+\mathcal{P})_2$ be positive definite. The converse seems to hold in most practical situations, but can be shown *not* to be universally true [Koiter (1976, Sect. 5, 1981, Sect 3)]. Approximating the Trefftz Criterion by $\delta\Delta(\mathcal{E}+\mathcal{P})_2=0$ leads, of course, to *linear* buckling equations, which greatly simplifies computing q_{cr} and $\underline{u}_{cr}$.

We shall not go into the details of postbuckling analysis and the computation of the snapping loads, q_s, of imperfect structures, but simply mention that the procedures are based on the pioneering work of Koiter (1945) which has been elaborated upon in several of the books and reviews cited at the beginning of this section. Essentially, one either incorporates the influence of $\Delta(\mathcal{E}+\mathcal{P})_3$ and $\Delta(\mathcal{E}+\mathcal{P})_4$ on the change in $\Delta(\mathcal{E}+\mathcal{P})$ near the point $(\underline{u}_{cr}, q_{cr})$ or else, equivalently, one does a perturbation analysis of the variational equilibrium equations. See, for example, Budiansky (1974) or Dym (1974). The ultimate goal of either approach is to compute the coefficients, a and b, in the expansion

$$q = q_{cr} + a\varepsilon + b\varepsilon^2 + \cdots , \tag{R.4}$$

which determines the shape of the q-δ curve near the critical point. Here, ε is a small parameter, typically, the dimensionless amplitude of a buckling mode. With these 2 coefficients in hand, approximate formulas can be given for the snapping load, q_s, of an imperfect beamshell with a *known* imperfection amplitude, $\bar{\varepsilon}$, in terms of q_{cr}, $\bar{\varepsilon}$, a, and b. For details, see Budiansky (1974) or Dym (1974, Sect. 6.5). A difficulty is the need for a priori knowledge of $\bar{\varepsilon}$ which, for design purposes, may have to be found statistically (Elisakoff, 1983).

1. Buckling equations

To get buckling equations for a beamshell, we calculate the change in potential energy due to an admissible $\underline{u}$. The final state of the beamshell is characterized by the 3-vector $\underline{\mathbf{y}}=\underline{\mathbf{y}}^{(0)}+\underline{u}$ or, more explicitly, by

$$\bar{\mathbf{y}} \equiv \bar{\mathbf{y}}^{(0)} + \mathbf{u}\,, \quad \beta = \beta^{(0)} + \beta^{(1)}\,. \tag{R.5}$$

All other quantities associated with the fundamental state and increments from it are denotes by the superscripts (0) and (1), respectively, while the final (total) state has no associated superscripts. In particular,

$$\mathbf{T}^{(1)} = \mathbf{T}^{(0)}(\cos\beta^{(1)} - 1) + \mathbf{B}^{(0)}\sin\beta^{(1)} \tag{R.6}$$

$$\mathbf{B}^{(1)} = \mathbf{B}^{(0)}(\cos\beta^{(1)} - 1) - \mathbf{T}^{(0)}\sin\beta^{(1)} \tag{R.7}$$

$$\begin{aligned} e^{(1)} &= \mathbf{T}\cdot\bar{\mathbf{y}}' - \mathbf{T}^{(0)}\cdot\bar{\mathbf{y}}^{(0)\prime} \\ &= \mathbf{T}^{(0)}\cdot\mathbf{u}' + (g^{(0)} + \mathbf{B}^{(0)}\cdot\mathbf{u}')\sin\beta^{(1)} + (1 + e^{(0)} + \mathbf{T}^{(0)}\cdot\mathbf{u}')(\cos\beta^{(1)} - 1) \end{aligned} \tag{R.8}$$

$$g^{(1)} = \mathbf{B}\cdot\bar{\mathbf{y}}' - \mathbf{B}^{(0)}\cdot\bar{\mathbf{y}}^{(0)\prime}$$
$$= \mathbf{B}^{(0)}\cdot\mathbf{u}' - (1 + e^{(0)} + \mathbf{T}^{(0)}\cdot\mathbf{u}')\sin\beta^{(1)} + (g^{(0)} + \mathbf{B}^{(0)}\cdot\mathbf{u}')(\cos\beta^{(1)} - 1) \tag{R.9}$$

$$k^{(1)} = \beta^{(1)\prime}\,. \tag{R.10}$$

For subsequent analysis, it is convenient to arrange $e^{(1)}$ and $g^{(1)}$ in ascending degrees of nonlinearity as

$$e^{(1)} = \sum_i e_i^{(1)}\ ,\quad g^{(1)} = \sum_i g_i^{(1)}\ , \tag{R.11}$$

where

$$e_1^{(1)} = \mathbf{T}^{(0)}\cdot\mathbf{u}' + g^{(0)}\beta^{(1)} \qquad (\text{R.12})_1$$

$$e_2^{(1)} = \mathbf{B}^{(0)}\cdot\mathbf{u}'\beta^{(1)} - (1/2)(1 + e^{(0)})(\beta^{(1)})^2 \qquad (\text{R.12})_2$$

$$e_3^{(1)} = -(1/6)g^{(0)}(\beta^{(1)})^3 - (1/2)\mathbf{T}^{(0)}\cdot\mathbf{u}'(\beta^{(1)})^2 \qquad (\text{R.12})_3$$

$$e_4^{(1)} = -(1/6)\mathbf{B}^{(0)}\cdot\mathbf{u}'(\beta^{(1)})^3 + (1/24)(1 + e^{(0)})(\beta^{(1)})^4\ , \qquad (\text{R.12})_4$$

etc.

$$g_1^{(1)} = \mathbf{B}^{(0)}\cdot\mathbf{u}' - (1 + e^{(0)})\beta^{(1)} \qquad (\text{R.13})_1$$

$$g_2^{(1)} = -\mathbf{T}^{(0)}\cdot\mathbf{u}'\beta^{(1)} - (1/2)g^{(0)}(\beta^{(1)})^2 \qquad (\text{R.13})_2$$

$$g_3^{(1)} = -(1/2)\mathbf{B}^{(0)}\cdot\mathbf{u}'(\beta^{(1)})^2 + (1/6)(1 + e^{(0)})(\beta^{(1)})^3 \qquad (\text{R.13})_3$$

$$g_4^{(1)} = (1/6)\mathbf{T}^{(0)}\cdot\mathbf{u}'(\beta^{(1)})^3 + (1/24)g^{(0)}(\beta^{(1)})^4\ , \qquad (\text{R.13})_4$$

etc. Only the linear and quadratic terms are used in the determination of the critical state, but the cubic and quartic terms are useful for postbuckling analyses.

The expression for the increment of strain energy is

$$\Delta\mathcal{E} = \int_0^L \{N^{(0)}e^{(1)} + Q^{(0)}g^{(1)} + M^{(0)}k^{(1)}$$
$$+ \tfrac{1}{2}[\phi,_{ee}^{(0)}(e^{(1)})^2 + \phi,_{gg}^{(0)}(g^{(1)})^2 + \phi,_{kk}^{(0)}(k^{(1)})^2] + R\}d\sigma\,. \tag{R.14}$$

Here, $\phi,_{ij}^{(0)}$ denotes the partial derivative of the dimensionless strain-energy density with respect to its ith and jth arguments, evaluated at the fundamental state and R includes all mixed and higher order terms. For quadratic, uncoupled strain-energy densities, $R \equiv 0$. For other cases, additional terms in the Taylor expansion can be taken as needed.

We consider, temporarily, only deformation-independent *dead loads*, $\mathbf{p} = \mathbf{d}(\sigma)$, so that $\mathcal{P}$ is linear in $\mathbf{u}$ and cancels the linear terms in $\Delta\mathcal{E}$. Of particular interest is the quadratic part of $\Delta(\mathcal{E} + \mathcal{P})$, which now reduces to

$$\Delta(\mathcal{E}+\mathcal{P})_2 = \Delta\mathcal{E}_2$$
$$= \int_0^L [N^{(0)}e_2^{(1)} + Q^{(0)}g_2^{(1)} + \tfrac{1}{2}(N^{(1)}e_1^{(1)} + Q^{(1)}g_1^{(1)} + M^{(1)}k^{(1)})]d\sigma, \tag{R.15}$$

where the linearized, incremental stress resultants and couples are *defined* by the expressions

$$N^{(1)} \equiv \phi,_{ee}^{(0)} e_1^{(1)} + \phi,_{eg}^{(0)} g_1^{(1)} + \phi,_{ek}^{(0)} k^{(1)} \tag{R.16}$$

$$Q^{(1)} \equiv \phi,_{gg}^{(0)} g_1^{(1)} + \phi,_{eg}^{(0)} e_1^{(1)} + \phi,_{gk}^{(0)} k^{(1)} \tag{R.17}$$

$$M^{(1)} \equiv \phi,_{kk}^{(0)} k^{(1)} + \phi,_{ek}^{(0)} e_1^{(1)} + \phi,_{gk}^{(0)} g_1^{(1)} . \tag{R.18}$$

To obtain buckling equations, we use (R.1) and take the variation of (R.15) which yields

$$\delta\Delta(\mathcal{E}+\mathcal{P})_2 = \int_0^L (N^{(0)}\delta e_2^{(1)} + Q^{(0)}\delta g_2^{(1)} + N^{(1)}\delta e_1^{(1)} + Q^{(1)}\delta g_1^{(1)} + M^{(1)}\delta k^{(1)})d\sigma = 0 . \tag{R.19}$$

Expressing $e_i^{(1)}$, $g_i^{(1)}$, and $k^{(1)}$ in terms of $\mathbf{u}$ and $\beta^{(1)}$, carrying out the variation, and integrating by parts, we arrive at

$$-\int_0^L \{[(M^{(1)})' + (1+e^{(0)})(Q^{(1)} + N^{(0)}\beta^{(1)}) - g^{(0)}(N^{(1)} - Q^{(0)}\beta^{(1)}) + \mathbf{F}^{(0)}\times\mathbf{e}_z)\cdot\mathbf{u}']\delta\beta^{(1)}$$
$$+ [(Q^{(1)} + N^{(0)}\beta^{(1)})\mathbf{B}^{(0)} + (N^{(1)} - Q^{(0)}\beta^{(1)})\mathbf{T}^{(0)}]'\cdot\delta\mathbf{u}\}d\sigma \tag{R.20}$$
$$+ \{[(Q^{(1)} + N^{(0)}\beta^{(1)})\mathbf{B}^{(0)} + (N^{(1)} - Q^{(0)}\beta^{(1)})\mathbf{T}^{(0)}]\cdot\delta\mathbf{u} + M^{(1)}\delta\beta^{(1)}\}_0^L = 0 .$$

From this variational equation, we obtain, by standard arguments, the following differential equations and boundary conditions for a beamshell under deformation-independent loads:

in the interior: $0 < \sigma < L$

$$(M^{(1)})' + (1+e^{(0)})(Q^{(1)} + N^{(0)}\beta^{(1)}) - g^{(0)}(N^{(1)} - Q^{(0)}\beta^{(1)}) + (\mathbf{F}^{(0)}\times\mathbf{e}_z)\cdot\mathbf{u}' = 0 \tag{R.21}$$

$$[(Q^{(1)} + N^{(0)}\beta^{(1)})\mathbf{B}^{(0)} + (N^{(1)} - Q^{(0)}\beta^{(1)})\mathbf{T}^{(0)}]' = 0 \tag{R.22}$$

at the edges: $\sigma = 0, L$

$$Q^{(1)} + N^{(0)}\beta^{(1)} = 0 \text{ or } \delta(\mathbf{B}^{(0)}\cdot\mathbf{u}) = 0 \tag{R.23}$$

$$N^{(1)} - Q^{(0)}\beta^{(1)} = 0 \text{ or } \delta(\mathbf{T}^{(0)}\cdot\mathbf{u}) = 0 \tag{R.24}$$

$$M^{(1)} = 0 \text{ or } \delta\beta^{(1)} = 0 . \tag{R.25}$$

By rearranging various terms, we may rewrite (R.21) in the form

$$(M^{(1)})' + (1+e^{(0)})Q^{(1)} + Q^{(0)}e_1^{(1)} - N^{(1)}g^{(0)} - N^{(0)}g_1^{(1)} = 0\,, \qquad \text{(R.26)}$$

which is identical to the linearized, incremental form of the nonlinear moment equilibrium condition, (O.11). Likewise, (R.22) is a linearized, incremental form of the force equilibrium condition, $\mathbf{F}' + \mathbf{p} = \mathbf{0}$, and can be rewritten as

$$(\mathbf{F}^{(1)})' = \mathbf{0} \ \text{ or } \ \mathbf{F}^{(1)} = \mathbf{C}\,. \qquad \text{(R.27)}$$

A component form of (R.22), which incorporates the equations of equilibrium in the fundamental state, is

$$(Q^{(1)})' + (\alpha^{(0)} + \beta^{(0)})' N^{(1)} + (\beta^{(1)})' N^{(0)} - \beta^{(1)} p_T^{(0)} = 0 \qquad \text{(R.28)}$$

$$(N^{(1)})' - (\alpha^{(0)} + \beta^{(0)})' Q^{(1)} - (\beta^{(1)})' Q^{(0)} + \beta^{(1)} p_B^{(0)} = 0\,. \qquad \text{(R.29)}$$

These equations are identical to the linearized, incremental form of (O.10). Note that the distributed load vector, $\mathbf{p}$, is constant but its components, $p_B = \mathbf{p}\cdot\mathbf{B}$ and $p_T = \mathbf{p}\cdot\mathbf{T}$, change due to the rotation of the unit vectors, $\mathbf{T}$ and $\mathbf{B}$.

We have derived the equations for the critical state by using the variational form of the energy criterion and have demonstrated that, as expected, it is equivalent to the method of *adjacent equilibrium*, the latter often being used to derive buckling equations directly, based on loss-of-uniqueness arguments. The advantages of the energy method are that it is grounded in the physics of the phenomenon and provides, via (R.1) and (R.15), a powerful tool for the approximate analysis of structural stability.

We now proceed to derive buckling equations if the load is a constant *fluid pressure*. The potential for this load was derived in Section J for moving boundaries, but, for simplicity, we shall assume that either the beamshell is closed or that $\mathbf{u} = \mathbf{0}$ at the edges, which covers most practical applications. The consequent vanishing of the edge potential simplifies the following analysis, but is not essential to its success. Thus, if $\mathbf{p} = p_0\mathbf{m}$,

$$\Delta\mathcal{P} = \tfrac{1}{2} p_0 \int_0^L [\mathbf{e}_z\cdot(\bar{\mathbf{y}}\times\bar{\mathbf{y}}' - \bar{\mathbf{y}}^{(0)}\times\bar{\mathbf{y}}^{(0)\prime})]\,d\sigma\,. \qquad \text{(R.30)}$$

Introducing the displacement vector and integrating by parts, we have

$$\Delta\mathcal{P} = p_0 \int_0^L [-\mathbf{m}^{(0)}\cdot\mathbf{u} + \tfrac{1}{2}\mathbf{e}_z\cdot(\mathbf{u}\times\mathbf{u}')]\,d\sigma \qquad \text{(R.31)}$$

which, with $\mathbf{u} = v\boldsymbol{\tau}^{(0)} - w\mathbf{n}^{(0)}$, has the component form

$$\Delta\mathcal{P} = p_0 \int_0^L [\lambda^{(0)} w - v w' + \tfrac{1}{2}(\alpha^{(0)} + \beta^{(0)})'(v^2 + w^2)]\,d\sigma\,. \qquad \text{(R.32)}$$

For a circular ring, (R.32) reduces to equation (4.65) of Budiansky (1974).

Returning to (R.31), carring out the variation, and integrating by parts, we have

$$\delta\Delta\mathcal{P} = -p_0 \int_0^L \mathbf{m}\cdot\delta\mathbf{u}\,d\sigma = p_0 \int_0^L (-\mathbf{m}^{(0)} + \mathbf{u}'\times\mathbf{e}_z)\cdot\delta\mathbf{u}\,d\sigma\,. \qquad \text{(R.33)}$$

Only the factor $p_0\,\mathbf{u}'\times\mathbf{e}_z$ enters $\delta\Delta\mathcal{P}_2$. Thus, by adding this term to the left side of (R.22), which is the incremental force equilibrium equation for buckling under a constant load, we obtain

$$[(Q^{(1)}+N^{(0)}\beta^{(1)})\mathbf{B}^{(0)}+(N^{(1)}-Q^{(0)}\beta^{(1)})\mathbf{T}^{(0)}+p_0\mathbf{e}_z\times\mathbf{u}]'=\mathbf{0}\,. \quad \text{(R.34)}$$

Expressing the pressure term in the basis $\{\mathbf{T}^{(0)},\mathbf{B}^{(0)}\}$—see $(\mathrm{O}.1)_3$—we have $p_T=-gp_0$ and $p_B=(1+e)p_0$. Introducing these expressions into (R.34), we obtain, after some manipulations, the following component form of the buckling equations:

$$(Q^{(1)})'+(\alpha^{(0)}+\beta^{(0)})N^{(1)}+N^{(0)}(\beta^{(1)})'+p_0e_1^{(1)}=0 \quad \text{(R.35)}$$

$$(N^{(1)})'-Q^{(1)}(\alpha^{(0)}+\beta^{(0)})'-Q^{(0)}(\beta^{(1)})'-p_0g_1^{(1)}=0\,. \quad \text{(R.36)}$$

Moment equilibrium, (R.21) or (R.26), is unchanged.

Potentials for other types of loads may be derived and applied without difficulty. As for edge potentials, we mention only those for linear and torsional springs given in Section J. These are frequently used to simulate elastic bodies attached to the ends of beamshells. The effects of such springs in the boundary conditions implied by $\delta\Delta(\mathcal{E}+\mathcal{P})_2=0$ are easily computed and need not be stated here.

In $(\mathrm{R}.12)_1$, $(\mathrm{R}.13)_1$, (R.16)-(R.18), (R.22) or (R.34), and (R.21) or (R.26), we have a 6th-order system of ordinary differential equations for $\mathbf{u}$ and $\beta^{(1)}$. The variable coefficients in this system depend on the fundamental state. The vector incremental equilibrium equations, (R.22) or (R.34), can be integrated once to obtain a 4th order system with 2 unknown constants of integration.[23] For deformation-independent (i.e., constant) loads, the system can be further reduced to 2nd order plus 2 quadratures by first integrating $(\mathrm{R}.22)_2$ and then substituting the result into (R.21), so obtaining

$$Q^{(1)}+N^{(0)}\beta^{(1)}=\mathbf{C}\cdot\mathbf{B}^{(0)} \quad \text{(R.37)}$$

$$N^{(1)}-Q^{(0)}\beta^{(1)}=\mathbf{C}\cdot\mathbf{T}^{(0)} \quad \text{(R.38)}$$

$$(M^{(1)})'+\mathbf{C}\cdot\mathbf{m}^{(0)}+(\mathbf{F}^{(0)}\times\mathbf{e}_z)\cdot\mathbf{u}'=0\,. \quad \text{(R.39)}$$

We adopt the common assumption that the strain-energy density is uncoupled, which simplifies some of the algebra. Then, using (R.12), (R.13), and (R.16)-(R.18) in (R.37) and (R.38), we obtain

$$\begin{aligned}\mathbf{u}'=\mathbf{m}^{(0)}\beta^{(1)}&+(\phi,_{gg}^{(0)})^{-1}(\mathbf{C}\cdot\mathbf{B}^{(0)}-N^{(0)}\beta^{(1)})\mathbf{B}^{(0)}\\&+(\phi,_{ee}^{(0)})^{-1}(\mathbf{C}\cdot\mathbf{T}^{(0)}+Q^{(0)}\beta^{(1)})\mathbf{T}^{(0)}\,.\end{aligned} \quad \text{(R.40)}$$

Finally, substitution of (R.16)-(R.18) and (R.40) into (R.39) yields

---

[23]If there is fluid pressure on a *circular* beamshell, the equilibrium and energy integrals (O.33) and (O.35) can be used to reduce the system to 2nd order, as shown by Antman & Dunn (1980). In particular, see their equation (4.6).

$$
\begin{aligned}
&[\phi,^{(0)}_{kk}(\beta^{(1)})']' + [-\mathbf{F}^{(0)}\cdot(\underline{\mathbf{y}}^{(0)})' + (\phi,^{(0)}_{ee})^{-1}(Q^{(0)})^2 + (\phi,^{(0)}_{gg})^{-1}(N^{(0)})^2]\beta^{(1)} \\
&\qquad + \mathbf{C}\cdot[\mathbf{m}^{(0)} + (\phi,^{(0)}_{ee})^{-1}Q^{(0)}\mathbf{T}^{(0)} - (\phi,^{(0)}_{gg})^{-1}N^{(0)}\mathbf{B}^{(0)}] = 0\,.
\end{aligned}
\tag{R.41}
$$

This is a 2nd-order differential equation for $\beta^{(1)}$ with (generally) variable coefficients. Once (R.41) is solved, computing **u** from (R.40) reduces to quadratures.

As a simple (though not very practical) example, we calculate the critical pressure for a materially uniform circular tube or ring of radius R under a deformation-independent uniform normal load, $\mathbf{p}=p_a\mathbf{b}$, where p_a is a constant. Note that the fundamental state is axisymmetric. Taking $\alpha=\sigma/R$ as the independent variable and imposing *periodicity* in place of boundary conditions, we obtain from (R.41) an eigenvalue problem which reduces to

$$
p_aR^3[1 + e^{(0)} + \underline{(\phi,^{(0)}_{gg})^{-1}p_aR}] = m^2\phi,^{(0)}_{kk}\ , \quad m = 2, 3, \cdots , \tag{R.42}
$$

where the underlined term represents shear effects. The condition $m\geq 2$ stems from the periodicity requirement.[24] The critical load, $(p_a)_{\rm cr}$, is the solution of (R.42) for $m=2$. If $e^{(0)}=0$ and shear effects are small, then, to a 1st approximation, we recover the results of Smith & Simitses (1969):

$$
(p_a)_{\rm cr} = 4R^{-3}\phi,^{(0)}_{kk}[1 - 4R^{-2}(\phi,^{(0)}_{gg})^{-1}\phi,^{(0)}_{kk}]\,. \tag{R.43}
$$

If shear effects are totally neglected in (R.43) and $\phi,^{(0)}_{kk}$ is identified with the bending stiffness of the beamshell—see (N.8), (N.9), and (N.17)—we have

$$
(p_a)_{\rm cr} = \frac{1}{3}\left[\frac{Eh^3}{(1-\nu^2)R^3}\right], \tag{R.44}
$$

which is the classical result.

For other types of pressure loads, a different coefficient is obtained in (R.44): $1/4$ for fluid pressure [Carrier (1947), Budiansky (1974)]; $3/8$ for constant, centrally directed pressure; and $3/16$ for inverse-square, centrally directed pressure [Sills & Budiansky (1978)]. The latter authors have shown that only the inverse-square case is imperfection sensitive. The effect of centrifugal forces on the buckling of rings under fluid pressure has been studied by Wang (1984b) who shows that the buckling pressure is increased by $1/3$ times the centrifugal loading.

Although the buckling and initial postbuckling of closed circular beamshells is fairly well understood, less is known about deeply postbuckled states. For this analysis, the equations and methods presented in subsection O.2.b apply. Antman (1970, 1972) has studied the problem qualitatively. Quantitative results for inextensional elasticas have been obtained by Lévy (1883) and Carrier (1947). The solutions involve elliptic functions and integrals but the constants of integration are left in implicit, transcendental form. Carrier's study, which includes methods for obtaining these constants, has been used to assess

---

[24]The ring should also be prevented from rigid body motion which, as pointed out by Singer & Babcock (1970), is unstable.

approximate methods. Elastica-like solutions have also been used by DaDeppo & Schmidt (1969) to analyize steep arches under concentrated loads.

2. Shallow beamshells

With the shallow shell approximation we can obtain *complete solutions* to important problems. Consequently, the location of limit points, the study of bifurcations, and the relation between the two can be studied in detail.

Depending on circumstances, a transversely or normally loaded, shallow cylindrical beamshell may be treated either as a special case of the theory of shallow shells of weak curvature—the DMV (Donnell-Mushtari-Vlasov) theory—or as a special case of Marguerre's (1938) general nonlinear theory of shallow shells. Detailed analyses of the shallow shell approximation may be found in Sanders (1963, pp. 33-34) and Koiter (1966, Sect. 9). Equations for shearable shallow shells are given by Reissner (1969, 1982). Koiter (1967c) has discussed the applicability of *shallow buckling modes* to stability analysis and Budiansky (1974, Sect. D) has presented equations for the initial postbuckling analysis of shallow shells.

"Shallowness" with respect to a plane (conveniently taken to be the xy-plane) is characterized by the requirement that

$$\alpha^2, (\alpha+\beta)^2 \ll 1 . \tag{R.45}$$

Hence, nonlinear shallow shell theory is, perforce, a *moderate rotation* theory. If, in addition, the Kirchhoff hypothesis is adopted and the deformed position represented as $\bar{\mathbf{y}} \equiv \bar{x}\mathbf{e}_x + \bar{y}\mathbf{e}_y$, then the bending and Lagrangian extensional strain, within these approximations, are given by

$$k = (\bar{y} - y),_{xx} \ , \ \varepsilon = (\bar{x} - x),_x + \tfrac{1}{2}(\bar{y},_x^2 - y,_x^2) , \tag{R.46}$$

where it has also been assumed that $|(\bar{x} - x),_x| \ll 1$. Introducing (R.46) and $\mathbf{p} = p_y\mathbf{e}_y$ into the Principle of Stationary Potential Energy, (Q.13), and taking variations, we find, because $\bar{x}$ does not appear in the expression for k, that

$$\phi,_\varepsilon = N = \text{constant} \ , \ 0 < \sigma < L . \tag{R.47}$$

This is the main consequence of the shallow beamshell approximation.

Similarly, in a shell of *weak curvature* (such that $L_* / R_* \ll 1$, where L_* is a characteristic wavelength of the deformation pattern and R_* is the minimum radius of curvature of the undeformed reference curve), the *tangential* displacement component can be eliminated from k. This is the well-known Donnell approximation. If $\mathbf{p} = -p_0\mathbf{m}$, we find, as with the case $\mathbf{p} = p_y\mathbf{e}_y$, that $N =$ constant. This result still holds if shear deformations are included (Reissner & Wan, 1969, 1982).

To examine some of the implications of shallow shell theory, we first note from (C.6), (C.7), (O.2), and (O.8) that

$$N = H\cos(\alpha+\beta) + V\sin(\alpha+\beta) \ , \ Q = -H\sin(\alpha+\beta) + V\cos(\alpha+\beta) \ , \tag{R.48}$$

where H and V are given by the integrated *equilibrium equations*, (O.6). For transversely loaded beamshells, $p_x = 0$ and H is a constant. It thus follows from (R.48) and (R.47) that, in a shallow beamshell, the V term in $(R.48)_1$ can be neglected *insofar as equilibrium is concerned.*[25]

We now incorporate the above order of magnitude considerations *directly* into the equilibrium and kinematic relations to derive the equations of a transversely loaded, shallow beamshell. First, we shall derive a nonlinear, Cartesian form for the exact beamshell equations and, only then, introduce the shallowness approximation, (R.45). We use the strain-stress relations $e = \Psi,_N$ and $g = \Psi,_Q$ and the uncoupled, quadratic mixed-energy density, (N.18), to express e and g in terms of N and Q.[26] Then, the introduction of these relations along with (R.48) into to the moment equilibrium condition, (O.11), yields

$$\begin{aligned} &M' - [H + (A_e - A_s)(H^2 - V^2)\cos(\alpha+\beta)]\sin(\alpha+\beta) \\ &\qquad + \{\cos(\alpha+\beta) + (A_e - A_s)H[\cos^2(\alpha+\beta) - \sin^2(\alpha+\beta)]\}V = 0, \end{aligned} \tag{R.49}$$

where $A_e \equiv (E_e h)^{-1}$ and $A_s \equiv (E_s h)^{-1}$. This is a differential equation for $(\alpha+\beta)$ involving the unknown constants, H and $V_0 \equiv V(0)$, which are to be found from the boundary conditions. If these are displacement boundary conditions then, by virtue of (C.6), (C.7), and (K.1), they can be written in the form

$$\begin{aligned} (\mathbf{u}_L - \mathbf{u}_0)\cdot\mathbf{e}_x = \int_0^L [&\cos(\alpha+\beta) - \cos\alpha - \Psi,_N\cos(\alpha+\beta) \\ &+ \Psi,_Q\sin(\alpha+\beta)]d\sigma \end{aligned} \tag{R.50}$$

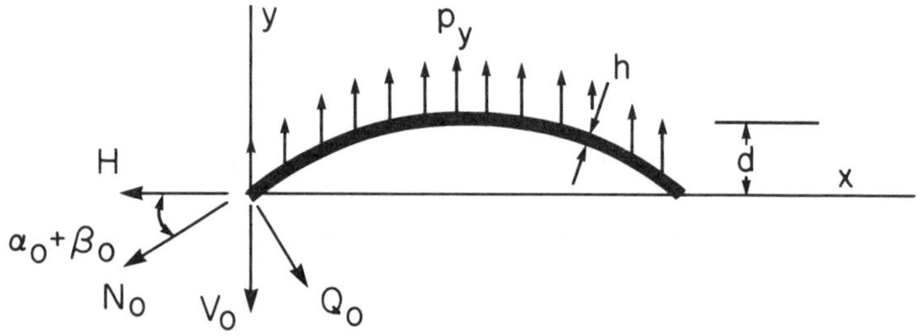

Fig. 4.14. Load and geometrical conventions for a shallow beamshell.

---

[25]Another implication of (R.48) is that if H *is* to influence equilibrium, then $|V/H| = O(\alpha+\beta)$. This is indeed the case in the stability analysis of shallow beamshells. See Fig. 4.14.

[26]The use of a quadratic Ψ in the 2nd-order terms implies that to the same degree of approximation, a *cubic* in Φ (if available) can be used with the first order terms. This should be borne in mind for later developments. The choice of an uncoupled strain-energy density is not essential, but it simplifies things considerably.

$$(\mathbf{u}_L - \mathbf{u}_0)\cdot\mathbf{e}_y = \int_0^L [\sin(\alpha+\beta) - \sin\alpha - \Psi_{,N}\sin(\alpha+\beta) - \Psi_{,Q}\cos(\alpha+\beta)]d\sigma\,. \tag{R.51}$$

This last boundary condition may be replaced by a *simple static equation* of overall moment equilibrium about one or the other end of the beamshell. This will determine V_0 or V_L in terms of other endpoint variables. Introducing N and Q from (R.48) and the solution $(\alpha+\beta)$ of (R.49) into (R.50), we get an equation for the determination of H.

If the mixed-energy density, Ψ, is quadratic, then (R.50) reduces to

$$(\mathbf{u}_L - \mathbf{u}_0)\cdot\mathbf{e}_x = \int_0^L \{\cos(\alpha+\beta) - \cos\alpha + H[A_e\cos^2(\alpha+\beta) + A_s\sin^2(\alpha+\beta)] + V(A_e - A_s)\sin(\alpha+\beta)\cos(\alpha+\beta)\}d\sigma\,. \tag{R.52}$$

At this stage, we introduce the shallowness approximation, so that (R.49) and (R.52) reduce to

$$E_b h^3\beta'' - H[1+(A_e - A_s)H](\alpha+\beta) + V[1+(A_e - A_s)H] = 0 \tag{R.53}$$

$$A_e LH = (\mathbf{u}_L - \mathbf{u}_0)\cdot\mathbf{e}_x + \int_0^L (\alpha + \tfrac{1}{2}\beta)\beta d\sigma\,. \tag{R.54}$$

These are the final nonlinear shallow beamshell equations.[27] [28] They differ from the classical equations only by the additional $(A_e - A_s)H$-terms in (R.53). We note that (R.53) is a linear, constant coefficient equation with a variable, nonhomogeneous part. The nonlinearity stems from the boundary condition, (R.54).

When the stability of a shallow beamshell (or arch) is to be investigated, the loading is described in terms of a varying load parameter, q, and the total angle to the deformed reference curve, $\alpha+\beta$, is determined from the buckling equations. The variation of q with a suitable deformation measure, δ, defines a *fundamental equilibrium path*, and the limit point value(s), q_s, can be sought. The superposition of an asymmetrical deformation on a symmetric response to symmetric loads defines a 2nd (bifurcated) equilibrium path, whose location can be found either by a direct analysis of the equations, as done by Masur & Lo (1972) for the circular shallow arch, or else by solving the buckling equations (R.40)-(R.41), as done by Schreyer & Masur (1966) in their complete analysis of the clamped shallow arch under uniform pressure. These later results are summarized and extended by Dym (1974). Here are some of the highlights.

Define the rise ratio

---

[27]The last term in (R.52) was deleted for consistency. If $|H/V| = O(1)$, then it might be retained to improve the accuracy of the calculation of H.

[28]If available, consideration should be given to using a cubic form for Ψ in (R.50) before applying the shallowness approximation.

$$\hat{\lambda} \equiv \frac{2d}{h}, \tag{R.55}$$

where d is defined in Fig. 4.14. If $\hat{\lambda} < 2.85$ the arch is stable. If $2.85 < \hat{\lambda} < 5.74$, the arch has a symmetric limit point instability where it snaps. If $5.74 < \hat{\lambda}$, the arch will bifurcate unsymmetrically. Finally, if $5.02 < \hat{\lambda} < 5.74$, there is an unsymmetrical bifurcation *beyond* the limit point that does not influence the physical behavior of the arch. *Schematic diagrams* of equilibrium paths, some with limit points and bifurcations, are shown in Fig. 4.15 for several values of $\hat{\lambda}$. The (nondimensional) area swept out by the deforming arch—the linear part of (R.31)—is used as the deformation measure.

Dym (1974) presents a detailed initial postbuckling analysis of this problem and finds that, in (R.4), $a = 0$ and $b = 1.61(1 - .199\hat{\lambda}^2)^{-1}$. Hence, all the bifurcation points for $\hat{\lambda} > 5.02$ are *unstable*.

The shallow arch under a concentrated, central load has also been studied extensively. Early results, starting with those of Navier (1833), are reported by Timoshenko & Gere (1961). Gjelsvik & Bodner (1962) have made a more detailed (though approximate) study of the problem and have proposed an energy criterion for buckling. Schreyer & Masur (1966) have investigated asymmetric bifurcations. By presenting a complete solution, they showed that a shallow clamped arch under a concentrated load is stable if $\hat{\lambda} < 3.20$ and snaps symmetrically if $\hat{\lambda} > 3.20$; all bifurcations lie on the unstable portions of the fundamental equilibrium path.

As a final note on the stability of shallow, shearable beamshells, we observe that the only place where *shear deformations* appear as 1st-order terms is in (R.51), which is used to calculate the displacements in the y-direction. Normally, these play a secondary role in the analysis (compared with u_x and u_y). However, whenever compatibility of deformations in this direction is significant, we should include shear-deformation effects. Similar remarks apply to axishells and unishells.

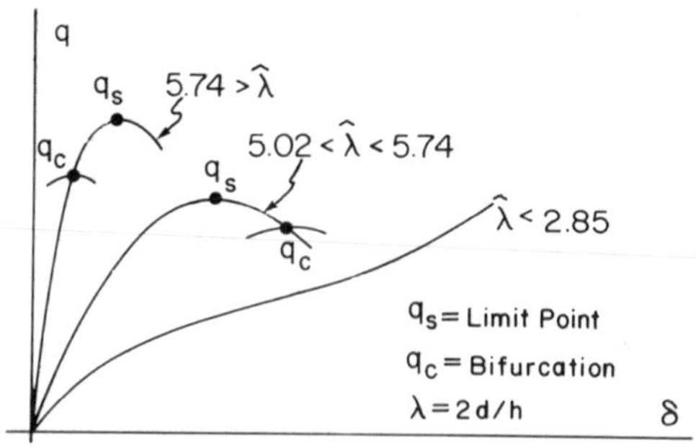

Fig. 4.15. Equilibrium paths for a shallow, pressure-loaded arch.

S. Some Remarks on Failure Criteria and Stress Calculations

When we contemplate a structure such as a beamshell, we may want to know:

(a) If it will fail or how close it is to failure under a given external disturbance, such as a dead load or a heat flux.[29]

(b) The position of particles during the deformation process.

(c) Useful global parameters such as natural frequencies, overall stiffnesses, etc.

Having *modeled* a loaded structure as a beamshell and solved the associated equations for the spatial and temporal variations of average quantities, such as N, M, Q, β, or $\bar{y}$, we should be able to apply these predictions to the physical problem at hand.

(Global or *local) failure criteria* are externally imposed design requirements. A global failure criterion involves *gross* quantities such as buckling loads, limit loads, or critical frequencies, which are usually predicted quite well by beamshell models. But, even *local* failure criteria involving, say, yield stresses or a yield surface, can often be expressed adequately in terms of the local static variables, N, M, and Q, the local geometry, $\mathcal{G}$ (thickness h, curvature b, etc.), and the local material, $\mathcal{M}$, of the beamshell. Typically, these quantities are combined into a *failure envelope*, $F(N, M, Q, \mathcal{G}, \mathcal{M})$. If, at any point, F attains or exceeds some critical value, F_c, then the beamshell has failed at this point. The nature of the failure envelope and F_c are usually determined by experiment, supplemented, as necessary, by analysis. We call this *the direct approach* to failure analysis.

Much of what we know about failure comes from experiments on plates or shells. Thus, it makes "physical sense" to form failure envelopes for the *stress resultants*, thereby employing measured, rather than inferred, data such as *allowable stresses*, which require one to assume a certain distribution of stress through the thickness of the specimens. One of the simpler engineering failure criterion, $|s| = s_c$, employs the *extreme fiber-stress formula*,

$$s = \frac{1}{h}\left[N \pm \frac{6M}{h}\right] . \tag{S.1}$$

Here, s_c is usually taken as the *engineering yield stress*[30], though other allowable stresses are occasionally used. This criterion, which has received

---

[29]Failure is usually assessed by applying a suitable *failure criterion* based on experiments, engineering codes, or theoretical analyses.

[30]Some yield criteria are multidimensional and expressed in terms of yield surfaces, in which case the existence of stresses in the z-direction (which follows from the vanishing of strains in that direction) should be accounted for. For example, a J_2 (von Mises) yield criterion produces a modified yield stress, s_c^*.

substantial experimental verification, applies to homogeneous beamshells in which the stress-strain relations are linear, the strains small, and N and M referred to the midsurface. Note that, under these same conditions, we may also regard $|s| = s_c$ as a failure envelope, expressible, via (S.1), directly in terms of N and M and not requiring any assumption on the distribution of stresses through the thickness of the beamshell. A similar situation occurs in sandwich shells with weak cores where, in the failure criterion, $|s| = s_c$,

$$s = \frac{1}{h_1}\left[\frac{N}{2} + \frac{M}{h}\right], \tag{S.2}$$

h_1 being the thickness of the cover sheet. The same remarks apply to (S.2) as to (S.1).

Some of the advantages of the direct approach become evident in more complex situations. For example, in curved sandwich beamshells, failure by separation of the cover sheet from the core results from local instability effects plus material failure of the adhesive or the core due to combined stresses. Here, an experimental determination of a *failure envelope* might be more useful than repeated, detailed analyses based on additional assumptions.

Nevertheless, there are cases in which a failure-envelope approach is inadequate and it is necessary to produce an approximate stress distribution through the thickness of the beamshell. This is especially true if failure can originate in the interior of the beamshell. If the strains are small, there is not much sense in trying to construct models of beamshells with elaborate thickness distributions—any "improved accuracy" is swamped by errors inherent in beamshell theory itself. Thus, it suffices to consider the widely-accepted *linear-strain-distribution model* in which strains parallel to the deformed reference curve, $\bar{C}$, when referred to the undeformed shape, C, vary linearly in the **b**-direction.[31] For homogeneous materials with linear stress-strain relations, this model leads to the well-known formula for the engineering stress

$$s = \frac{N}{h} + \frac{3M}{h_-^2 + h_+^2}\zeta, \tag{S.3}$$

where ζ is distance along **b** from C and $h_\pm$ are the values of ζ at the upper and lower faces of the shell.

In nonhomogeneous beamshells too, the linear distribution of strains plus the local conversion from strains to stresses determines the stress distribution

[31]Both the well-known Euler-Bernoulli assumption and the plane stress assumption lead to such a distribution (see subsection N.3 for further discussion). Curvature effects are sometimes added, but are important only in thick beamshells. See Timoshenko & Goodier (1970, Sections 29 and 33) who also show that the Wrinkler approximation—which strictly assumes that normal sections remain normal and thereby includes curvature effects—yields excellent results in thick, curved beamshells under end moments. The assumption of a linear distribution of strain through the thickness has also been used if transverse shearing-strain effects are not negligible and thickness averages are needed.

through the thickness to sufficient accuracy for practical purposes. The transverse shearing stresses are estimated by integrating the differential equations of equilibrium in the thickness direction. If there are no shear stresses applied to the faces of the shell, then a linear variation of s through the thickness leads to a parabolic distribution of shear stress with a maximum of $3Q/2h$.

1. Some large-strain refinements

If the strains are large, their variation through the thickness may be far from linear. Nevertheless, the latter assumption can be useful in the following cases:

(a) Small bending strains superimposed on large extensional strains If $e \gg hk$, then the moment arm of the stress distribution is mainly affected. An approximate correction is obtained by replacing M by $\lambda_\zeta^{-1} M$, where λ_ζ is the average stretch in the thickness direction (derived from the material model at hand). This correction is inadequate if the extensional and bending strains are of the same order of magnitude. In the latter case, the unsymmetrical distribution of strain must be accounted for. For further discussion on this point, see Libai & Simmonds (1981).

(b) Effects of shearing strains A crude approximation can be obtained by assuming a linear distribution in the *non-material* $\mathbf{\bar{b}}$-direction. The elementary correction replaces N by $N^* = N\cos\gamma + Q\sin\gamma$. This approximation is useful in beamshells with low shear moduli, e.g., sandwich shells with weak cores (see Fig. 4.16), provided that the shearing deformations are not very large. If these conditions do not obtain, a failure envelope may have to be determined or a more refined analysis used.

(c) Use of 3-dimensional models Approximate, 3-dimensional, large-strain models are occasionally used for the derivation of shell constitutive relations and are obtained by averaging, assuming thickness distributions, variational

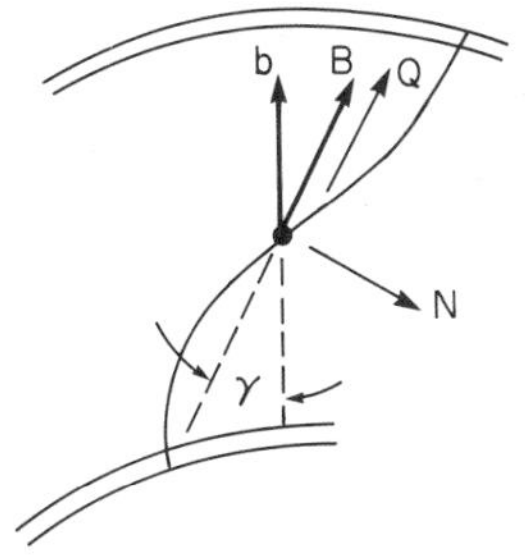

Fig. 4.16. Deformed cross section of a sandwich beamshell.

techniques, etc. The thickness variations of the associated 3-dimensional stresses follow from 3-dimensional stress-strain relations and can be expected, in general, to contain larger relative errors than the 2-dimensional constitutive relations derived from the models. For example, it follows from subsection N.3 that in a neo-Hookean material that is stretched and bent, we have, upon neglecting transverse shear strain effects, the following estimate of the dominant component of the 1st Piola-Kirchhoff stress tensor:

$$S_\sigma = \frac{\partial W}{\partial \varepsilon} = 2\frac{\partial W}{\partial G_\sigma} = \frac{E}{3}\left[1 - \frac{1}{G_\sigma^2} - \frac{4\Gamma^2}{G_\sigma^2}\right] \approx \frac{E}{3}\left[1 - \frac{1}{\lambda^4} + \frac{2\bar{\kappa}\zeta}{\lambda^6}\right] . \tag{S.4}$$

For the distribution of the *true* (Cauchy) stresses in a large-strain model *including curvature effects*, see Fig. 1 of Libai & Simmonds (1981). The introduction of shear strain effects can modify these distributions, as discussed in subsection N.3.

2. Determination of the deformed configuration

We remind the reader that points on $\bar{\mathcal{C}}$ are *not* the deformed images of material points on $\mathcal{C}$, but rather represent through-the-thickness averages of the deformed positions of particles. Nevertheless, the differences are slight and can be disregarded for most practical purposes. Hence, in applications, we can take $\bar{\mathcal{C}}$ as the deformed image of $\mathcal{C}$. Similarly, β can be regarded (in the sense of a thickness average) as the angle of rotation of a cross section and $\lambda_\zeta h$ as the thickness of the deformed beamshell. Beamshell theory cannot improve upon these approximations—to do so 3-dimensional elasticity theory must be invoked.

T. Thermodynamics

Thermal effects in shells need to be considered for at least four reasons. First, constrained shells, when heated, may experience large stresses. Second, a complete analysis of crack growth should involve heat flow (Gurtin, 1979). Third, the coupling between deformation and temperature, which results from the time rate of dilation appearing in the energy equation, produces a slight damping of free vibrations. (Although normally swamped by air and joint damping, thermal damping could be important in large, thin, homogeneous structures in outer space). And fourth, as first emphasized by Duhem (1911) and later by Ericksen (1966, 1967), Koiter (1969, 1971), and Gurtin (1973a,b, 1974, 1975), among others, a comprehensive definition of stability needs thermodynamics. See Section U for more details.

The approach in this section was motivated by the work of Coleman & Noll (1963), Truesdell (1969, 1984), Müller (1969), and Green & Naghdi (1979). The last authors are concerned with the reconciliation of a direct approach to the thermodynamics of a 2-dimensional Cosserat continuum with a 3-dimensional approach in which *all dependent variables are represented by truncated power series in the thickness coordinate*. In contrast, *no such expansions are used in what follows.* Rather, certain 2-dimensional thermodynamic variables (an entropy density resultant ι, a heat influx resultant q, an entropy flux resultant j, an "effective" temperature T, and an "effective" transverse temperature gradient Δ) fall out naturally from the integral equations of 3-dimensional thermodynamics, while other secondary quantities are a consequence of Conservation of Energy, the Clausius-Duhem inequality, and the assumed form of the 2-dimensional constitutive relations. Our results for beamshells are a special case of those developed by Simmonds (1984) for general shells.

Just as the 3-dimensional integral equations of motion (II.A.4) and (II.A.5), when specialized to a beamshell, imply (B.2) and (B.3), so (II.C.1) and (II.D.2) imply, for a beamshell, that

$$\mathcal{Q} = \int_{\partial A} \nu ds + \int_A n dA \tag{T.1}$$

$$\mathcal{H} = \int_A \eta dA \text{ and } \mathcal{J} = \int_{\partial A} \frac{\nu}{\theta} ds + \int_A \frac{n}{\theta} dA , \tag{T.2}$$

where, as in (B.2) and (B.3), the dimensions of all 3-dimensional variables have been multiplied by the z unit of length. Thus, for example, νds is the heat *influx* across the image of the differential of arc length ds.

From (II.C.2) and Fig. 4.1, it follows that, on the edges, $\nu ds = \pm \mathbf{c} \cdot (\mathbf{e}_z \times \mathbf{r},_\zeta) d\zeta$, while on the faces, $\nu ds = \pm \mathbf{c} \cdot (\mathbf{r},_\sigma \times \mathbf{e}_z)$. Thus, if we set

$$\mathbf{c} = c^\sigma \mathbf{r},_\sigma + c^\zeta \mathbf{r},_\zeta , \tag{T.3}$$

then, with the aid of (B.4), (T.1) and (T.2) reduce to

$$\mathcal{Q} = q \Big|_a^b + \int_a^b s d\sigma \tag{T.4}$$

$$\mathcal{H} = \int_a^b \iota d\sigma \text{ and } \mathcal{J} = j \Big|_a^b + \int_a^b v d\sigma , \tag{T.5}$$

where

$$q = \int_{-H}^{H} c^\sigma \mu d\zeta \ , \ s = c^\zeta \mu \Big|_{-H}^{H} + \int_{-H}^{H} n \mu d\zeta \tag{T.6}$$

$$\iota = \int_{-H}^{H} \eta \mu d\zeta \ , \ j = \int_{-H}^{H} (c^\sigma / \theta) \mu d\zeta \tag{T.7}$$

$$v = (c^\zeta \mu / \theta)_{-H}^{H} + \int_{-H}^{H} (n/\theta) \mu d\zeta . \tag{T.8}$$

We call q and j the *heat and entropy influx resultants*, respectively, and ι the *entropy density resultant*; s and v are, respectively, the *heating* and the *supply of entropy*.

If the heat influx ν in (II.D.2) is replaced by $\mathbf{c} \cdot \mathbf{n}$ and the Divergence Theorem applied, we obtain a volume integral with integrand $\nabla \cdot (\mathbf{c}/\theta)$. In place of this integrand, Müller (1969) introduced an entropy flux j and then *derived* $j = \nabla\cdot(\mathbf{c}/\theta)$ as a constitutive relation. It is therefore interesting that an entropy influx arises naturally in the descent from 3 dimensions.

Observe that, if we introduce a *mean reciprocal temperature* and a *transverse temperature gradient* by setting

$$\frac{1}{T} \equiv \frac{1}{2}\left[\frac{1}{\theta_-} + \frac{1}{\theta_+}\right], \quad \Delta \equiv \frac{T^2}{2H}\left[\frac{1}{\theta_-} - \frac{1}{\theta_+}\right], \tag{T.9}$$

then the supply of entropy, v, defined by (T.8), may be given the form

$$v = \frac{s + \Delta\chi}{T}, \tag{T.10}$$

where

$$\chi \equiv -\frac{H[(c^{\zeta}\mu)_+ + (c^{\zeta}\mu)_-]}{T} + \frac{1}{\Delta}\int_{-H}^{H}\left[\frac{T}{\theta} - 1\right] n\mu d\zeta . \tag{T.11}$$

The last term in (T.10) measures the supply of entropy due to the nonuniform distribution of temperature through the thickness and is thus analogous to the external distributed couple, l, defined by (B.19).[32] Except in special cases, the distribution through the thickness of neither the temperature θ nor the deformed position $\bar{\mathbf{x}}$ can be inferred exactly from beamshell variables, so that neither χ nor l can be specified precisely. However, in many practical problems, these terms are negligible.

Observe also that if $\theta = T$, then, by $(T.6)_1$ and $(T.7)_2$, $j = q/T$. This suggests that we work with a new unknown, w, by setting

$$j \equiv \frac{q}{T} + \Delta w . \tag{T.12}$$

Again, we emphasize that the equations of motion, (B.13) and (B.17), and the Clausius-Duhem inequality, (II.D.1), with $\mathcal{H}$ and $\mathcal{J}$ given by (T.5), are *exact;* the approximate nature of beamshell theory enters when we take the apparent external mechanical power, $\mathcal{W}$, to be the actual external mechanical power and when we approximate (or ignore) l and χ.

To obtain a complete thermodynamic theory of beamshells, we *postulate* the following law of *Conservation of Energy* (1st Law of Thermodynamics):

$$\int_{t_1}^{t_2}(Q + \mathcal{W})dt = (\mathcal{K} + I)\Big|_{t_1}^{t_2}, \tag{T.13}$$

where I is the *internal energy*. By (G.1), (T.13) is equivalent to the *reduced equation of energy balance*

---

[32] If θ^{-1} is linear in ζ, then $-T\chi = H[(c^{\zeta}\mu)_+ + (c^{\zeta}\mu)_-] + \int_{-H}^{H} \zeta n\mu d\zeta$.

$$\int_{t_1}^{t_2}(Q+\mathcal{D})dt = I\,\Big|_{t_1}^{t_2} . \qquad \text{(T.14)}$$

We call a beamshell *thermoelastic* if there exists an *internal energy density* i, depending on

$$\Lambda \equiv (e, g, k, T, \Delta) \qquad \text{(T.15)}$$

and its gradient, $\Lambda' \equiv (e', g', k', T', \Delta')$, such that

$$I = \int_a^b i d\sigma . \qquad \text{(T.16)}$$

It is convenient to introduce the *free-energy density*, Φ, by setting

$$i = \Phi + \iota T + f\Delta , \qquad \text{(T.17)}$$

where the unknown f may be called the *entropy density couple*. Both Φ and f are assumed to be functions of Λ and Λ'. Inserting (T.17) into (T.16), the resulting expression into (T.14), and using (K.4), (K.5), and (T.4), we have

$$\int_{t_1}^{t_2}[q\,\Big|_a^b + \int_a^b(N\dot{e} + Q\dot{g} + M\dot{k} + s)d\sigma]dt = \int_a^b(\Phi + \iota T + f\Delta)d\sigma\,\Big|_{t_1}^{t_2} . \qquad \text{(T.18)}$$

For sufficiently smooth fields, (T.18) implies the *reduced differential equation of energy balance*

$$q' + \Xi\cdot\dot{\Lambda} + s = \dot{\Phi} + \dot{\iota}T + \dot{f}\Delta , \qquad \text{(T.19)}$$

where

$$\Xi \equiv (N, Q, M, -\iota, -f) . \qquad \text{(T.20)}$$

Likewise, for sufficiently smooth fields, (II.D.1) and (T.5) imply the *differential Clausius-Duhem inequality*,

$$\dot{\iota} \geq j' + v , \qquad \text{(T.21)}$$

which, with the aid of (T.10) and (T.12) and upon multiplication by T, may be given the form

$$T\dot{\iota} \geq q' + T^{-1}(-qT' + T^2 w\Delta') + s + \Delta(Tw' + \chi) . \qquad \text{(T.22)}$$

Using the reduced equation of energy balance, (T.19), we obtain

$$\Xi\cdot\dot{\Lambda} - \dot{\Phi} - \dot{f}\Delta \geq T^{-1}(-qT' + T^2 w\Delta') + \Delta(Tw' + \chi) . \qquad \text{(T.23)}$$

We now assume that Φ, Ξ, q, and w are given by differentiable constitutive functions of the form

$$\Phi = \Phi(\Lambda, \Lambda') , \ldots , w = w(\Lambda, \Lambda') . \qquad \text{(T.24)}$$

Thus, in particular,

$$\dot{\Phi} = \Phi_{,\Lambda}\cdot\dot{\Lambda} + \Phi_{,\Lambda'}\cdot\dot{\Lambda}' , \qquad \text{(T.25)}$$

where $\Phi,_\Lambda$ is short for $(\Phi,_e, \cdots, \Phi,_\Delta)$, etc. (Though it is customary *not* to include strain gradients in the parameter list of the constitutive functions, it actually *simplifies* the form of the resulting equations to do so. As will be seen, the same final form for Φ emerges as if the strain gradients had not been included.)

We now wring a number of consequences from (T.23). First, since $\dot{f}$ may be expressed in a form similar to (T.25) and

$$w' = w,_\Lambda \cdot \Lambda' + w,_{\Lambda'} \cdot \Lambda'' , \tag{T.26}$$

(T.23) can be rewritten as

$$\begin{aligned}(\tilde{\Xi} - \tilde{\Phi},_\Lambda) \cdot \dot{\Lambda} - \tilde{\Phi},_{\Lambda'} \cdot \dot{\Lambda}' \\ \geq T^{-1}(-qT' + T^2 w \Delta') + \Delta[T(w,_\Lambda \cdot \Lambda' + w,_{\Lambda'} \cdot \Lambda'') + \chi] ,\end{aligned} \tag{T.27}$$

where $\tilde{\Xi} = (N, Q, M, -\iota, 0)$ and $\tilde{\Phi} = \Phi + f\Delta$ is a modified free-energy density.

At any σ and t,

$$\Upsilon \equiv (\Lambda, \dot{\Lambda}, \Lambda', \dot{\Lambda}', \Lambda'') \tag{T.28}$$

can be assigned arbitrarily, provided that $\mathbf{p}$, l, and s are chosen so that the differential equations of momentum and reduced energy balance, (E.1), (E.2), and (T.19), are satisfied. By virtue of the *form* of the constitutive equations, (T.24), it follows that a choice of Υ fixes every term in (T.27) *except* χ.[33] Thus, there must be a relation involving χ and at least some of the components of Υ; otherwise, by (T.11), we could give χ any value by prescribing suitable heat fluxes along the faces of the beamshell and thus violate (T.27).

We write the relation involving χ in the form

$$Tw' + \chi = A . \tag{T.29}$$

If we regard A as a new unknown and adopt the *Principle of Equipresence* (Truesdell & Toupin, 1960, Sect. 293), then A must be a function of Λ and Λ'. With this understanding, the right side of (T.27) reduces to $T^{-1}(-qT' + T^2 w\Delta') + \Delta A$—a function of Λ and Λ' only. Because the coefficients of $\dot{\Lambda}$ and $\dot{\Lambda}'$ on the left side of (T.27) are independent of $\dot{\Lambda}$ and $\dot{\Lambda}'$, and because $\dot{\Lambda}$ and $\dot{\Lambda}'$ may be prescribed arbitrarily at any σ and t, we conclude that, to not violate the inequality, we must have

$$\tilde{\Xi} = \tilde{\Phi},_\Lambda \ , \ \ \tilde{\Phi},_{\Lambda'} = \mathbf{0} . \tag{T.30}$$

Thus, $\tilde{\Phi}$ depends on e, g, k, and T only. But obviously, if we are to account for so simple a phenomenon as the build up of stress couples in a plate heated uniformly on one face and constrained to remain flat, we need a constitutive equation for M that depends on Δ; i.e., $\tilde{\Phi}$ is *not* a suitable free-energy density. We

---

[33]If $n = 0$, then, from $(T.6)_2$ and (T.11), $s = (c^\zeta\mu)_+ - (c^\zeta\mu)_-$ and $\chi = -H[(c^\zeta\mu)_+ + (c^\zeta\mu)_-]$. Thus, satisfaction of (T.19) fixes the sum of the heat flow through the faces of the beamshell, but leaves the difference at our disposal.

therefore return to (T.23) and take the relation for χ to be of the form

$$\dot{f} + Tw' + \chi = B \, , \tag{T.31}$$

where

$$B = B(\Lambda, \Lambda') \, . \tag{T.32}$$

By arguments similar to those used before, we conclude that (T.27) will remain inviolate if and only if $\Phi = \hat{\Phi}(\Lambda)$ and $\Xi = \Phi,_\Lambda$, i.e., if and only if

$$\Phi = \Phi(e, g, k, T, \Delta) \tag{T.33}$$

and

$$N = \Phi,_e \, , Q = \Phi,_g \, , M = \Phi,_k \, , \iota = -\Phi,_T \, , f = -\Phi,_\Delta \, . \tag{T.34}$$

With these relations, (T.19) and (T.23) reduce to

$$q' + s = \dot{\iota} T + \dot{f} \Delta \tag{T.35}$$

$$-qT' + T^2 w \Delta' + T \Delta B \leq 0 \, . \tag{T.36}$$

It might appear as if the decision to include all of

$$Tw' = T(w,_\Lambda \cdot \Lambda' + w,_{\Lambda'} \cdot \Lambda'') \tag{T.37}$$

in (T.31) was arbitrary. But this is not so. Consider the second term on the right of (T.37). Had it been left out of (T.31), then, because Λ'' can be prescribed arbitrarily at any σ and t and since w is independent of Λ'', we would have concluded that, to keep (T.27) inviolate, $w_{\Lambda'} = 0$. But this is a contradiction if $w \neq 0$, because (T.12) implies that w must change sign if σ is replaced by $L - \sigma$. The first term on the right of (T.37), being a function of Λ and Λ' only, could, if omitted, be reintroduced by redefining B as $B - Tw,_\Lambda \cdot \Lambda' \equiv C(\Lambda, \Lambda')$, say.

For simplicity, we assume that the constitutive relations for q, w, and B do *not* depend on the strain gradients e', g', and k'. Furthermore, we set $T = T_0 + \tau$, where T_0 is a constant, and assume that, in the reference state, $N, Q, M, e, g, k, \iota, f, \tau, \Delta, B, \tau'$, and Δ' are 0.

If σ is replaced by $L - \sigma$ or $\mathbf{b}$ by $-\mathbf{b}$, certain quantities in the constitutive relations change sign, as indicated in Table 4.2.

Table 4.2
Sign Changes and Dimensions of Constitutive Variables

	(C.4)	(K.1)	(K.6)	(T.6)	(T.9)	(T.12)	(T.17)	(T.31)
$\sigma, \mathbf{b}$	α'	e, g	k	q	$\tau, \tau', \Delta, \Delta'$	w	Φ	B
$L-\sigma, \mathbf{b}$	α'	$e, -g$	k	$-q$	$\tau, -\tau', \Delta, -\Delta'$	$-w$	Φ	B
$\sigma, -\mathbf{b}$	$-\alpha'$	$e, -g$	$-k$	q	$\tau, \tau', -\Delta, -\Delta'$	$-w$	Φ	$-B$
	L^{-1}	$1, 1$	L^{-1}	MLT^{-3}	$\Theta, L^{-1}\Theta, L^{-1}\Theta, L^{-2}\Theta$	$ML^2T^{-3}\Theta^{-2}$	MT^{-2}	$MLT^{-3}\Theta^{-1}$

The first row indicates the equation that defines the quantity and the last row indicates the dimensions of the quantity in terms of MASS (M), LENGTH (L), TIME (T), and TEMPERATURE (Θ). Invariance requirements and dimensional similarity may now be used to limit the possible forms of the constitutive relations for Φ, q, w, and B.

To be specific, we consider only the simplest case in which N, Q, M, ι, f, q, w, and B are linear in the state variables and the base-reference deviation $d=0$. This implies that

$$\Phi = \tfrac{1}{2}h(E_{11}e^2 + h^2E_{13}\alpha' ek - T_0^{-1}E_{14}e\tau - hT_0^{-1}E_{15}\alpha' e\Delta + E_{22}g^2 + h^2E_{33}k^2 - h^2T_0^{-1}E_{34}\alpha' k\tau - hT_0^{-1}E_{35}k\Delta - T_0^{-2}E_{44}\tau^2 - T_0^{-2}E_{55}\Delta^2) \tag{T.38}$$

$$q = h(q_0\tau' + h^2q_1\alpha'\Delta' + \underline{T_0q_2\alpha' g}) \tag{T.39}$$

$$w = -h^3(w_0\Delta' + w_1\alpha'\tau') - \underline{hT_0w_2g} \tag{T.40}$$

$$B = -h[B_0\Delta + \underline{h(B_1\alpha'\tau + T_0B_2k)}]\,, \tag{T.41}$$

where h is the beamshell thickness and $E_{11}, \cdots, B_2$ depend, at most, on σ and α'^2. The E_{ij}'s are "Young's moduli" and the q_i's are thermal conductivities. Substituting (T.39)-(T.41) into (T.36), we find that choosing all unknowns to be 0 except: g and τ' implies that $q_2=0$; g and Δ' implies that $w_2=0$; k and Δ implies that $B_2=0$; and τ and Δ implies that $B_1=0$. Thus, all the underlined terms in (T.39)-(T.41) must vanish. An examination of the resulting simplified form of (T.36) then shows that this inequality will always be satisfied provided

$$q_0, w_0, B_0 \geq 0 \tag{T.42}$$

$$4(T_0+\tau)^2q_0w_0 \geq \alpha'^2[q_1 + (T_0+\tau)^2w_1]^2\,. \tag{T.43}$$

For a flat beamshell, homogeneous in the thickness direction, (T.38)-(T.43), upon the neglect of shear-strain effects, reduce to

$$\Phi = \frac{C}{2}\left[e^2 + \frac{h^2k^2}{12} - 2\tilde{\alpha}\left(e\tau - \frac{h^2k\Delta}{12}\right)\right] - \frac{1}{2}\left(\frac{cm}{T_0}\right)\left(\tau^2 + \frac{h^2\Delta^2}{12}\right) \tag{T.44}$$

$$q = hq_0\tau'\,,\quad w = -h^3w_0\Delta'\,,\quad B = -hB_0\Delta\,, \tag{T.45}$$

where $(1-\nu^2)C \equiv Eh$, $\tilde{\alpha}$ is the *coefficient of thermal expansion*, c is the *specific heat at constant volume*, and q_0 is the *thermal conductivity*.

If ι, f, q, w, and B are computed from $(T.34)_4$, $(T.34)_5$, (T.44), and (T.45), and the resulting expressions substituted into (T.35) and (T.31), the equations we obtain, *if linearized*, read

$$(hq_0\tau')' + s = (cm\tau + T_0C\tilde{\alpha}e)^{\bullet} \tag{T.46}$$

$$[h^2/(12T_0)](cm\Delta - T_0 C\tilde{\alpha}k)^{\bullet} - T_0(h^3 w_0 \Delta')' + \chi = -hB_0\Delta . \qquad \text{(T.47)}$$

The homogeneous parts of these equations—with one exception—agree with equations (10.13) and (10.14) of Green & Naghdi (1979) for a plate, provided that, in (T.47), we take $12T_0^2 w_0 = q_0$ and $T_0 B_0 = q_0$, and that in Green & Naghdi's equations we set $(1-\nu)\gamma_3 = -\nu e$, $(1+\nu)\alpha^* = \alpha$, $\kappa = q_0$, and $\rho^* h = m$, where γ_3 is the transverse normal strain. The exception we mentioned above is that our coefficient of the bending strain in (T.47)—the factor $T_0 C\tilde{\alpha}$—does not agree with that of Green & Naghdi. The reason is that, in their equation (6.6), they introduce the assumption that the deformed position is a *linear* function of ζ (θ^3 in their notation). This leads to a transverse normal strain which is *independent* of ζ. However, for a flat beamshell whose reference surface lies in the xz-plane (so that $x = \sigma$ and $y = \zeta$), i.e., for a plate in a state of plane strain in the z-direction and in a state of plane stress in the y-direction, $(1-\nu)\gamma_3 = -\nu\gamma_\sigma$, where γ_σ is the 3-dimensional strain in the x-direction. But, to a 1st approximation, $\gamma_\sigma = e(\sigma) + \zeta k(\sigma)$. Thus, if bending and stretching are of equal importance (or if bending effects dominate, which is often the case in beamshell problems), any approximation to the transverse normal strain needs, at least, a term linear in the transverse coordinate.[34]

The homogeneous forms of (T.46) and (T.47) also agree with equations $(20)_1$ and $(20)_2$ of Jones (1966) if we set his $T_y = h^3\Delta/12$, replace C by E, i.e., ignore the factor $1-\nu^2$ (which is conventional in going from linear beamshell theory to linear beam theory), and neglect the dimensionless factor,

$$\frac{2(1+\nu)E\tilde{\alpha}^2 T_0}{(1-2\nu)\rho c}, \qquad \text{(T.48)}$$

which, as Dawson (1976, p.167) notes, is of the order of magnitude 10^{-2} for most solids near room temperature. [Alternatively, we may simply take our c equal to Jones' c multiplied by (1 – the above parameter).]

The nonhomogeneous terms in Green & Naghdi's equations (10.13) and (10.14) come from assuming that the only external heating is from radiation at the faces—a condition they express in linearized form. The nonhomogeneous terms in (T.46) and (T.47) reduce to these provided we note that "the ambient temperature[s] of the surroundings at the ... surface of the plate" must be measured relative to T_0.

U. The Thermodynamic Theory of Stability of Equilibrium

In this section, we indicate how thermodynamics can serve to illuminate and to justify the mechanical (or static) theory of elastic stability set forth in Section R. We do this by adopting to elastic beamshells ideas first propounded

[34]However, if membrane effects dominate, the assumption that the deformed position varies linearily with ζ is justified.

by Duhem (1911) and later refined and extended by Ericksen (1966, 1967, 1984), Koiter (1969, 1971, 1981), and Gurtin (1973a, 1973b, 1974, 1975). Here, we follow Koiter's presentation.

Consider a static, isothermal, equilibrium configuration in which $\bar{y} = \bar{y}_0(\sigma)$ and $\theta = T_0$, a constant. Let

$$\underline{u} \equiv \bar{y}(\sigma, t) - \bar{y}_0(\sigma) \ , \ \tau \equiv T(\sigma, t) - T_0 \, , \tag{U.1}$$

and $\underline{e}$, g, and $\underline{k}$ denote, respectively, increments, from their *equilibrium values*, of position, temperature, and strain and assume that the *incremental* external loads are derivable from an incremental potential, $\mathcal{P}[\underline{u}]$. Further, assume that the beamshell resides in a *passive* thermal environment in the sense that the ambient temperature remains fixed at T_0 and that there are no internal sources of heat ($n = 0$). The aim of our analysis, following Koiter, is to construct a *Lyapunov functional*, $\mathcal{V}[\underline{u}, \dot{\underline{u}}, \tau, \Delta]$, such that $\dot{\mathcal{V}} \le 0$ and thus to conclude—we defer to Koiter's papers for the analysis—that the L_2 norms,

$$\| \underline{u} \|_t^2 \equiv M^{-1} \int_0^L m \, | \underline{u}(\sigma, t) |^2 d\sigma \tag{U.2}$$

$$\| \dot{\underline{u}} \|_t^2 \equiv M^{-1} \int_0^L m \, | \dot{\underline{u}}(\sigma, t) |^2 d\sigma \tag{U.3}$$

$$\| \tau \|_t^2 \equiv M^{-1} \int_0^L m \tau^2(\sigma, t) d\sigma \tag{U.4}$$

$$\| \Delta \|_t^2 \equiv M^{-1} \int_0^L m \Delta^2(\sigma, t) d\sigma \, , \tag{U.5}$$

can be made as uniformly small as we please by taking the L_2 norms ($\| \underline{u} \|_0^2$, $\| \dot{\underline{u}} \|_0^2$, $\| \tau \|_0^2$, $\| \Delta \|_0^2$) of any initial disturbance sufficiently small. Here, $M \equiv \int_0^L m(\sigma) d\sigma$ is the total mass of the beamshell. For simplicity, we shall restrict ourselves to classical beamshell theory.

To obtain the differential inequality $\dot{\mathcal{V}} \le 0$, we assume first that the Clausius-Duhem inequality (II.D.1) and the energy balance (T.13)—with the external mechanical power, $\mathcal{W} = -\dot{\mathcal{P}}$—involve sufficiently smooth fields to be written in the differential forms $\dot{\mathcal{H}} \ge \mathcal{J}$ and $\mathcal{Q} = (\mathcal{K} + \mathcal{P} + \mathcal{I})^{\bullet}$. By multiplying the first equation by T_0 and adding it to the second, we have, upon rearranging terms,

$$\dot{\mathcal{V}} \le \mathcal{Q} - T_0 \mathcal{J} \, , \tag{U.6}$$

where

$$\mathcal{V} \equiv \mathcal{K} + \mathcal{P} + \mathcal{I} - T_0 \mathcal{H} \, . \tag{U.7}$$

We assume that, if the local temperature θ is greater than the ambient temperature T_0 at any point on the edges or faces of the beamshell, then heat flows outward. This implies

$$\left[\left(\frac{1}{T_0}-\frac{1}{\theta}\right)c^{\zeta}\mu\right]_{-}^{+}\overset{<}{\underset{>}{=}}0\ ,\quad \left[\left(\frac{1}{T_0}-\frac{1}{\theta}\right)c^{\sigma}\mu\right]_{0}^{L}\overset{<}{\underset{>}{=}}0\ . \tag{U.8}$$

Thus, by $(T.2)_2$, (T.4), (T.6), and (T.7),

$$\mathcal{Q}-T_0\mathcal{J}=T_0\left\{\int_{-H}^{H}\left[\left(\frac{1}{T_0}-\frac{1}{\theta}\right)c^{\sigma}\mu\right]_{0}^{L}d\zeta\right.$$
$$\left.+\int_0^L\left[\left(\frac{1}{T_0}-\frac{1}{\theta}\right)c^{\zeta}\mu\right]_{-}^{+}d\sigma\right\}\leq 0\ . \tag{U.9}$$

We now introduce the free energy via (T.16) and (T.17) and, taking note of $(T.5)_1$, $(T.34)_4$, $(T.34)_5$, and (U.1), set

$$I-T_0\mathcal{H}=\int_a^b[\Phi(\Lambda)-\tau\Phi,_T(\Lambda)-\Delta\Phi,_\Delta(\Lambda)]d\sigma\ , \tag{U.10}$$

where Λ is defined by (T.15). It was Ericksen's happy notion to construct a finite Taylor expansion of the free-energy density, Φ, *not* about the ambient, equilibrium temperature, T_0, but about the local temperature, T (and, in our case, the local transverse temperature gradient, Δ). Thus, assuming obvious smoothness, we may set

$$\Phi(\underline{e},\underline{g},\underline{k},T_0,0)=\Phi(\Lambda)-\tau\Phi,_T(\Lambda)-\Delta\Phi,_\Delta(\Lambda)$$
$$-\frac{1}{2T^*}(c^*_{\tau\tau}\tau^2+2c^*_{\tau\Delta}\tau\Delta+c^*_{\Delta\Delta}\Delta^2)\ , \tag{U.11}$$

where

$$c^*_{\tau\tau}\equiv-\Phi,_{TT}(\Lambda^*)\ ,\quad c^*_{\tau\Delta}\equiv-\Phi,_{T\Delta}(\Lambda^*)\ ,\quad c^*_{\Delta\Delta}\equiv-\Phi,_{\Delta\Delta}(\Lambda^*)\ , \tag{U.12}$$

and $\Lambda^*=(e, g, k, T^*, \Delta^*)$, with T^* and Δ^* some unknown values between T_0 and T and 0 and Δ, respectively. Inserting (U.11) into (U.10) and the resulting expression into (U.7), the Lyapunov functional takes the form

$$\mathcal{V}=\Pi+\mathcal{K}+\mathcal{C}^*\ , \tag{U.13}$$

where

$$\Pi\equiv\mathcal{P}+\mathcal{E} \tag{U.14}$$

is the *total (incremental) potential energy functional*,

$$\mathcal{E}\equiv\int_0^L\Phi(\underline{e},\underline{g},\underline{k},T_0,0)d\sigma \tag{U.15}$$

is the (incremental) *strain-energy density of isothermal deformation* at temperature T_0, and

$$C^* \equiv \tfrac{1}{2}\int_0^L T^{*-1}(c^*_{\tau\tau}\tau^2 + 2c^*_{\tau\Delta}\tau\Delta + c^*_{\Delta\Delta}\Delta^2)d\sigma\,. \tag{U.16}$$

In analogy with the terminology of 3-dimensional thermodynamics, we call the matrix, C^*, the *specific heat matrix at constant strain.* {If Φ has the quadratic form (T.44), then $C^* = \text{diag}[cm/T_0, cmh^2/(12T_0)]$.} Thus, if this matrix is positive definite (in analogy with the conventional assumption in 3-dimensions that the scalar specific heat is positive) and continuous, and if the incremental fields are continuous, then $\mathcal{K}$ and C^* are non-negative definite and one sees, at least intuitively, that the positive definiteness of the Lyapunov functional $\mathcal{V}$ depends on the positive definiteness of Π, *which is precisely the functional considered in Section R on the mechanical theory of stability.* We refer the reader to the papers of Ericksen, Koiter, and Gurtin cited at the beginning of this section for statements and proofs of theorems on stability and instability in terms of Π.

References

[A letter in brackets indicates the section where the reference appears.]

Antman, S. S. (1968). General solutions for plane extensible elasticae having nonlinear stress-strain laws. *Q. Appl. Math.* **26,** 35-47. [O]

Antman, S. S. (1969). Equilibrium states of nonlinearly elastic rods. *In* "Bifurcation Theory and Nonlinear Eigenvalue Problems" (J. B. Keller & S. S. Antman, eds.), pp. 331-358. Benjamin, New York. [O]

Antman, S. S. (1970). The shape of buckled nonlinearly elastic rods. *ZAMP* **21,** 422-438. [R]

Antman, S. S. (1972). The theory of rods. *In* "Encyclopedia of Physics," 2nd. ed., (S. Flügge, ed.). pp. 641-703. Springer-Verlag, Berlin and New York. [H, O, R]

Antman, S. S. (1974a). Qualitative theory of the ordinary differential equations of nonlinear elasticity. *Mechanics Today* **1,** 58-101. [K]

Antman, S. S. (1974b). Kirchoff's problem for nonlinearly elastic rods. *Q. Appl Math.* **32,** 221-240. [O]

Antman, S. S. (1976). Ordinary differential equations of non-linear elasticity I: foundations of the theories of non-linearly elastic rods and shells. *Arch. Rat. Mech. Anal.* **61,** 307-351. [J]

Antman, S. S., and Dunn, J. E. (1980). Qualitative behavior of buckled nonlinearly elastic arches. *J. Elasticity* **10,** 225-239. [R]

Atluri, S. N. (1984). Alternate stress and conjugate strain measures, and mixed variational formulations involving rigid rotations, for computational analyses of finitely deformed solids, with application to plates and shells—I. *Computers & Struct.* **18,** 93-116. [Q]

Babcock, C. D. (1983). Shell stability. *J. Appl. Mech.* **50,** 935-940. [R]

Benson, R. C. (1981). The deformation of a thin, incomplete, elastic ring in a frictional channel. *J. Appl. Mech.* **48,** 895-899. [H]

Berger, N. (1973). Estimates for stress derivatives and error in interior equations for shells of variable thickness with applied forces. *SIAM J. Appl. Math.* **24,** 97-120. [N]

Bickley, W. G. (1934). The heavy elastica. *Phil. Mag.* **17,** 603-622. [O]
Biezeno, C. B., and Grammel, R. (1939). "Technische Dynamik," vol. 1. Springer-Verlag, Berlin. [R]
Bolotin, V. V. (1963). "Nonconservative Problems in the Theory of Elastic Stability." Macmillan, New York. [R]
Brush, D. O., and Almroth, B. O. (1975). "Buckling of Bars, Plates, and Shells." McGraw-Hill, New York, etc. [R]
Buchanan, G. R., Huang, J.C., and Cheng, T.K.M. (1970). Effect of shear on nonlinear behavior of elastic bars. *J. Appl. Mech.* **37,** 212-215. [O]
Buck, R. C. (1978). "Advanced Calculus." McGraw-Hill, New York, etc. [J]
Budiansky, B. (1968). Notes on nonlinear shell theory. *J. Appl. Mech.* **35,** 393-401. [L]
Budiansky, B. (1974). Theory of buckling and post-buckling behavior of elastic structures. *Adv. Appl. Mech.* **14,** 1-65. [R]
Budiansky, B., and Hutchinson, J. W. (1979). Buckling: progress and challange. *In* "Trends in Solid Mechanics" (Koiter birthday volume, J. F. Besseling & A. M. A. Van der Heijden, eds.), pp. 93-116. Delft Univ. Press, The Netherlands. [R]
Budiansky, B., and Payton, R. (1961). Studies on the dynamic response of shell structures and materials to a pressure pulse. AVCO AF SWC-TR-61-31 (II). [P]
Budiansky, B., and Sanders, J. L., Jr. (1963). On the "best" first-order linear shell theory. *In* "Progress in Applied Mechanics" (Prager Anniversary Volume, D. C. Drucker, ed.), pp. 129-140. Macmillan, New York. [L]
Buffler, H. (1984). Pressure loaded structures under large deformations. *ZAMM* **64,** 287-295. [J]
Bushnell, D. (1985). Static collapse: a survey of methods and modes of behavior. *Finite Elements in Anal. Design.* **1,** 165-205. [R]
Carrier. G. F. (1947). On the buckling of elastic rings. *J. Math. & Phys.* **26,** 94-103. [R]
Coleman, B. D., and Noll, W. (1963). The thermodynamics of elastic materials with heat conduction and viscosity. *Arch. Rat. Mech. Anal.* **51,** 1-53. [T]
Cowper, G. R. (1966). The shear coefficient in Timoshenko's beam theory. *J. Appl. Mech.* **33,** 335-340. [B, N]
DaDeppo, D. A., and Schmidt, R. (1969). Sidesway buckling of deep circular arches under a concentrated load. *J. Appl. Mech.* **36,** 325-327. [R]
Dawson, T. H. (1976). "Theory and Practice of Solid Mechanics." Plenum, New York and London. [T]
Duhem, P. (1911). "Traité d' Energétique ou Thermodynamique Générale," vols. 1 & 2. Gauthier-Villars, Paris. [R, T, U]
Dym, C. L. (1974). "Stability Theory and its Application to Structural Mechanics." Noordhoff, Leyden. [R]
Elisakoff, I. (1983). "Probabilistic Methods in the Theory of Structures." Wiley, New York. [R]
Ericksen, J. L. (1966). A thermo-kinetic view of elastic stability theory. *Int. J. Solids Struct.* **2,** 573-580. [R, T, U]

Ericksen, J. L. (1967). Thermoelastic stability. *In* "Proc. 5th U. S. Nat. Cong. Appl. Mech.," pp. 187-193. Am Soc. Mech. Engrs., New York. [R, T, U]
Ericksen, J. L. (1984). Thermodynamics and stability of equilibrium. *In* "Rational Thermodynamics," 2nd ed. (C. Truesdell, author and ed.), pp. 503-508. Springer-Verlag, New York, etc. [U]
Ericksen, J. L., and Truesdell, C. A. (1958). Exact theory of stress and strain in rods and shells. *Arch. Rat. Mech. Anal.* **1,** 295-323. [M]
Euler, L. (1744). "Methodus Inveniendi Lineas Curvas Maximi Minimive Proprietate Gaudentes." Lausanne and Geneva. [R]
Fisher, D. H. (1986). Private communication. [A]
Fisher, D. H. (1987). Configuration dependent pressure potentials. *J. Elasticity* (to appear). [J]
Fraeys de Veubeke, B. (1972). A new variational principle for finite elastic displacements. *Int. J. Engr. Sci.* **10,** 745-763. [Q]
Frisch-Fay, R. (1962). "Flexible Bars." Butterworth, London. [O]
Fukuchi, N., and Atluri, S. N. (1981) Finite deformation analysis of shells: a complementary energy-hybrid method. *In* "Nonlinear Finite Element Analysis of Plates and Shells" (T. J. R. Hughes, ed.), AMD-vol. 48, pp. 233-247. Am. Soc. Mech. Engrs. [Q]
Gelfand, I. M., and Fomin, S. V. (1963). "Calculus of Variations" (Transl. from the Russian by R. Silverman). Prentice-Hall, Englewood Cliffs, New Jersey. [J]
Gjelsvik, A., and Bodner, S. R. (1962). Energy criterion and snap buckling of arches. *ASCE J. Engr. Mech. Div.* **88,** 87-134. [R]
Gorski, W. (1976). A review of the literature and a bibliography on finite elastic deflections of bars. *Civil Engr. Trans. I. E. Australia* **CE18,** 74-85. [O]
Green, A. E., and Naghdi, P. M. (1979). On thermal effects in the theory of shells. *Proc. R. Soc. London* **A365,** 161-190. [T]
Green, A. E., and Zerna, W. (1968). "Theoretical Elasticity," 2nd ed., Oxford Univ. Press. [N]
Gurtin, M. E. (1973a). Thermodynamics and the potential energy of an elastic body. *J. Elasticity* **3,** 23-26. [U]
Gurtin, M. E. (1973b). Thermodynamics and the energy criterion for stability. *Arch. Rat. Mech. Anal.* **52,** 93-103. [U]
Gurtin, M. E. (1974). Modern continuum thermodynamics. *Mechanics Today* **1,** 168-213. [T, U]
Gurtin, M. E. (1975). Thermodynamics and stability. *Arch. Rat. Mech. Anal.* **59,** 63-96. [T, U]
Gurtin, M. E. (1979). Thermodynamics and the Griffith criterion for brittle fracture. *Int. J. Solids Struct.* **15,** 553-560. [T]
Hildebrand, F. B., Reissner, E. and Thomas, G. B. (1949). Notes on the foundations of the theory of small displacement of orthotropic shells. NACA Tech. Note No. 1833. [N]
Holden, J. T. (1972). On the finite deflection of thin beams. *Int. J. Solids Struct.* **8,** 1051-1055. [O]
Hummel, F. H., and Morton, W. B. (1927). On the large bending of thin flexible

strips and the measurement of their elasticity. *Phil. Mag.* **4** (Series 7), 348-357. [O]

Hutchinson, J. W. (1974). Plastic buckling. *Adv. Appl. Mech.* **14,** 67-143. [R]

Hutchinson, J. W., and Koiter, W. T. (1970). Postbuckling theory. *Appl. Mech. Rev.* **23,** 1353-1366. [R]

Irvine, H. M. (1981). "Cable Structures." MIT Press, Cambridge. [O]

John, F. (1965a). A priori estimates applied to nonlinear shell theory. *Proc. Sympos. Appl. Math., vol. 17,* pp. 102-110. Am. Math Soc. [N]

John, F. (1965b). Estimates for the derivatives of the stresses in a thin shell and interior shell equations. *Comm. Pure Appl. Math.* **18,** 235-267. [N]

John, F. (1969). Refined interior shell equations. *In* "Theory of Thin Shells," Proc. I.U.T.A.M. Sympos., Copenhagen, 1967 (F.I. Niordson, ed.), pp. 1-14. Springer-Verlag, Berlin, etc. [N]

John, F. (1971). Refined interior equations for thin elastic shells. *Comm. Pure Appl. Math.* **24,** 583-615. [N]

Jones, J. P. (1966). Thermoelastic vibrations of beams. *J. Acoustical Soc. Am.* **39,** 542-548. [B,T]

Keller, J. B., and Ting, L. (1966). Periodic vibrations of systems governed by nonlinear partial differential equations. *Comm. Pure Appl. Math.* **19,** 371-420. [P]

Koiter, W. T. (1945). "Over de Stabiliteit van het Elastisch Evenwicht." Thesis, Delft. Amsterdam. [R]

Koiter, W. T. (1960). A consistent first approximation in the general theory of thin elastic shells. *In* "The Theory of Thin Elastic Shells," Proc. I.U.T.A.M. Sympos., Delft, 1959 (W. T. Koiter, ed.), pp. 12-33. North-Holland Publ., Amsterdam. [L, N]

Koiter, W. T. (1966). On the nonlinear theory of thin elastic shells, I-III. *Proc. Kon. Ned. Ak. Wet.* **B69,** 1-54. [R]

Koiter, W. T. (1967a), "On the Stability of Equilibrium." NASA TT-F-10. [R]

Koiter, W. T. (1967b). General equations of elastic stability for thin shells. *In* "Proceedings of the Symposium on the Theory of Shells to Honor L. H. Donnell" (D. Muster, ed.), pp. 187-228. Univ. of Houston Press, Houston. [R]

Koiter. W. T. (1967c). A sufficient condition for the stability of shallow shells. *Proc. Kon. Ned. Ak. Wet.* **B70,** 367-375. [R]

Koiter. W. T. (1968). Purpose and achievements of research in elastic stability. *In* "Recent Advances in Engineering Science," Proc. 4th Tech. Meeting Soc. Engr. Sci., Raleigh, N. C., 1966, pp. 197-218. [R]

Koiter, W. T. (1969). On the thermodynamic backround of elastic stability theory. *In* "Problems of Hydrodynamics and Continuum Mechanics" (The Sedov Anniversary Volume), pp. 423-433. SIAM, Philadelphia. [T]

Koiter, W. T. (1970). "The Stability of Elastic Equilibrium." Air Force Flight Dynamics Lab. Rept. TR-70-25. [R]

Koiter, W. T. (1971). Thermodynamics of elastic stability. *In* "Proc. 3rd Canadian Cong. Appl. Mech., Calgary," pp. 29-37. [R, T, U]

Koiter, W. T. (1976). Current trends in the theory of buckling. *In* "Proc.

I.U.T.A.M. Sympos. on Buckling of Structures, Cambridge, USA, 1974," pp. 1-16. [R]

Koiter, W. T. (1981). Elastic stability, buckling and post-buckling behaviour. *In* "Proc. I.U.T.A.M. Sympos. on Finite Elasticity, Lehigh, 1980" (D. E. Carlson & R. T. Shields, eds.), pp. 13-24. [R, U]

Lagrange, J. L. (1788). "Méchanique Analitique." Paris. [R]

Lee, S. Y. (1971). On the finite deflection dynamics of thin elastic beams. *J. Appl. Mech.* **38,** 961-963. [P]

Lee, S. J., and Shield, R. T. (1980). Variational principles in finite elastostatics. *ZAMP* **31,** 437-453. [Q]

Leipholz, H. H. E. (1980). "Stability of Elastic Systems." Noordhoff, Leyden. [R]

Lévy, M. (1883). Sur un nouveau cas intégrable de la problème de l'élastique et l'une de ses applications. *Compt. Rend. Acad. Sci., Paris* **97,** 694-697. [R]

Libai, A. (1962). On the nonlinear elastokinetics of shells and beams. *J. Aerospace Sci.* **29,** 1190-1195, 1209. [P]

Libai, A. (1984). Strongly nonlinear bending, extensional and shearing deformations of ringlike structures. Technion, Israel Institute of Technology, Dept. of Aeronautical Engr. Rept. No. 552. [O]

Libai, A. and Simmonds, J. G. (1981). Large-strain constitutive laws for the cylindrical deformation of shells. *Int. J. Non-Linear Mech.* **16,** 91-103. [N, O, S]

Libai, A., and Simmonds, J. G. (1983a). Nonlinear elastic shell theory. *Adv. Appl. Mech.* **23,** 271-371. [L, T]

Libai, A., and Simmonds, J. G. (1983b). Highly non-linear cylindrical deformations of rings and shells. *Int. J. Non-Linear Mech.* **18,** 181-197. [N, O]

Mansfield. L., and Simmonds, J. G. (1987). The reverse spaghetti problem: drooping motion of an elastica issuing from a horizontal guide. *J. Appl. Mech.* **54,** 147-150. [P]

Marguerre, K. (1938). Zur Theorie der gekrümmten Platte grosser Formänderung. *In* "Proc. 5th Int. Cong. Appl. Mech.," pp. 93-101. Wiley, New York. [R]

Masur, E. F., and Lo, D. L. C. (1972). The shallow arch: general buckling, post buckling and imperfection analysis. *J. Struct. Mech.* **1,** 91-112. [R]

Müller, I. (1969). On the entropy inequality. *Arch. Rat. Mech. Anal.* **26,** 118-141. [T]

Namias, V. (1985). Load-supporting fluid-filled cylindrical membranes. *J. Appl. Mech.* **52,** 913-918. [O]

Navier, C. H. (1833). "Résumé des Leçons sur l'Application de la Méchanique," 2nd ed. Paris. [R]

Nemat-Nasser, S. (1974). General variational principles in nonlinear and linear elasticity with applications. *Mechanics Today* **1,** 214-261. [Q]

Newman, B. G. (1975). Shape of a towed boom of logs. *Proc. Roy. Soc. London* **A346,** 329-348. [O]

Niordson, F. I. (1971). A note on the strain energy of elastic shells. *Int. J. Solids Struct.* **7,** 1573-1579. [N]

Novozhilov, V. V. (1953). "Foundations of the Nonlinear Theory of Elasticity." Graylock Press, Rochester, New York. [K]
Oden, J. T., and Reddy, J. N. (1983). "Variational Methods in Theoretical Mechanics," 2nd ed. Springer-Verlag, Berlin, etc. [J, Q]
Parikh, B. B., Schmidt, R., and DaDeppo, D. A. (1972). Finite deflections of beams with restrained ends. *J. Indust. Math. Soc.* **22,** 77-82. [O]
Pielorz, A., and Nadolski, W. (1986). Non-linear vibration of a cantilever beam of variable cross-section. *ZAMM* **66,** 147-154. [P]
Pietraszkiewicz, W. (1979). "Finite Rotations and Lagrangian Description in the Non-Linear Theory of Shells." Polish Scientific Pub., Warsaw. [K]
Pflüger, A. (1975). "Stabilitätsprobleme der Elastostatik," 3rd ed. Springer-Verlag, Berlin, and New York. [O]
Prescott, J. (1942). Elastic waves and vibrations of thin rods. *Phil. Mag.* **33,** 703-754. [B]
Rankine, W. J. W. (1858). "A Manual of Applied Mechanics." Richard Griffin and Co., pp. 342-344. [O]
Reissner, E. (1962). Variational considerations for elastic beams and shells. *Proc. ASCE Eng. Mech. Div.* **8,** 23-57. [K]
Reissner, E. (1972). On one-dimensional finite-strain beam theory. *ZAMP* **23,** 795-804. [K]
Reissner, E. (1984). Formulation of variational theorems in geometrically non-linear elasticity. *J. Engr. Mech.* **110,** 1377-1390. [Q]
Reissner, E., and Wan, F. Y. M. (1969). On the equations of linear shallow shell theory. *Studies Appl. Math.* **48,** 133-145. [R]
Reissner, E., and Wan, F. Y. M. (1982). A note on the linear theory of shallow shear-deformable shells. *ZAMP* **33,** 425-427. [R]
Rohde, F. V. (1953). Large deflections of a cantilever beam with uniformly distributed load. *Q. Appl. Math.* **11,** 337-338. [O]
Romano, G. (1972). Potential operators and conservative systems. *Meccanica* **7,** 141-146. [J]
Sanders, J. L., Jr. (1959). An improved first-approximation theory for thin shells. NASA Rep. No. 24. [L]
Sanders, J. L., Jr. (1963). Nonlinear theories for thin shells. *Q. Appl. Math.* **21,** 21-36. [R]
Schreyer, H. L., and Masur, E. F. (1966). Buckling of shallow arches. *ASCE J. Engr. Mech. Div.* **92,** 1-19. [R]
Sewell, M. J. (1967). On configuration-dependent loading. *Arch. Rat. Mech. Anal.* **23,** 327-351. [J]
Sills, L. B., and Budiansky, B. (1978). Postbuckling ring analysis. *J. Appl. Mech.* **45,** 208-210. [J, R]
Simitses, G. J. (1976). "An Introduction to the Elastic Stability of Structures." Prentice-Hall, Englewood Cliffs, New Jersey. [R]
Simmonds, J. G. (1966). A set of simple, accurate equations for circular cylindrical elastic shells. *Int. J. Solids Stuct.* **2,** 525-541. [P]
Simmonds, J. G. (1979). Accurate nonlinear equations and a perturbation solution for the free vibrations of a circular elastic ring. *J. Appl. Mech.* **46,**

156-160. [P]

Simmonds, J. G. (1984). The nonlinear thermodynamical theory of shells: descent from 3-dimensions without thickness expansions. *In* "Flexible Shells" (E. L. Axelrad & F. A. Emmerling, eds.), pp. 1-11. Springer-Verlag, Berlin, etc. [T]

Simmonds, J. G. (1985a). A new displacement form for the nonlinear equations of motion of shells of revolution. *J. Appl. Mech.* **52,** 507-509. [L]

Simmonds, J. G. (1985b). The strain energy density of rubber-like shells. *Int J. Solids Struct.* **21,** 67-77. [N]

Simmonds, J. G., and Danielson, D. A. (1970). Nonlinear shell theory with a finite rotation vector. *Proc. Kon. Ned. Ak. Wet.* **B73,** 460-478. [K]

Simmonds, J. G., and Danielson, D. A. (1972). Nonlinear shell theory with finite rotation and stress-function vectors. *J. Appl. Mech.* **39,** 1085-1090. [K, Q]

Simmonds, J. G., and Mann, J. E., Jr. (1986). "A First Look at Perturbation Theory." Krieger, Melbourne, Florida. [O]

Singer, J., and Babcock, C. D. (1970). On the buckling of rings under constant directional and centrally directed pressure. *J. Appl. Mech.* **37,** 215-218. [R]

Smith, C. V., and Simitses, G. J. (1969). Effect of shear and load behavior on ring stability. *ASCE J. Eng. Mech. Div.* **95,** 559-569. [R]

Stackgold, I. (1979). "Green's Functions and Boundary Value Problems." Wiley, New York, etc. [J]

Stein. M. (1968). Some recent advances in the investigation of shell buckling. *AIAA J.* **6,** 2339-2345. [R]

Stephen, N. G. (1981). Considerations on second order beam theories. *Int. J. Solids Struct.* **17,** 325-333. [B]

Thompson, J. M. T., and Hunt, G. W. (1973). "A General Theory of Elastic Stability." Wiley, London, etc. [R]

Timoshenko, S. P. (1921). On the correction for shear of the differential equation for transverse vibrations of prismatic bars. *Phil. Mag.* **41,** 744-746. [O]

Timoshenko, S. P. (1922). On the transverse vibrations of bars of uniform cross section. *Phil. Mag.* **43,** 125-131. [O]

Timoshenko, S. P., and Gere, J. M. (1961). "Theory of Elastic Stability," 2nd ed. McGraw-Hill, New York, etc. [R]

Timoshenko, S. P., and Goodier, J. N. (1970). "Theory of Elasticity," 3rd ed. McGraw-Hill, New York, etc. [H, S]

Truesdell, C. A. (1969). "Rational Thermodynamics." McGraw-Hill, New York. [T]

Truesdell, C. A. (1984). "Rational Thermodynamics," 2nd ed. Springer-Verlag, New York, etc. [T]

Truesdell, C. A., and Toupin, R. A. (1960). The classical field theories. *In* "Encyclopedia of Physics" (S. Flügge, ed.), vol. III/1. Springer-Verlag, Berlin and New York. [T]

Tvergaard, V. (1976). Buckling behaviour of plate and shell structures. *In* "Theoretical and Applied Mechanics" (W. T. Koiter, ed.), pp. 233-247. North-Holland. [R]

Vainberg, M. M. (1964). "Variational Methods for the Study of Nonlinear Operators." Holden-Day, San Francisco. [J]
van der Heijden, (1973). On the influence of the bending stiffness in cable analysis. *Proc. Kon. Ned. Ak. Wet.* **B76,** 217-229. [O]
Verma, G. R. (1972). Non-linear vibrations of beams and membranes. *ZAMP* **23,** 805-814. [P]
Wang, C. Y. (1981a). Large deformations of a heavy cantilever. *Q. Appl. Math.* **39,** 261-273. [O]
Wang, C. Y. (1981b). Unfolding a curved elastic sheet. *J. Mech. Engr. Sci.* **23,** 217-219. [O]
Wang, C. Y. (1981c). Winding a long elastic sheet. *Acta Mech.* **39,** 297-301. [O]
Wang, C. Y. (1981d). Equilibrium of a heavy, naturally curved sheet on an inclined plane. *ZAMM* **61,** 267-269. [O]
Wang, C. Y. (1984a). The filling of a long membrane container. *J. Struct. Mech.* **12,** 1-11. [O]
Wang, C. Y. (1984b). Stability of a rotating ring under external pressure. *J. Appl. Mech.* **51,** 439-440. [R]
Wang, C. Y. (1986a). The lifted beam. *Acta Mech.* **65,** 145-152. [O]
Wang, C. Y. (1986b). A critical review of the heavy elastica. *Int. J. Mech. Sci.* **28,** 549-559. [O]
Wang, C. Y., and Watson, L. T. (1981a). Equilibrium of heavy elastic cylindrical shells. *J. Appl. Mech.* **48,** 582-586. [O].
Wang, C.Y., and Watson, L. T. (1981b). The fluid-filled cylindrical membrane container. *J. Engr. Math.* **15,** 81-88. [O]
Wang, C. Y., and Watson, L. T. (1983). Free rotation of a circular ring about a diameter. *ZAMP* **34,** 13-24. [O]
Wang, C. Y., and Shodja, H. M. (1984). Equilibrium of a tip weighted curved sheet on an inclined plane. *Acta Mech.* **53,** 173-181. [O]
Wang, T. M. (1969). Non-linear bending of beams with uniformly distributed loads. *Int. J. Non-lin. Mech.* **4,** 389-395. [O]
Watson, L. T. (1981). Engineering applications of the Chow-York Algorithm. *Appl. Math. and Computation* **9,** 111-133. [O]
Watson, L. T., and Wang, C. Y. (1982). Overhang of a heavy elastic sheet. *ZAMP* **33,** 17-23. [O]
Watson, L. T., and Wang, C. Y. (1983). The periodically supported heavy elastic sheet. *J. Engr. Mech.* **109,** 811-820. [O]
Weinitschke, H. J. (1985). On the calculation of limit and bifurcation points of stability problems in elastic shells. *Int. J. Solids Struct.* **21,** 79-95. [R]
Wempner, G. A. (1969). Finite elements, finite rotations and small strain of flexible shells. *Int. J. Solids Struct.* **5,** 117-153. [K]
Yamaki, N. (1984). "Elastic Stability of Circular Cylindrical Shells." North-Holland. [Q]
Ziegler, H. (1967). Some developments in the theory of stability. *In* "Proc. Can. Cong. Appl. Mech., Quebec," vol. 3, pp. 233-250. [R]

Chapter V

Torsionless, Axisymmetric Motion of Shells of Revolution (Axishells)

A *true* shell has double curvature: if undeformed, its Gaussian curvature vanishes nowhere, except possibly along certain curves (such as the crown of a toroid) or at isolated points (such as the apex of a very flat dome). Shells of revolution, with the exception of conical and cylindrical shells, are the simplest, useful class of true shells. It is a corollary of Gauss' *Theorema egregium* (Struik, 1961) that if a surface is bent, its Gaussian curvature will change only if the surface is simultaneously stretched. Thus, a shell will be relatively stiff if, by virtue of symmetry, loading, or boundary conditions, inextensional deformation is impossible.[1] Such is the case with a shell of revolution that undergoes torsionless, axisymmetric deformation without rigid body motion. This feature of *axishells* is in contrast to the behavior of beamshells where near inextensional bending is typical.

In beamshells there is 1 extensional strain, 1 shearing strain, and 1 bending strain; an arbitrary prescription of these 3 fields always yields some displacement field. However, in an axishell there are 6 strains: 2 extensional strains, 1 shearing strain, 2 bending strains, and 1 *normal* bending strain; these fields cannot be prescribed independently but must satisfy *compatibility* conditions if an associated displacement field is to exist. There is 1 differential compatibility condition if the rotation is introduced as a dependent variable, but 3—2 are differential and 1 is algebraic—if the strains are the basic kinematic unknowns.

The normal bending strain—call it k_n—was introduced by Reissner (1969). If we assume that the stress couple vector has a component, M_n, along the normal to the deformed shell reference surface—a nonclassical assumption—then a term $M_n \dot{k}_n$ appears in the deformation power density. In the linear theory of axishells, Reissner (1969) has pointed out that the pair (M_n, k_n) is the static-geometric dual of the pair (g, Q), where g is the shear strain and Q

[1]The role of inextensionality in shell theory is emphasized in the book by Calladine (1983).

is the transverse shear stress resultant. Reissner (1969) also has pointed out—see his equation (24)—that the hoop and normal bending strains, k_θ and k_n, satisfy an algebraic relation. This implies that the rates (or, equivalently, the variations) of these strains are not independent, so that in an elastic shell, where the variation of the strain-energy density is equal to the virtual work of the stress resultants and couples, we cannot infer independent stress-strain relations for the hoop and normal stress couples *unless* we introduce a Lagrange multiplier. Remarkably, this multiplier appears neither in the equations of motion nor in the boundary conditions.[2] The situation is reminiscent of the observation of Koiter (1964) that in the couple-stress theory of 3-dimensional elasticity, the linear invariant of the couple-stress tensor likewise appears neither in the field equations nor in the boundary conditions. As there is no way to determine our multiplier within the framework of our theory, we shall set it to 0, as did Koiter with his linear invariant.

A. Geometry of the Undeformed Shell

A shell of revolution is a body whose reference shape is generated by rotating a plane region—its *meridional section*—about a fixed (vertical) axis. See Fig. 5.1a. Mathematically, the meridional section is the image of a rectangular region; the images of the top and bottom of the rectangular region generate the *faces* of the shell and the images on the ends of the rectangular region generate the *edges* of the shell.

For a quantitative description, let (r, θ, z) be a set of circular cylindrical coordinates in a fixed, right-handed Cartesian frame $Oxyz$. Denote by $\{\mathbf{e}_r, \mathbf{e}_\theta, \mathbf{e}_z\}$ the standard set of orthonormal base vectors associated with (r, θ, z), where

$$\mathbf{e}_r = \mathbf{e}_x \cos\theta + \mathbf{e}_y \sin\theta \;, \quad \mathbf{e}_\theta = -\mathbf{e}_x \sin\theta + \mathbf{e}_y \cos\theta \,. \tag{A.1}$$

The reference shape may then be expressed in the vector-parametric form

$$\mathbf{x}(\sigma, \theta, \zeta) = R(\sigma, \zeta)\mathbf{e}_r(\theta) + Z(\sigma, \zeta)\mathbf{e}_z \,, \; 0 \le \sigma \le L \,, \; 0 \le \theta < 2\pi \,, \; -H \le \zeta \le H \,, \qquad \text{L}$$

where, for convenience, L and H have dimensions of LENGTH (though σ and ζ need not measure arc length). The faces of the shell are given by $\mathbf{x}(\sigma, \theta, \pm H)$ and the edges by $\mathbf{x}(0, \theta, \zeta)$ and $\mathbf{x}(L, \theta, \zeta)$. We call $\mathbf{x}(\sigma, 0, 0)$ the parametric equation of the *base curve*, $\mathcal{B}$. See Fig. 5.1b. The functions $R(\sigma, \zeta)$ and $Z(\sigma, \zeta)$ are usually dictated by the construction of the shell and often have a simple form. Thus, if the shell is materially isotropic, homogeneous, and of constant thickness, h, it is natural to take

---

[2]Equivalently, by introducing a certain combination of stress couples as a new unknown in the deformation power density, we may eliminate k_n from the axishell equations. For details, see Sections L and O.

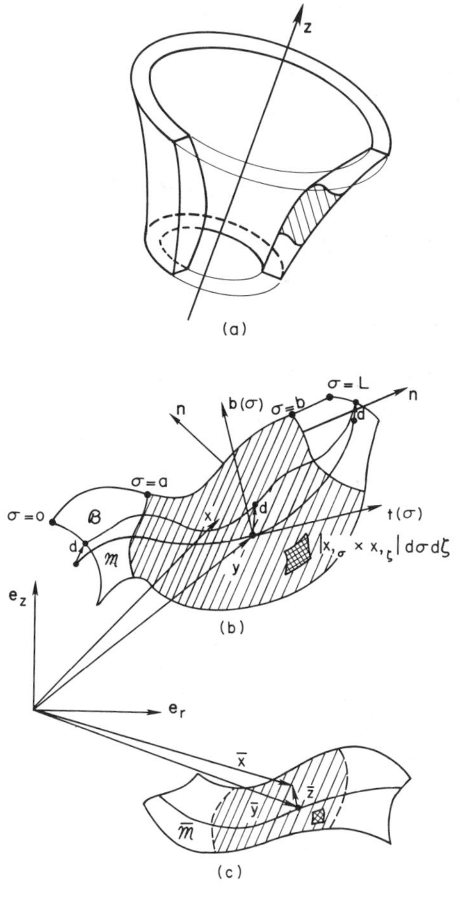

Fig. 5.1. Geometry of a shell of revolution: (a) perspective view; (b) undeformed and (c) deformed meridional section $\theta =$ constant.

$$\mathbf{x}(\sigma, \theta, \zeta) = R(\sigma)\mathbf{e}_r(\theta) + Z(\sigma)\mathbf{e}_z + \zeta[-Z'(\sigma)\mathbf{e}_r(\theta) + R'(\sigma)\mathbf{e}_z] , \qquad 0 \le \sigma \le L \ , \ -h/2 \le \zeta \le h/2 . \tag{A.3}$$

Here, $R(\sigma) \equiv R(\sigma, 0)$, $Z(\sigma) \equiv Z(\sigma, 0)$, and, now, σ and ζ are, respectively, arc length along and distance normal to $\mathcal{B}$. In this case, σ and ζ are called *shell coordinates*. More generally, if the shell is made of layers of different (possibly inhomogeneous) materials, each of variable thickness, then it is natural to choose $\mathbf{x}(\sigma, \theta, \zeta)$ so that $\zeta =$ constant along interfaces. As with beamshells, we conform with Dawn Fisher's suggestion that leaving more latitude in the specification of $\mathbf{x}(\sigma, \theta, \zeta)$ not only simplifies the descent from 3-dimensions, but also prevents us from making a premature (and what can prove to be an unnatural) choice of a reference surface, especially if the shell is nonhomogeneous or of variable thickness.

B. Integral Equations of Motion

Integral equations of motion for shells of revolution may be obtained by specializing (II.A.4) and (II.A.5), the integral equations of motion for a general 3-dimensional body. Let the frame $Oxyz$ be inertial and take the arbitrary volume V of Section II.A to be a curvilinear panel of angular width $2\theta_0$ and meridional section

$$\partial V_\theta \equiv \{\, \mathbf{x}(\sigma, 0, \zeta) |\ 0 \leq a \leq \sigma \leq b \leq L \ , -H \leq \zeta \leq H \,\} \ , \tag{B.1}$$

as indicated by the shaded region in Fig. 5.1b.

It follows from (A.2) that in (II.A.4) and (II.A.5),

$$dV = |(\mathbf{x},_\sigma \times \mathbf{x},_\theta)\cdot\mathbf{x},_\zeta| d\sigma d\theta d\zeta \equiv \mu(\sigma, \zeta) r(\sigma) d\sigma d\theta d\zeta \, . \tag{B.2}$$

The factor r, which will be specified presently, is introduced so that μ will be dimensionless (i.e., so that $rd\theta$ will have the dimension of LENGTH) and so that the equations to follow will resemble those of a beamshell as closely as possible.

On the edges of the panel,

$$\mathbf{n}dA = \pm \mathbf{x},_\theta \times \mathbf{x},_\zeta d\theta d\zeta \ , \ \sigma = {}^b_a ; \tag{B.3}$$

on the sides of the panel,

$$\mathbf{n}dA = \pm \mathbf{x},_\sigma \times \mathbf{x},_\zeta d\sigma d\zeta \ , \ \theta = \pm\theta_0 \ ; \tag{B.4}$$

and on the faces of the panel,

$$\mathbf{n}dA = \pm \mathbf{x},_\sigma \times \mathbf{x},_\theta d\sigma d\theta \ , \ \zeta = \pm H \ . \tag{B.5}$$

Let the stress vector $\mathbf{S}$ of Section II.B be represented in the form

$$\mathbf{S} = \mathbf{S}^\sigma(\mathbf{x},_\sigma\cdot\mathbf{n}) + \mathbf{S}^\theta(\mathbf{x},_\theta\cdot\mathbf{n}) + \mathbf{S}^\zeta(\mathbf{x},_\zeta\cdot\mathbf{n}) \, . \tag{B.6}$$ [3]

It now follows from (B.3)-(B.6) that on the edges of the panel,

$$\mathbf{S}dA = \mathbf{S}^\sigma \mu r d\theta d\zeta \ , \ \sigma = {}^b_a ; \tag{B.7}$$

on the sides of the panel,

$$\mathbf{S}dA = \mathbf{S}^\theta \mu r d\sigma d\zeta \ , \ \theta = \pm\theta_0 \ ; \tag{B.8}$$

and on the faces of the panel,

$$\mathbf{S}dA = \mathbf{S}^\zeta \mu r d\sigma d\theta \ , \ \zeta = \pm H \ . \tag{B.9}$$

Adding the three contributions to the contact force in (II.A.4) coming from the edges, sides, and faces of our arbitrary curvilinear panel, we have, from (B.7)-

[3]By (II.E.1) and (II.E.3), $\mathbf{S}^\sigma$, $\mathbf{S}^\theta$, and $\mathbf{S}^\zeta$ are the vector contravariant components of the first Piola-Kirchhoff stress tensor $\mathbf{P}$.

(B.9),

$$\int_{\partial V} \mathbf{S}dA = \int_{-\theta_0}^{\theta_0} \Big(\int_-^+ \mathbf{S}^\sigma \mu d\zeta\Big)_a^b r d\theta + \int_a^b \Big(\int_-^+ \mathbf{S}^\theta \mu\Big)_{-\theta_0}^{\theta_0} r d\sigma$$
$$+ \int_{-\theta_0}^{\theta_0} \int_a^b (\mathbf{S}^\zeta \mu)_-^+ r d\sigma d\theta\,. \tag{B.10}$$

Here and elsewhere, $\int_-^+$ is short for $\int_{-H}^{H}$. By (II.A.4) and (B.2), the panel is subject to the net body force,

$$\int_V \mathbf{f}\, dV = \int_{-\theta_0}^{\theta_0} \int_a^b \Big(\int_-^+ \mathbf{f} \mu d\zeta\Big) r d\sigma d\theta \tag{B.11}$$

and has linear momentum,

$$\int_V \rho \dot{\bar{\mathbf{x}}} dV = \int_{-\theta_0}^{\theta_0} \int_a^b \Big(\int_-^+ \rho \bar{\mathbf{x}} \mu d\zeta\Big)^{\boldsymbol{\cdot}} r d\sigma d\theta\,. \tag{B.12}$$

To express the right sides of (B.10)-(B.12) in terms of shell variables, we first let

$$m \equiv \int_-^+ \rho \mu d\zeta = m(\sigma) \tag{B.13}$$

denote a mass per unit undeformed area and let

$$\bar{\mathbf{y}} \equiv \int_-^+ \rho \bar{\mathbf{x}} \mu d\zeta / m = \bar{\mathbf{y}}(\sigma, \theta, t) \equiv \bar{r}(\sigma, t)\mathbf{e}_r(\theta) + \bar{z}(\sigma, t)\mathbf{e}_z \tag{B.14}$$

denote a *weighted deformed position.* The interpretation of $\bar{\mathbf{y}}$ for an shell of revolution is analogous to that for a beamshell given in the paragraph following (IV.B.9) and the digression to follow parallels closely that discussion.

We call

$$\mathbf{y}(\sigma, \theta) \equiv \bar{\mathbf{y}}(\sigma, \theta, 0) = \bar{r}(\sigma, 0)\mathbf{e}_r(\theta) + \bar{z}(\sigma, 0)\mathbf{e}_z \equiv r(\sigma)\mathbf{e}_r(\theta) + z(\sigma)\mathbf{e}_z \tag{B.15}$$

the parametric equation of *the initial (or undeformed) reference surface* $\mathcal{S}$ and call (B.14) the parametric equation of its deformed image, $\bar{\mathcal{S}}$. As the notation indicates, we now take the unspecified factor $r(\sigma)$ in (B.2) to be the radial distance to a point on the reference surface, $\mathcal{S}$. Arc length along a *meridian,* $\mathcal{M}$, of $\mathcal{S}$ is given by

$$\tilde{\sigma} \equiv \int_0^\sigma |\mathbf{y}_{,\sigma}(u, \theta)|\, du \equiv \psi(\sigma)\,. \tag{B.16}$$

The length of $\mathcal{M}$ is $\tilde{L} \equiv \psi(L)$.

For fixed σ and θ, (B.14) locates the *center of mass* of a curved, differential column of variable cross section capped by the shell faces, $\zeta = \pm H$. Our method of defining $\mathbf{y}$ and its deformed image, $\bar{\mathbf{y}}$, is thus equivalent to defining a *surface of centers of mass* for an undeformed and deformed shell of revolution, respectively.

We emphasize that, in general, $\bar{\mathbf{y}}$ *is not the deformed position of some particle in the shell* and remind the reader that (B.14) comes from the inertial terms in the 3-dimensional equations of motion. Therefore, in statics (where t may be regarded as a parameter measuring the magnitude of the loads) other reasonable relations between $\bar{\mathbf{y}}$ and $\bar{\mathbf{x}}$ may be defined, e.g., $\bar{\mathbf{y}}=\bar{\mathbf{x}}(\sigma, \theta, 0, t)$ (or, what comes to the same thing, we may, in a static problem, chose ρ at our convenience). This point must be recognized if one attempts to derive constitutive relations for a shell by postulating a specific form for $\mathbf{x}$ and then averaging in some way through the thickness.

There are theoretical and practical advantages in taking, as the independent spatial variable, arc length along $\mathcal{M}$, which σ, in general, is not. To do so, we note that the arc length function $\psi(\sigma)$, defined in (B.16), is monotonically increasing and differentiable; hence so is its inverse, ψ^{-1}. Thus, if we make the change of variable $\sigma=\psi^{-1}(\tilde{\sigma})$, we need merely replace σ by $\tilde{\sigma}$, L by $\tilde{L}$, and μ by $\tilde{\mu}$, in (A.2)-(B.15), (B.16) reducing to the identity $\tilde{\sigma}\equiv\psi(\psi^{-1}(\tilde{\sigma}))$. We may then drop the tildes.

Returning to the task of expressing the right sides of (B.10)-(B.12) in terms of shell variables, we let

$$\mathbf{N}_\sigma \equiv \int_-^+ \mathbf{S}^\sigma \mu d\zeta = \mathbf{N}_\sigma(\sigma, \theta, t) \tag{B.17}$$

denote the *meridional stress resultant* (force per unit length of a parallel, $r=$constant),

$$\mathbf{N}_\theta \equiv r\int_-^+ \mathbf{S}^\theta \mu d\zeta = \mathbf{N}_\theta(\sigma, \theta, t) \tag{B.18}$$

denote the *hoop (circumferential) stress resultant* (force per unit length of $\mathcal{M}$), and

$$\mathbf{p} \equiv \int_-^+ \mathbf{f}\mu d\zeta + (\mathbf{S}^\zeta \mu)_-^+ = \mathbf{p}(\sigma, \theta, t) \tag{B.19}$$

denote the *external force per unit area* of the reference surface, $\mathcal{S}$. Substituting (B.13), (B.14), and (B.17)-(B.19) into the right sides of (B.10)-(B.12) and the resulting expressions into (II.A.4), we have for the balance of linear momentum,

$$\mathbf{L} \equiv \int_{t_1}^{t_2}[\int_{-\theta_0}^{\theta_0} (r\mathbf{N}_\sigma)_a^b d\theta + \int_a^b (\mathbf{N}_\theta)_{-\theta_0}^{\theta_0} d\sigma + \int_{-\theta_0}^{\theta_0}\int_a^b \mathbf{p} r d\sigma d\theta]dt \\ - \int_{-\theta_0}^{\theta_0}\int_a^b m\mathbf{v} r d\sigma d\theta \,\Big|_{t_1}^{t_2} = \mathbf{0}, \tag{B.20}$$

where $\mathbf{v}\equiv\dot{\bar{\mathbf{y}}}$.

To obtain the integral form of the equation of balance of rotational momentum for a shell of revolution, we first set

$$\bar{\mathbf{x}} = \bar{\mathbf{y}} + \bar{\mathbf{z}} \tag{B.21}$$

and note that (B.14) implies

$$\int_{-}^{+} \rho \bar{\mathbf{z}} \mu d\zeta = \mathbf{0} \,. \tag{B.22}$$

The discrepancy between the base curve, $\mathcal{B}$, and the reference meridian, $\mathcal{M}$, is measured by the *base-reference deviation*

$$\mathbf{d}(\sigma) \equiv \bar{\mathbf{z}}(\sigma, 0, 0) \,. \tag{B.23}$$

If $\mathbf{x}(\sigma, \theta, \zeta)$ has the form (A.3), it follows from (B.13) and (B.14) that

$$\mathbf{d} = d\,[-Z'(\sigma)\mathbf{e}_r(\theta) + R'(\sigma)\mathbf{e}_z] \,. \tag{B.24}$$

Here, $d = \int_{-h/2}^{h/2} \rho\zeta\mu d\zeta / m$ and $\mu = (1 - 2\zeta M + \zeta^2 G)R/r$, where M and G are, respectively, the mean and Gaussian curvatures of the base reference surface with the vector-parametric equation $\mathbf{x}(\sigma, \theta, 0)$. If the shell of revolution is also homogeneous and of constant thickness, h, then $d = (Mh^2/6)(1 + Gh^2/12)^{-1}$ and, as we shall see, d/h can then be set to zero within the error in shell theory coming from the constitutive relations. In such a case, the midcurve, the base curve, and the undeformed reference meridian can be regarded as virtually coincident and the extensive literature based on using the midcurve as the undeformed reference meridian is applicable.

Turning to balance of rotational momentum, we insert (B.7)-(B.9), (B.21), and the definitions (B.17) and (B.18) into (II.A.5) to express the net moment about the origin of the contact forces on the panel as

$$\begin{aligned} \int_{\partial V} \bar{\mathbf{x}} \times \mathbf{S} dA &= \int_{-\theta_0}^{\theta_0} [r(\bar{\mathbf{y}} \times \mathbf{N}_\sigma + \int_{-}^{+} \bar{\mathbf{z}} \times \mathbf{S}^\sigma \mu d\zeta)]_a^b d\theta \\ &+ \int_a^b (\bar{\mathbf{y}} \times \mathbf{N}_\theta + r \int_{-}^{+} \bar{\mathbf{z}} \times \mathbf{S}^\theta \mu d\zeta)_{-\theta_0}^{\theta_0} d\sigma \\ &+ \int_{-\theta_0}^{\theta_0} \int_a^b [\bar{\mathbf{y}} \times (\mathbf{S}^\zeta \mu)_{-}^{+} + (\bar{\mathbf{z}} \times \mathbf{S}^\zeta \mu)_{-}^{+}] r d\sigma d\theta \,. \end{aligned} \tag{B.25}$$

The panel is subject to the net body couple

$$\int_V \bar{\mathbf{x}} \times \mathbf{f} dV = \int_{-\theta_0}^{\theta_0} \int_a^b (\bar{\mathbf{y}} \times \int_{-}^{+} \mathbf{f} \mu d\zeta + \int_{-}^{+} \bar{\mathbf{z}} \times \mathbf{f} \mu d\zeta) r d\sigma d\theta \,. \tag{B.26}$$

Its rotational momentum about the origin, in view of (B.2), (B.13), and (B.22), is

$$\int_V \bar{\mathbf{x}} \times \rho \dot{\bar{\mathbf{x}}} dV = \int_{-\theta_0}^{\theta_0} \int_a^b (\bar{\mathbf{y}} \times m\mathbf{v} + \int_{-}^{+} \bar{\mathbf{z}} \times \rho \dot{\bar{\mathbf{z}}} \mu d\zeta) r d\sigma d\theta \,. \tag{B.27}$$

Inserting the right sides of (B.25)-(B.27) into (II.A.5) and noting (B.19), we may write the resulting expression in the form

$$\mathbf{R} \equiv \int_{t_1}^{t_2} \{ \int_{-\theta_0}^{\theta_0} [r(\bar{\mathbf{y}} \times \mathbf{N}_\sigma + \mathbf{M}_\sigma)]_a^b d\theta + \int_a^b (\bar{\mathbf{y}} \times \mathbf{N}_\theta + \mathbf{M}_\theta)_{-\theta_0}^{\theta_0} d\sigma$$
$$+ \int_{-\theta_0}^{\theta_0} \int_a^b (\bar{\mathbf{y}} \times \mathbf{p} + \mathbf{l}) r d\sigma d\theta \} dt - \int_{-\theta_0}^{\theta_0} \int_a^b (\bar{\mathbf{y}} \times m\mathbf{v} + I\boldsymbol{\omega}) r d\sigma d\theta \,\Big|_{t_1}^{t_2} = \mathbf{0}, \tag{B.28}$$

where

$$\mathbf{M}_\sigma \equiv \int_-^+ \bar{\mathbf{z}} \times \mathbf{S}^\sigma \mu d\zeta = \mathbf{M}_\sigma(\sigma, \theta, t), \tag{B.29}$$

is the *meridional stress couple,*

$$\mathbf{M}_\theta \equiv r \int_-^+ \bar{\mathbf{z}} \times \mathbf{S}^\theta \mu d\zeta = \mathbf{M}_\theta(\sigma, \theta, t) \tag{B.30}$$

is the *hoop (circumferential) stress couple,*

$$\mathbf{l} \equiv \int_-^+ \bar{\mathbf{z}} \times \mathbf{f} \mu d\zeta + (\bar{\mathbf{z}} \times \mathbf{S}^\zeta \mu)_{-\theta_0}^{\theta_0} = \mathbf{l}(\sigma, \theta, t) \tag{B.31}$$

is the *external couple per unit area* of $\mathcal{S}$,

$$I \equiv \int_-^+ \bar{\mathbf{z}} \cdot \bar{\mathbf{z}} \rho \mu d\zeta = I(\sigma) \tag{B.32}$$

is the *initial moment of inertia per unit area* of $\mathcal{S}$, and

$$\boldsymbol{\omega} \equiv \int_-^+ \bar{\mathbf{z}} \times \rho \dot{\bar{\mathbf{z}}} \mu d\zeta / I = \boldsymbol{\omega}(\sigma, \theta, t) \tag{B.33}$$

is the *spin.*

The same observations apply to the definitions (B.32) and (B.33) as we made for the beamshell following equation (IV.B.21).

C. Differential Equations of Motion

If the fields $\mathbf{N}_\sigma, \mathbf{N}_\theta, \mathbf{v}$, etc. that appear in (B.20) and (B.28) are sufficiently smooth in space and time, then these global integral equations imply local differential equations which we obtain by using the Fundamental Theorem of Calculus to express terms of the form $f|_a^b, g|_{-\theta_0}^{\theta_0}$ or $h|_{t_1}^{t_2}$ as $\int_a^b f_{,\sigma} d\sigma, \int_{-\theta_0}^{\theta_0} g_{,\theta} d\theta$ or $\int_{t_1}^{t_2} \dot{h} dt$. In this way we reduce (B.20) and (B.28) to expressions of the form $\int_{t_1}^{t_2} \int_{-\theta_0}^{\theta_0} \int_a^b \mathbf{w} d\sigma d\theta dt = \mathbf{0}$. As the intervals (a, b), $(-\theta_0, \theta_0)$, and (t_1, t_2) are arbitrary (save that $0 \le a \le b \le L$, etc.), it follows that, if $\mathbf{w}$ is continuous, then $\mathbf{w} \equiv \mathbf{0}$. That is,

$$(r\mathbf{N}_\sigma)_{,\sigma} + \mathbf{N}_{\theta,\theta} + r\mathbf{p} = mr\dot{\mathbf{v}} \tag{C.1}$$

$$(r\mathbf{M}_\sigma)_{,\sigma} + \mathbf{M}_{\theta,\theta} + r\bar{\mathbf{y}}_{,\sigma} \times \mathbf{N}_\sigma + \bar{\mathbf{y}}_{,\theta} \times \mathbf{N}_\theta + r\mathbf{l} = Ir\dot{\boldsymbol{\omega}}, \tag{C.2}$$

where we have used (C.1) in arriving at (C.2).

D. Differential and Integral Equations of Torsionless, Axisymmetric Motion

We now assume that the shell is circumferentially complete ($0 \le \theta < 2\pi$), that all external disturbances, boundary conditions, and material properties are independent of θ, and that each meridional cross section of the shell undergoes the same planar motion. For conciseness, we shall refer to shells of revolution undergoing such motion as *axishells*. Our assumptions imply that

$$(\mathbf{N}_\theta, \bar{\mathbf{y}},_\theta, \mathbf{M}_\sigma, \mathbf{l}, \boldsymbol{\omega}) = (N_\theta, \bar{r}, -M_\sigma, -l, -\omega)\mathbf{e}_\theta \tag{D.1}$$

$$(\mathbf{N}_\sigma, \mathbf{p}, \mathbf{v}, \mathbf{M}_\theta)\cdot\mathbf{e}_\theta = 0 \tag{D.2}$$

so that (C.1) reduces to

$$(r\mathbf{N}_\sigma)' - N_\theta \mathbf{e}_r + r\mathbf{p} = mr\dot{\mathbf{v}} \,, \tag{D.3}$$

where, henceforth, a prime will denote differentiation with respect to σ.

To simplify (C.2), note from (D.2) that $\mathbf{M}_\theta,_\theta\cdot\mathbf{e}_\theta = \mathbf{M}_\theta\cdot\mathbf{e}_r$. Furthermore, we may set $(\bar{\mathbf{y}}' \times \mathbf{N}_\sigma)\cdot\mathbf{e}_\theta = -\mathbf{m}\cdot\mathbf{N}_\sigma$, where

$$\mathbf{m} \equiv \bar{\mathbf{y}}' \times \mathbf{e}_\theta \,. \tag{D.4}$$

With these observations, (C.2) reduces to

$$(rM_\sigma)' - \mathbf{M}_\theta\cdot\mathbf{e}_r + r\mathbf{N}_\sigma\cdot\mathbf{m} + rl = Ir\dot{\omega} \,. \tag{D.5}$$

Let us return to the integral equations of motion, (B.20) and (B.28). In view of (D.1) and (D.2), the only remaining dependence on θ in the integral equations is through the unit vectors $\mathbf{e}_r$ and $\mathbf{e}_\theta$, which are smooth functions of θ. Thus, because $f|_{-\theta_0}^{\theta_0} = 0$ implies that $f,_\theta = 0$, if θ_0 is arbitrary and f is differentiable, we may write (B.20) and (B.28) in the form

$$\mathbf{L} \equiv \oint_{\partial\Omega} r(m\mathbf{v}d\sigma + \mathbf{N}_\sigma dt) + \int_\Omega (N_\theta \mathbf{e}_r + r\mathbf{p})d\sigma dt = \mathbf{0} \tag{D.6}$$

$$R \equiv \oint_{\partial\Omega} r(I\omega d\sigma + M_\sigma dt) + \int_\Omega (\mathbf{M}_\theta\cdot\mathbf{e}_r + r\mathbf{N}_\sigma\cdot\mathbf{m} + rl)d\sigma dt = 0 \,, \tag{D.7}$$

where Ω is a rectangle $(a, b)\times(t_1, t_2)$ in the σt-plane and $\partial\Omega$ is its boundary. (See Fig. 3.3a). Note that (D.6) and (D.7) also follow *formally* from (D.3) and (D.5) upon integrating over Ω and applying the Divergence Theorem to eliminate certain space and time derivatives. As with the birod, (D.6) and (D.7) actually hold for *any* region Ω in the σt-plane, providing only that $\partial\Omega$ is piecewise smooth and is cut in no more than 2 points by any line σ = constant.

Suppose that the initial velocity and spin are prescribed vector and scalar functions, $\mathbf{g}$ and w, on $0 < z < L$. If we take Ω to be the rectangle $(a, b)\times(0, t_2)$ and let $t_2 \to 0$, then (D.6) and (D.7) reduce to the *natural initial conditions*

$$\int_a^b rm(\mathbf{g} - \overset{0}{\mathbf{v}})d\sigma = \mathbf{0} \ , \ \int_a^b rI(w - \overset{0}{\omega})d\sigma = 0 \ , \tag{D.8}$$

where $\overset{0}{\mathbf{v}}$ denotes the limit of $\mathbf{v}(\sigma, t)$ as $t \to 0$ through positive values, with a similar definition of $\overset{0}{\omega}$. Likewise, let $\hat{\mathbf{N}}_\sigma^0, \hat{M}_\sigma^0$ and $\hat{\mathbf{N}}_\sigma^L, \hat{M}_\sigma^L$ denote prescribed values of $\mathbf{N}_\sigma$ and M_σ at $\sigma=0$ and $\sigma=L$. If we take Ω to be the rectangle $(0, b)\times(t_1, t_2)$ or $(a, L)\times(t_1, t_2)$ and let, respectively, $b \to 0$ from the right and $a \to L$ from the left, then (D.6) and (D.7) reduce to the *natural boundary conditions*

$$\begin{gathered}\int_{t_1}^{t_2} r_0(\mathbf{N}_\sigma^0 - \hat{\mathbf{N}}_\sigma^0)dt = \int_{t_1}^{t_2} r_L(\hat{\mathbf{N}}_\sigma^L - \mathbf{N}_\sigma^L)dt = \mathbf{0} \\ \int_{t_1}^{t_2} r_0(M_\sigma^0 - \hat{M}_\sigma^0)dt = \int_{t_1}^{t_2} r_L(\hat{M}_\sigma^L - M_\sigma^L)dt = 0 \ ,\end{gathered} \tag{D.9}$$

where the subscript or superscript 0 denotes the limit as $\sigma \to 0$ from the right, with an analogous definition of the subscript or superscript L.

E. Initial and Spin Bases

From the parametric representation (B.15) for the reference surface, $\mathcal{S}$, it follows that the unit tangent to the reference meridian, $\mathcal{M}$, is given by

$$\mathbf{t} = r'(\sigma)\mathbf{e}_r + z'(\sigma)\mathbf{e}_z \equiv \cos\alpha(\sigma)\mathbf{e}_r + \sin\alpha(\sigma)\mathbf{e}_z \ , \tag{E.1}$$

where α is the angle the tangent to $\mathcal{M}$ makes with the r-axis. A unit normal to $\mathcal{M}$ is given by

$$\mathbf{b} \equiv \mathbf{t} \times \mathbf{e}_\theta = -\sin\alpha\mathbf{e}_r + \cos\alpha\mathbf{e}_z \ . \tag{E.2}$$

We call $\{\mathbf{t}, \mathbf{b}\}$ the *initial basis.*

Let $\{\mathbf{T}, \mathbf{B}\}$ denote a rigid, orthonormal basis that rotates in the rz-plane with spin ω relative to $\{\mathbf{t}, \mathbf{b}\}$ and coincides with this latter basis when $t=0$. Then, with

$$\beta \equiv \int_0^t \omega dt \tag{E.3}$$

denoting the *rotation* of the axishell, we have

$$\mathbf{T} = \cos\beta\mathbf{t} + \sin\beta\mathbf{b} = \cos(\alpha + \beta)\mathbf{e}_r + \sin(\alpha + \beta)\mathbf{e}_z \tag{E.4}$$

$$\mathbf{B} = -\sin\beta\mathbf{t} + \cos\beta\mathbf{b} = -\sin(\alpha + \beta)\mathbf{e}_r + \cos(\alpha + \beta)\mathbf{e}_z \ , \tag{E.5}$$

where $\alpha + \beta$ is the angle $\mathbf{T}$ makes with the r-axis. We call $\{\mathbf{T}, \mathbf{B}\}$ the *spin basis.*

The geometry here is identical to that in Fig. 4.2 for the beamshell, save that $\mathbf{e}_x$ and $\mathbf{e}_y$ are to be replaced by $\mathbf{e}_r$ and $\mathbf{e}_z$ and $\mathcal{C}$ and $\overline{\mathcal{C}}$ by $\mathcal{M}$ and $\overline{\mathcal{M}}$.

*F. Jump Conditions and Propagation of Singularities

Let the curve $\sigma = s(t)$ be a candidate for a shockline in the σt-plane. If we require that there be no jumps in displacement or rotation across a shock, then a virtual repetition of the arguments made for the birod in Section III.D shows that (D.6) and (D.7) imply that, at a shock, the following conditions must hold:

$$[\bar{\mathbf{y}}] = \mathbf{0} \ , \ [\beta] = 0 \tag{F.1}$$

$$[\mathbf{N}_\sigma] = mc^2[\bar{\mathbf{y}}'] \ , \ [M_\sigma] = Ic^2[\beta'] \, , \tag{F.2}$$

where $c \equiv |\dot{s}(t)|$. In an initial/boundary value problem, (F.1) and (F.2) serve as boundary conditions for the classical solutions of the field equations that hold on either side of the shock. As we shall see, the condition $c^2 \geq 0$ implies certain restrictions of the form of the constitutive relations.

G. The Weak Form of the Equations of Motion

Assume that the fields $(\mathbf{N}_\sigma, N_\theta, \mathbf{p}, \mathbf{v}, M_\sigma, \mathbf{M}_\theta, l, \omega)$ are classical solutions, i.e., that they satisfy (D.3) and (D.5), and assume that $\mathbf{V}$ and Ω are arbitrary vector and scalar fields, sufficiently smooth for the following operations to make sense. Take the dot product of (D.3) with $\mathbf{V}$, the product of (D.5) with Ω, add, and integrate over $(a, b) \times (t_1, t_2)$. Integrating by parts to eliminate the spatial derivative on $r\mathbf{N}_\sigma$ and rM_σ and the time derivative on $\mathbf{v}$ and ω, we obtain the *weak form of the equations of motion:*

$$\begin{aligned} &\int_{t_1}^{t_2} [r(\mathbf{N}_\sigma \cdot \mathbf{V} + M_\sigma \Omega)]_a^b dt - \int_a^b r(m\mathbf{v}\cdot\mathbf{V} + I\omega\Omega)_{t_1}^{t_2} d\sigma \\ &- \int_{t_1}^{t_2} \int_a^b [r\mathbf{N}_\sigma \cdot \mathbf{V}' + N_\theta \mathbf{e}_r \cdot \mathbf{V} + rM_\sigma \Omega' + (\mathbf{M}_\theta \cdot \mathbf{e}_r - r\mathbf{N}_\sigma \cdot \mathbf{m})\Omega \\ &+ r(m\mathbf{v}\cdot\dot{\mathbf{V}} + I\omega\dot{\Omega} + \mathbf{p}\cdot\mathbf{V} + l\Omega)] d\sigma dt = 0 \ , \ \forall \, \mathbf{V} \, , \Omega \, . \end{aligned} \tag{G.1}$$

Because integration is a smoothing operation, there may exist fields $(\mathbf{N}_\sigma^*, \cdots, \omega^*)$ too rough to satisfy the differential equations of motion, yet such that, if $\mathbf{V}$ and Ω are sufficiently smooth (to compensate for the roughness), (G.1) is satisfied. Such fields are called *weak solutions.* Note that (G.1) implies nothing about boundary or initial conditions. As with beamshells, the weak solutions of (G.1) satisfy the integral equations of motion, (D.6) and (D.7). Conversely, as we may infer from Antman & Osborn (1979), solutions of (D.6) and (D.7) satisfy (G.1) and are compatible with the jump conditions discussed in the preceding section.

H. The Mechanical Work Identity

In (G.1), let $(\mathbf{V}, \Omega) = (\mathbf{v}, \omega)$, the actual velocity and spin fields. Then, the resulting equation reduces to *the Mechanical Work Identity*

$$\int_{t_1}^{t_2} \mathcal{W} dt \equiv \mathcal{K}\Big|_{t_1}^{t_2} + \int_{t_1}^{t_2} \mathcal{D} dt \,, \tag{H.1}$$

where

$$\mathcal{W} \equiv [r(\mathbf{N}_\sigma \cdot \mathbf{v} + M_\sigma \omega)]_a^b + \int_a^b (\mathbf{p} \cdot \mathbf{v} + l\omega) r d\sigma \tag{H.2}$$

is the *apparent external mechanical power*,

$$\mathcal{K} \equiv \tfrac{1}{2} \int_a^b (m\mathbf{v} \cdot \mathbf{v} + I\omega^2) r d\sigma \tag{H.3}$$

is the *kinetic energy*, and

$$\mathcal{D} \equiv \int_a^b [\mathbf{N}_\sigma \cdot (\mathbf{v}' - \omega \mathbf{m}) + N_\theta r^{-1} \mathbf{e}_r \cdot \mathbf{v} + M_\sigma \omega' + \mathbf{M}_\theta \cdot \mathbf{e}_r r^{-1} \omega] r d\sigma \tag{H.4}$$

is the *deformation power*.

I. Mechanical Boundary Conditions

This section is quite similar to Section IV.H on beamshells, which the reader should consult, along with Section III.G, for motivation.

Let

$$y \equiv [\bar{\mathbf{y}}_0, \beta_0, \bar{\mathbf{y}}_L, \beta_L]^T \tag{I.1}$$

denote the 6-component *edge kinematic vector*, where it is understood that $\bar{\mathbf{y}}$ is to be represented by a pair of scalars (which depend on what basis we choose in the rz-plane). We shall assume that y and $v \equiv \dot{y}$ satisfy $m\,(\leq 6)$ *independent, nonholonomic constraints* of the form

$$Av - \mathcal{B} = 0 \,, \tag{I.2}$$

where A and $\mathcal{B}$ are, respectively, $m \times 6$ and $m \times 1$ matrices that depend on y and t only. Clearly, (I.2) is not the most general constraint we could impose. For example, thermal effects might have been included or "=" could have been replaced by "≤" (unilateral constraints). Still, (I.2) is far more general than the conditions that $\bar{\mathbf{y}}$ or β be prescribed on one or the other edge that one often sees discussed.

By m independent constraints we mean that $\operatorname{rank} A = m$. Thus, the solutions of (I.2) may be expressed as

$$v = \sum_{1}^{6-m} q_k \mathcal{M}_k(y, t) + \mathcal{H}(y, t) \,, \tag{I.3}$$

where the q_k's are unknowns, the $\mathcal{M}_k$'s are any 6–m linearly independent vectors that span the null space of A (i.e., $A\mathcal{M}_k = 0$), and $\mathcal{H}$ is determined uniquely by the conditions $A\mathcal{H} = \mathcal{B}$ and $\mathcal{H}^T\mathcal{M}_k = 0$, $k = 1, 2, \cdots 6{-}m$.

Now let

$$\mathcal{N} \equiv [-r_0\mathbf{N}_\sigma^0, -r_0M_\sigma^0, r_L\mathbf{N}_\sigma^L, r_LM_\sigma^L]^T \tag{I.4}$$

denote the *edge load vector* conjugate to v, where $\mathbf{N}_\sigma^0 \equiv \mathbf{N}_\sigma(0, \theta, t)$, $r_0 \equiv r(0)$, etc. and it is understood in (I.4) that $\mathbf{N}_\sigma$ represents an ordered pair of scalars. With the aid of (I.3) and (I.4), the rate of work of the edge loads may be written

$$[r(\mathbf{N}_\sigma \cdot \mathbf{V} + M_\sigma\Omega)]_0^L = \mathcal{N}^T v = \sum_1^{6-m} \mathcal{N}^T\mathcal{M}_k q_k + \mathcal{N}^T\mathcal{H}. \tag{I.5}$$

We call the $6-m$ scalars $\mathcal{N}^T\mathcal{M}_k$ the *generalized applied edge loads* and the term $\mathcal{N}^T\mathcal{H}$ the *power of the forces of constraint.*

As with beamshells, we postulate for axishells *the Principle of Mechanical Boundary Conditions,* namely, on the boundary of the deformed initial surface there must be specified

- $m \le 6$ kinematic constraints (possibly nonholonomic)
- $6-m$ generalized applied forces of the form $\mathcal{N}^T\mathcal{M}_k$.

An example will illustrate the ideas of this section.

Problem: A circular cylindrical shell with reference surface of radius a is fixed at one end ($\sigma = 0$) to an immovable wall. The other end ($\sigma = L$) is in contact with a rough, rigid ball of radius $R > a$. Beginning at $t = 0$, the ball is moved along the axis of the cylinder toward the fixed end at a constant speed, v_0. Determine the boundary conditions, ignoring the thickness of the shell. (See Fig. 5.2)

Solution: The upper edge of the shell is free to rotate but its coordinates $(\bar{r}_L, \bar{z}_L)$ must satisfy the (holonomic) constraint

$$\bar{r}_L^2 + (\bar{z}_L - D + v_0t)^2 = R^2\,, \tag{I.6}$$

where $D \equiv \sqrt{R^2 - a^2} + L$. At the fixed end,

$$\bar{r}_0 = a\ ,\ \ \bar{z}_0 = \beta_0 = 0\,. \tag{I.7}$$

Hence, there are $m = 4$ constraints of the form (I.2), namely,

$$\bar{r}_L\dot{\bar{r}}_L + (\bar{z}_L - D + v_0t)(\dot{\bar{z}}_L + v_0) = 0 \tag{I.8}$$

$$\dot{\bar{r}}_0 = \dot{\bar{z}}_0 = \dot{\beta}_0 = 0\,. \tag{I.9}$$

We choose as *degrees of freedom*, $q_1 = (\bar{z}_L - D + v_0t)^{-1}$ and $q_2 = \dot{\beta}_L$, so that using (I.8) and (I.9) to express v in the form (I.3), we have

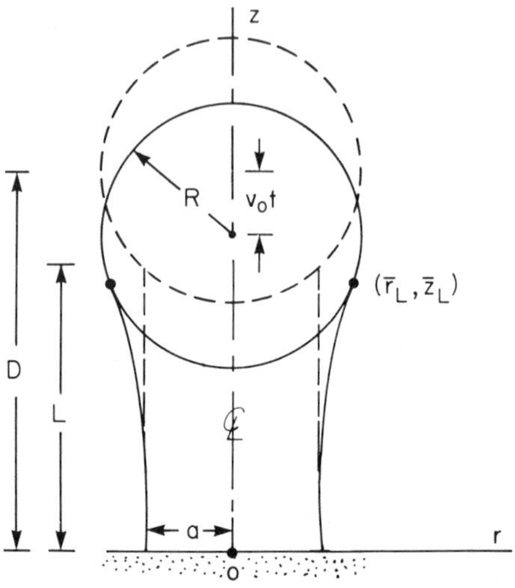

Fig. 5.2. Ball moving into a cylindrical shell.

$$v = [\dot{\bar{r}}_0, \dot{\bar{z}}_0, \dot{\beta}_0, \dot{\bar{r}}_L, \dot{\bar{z}}_L, \dot{\beta}_L]^T$$
$$= [0, 0, 0, (\bar{z}_L - D + v_0 t)q_1, -\bar{r}_L q_1 - v_0, q_2]^T . \tag{I.10}$$

Let $\mathbf{N}_\sigma = H\mathbf{e}_r + V\mathbf{e}_z$ and $M_\sigma = M$ so that

$$\mathcal{N} = a[-H_0, -V_0, -M_0, H_L, V_L, M_L]^T . \tag{I.11}$$

Then, with $P_L \equiv H_L(\bar{z}_L - D + v_0 t) - V_L \bar{r}_L$ denoting R times the component of $-\mathbf{N}_\sigma^L$ tangent to the ball, we have

$$\mathcal{N}^T v = aP_L q_1 + M_\sigma^L q_2 - V_L v_0 . \tag{I.12}$$

Thus, the boundary conditions are the four constraints (I.8) and (I.9), or their original, undifferentiated forms, (I.6) and (I.7), plus prescribed values of the generalized edge loads, P_L and M_L.

To pin down these last two conditions, let us assume that the roughness of the ball creates no edge moment but only a force on the edge of the shell, tangent to the ball and proportional to the normal force of contact between the ball and the edge, $R^{-1}[H_L \bar{r}_L + V_L(z_L - D + v_0 t)]$. Then, the generalized force boundary conditions become

$$(H_L - \mu V_L)(\bar{z}_L - D + v_0 t) = (\bar{V}_L + \mu H_L)\bar{r}_L \ , \ M_L = 0 , \tag{I.13}$$

where μ is the coefficient of friction. These conditions are valid so long as the slope of the deformed cylindrical shell at the point of contact is less than or equal to that of the ball. □

For a discussion of classical and typical boundary conditions, see subsections IV.H.2 and IV.H.3; the reader should have no difficulties in making the slight adjustments in viewpoint and notation necessary for axishells.

J. The Principle of Virtual Work

This principle is obtained from the weak form of the equations of motion, (G.1), as follows. First, we set $a=0, b=L, t_1=0$, and $t_2=T$, where $T>0$ is arbitrary. Next, we introduce the subclass of test functions $\mathcal{T}_H$ whose elements $(\mathbf{V}, \Omega)$ satisfy the *final conditions*

$$\mathbf{V}(\sigma, T)=\mathbf{0} \ , \ \Omega(\sigma, T)=0 . \tag{J.1}$$

Finally, we impose the natural initial conditions (D.8) so that, altogether, the *left side* of (G.1) takes the form

$$J[\mathcal{S}; \mathbf{V}, \Omega] \equiv \int_0^T \mathcal{N}^T \mathcal{V} dt + J_R . \tag{J.2}$$

On the left side of this equation, $\mathcal{S} \equiv (\mathbf{N}_\sigma, N_\theta, M_\sigma, \mathbf{M}_\theta, \mathbf{v}, \omega)$, while on the right side, $\mathcal{V}=[\mathbf{V}_0, \Omega_0, \mathbf{V}_L, \Omega_L]^T$ is the (6-dimensional) edge vector of test functions and

$$J_R \equiv \int_0^L r(m\mathbf{g}\cdot\overset{0}{\mathbf{V}} + Iw\overset{0}{\Omega})d\sigma$$
$$-\int_0^T\int_0^L [r\mathbf{N}_\sigma\cdot\mathbf{V}' + N_\theta\mathbf{e}_r\cdot\mathbf{V} + rM_\sigma\Omega' + (\mathbf{M}_\theta\cdot\mathbf{e}_r - r\mathbf{N}_\sigma\cdot\mathbf{m})\Omega \tag{J.3}$$
$$-r(m\mathbf{v}\cdot\dot{\mathbf{V}} + I\omega\dot{\Omega} + \mathbf{p}\cdot\mathbf{V} + l\Omega)]d\sigma dt ,$$

where $\overset{0}{\mathbf{V}} \equiv \mathbf{V}(\sigma, 0+)$ and $\overset{0}{\Omega} \equiv \Omega(\sigma, 0+)$.

Suppose that there are edge constraints of the form $A v = \mathcal{B}$ and natural boundary conditions as in (D.9). To enforce this condition, we introduce $m \times 1$ Lagrange multipliers $\mathcal{L}$ and $\tilde{\mathcal{L}}$ into the functional J to obtain the new functional

$$J^*[\mathcal{S}, \tilde{\mathcal{L}}; \mathbf{V}, \Omega, \mathcal{L}] \equiv \int_0^T [(\hat{\mathcal{N}}^T + \tilde{\mathcal{L}}^T A)\mathcal{V} + \mathcal{L}^T(A v - \mathcal{B})]dt + J_R , \tag{J.4}$$

where $\hat{\mathcal{N}}$ is the prescribed part of the edge-load vector. The first form of the Principle of Virtual work states that if the *essential* initial conditions

$$\bar{\mathbf{y}}(\sigma, 0) = \mathbf{f}(\sigma) \ , \ \beta(\sigma, 0) = 0 \tag{J.5}$$

are satisfied and if $J^*=0$ for all $(\mathbf{V}, \Omega)$ in $\mathcal{T}_H$ and all $\mathcal{L}$, then $\mathcal{S}$ is a weak solution which, if sufficiently smooth, is also a classical solution that satisfies the additional initial and boundary conditions

$$\mathbf{v}(\sigma, 0) = \mathbf{g}(\sigma) \ , \ \omega(\sigma, 0) = w(\sigma) \tag{J.6}$$

$$\hat{\mathcal{N}}^T + \tilde{\mathcal{L}}^T A = \mathcal{N}^T \ , \ A v = \mathcal{B} . \tag{J.7}$$

The second form of the Principle of Virtual Work requires the introduction of the subclass of test functions, $\mathcal{T}_H$, whose elements, in addition to satisfying the final conditions (J.1), also satisfy $A\mathcal{V}=0$ *explicitly*, i.e., on the edges of the axishell, $\mathcal{V}$ has the form

$$\mathcal{V} = \sum_{1}^{6-m} r_k \mathcal{M}_k \,, \tag{J.8}$$

where the r_k are arbitrary. Thus, let

$$\bar{J}[\mathcal{S}, r_k; \mathbf{V}, \Omega, \mathcal{L}] \equiv \int_0^T [\sum_{1}^{6-m} \hat{\mathcal{N}}^T r_k \mathcal{M}_k + \mathcal{L}^T(A\,v - \mathcal{B})]dt + J_R \,. \tag{J.9}$$

If the essential initial conditions (J.5) are satisfied and if $\bar{J}=0$ for all $(\mathbf{V}, \Omega)$ in $\mathcal{T}_H$ and all $\mathcal{L}$, then $\mathcal{S}$ is a weak solution which, if sufficiently smooth, is also a classical solution satisfying the natural initial conditions (D.8) and the boundary conditions

$$\hat{\mathcal{N}}^T \mathcal{M}_k = \mathcal{N}^T \mathcal{M}_k \,,\; k = 1, 2, \cdots, 6-m \;,\; A\,v = \mathcal{B} \,. \tag{J.10}$$

K. Load Potentials

The theory of load potentials for axishells is virtually identical to that for beamshells, discussed in detail in Section IV.J, so we may be brief.

By definition, there exists a load potential $\mathcal{P}(\bar{\mathbf{y}}, \beta)$ if the power of the external loads acting over the entire axishell,

$$\mathcal{W} = <\mathcal{F}(\bar{w}), \dot{\bar{w}}> \equiv [r(\mathbf{N}_\sigma \cdot \dot{\bar{\mathbf{y}}} + M_\sigma \dot{\beta})]_0^L + \int_0^L (\mathbf{p} \cdot \dot{\bar{\mathbf{y}}} + l\dot{\beta}) r d\sigma \,, \tag{K.1}$$

is equal to the time derivative of $\mathcal{P}$. In (K.1), $\mathcal{F}$ is an external load operator that sends any admissible kinematic element,

$$\bar{w} \equiv (\bar{\mathbf{y}}, \beta, \bar{\mathbf{y}}_0, \beta_0, \bar{\mathbf{y}}_L, \beta_L) \,, \tag{K.2}$$

into some load element,

$$\mathcal{F}(\bar{w}) \equiv (\mathbf{p}, l, \mathbf{N}_\sigma^0, M_\sigma^0, \mathbf{N}_\sigma^L, M_\sigma^L) \,, \tag{K.3}$$

and $<\mathcal{F}(\bar{w}), \dot{\bar{w}}>$ is an inner product. Each element, $\bar{w}$, of the admissible set, $\mathcal{A}$, satisfies edge constraints of the form $A\,v = \mathcal{B}$ and is sufficiently smooth for $\mathcal{F}(\bar{w})$ to exist.

We next present three common potential loads and derive the associated potential for each, more or less by inspection. Finally, we close with a general test for potentiality and a formula for computing $\mathcal{P}$ if it exists.

1. Self-weight (gravity loading)

Assume that the only external load is given by $\mathbf{p} = -mg\mathbf{e}_z$, where g is the acceleration of gravity. Then (K.1) reduces to

$$\mathcal{W} = -mg\int_0^L \mathbf{e}_z \cdot \dot{\bar{\mathbf{y}}} r d\sigma = -mg\int_0^L \dot{\bar{z}} r d\sigma = -(mg\int_0^L \bar{z} r d\sigma)^{\boldsymbol{\cdot}} \equiv -\dot{\mathcal{P}} \,. \tag{K.4}$$

2. Centrifugal loading

Assume that the only external load is given by $\mathbf{p}=m\bar{r}\Omega^2\mathbf{e}_r$, where Ω is the constant angular velocity of the axishell about the z-axis. Then (K.1) reduces to

$$\mathcal{W}=m\Omega^2\int_0^L \bar{r}\mathbf{e}_r\cdot\dot{\bar{\mathbf{y}}}r\,d\sigma=m\Omega^2\int_0^L \bar{r}\dot{\bar{r}}\,r\,d\sigma=(\tfrac{1}{2}m\Omega^2\int_0^L \bar{r}^2 r\,d\sigma)^{\bullet}=-\dot{\mathcal{P}}\,. \quad \text{(K.5)}$$

3. Arbitrary normal pressure

The deformed image of a differential element of area, $dA=rd\sigma d\theta$, of the reference surface is $d\bar{A}=\bar{r}\,|\bar{\mathbf{y}}'|\,d\sigma d\theta$ so that, if $p(\bar{r},\bar{z})$ is a normal pressure reckoned per unit area of the deformed reference surface, $\bar{\mathfrak{Z}}$, then the force on $d\bar{A}$ will be $-p\bar{\mathbf{b}}d\bar{A}=-p\mathbf{m}\bar{r}d\sigma d\theta$, where $\bar{\mathbf{b}}$ is a unit normal to $\bar{\mathfrak{Z}}$ and $\mathbf{m}=\bar{\mathbf{y}}'\times\mathbf{e}_\theta=|\bar{\mathbf{y}}'|\bar{\mathbf{b}}$. Hence, (K.1) reduces to

$$\mathcal{W}=-\int_0^L \bar{r}p\mathbf{m}\cdot\dot{\bar{\mathbf{y}}}d\sigma=\int_0^L \bar{r}p(\bar{r}'\dot{\bar{z}}-\bar{z}'\dot{\bar{r}})d\sigma\,. \quad \text{(K.6)}$$

Let

$$q\equiv\int_0^{\bar{r}}\tau p(\tau,\bar{z})d\tau \quad \text{(K.7)}$$

so that $\bar{r}p=q_{,\bar{r}}$. Then, an integration by parts converts (K.6) into

$$\mathcal{W}=q\dot{\bar{z}}\,\Big|_0^L-(\int_0^L q\bar{z}'d\sigma)^{\bullet}\,. \quad \text{(K.8)}$$

Thus, an arbitrary normal pressure that depends on $\bar{r}$ and $\bar{z}$ only has a potential

$$\mathcal{P}=\Gamma+\int_0^L q\bar{z}'d\sigma \quad \text{(K.9)}$$

if there exists an *edge potential* such that $\dot{\Gamma}=-q\dot{\bar{z}}\,\Big|_0^L$. If the pressure is constant, i.e., if $p=p_0$, the load potential is

$$\mathcal{P}=\Gamma+p_0\bar{V}/2\pi\,, \quad \text{(K.10)}$$

where

$$\bar{V}=\pi\int_0^L \bar{r}^2\bar{z}'d\sigma \quad \text{(K.11)}$$

is the volume enclosed by the deformed surface of revolution $\bar{\mathfrak{Z}}$ and planes perpendicular to the z-axis passing through any deformed edges. If the pressure is hydrostatic, i.e., if $p=-\rho g\bar{z}$, then the load potential is

$$\mathcal{P}=\Gamma-\rho g\bar{P}/2\pi\,, \quad \text{(K.12)}$$

where

$$\bar{P}=\pi\int_0^L \bar{r}^2\bar{z}\bar{z}'d\sigma \quad \text{(K.13)}$$

is the centroid of $\bar{V}$.

In general, a necessary condition that $\mathcal{F}$ be potential, as we showed in Section IV.J, is that it be symmetric; i.e.,

$$<\mathcal{F}'(\overline{w})u, v> = <\mathcal{F}'(\overline{w})v, u> \quad , \quad \forall\, u, v \in \mathcal{A}_\delta \; , \tag{K.14}$$

where $\mathcal{A}_\delta$ is the linear vector space of admissible variations associated with $\mathcal{A}$. If the admissible set, $\mathcal{A}$, is convex, then (K.14) is also a sufficient condition that there exist a load potential, $\mathcal{P}$, which can then be computed from the formula

$$\mathcal{P}[\overline{w}] = -\int_0^1 <\mathcal{F}(w + \tau(\overline{w} - w)), \overline{w} - w> d\tau \; , \tag{K.15}$$

where the inner product is defined by (K.1).

L. Strains

To measure the deviation of $\overline{\mathbf{y}}'$ from $\mathbf{T}$, we introduce a *meridional strain*, e_σ, and a *shear strain*, g, by setting

$$\overline{\mathbf{y}}' = (1 + e_\sigma)\mathbf{T} + g\mathbf{B} \; . \tag{L.1}$$

Thus, from (D.4),

$$\mathbf{m} = (1 + e_\sigma)\mathbf{B} - g\mathbf{T} \tag{L.2}$$

so that, in (H.4),

$$\mathbf{v}' - \omega\mathbf{m} = \dot{e}_\sigma\mathbf{T} + \dot{g}\mathbf{B} \equiv \overset{*}{\overline{\mathbf{y}}}{}' \; , \tag{L.3}$$

the asterisk denoting time rate-of-change with respect to the spin basis, $\{\mathbf{T}, \mathbf{B}\}$.

If we now introduce the component representations

$$\mathbf{N}_\sigma \equiv N_\sigma\mathbf{T} + Q\mathbf{B} \quad , \quad \mathbf{M}_\theta \equiv M_\theta\mathbf{T} + M_n\mathbf{B} \; , \tag{L.4}$$

(H.4) may be expressed as

$$\mathcal{D} = \int_a^b \tau{:}\dot{\varepsilon} r d\sigma \; , \tag{L.5}$$

where

$$\tau{:}\dot{\varepsilon} \equiv N_\sigma \dot{e}_\sigma + N_\theta \dot{e}_\theta + Q\dot{g} + M_\sigma \dot{k}_\sigma + M_\theta \dot{k}_\theta + M_n \dot{k}_n \tag{L.6}$$

is the *deformation power density*. In (L.6),

$$e_\theta \equiv \frac{\mathbf{e}_r \cdot (\overline{\mathbf{y}} - \mathbf{y})}{r} = \frac{\overline{r} - r}{r} \tag{L.7}$$

is the *hoop (circumferential) strain*,

$$k_\sigma \equiv \beta' \tag{L.8}$$

is the *meridional bending strain*,

$$k_\theta \equiv \frac{\sin(\alpha+\beta)-\sin\alpha}{r} \tag{L.9}$$

is the *hoop (circumferential) bending strain*, and

$$k_n \equiv \frac{\cos(\alpha+\beta)-\cos\alpha}{r} \tag{L.10}$$

is the *normal bending strain* associated with the (non-classical) normal component of the stress couple, M_n.

By (E.4), (E.5), (L.1), $(L.4)_1$, and (L.8), the jump conditions (F.2) imply that

$$[N_\sigma]=mc^2[e_\sigma]\ ,\ [Q]=mc^2[g]\ ,\ [M_\sigma]=Ic^2[k_\sigma]\,, \tag{L.11}$$

while (D.3), the vector equation of balance of linear momentum, takes the component form

$$(rN_\sigma)'-N_\theta\cos(\alpha+\beta)-r(\alpha+\beta)'Q+rp_T=rm(\dot v_T-\omega v_B) \tag{L.12}$$

$$(rQ)'+r(\alpha'+k_\sigma)N_\sigma+N_\theta\sin(\alpha+\beta)+rp_B=rm(\dot v_B+\omega v_T)\,, \tag{L.13}$$

where we have set

$$\mathbf{p}=p_T\mathbf{T}+p_B\mathbf{B}\ \text{ and }\ \mathbf{v}=v_T\mathbf{T}+v_B\mathbf{B}\,. \tag{L.14}$$

Note also that, with (E.4), (E.5), and $(L.4)_2$, the equation of balance of rotational momentum of an axishell, (D.5), takes the form

$$(rM_\sigma)'-M_\theta\cos(\alpha+\beta)+M_n\sin(\alpha+\beta)+r[(1+e_\sigma)Q-gN_\sigma]+rl=Ir\ddot\beta\,. \tag{L.15}$$

To obtain a form of the axishell equations in which neither M_n nor k_n appears, we differentiate (L.9) and (L.10) with respect to time to get $\dot k_n=-\tan(\alpha+\beta)\dot k_\theta$. This enables us to set

$$M_\theta\dot k_\theta+M_n\dot k_n=\overline{M}_\theta\dot k_\theta\,, \tag{L.16}$$

where

$$\overline{M}_\theta\equiv\sec(\alpha+\beta)\mathbf{M}_\theta\cdot\mathbf{e}_r=M_\theta-\tan(\alpha+\beta)M_n\,. \tag{L.17}$$

Thus, the deformation power density and the balance of rotational momentum—the only basic equations in which M_n and k_n appear—take the respective forms

$$\tau{:}\dot\varepsilon=N_\sigma\dot e_\sigma+N_\theta\dot e_\theta+Q\dot g+M_\sigma\dot k_\sigma+\overline{M}_\theta\dot k_\theta \tag{L.18}$$

$$(rM_\sigma)'-\overline{M}_\theta\cos(\alpha+\beta)+r[(1+e_\sigma)Q-gN_\sigma]+rl=Ir\ddot\beta\,. \tag{L.19}$$

M. Compatibility Conditions

Given e_σ, e_θ, g, and β, how do we compute $\bar{\mathbf{y}}$? From (L.1), (D.1), and (L.7), we may compute $\bar{\mathbf{y}},_\sigma$ and $\bar{\mathbf{y}},_\theta$ in terms of e, g, and β. As $\mathcal{S}$ has no holes, $\bar{\mathbf{y}}$ exists if and only if $\bar{\mathbf{y}},_{\theta\sigma} = \bar{\mathbf{y}},_{\sigma\theta}$. But (E.4) and (E.5) imply that

$$\mathbf{T},_\theta = \cos(\alpha + \beta)\mathbf{e}_\theta \ , \ \mathbf{B},_\theta = -\sin(\alpha + \beta)\mathbf{e}_\theta \ . \tag{M.1}$$

Thus, $\bar{\mathbf{y}},_{\theta\sigma} = \bar{\mathbf{y}},_{\sigma\theta}$ is equivalent to the single scalar *strain-rotation compatibility condition*

$$[r(1 + e_\theta)]' - (1 + e_\sigma)\cos(\alpha + \beta) + g\sin(\alpha + \beta) = 0 \tag{M.2}$$

(Reissner, 1963a).

With the aid of (L.9) and (L.10), (M.2) can be expressed in terms of the strains alone as

$$(re_\theta)' - e_\sigma\cos\alpha + g\sin\alpha - r[(1 + e_\sigma)k_n - gk_\theta] = 0 \ . \tag{M.3}$$

Following Reissner (1974), we call (M.3) the *intrinsic form* of the strain-rotation compatibility condition for an axishell because the angle of rotation β does not appear. It is important to note that, if we wish to eliminate β as an unknown, then we must append to (M.3) the algebraic identity (Reissner, 1969)

$$(rk_\theta + \sin\alpha)^2 + (rk_n + \cos\alpha)^2 - 1 = 0 \ , \tag{M.4}$$

which follows immediately from (L.9) and (L.10). The implications of (M.4) have already been mentioned in the introduction to this chapter and will be discussed further in Section O on strain-energy densities. In *linearized* static problems with no distributed couples, (L.15) and (M.3) display the static-geometric duality:

$$M_\sigma <> e_\theta \ , \ M_\theta <> e_\sigma \ , \ M_n <> g \ , \ Q <> -k_n \ . \tag{M.5}$$

The static dual of (M.4) is the *linearized* equation of overall vertical equilibrium in the absence of vertical surface or edge loads: $\mathbf{N}_\sigma \cdot \mathbf{e}_z = N_\sigma \sin\alpha + Q\cos\alpha = 0$.

When (M.2) is satisfied, (L.1) yields

$$\bar{\mathbf{y}} = \int [(1 + e_\sigma)\mathbf{T} + g\mathbf{B}]d\sigma \tag{M.6}$$

or, in more explicit form,

$$\begin{aligned}\bar{\mathbf{y}}(\sigma, \theta, t) &\equiv \bar{r}(\sigma, t)\mathbf{e}_r(\theta) + \bar{z}(\sigma, t)\mathbf{e}_z \\ &= r(\sigma)[1 + e_\theta(\sigma, t)]\mathbf{e}_r(\theta) \\ &\quad + <\bar{z}(0, t) + \int_0^\sigma \{[1 + e_\sigma(s, t)]\sin\bar{\alpha}_T(s, t) + g(s, t)\cos\bar{\alpha}_T(s, t)\}ds > \mathbf{e}_z \ ,\end{aligned} \tag{M.7}$$

where $\bar{\alpha}_T \equiv \alpha + \beta$

In addition to (M.3) and (M.4), there exists another vector compatibility condition among the bending strains. If we set

$$\mathbf{K}_\theta \equiv k_\theta \mathbf{T} + k_n \mathbf{B} \,, \tag{M.8}$$

then it is readily verified, using (L.8)-(L.10), that

$$(-r\mathbf{K}_\theta)' + k_\sigma \mathbf{e}_r + rk_\sigma \mathbf{K}_\theta \times \mathbf{e}_\theta = \mathbf{0} \,. \tag{M.9}$$

If *linearized,* this equation becomes the geometric dual of the static version of (D.3) in the absence of surface loads. Specifically, the duality indicated in (M.5) is supplemented by the additional relations

$$N_\sigma < > -k_\theta \;\; , \; N_\theta < > -k_\sigma \,. \tag{M.10}$$

Setting

$$\mathbf{e}_r = \mathbf{T}\cos(\alpha + \beta) - \mathbf{B}\sin(\alpha + \beta) = \mathbf{T}(\cos\alpha + rk_n) - \mathbf{B}(\sin\alpha + rk_\theta) \,, \tag{M.11}$$

we can reduce (M.9) to the two scalar compatibility conditions of Reissner (1969):

$$(-rk_\theta)' + k_\sigma \cos\alpha + (\alpha' + k_\sigma) rk_n = 0 \tag{M.12}$$

$$(rk_n)' + k_\sigma \sin\alpha + (\alpha' + k_\sigma) rk_\theta = 0 \,. \tag{M.13}$$

(Reissner's bending strains, which he denotes by κ_ξ, κ_θ, and λ, are equal to $-k_\sigma$, $-k_\theta$, and $-k_n$, respectively.)

We note that (M.4), (M.12), and (M.13) are not independent of each other because (M.4) is a first integral of (M.12) and (M.13). Hence, one of the latter equations should be deleted. The obvious candidate is (M.13) in view of the secondary importance of the normal bending strain, k_n, in axishell theory. In classical theory (where k_n is not used), we can further eliminate k_n between (M.4) and (M.12) to arrive at a single compatibility condition in k_σ and k_θ; we cannot retain both (M.12) and (M.13) without adjoining (M.4) too.

N. The Mechanical Theory of Axishells

If thermal effects are negligible, we may cast all of the approximations made in the theory of axishells into one postulate, namely, *there exists an internal energy $\mathcal{E}$, depending on the unknowns only through the strains $(e_\sigma, e_\theta, g, k_\sigma, k_\theta, k_n)$, such that*

$$\int_{t_1}^{t_2} \mathcal{D}\, dt = \mathcal{E} \Big|_{t_1}^{t_2} \,, \tag{N.1}$$

or, in view of (H.1), such that

$$\int_{t_1}^{t_2} \mathcal{W}\, dt = (\mathcal{K} + \mathcal{E})_{t_1}^{t_2} \,. \tag{N.2}$$

When we come to thermal effects in Section X, the postulate that generalizes (N.2) is called the First Law of Thermodynamics or the Law of Conservation of Energy. We therefore may call (N.2) the *Postulate of Conservation of*

Mechanical Energy. While the equations of motion for a shell of revolution and the Mechanical Work Identity (H.1) are *exact* consequences of 3-dimensional Continuum Mechanics, (N.1) or (N.2) are *not.* This was explained in Section III.F on birods. However, as we have emphasized in Chapter I and elsewhere, we have introduced our approximations in precisely the same place that the approximations of 3-dimensional continuum mechanics appear, namely, in the internal energy, which, ultimately, must be determined experimentally. Indeed, one often does experiments on special shells (plates, circular cylinders, etc.) to infer information about the 3-dimensional structure of the internal energy.

O. Elastic Axishells and Strain-Energy Densities

So far, our field equations for axishells consist of a vector equation of balance of linear momentum, (D.3), a scalar equation of balance of rotational momentum, (D.5) or (L.15), a vector position-strain relation, (L.1), three scalar bending strain-rotation relations, (L.8)-(L.10), and a scalar compatibility condition, (M.2). These pointwise conditions are to be supplemented by mechanical boundary conditions, as discussed in section I, and, possibly, by jump conditions, as discussed in section F.

The field equations are equivalent to 9 scalar equations for 15 unknowns: the 6 stresses, $(N_\sigma, N_\theta, Q, M_\sigma, M_\theta, M_n)$, the 6 strains, $(e_\sigma, e_\theta, g, k_\sigma, k_\theta, k_n)$, the rotation β, and the 2 components of $\bar{\mathbf{y}}$, $\bar{r}$ and $\bar{z}$. To obtain a complete set of field equations for a *mechanical theory* of axishells, we need 6 additional constitutive relations. While our results to this point apply regardless of the shell material—we could be talking about water bells—we now restrict attention to elastic shells.

An axishell is said to be *elastic* if there exists a *strain (= internal)–energy density*, Φ, reckoned per unit area of S and depending on σ and the pointwise values of $(e_\sigma, e_\theta, g, k_\sigma, k_\theta, k_n)$ only, such that

$$\tau{:}\dot{\varepsilon} = \dot{\Phi} \ , \quad 0<\sigma<L \, , 0<t \ , \quad \forall\, \bar{\mathbf{y}} \, , \beta \, . \tag{O.1}$$

To deduce stress-strain relations from (O.1), we replace the left side by $N_\sigma \dot{e}_\sigma + \cdots + M_n \dot{k}_n$ and apply the chain rule to the right to obtain $\Phi_{,e_\sigma} \dot{e}_\sigma + \cdots + \Phi_{,k_n} \dot{k}_n$. However, at a fixed but arbitrary arc length, σ^*, and time, t^*, the sextuple of rates $(\dot{e}_\sigma, \cdots, \dot{k}_n)$ cannot be prescribed independently because the algebraic identity (M.4) implies that

$$(rk_\theta + \sin\alpha)\dot{k}_\theta + (rk_n + \cos\alpha)\dot{k}_n = 0 \, . \tag{O.2}$$

Replacing Φ in (O.1) by Φ plus half of an unknown Lagrange multiplier, $\lambda(\sigma, t)$, times (M.4), we may, by a standard argument, now treat the elements of the septuple $(\dot{e}_\sigma, \cdots, \dot{k}_n, \lambda)$ as if they were linearly independent and thus conclude that (O.1) implies (M.4) as well as the constitutive relations

$$N_\sigma = \Phi,_{e_\sigma} \ , \ N_\theta = \Phi,_{e_\theta} \ , \ Q = \Phi,_g \ , \ M_\sigma = \Phi,_{k_\sigma}$$
$$M_\theta = \Phi,_{k_\theta} + \lambda(rk_\theta + \sin\alpha) \ , \ M_n = \Phi,_{k_n} + \lambda(rk_n + \cos\alpha) \ . \tag{O.3}$$

If $(O.3)_{1,2,3}$ can be solved, uniquely, for e_σ, e_θ, and g in terms of $N_\sigma, N_\theta, Q, k_\sigma, k_\theta$, and k_n then we may, via the following Legendre transformation, introduce the *mixed-energy density*

$$\Psi \equiv \Phi - (N_\sigma e_\sigma + N_\theta e_\theta + Qg) \ . \tag{O.4}$$

From (O.3) and (O.4), we obtain the new constitutive relations

$$e_\sigma = -\Psi,_{N_\sigma} \ , \ e_\theta = -\Psi,_{N_\theta} \ , \ g = -\Psi,_Q \ , \ M_\sigma = \Psi,_{k_\sigma}$$
$$M_\theta = \Psi,_{k_\theta} + \lambda(rk_\theta + \sin\alpha) \ , \ M_n = \Psi,_{k_n} + \lambda(rk_n + \cos\alpha) \ . \tag{O.5}$$

Note from (L.9), (L.10), and $(O.5)_{4,5}$ that

$$-M_\theta \cos(\alpha+\beta) + M_n \sin(\alpha+\beta) = -\Psi,_{k_\theta}\cos(\alpha+\beta) + \Psi,_{k_n}\sin(\alpha+\beta) \ , \tag{O.6}$$

so that the unknown Lagrange multiplier, λ, does not enter the equation of balance of rotational momentum, (L.15); neither can λ enter any boundary conditions. Thus, as our theory is incapable of determining λ, we simply set it to 0.

An alternative to introducing a Lagrange multiplier is to first use (M.4) to express k_n in terms of k_θ. Then, in view of (L.9), (L.10), and (L.17), it follows that

$$\frac{d\psi}{dk_\theta} = \psi,_{k_\theta} + \psi,_{k_n}\frac{dk_n}{dk_\theta} = \overline{M}_\theta \ . \tag{O.7}$$

Hence, k_θ and the *modified* stress couple, $\overline{M}_\theta$, which are conjugates in the deformation power, could be taken as basic unknowns in our theory of axishells. However, we find that it is more convenient to retain both conjugate pairs (k_θ, M_θ) and (k_n, M_n) and, where appropriate, to take account of their interrelation.

The *classical theory* of axishells assumes that Ψ is independent of Q and k_n; the theory of *aximembranes* assumes, in addition, that Ψ is independent of k_σ and k_θ.

There are both theoretical and practical advantages in using the mixed-energy density, Ψ, in place of Φ, including greater symmetry among the field equations and variables and robustness in numerical approximations.

1. Quadratic strain-energy densities

In the small-strain theory of transversely homogeneous axishells with linear, isotropic stress-strain relations referred to the midsurface, the expression for Φ takes the well-known form

$$\Phi = \tfrac{1}{2}C(e_\sigma^2 + e_\theta^2 + 2\nu_e e_\sigma e_\theta) + \tfrac{1}{2}D(k_\sigma^2 + k_\theta^2 + 2\nu_b k_\sigma k_\theta) + \tfrac{1}{2}Kg^2 + \tfrac{1}{2}D_n k_n^2 . \tag{O.8}$$

Here,

$$C \equiv E_e h C_e \ , \ D \equiv E_b h^3 C_b \ , \ K \equiv E_s h C_s \ , \tag{O.9}$$

where E_e, E_b, and E_s are, respectively, Young's moduli in stretching, bending, and shear, and ν_e and ν_b are, respectively, Poisson ratios in stretching and bending. Conventionally, one takes $E_e = E_b = E_s = E$ and $\nu_e = \nu_b = \nu$. The coefficients C_e, C_b, and C_s are as in (IV.N.17)[4], so that

$$C = \frac{Eh}{1-\nu^2} \text{ and } D = \frac{Eh^3}{12(1-\nu^2)} . \tag{O.10}$$

The corresponding expression for the mixed-energy density is

$$\Psi = -\tfrac{1}{2}A_e(N_\sigma^2 + N_\theta^2 - 2\nu_e N_\sigma N_\theta) + \tfrac{1}{2}D(k_\sigma^2 + k_\theta^2 + 2\nu_b k_\sigma k_\theta) - \tfrac{1}{2}A_s Q^2 + \tfrac{1}{2}D_n k_n^2 , \tag{O.11}$$

where

$$A \equiv A_e \equiv \frac{1}{E_e h} \ , \ A_s \equiv \frac{1}{E_s h C_s} . \tag{O.12}$$

In classical theory, D_n and Kg^2 are set to 0 in Φ, and D_n and A_s are set to 0 in Ψ. As we shall see in Section R, the use of the classical form (O.11), with error estimates appended, yields (in static problems) the simplified form of the Reissner-Meissner-Reissner equations derived by Koiter (1980).

We may take (O.8) and (O.11) as approximations for isotropic, transversely *nonhomogeneous* axishells (such as shells constructed from isotropic layers), provided that the nonhomogeneity is *symmetric* with respect to the midsurface of the shell.[5] In this case, the elastic moduli E_i and ν_i will be different from each other.

If the isotropic layers in an axishell are not symmetrically arranged, the the base-reference deviation introduced in Section B induces *mixed extensional-bending terms* in Φ. However, if, following the example in subsection IV.N.1, we choose d to be the centroid of the *modulus-weighted cross section* (in static problems), then the mixed terms vanish. The procedure used in the aforementioned example can be extended to layered axishells by taking the 3-dimensional strain-energy density W for each layer to be of the form

---

[4]The value of C_s varies somewhat, according to the source quoted.

[5]Throughout *this subsection*, we neglect the effects of curvature on the strain-energy density, which is equivalent to taking the strain-energy density equal to that of a plate. For a detailed discussion of this approximation, see the subsections to follow and Section IV.N.

$$W_i = \frac{E_i}{2(1-\nu_i^2)}(\varepsilon_\sigma^2 + \varepsilon_\theta^2 + 2\nu_i \varepsilon_\sigma \varepsilon_\theta)\,, \tag{O.13}$$

where ε_σ and ε_θ are components of the 3-dimensional strain. These forms reflect the assumption of an approximately plane state of stress in the axishell. If we further take $\varepsilon_\sigma = e_\sigma + (s-d)k_\sigma$, $\varepsilon_\theta = e_\theta + (s-d)k_\theta$, and integrate W through the thickness to obtain Φ, specific expressions fall out for the coefficients of the various strain terms in Φ.

In the more general case of polar-orthotropic layers, the above procedure is normally used, with the appropriate form for W in each layer. Here, however, mixed extensional-bending terms in Φ are unavoidable, and the choice of d in static problems is a matter of convenience. For more details and more elaborate models, the reader should consult the literature on layered and composite plates and shells, including Stavsky & Hoff (1969), Bert (1975), Vinson & Sierakowski (1986), and Librescu (1987).

*2. *General strain-energy densities*

Our aims in the next 3 subsections are to estimate the error made by taking (O.8) as an approximation to the exact, 2-dimensional strain-energy density and to discuss the general form that Φ should take when the 3-dimensional strains are large, say O(1). Our main approach—which follows closely that of Section IV.N while accounting for some additional terms and the more complex geometry of axishells—is to use logical constraints and plausible assumptions to delimit and to simplify the possible forms of Φ.

For simplicity, we assume henceforth that the axishell consists of a single, transversely homogeneous elastic layer, and we use the shell coordinates defined in (A.3) (possibly offset by a distance d from the midsurface). Then, using the same dimensional arguments as with beamshells, we may set

$$\Phi = Eh\phi(e_\sigma, e_\theta, g, hk_\sigma, hk_\theta, hk_n, d/h, hb_\sigma, hb_\theta)\,, \tag{O.14}$$

where ϕ is a dimensionless function of its dimensionless arguments, E is a Young's modulus, h is the shell thickness, and b_σ and b_θ are the principal curvatures of $\mathcal{S}$, defined by

$$b_\sigma \equiv \mathbf{y}_{,\sigma\sigma}\cdot\mathbf{b} = \mathbf{t}'\cdot\mathbf{b} = \alpha'\ ,\ b_\theta \equiv r^{-2}\mathbf{y}_{,\theta\theta}\cdot\mathbf{b} = -r^{-1}\mathbf{e}_r\cdot\mathbf{b} = r^{-1}\sin\alpha\,. \tag{O.15}$$

In setting down the right side of (O.14), we have made 3 assumptions:

(1) The thickness, h, is the natural scale of LENGTH for rendering ϕ, k_σ, k_θ, k_n, b_σ, and b_θ nondimensional.

(2) ϕ does not depend on the derivatives of $h(\sigma)$, i.e., we have *locality.*

(3) ϕ depends on the geometry of $\mathcal{S}$ *through its principal curvatures only,* i.e., we have *geometric inhomogeneity.*

For further discussion of these 3 assumptions, see Section IV.N (especially the introduction and subsection IV.N.2). We also mention that the form of the right side of (O.14) assures the *local* nature of ϕ, i.e., no action-at-a-distance effects and

(4) That Φ vanishes as the shell thickness approaches 0.

To proceed, we now exploit the following additional conditions that we impose on an axishell and its strain-energy density, Φ.

(5) Φ is invariant under coordinate transformations on $\mathcal{S}$ and $\bar{\mathcal{S}}$.

(6) The material of the axishell is stable.

(7) The shell is thin.

(8) Φ has the small-strain quadricity property (if the strains are measured from the undeformed state).

(9) Φ is positive definite in the strains.

(10) Φ approaches infinity as the sum of the squares of the (nondimensionalized) strains approachs infinity.

The last 2 conditions will not be exploited explicitly in this subsection, but, nevertheless, must be kept in mind. In particular, (10) can be useful in analyizing singular behavior, such as occurs under point loads, even though the constitutive model can not be expected to be valid for very large strains.

We examine the implications of (5) by making 2 coordinate transformations. In the first, we replace σ by $L-\sigma$. This transformation reverses the directions of the vectors $\mathbf{t}, \bar{\mathbf{y}}'$, and $\mathbf{T}$; the differentiated vectors $\mathbf{t}'$ and $\mathbf{T}'$ are unaffected. In the second transformation, we replace θ by $-\theta$, which reverses the directions of $\mathbf{b}$ and $\mathbf{B}$ and changes the sign of the directed distance, d, from the reference curve, C, to the base curve, $\mathcal{B}$. The 2 transformations must leave Φ invariant.

Physically, the second transformation is equivalent to rotating our Cartesian reference frame $\{\mathbf{e}_x, \mathbf{e}_y, \mathbf{e}_z\}$ through 180° about $\mathbf{e}_x$, so that $\mathbf{e}_z$ reverses direction. This interpretation is equivalent to examining a *second* axishell whose movement is the mirror image, in the *xy*-plane, of that of the first. This interpretation implies that Φ should not change if α and β are replaced, simultaneously, by $-\alpha$ and $-\beta$.

Using (O.15) and the vectorial definitions

$$e_\sigma \equiv \bar{\mathbf{y}}'\cdot\mathbf{T} - 1 \;,\; re_\theta \equiv (\bar{\mathbf{y}} - \mathbf{y})\cdot\mathbf{e}_r \;,\; g \equiv \bar{\mathbf{y}}'\cdot\mathbf{B} \tag{O.16}$$

$$k_\sigma \equiv \mathbf{T}'\cdot\mathbf{B} - \mathbf{t}'\cdot\mathbf{b} \;,\; rk_\theta \equiv (\mathbf{b} - \mathbf{B})\cdot\mathbf{e}_r \;,\; rk_n \equiv (\mathbf{T} - \mathbf{t})\cdot\mathbf{e}_r \,, \tag{O.17}$$

we can establish the dependence of the curvatures and the strains on our 2 coordinate transformations, $\sigma \leftarrow L-\sigma$ and $\theta \leftarrow -\theta$. The results are summarized in Table 5.1, from which it follows that $\phi = \Phi/Eh$ must have the functional form

Table 5.1
Sign Changes of Constitutive Variables

$\sigma, \mathbf{b}$	b_σ, b_θ	d	e_σ, e_θ	g	k_σ, k_θ	k_n
$L-\sigma, \mathbf{b}$	b_σ, b_θ	d	e_σ, e_θ	$-g$	k_σ, k_θ	$-k_n$
$\sigma, -\mathbf{b}$	$-b_\sigma, -b_\theta$	$-d$	e_σ, e_θ	$-g$	$-k_\sigma, -k_\theta$	k_n

$$\begin{aligned}\phi = \phi[&e_\sigma, e_\theta, g^2, d^2/h^2;\\ &dk_\sigma, dk_\theta, h^2(k_\sigma^2, k_\sigma k_\theta, k_\theta^2, k_n^2);\\ &db_\sigma, db_\theta, h^2(b_\sigma k_\sigma, b_\sigma k_\theta, b_\theta k_\sigma, b_\theta k_\theta, b_\sigma^2, b_\theta^2, b_\sigma b_\theta)]\\ \equiv \phi(&\mathcal{X}; \mathcal{Y}; \mathcal{Z}),\end{aligned} \tag{O.18}$$

where the grouping of the arguments of ϕ is analogous to that for beamshells in (IV.N.24). Terms containing both g and k_n, such as $gk_n[d, h^2(k_\sigma, k_\theta, b_\sigma, b_\theta)]$, have not been included, being of marginal importance in shell theory

To examine the implications of *material stability*, we use the jump conditions, (F.2), together with the constitutive relations, (O.3) (with λ set to zero), so obtaining

$$[\Phi,_{e_\sigma}] = mc^2[e_\sigma] \;,\; [\Phi,_g] = mc^2[g] \;,\; [\Phi,_{k_\sigma}] = Ic^2[k_\sigma] \,, \tag{O.19}$$

where, as before, the boldface brackets denote jumps in the argument across a shock. In calculating jumps, one should note that, as a consequence of (F.1), $[e_\theta] \equiv [k_\theta] \equiv [k_n] \equiv 0$.

Material stability requires that Φ be such that (O.19) yields positive values of c^2 for arbitrary, infinitesimal jumps in e_σ, g, and k_σ. (See the discussion in subsections III.J.1 and IV.N.2.) This same equation can be used for calculation the shock speed c (if, indeed, a shock exists).

Small-strain quadricity assumes the existence of a stress-free state from which Φ and the strains are measured. If we further take the constitutive relations (O.3) to be linear, homogeneous, and nondegenerate for sufficiently small strains, then any polynomial approximation to ϕ must be of the form

$$\phi = e^T C e + O(\| e \|^3) \,, \tag{O.20}$$

where $e \equiv [e_\sigma, e_\theta, g, hk_\sigma, hk_\theta, hk_n]^T$, $\| e \|^2 \equiv e^T e$, and the matrix C may depend on the non-strain parameters on the right side of (O.18). We can also express the requirement of small-strain quadricity as

$$\underline{C} \| e \|^2 \le \phi \le \overline{C} \| e \|^2 \,, \tag{O.21}$$

for all $\| e \| \le e_0$, where $\underline{C}$, $\overline{C}$, and e_0 are positive constants.

**3. Elastic isotropy*

First, let us consider a general shell. We assume that its dimensionless strain-energy density, ϕ, depends on various material constants, the curvature of its reference, surface $\mathcal{S}$, the base-reference deviation, $d \equiv \mathbf{d}\cdot\mathbf{b}$, and the strains. Let $\mathbf{B}$, $\mathbf{E}$, and $\mathbf{K}$ denote 2-order surface tensors that represent, respectively, the curvature of $\mathcal{S}$ and the extensional and bending strains that it suffers, and let $\mathbf{g}$ and $\mathbf{k}$ denote surface vectors whose components represent transverse shearing and normal bending strains, respectively. If the shell is elastically isotropic, then ϕ can depend on $\mathbf{B}, \mathbf{E}, \mathbf{K}, \mathbf{g}$, and $\mathbf{k}$ only through the various independent scalar invariants that we can form among them. Moreover, the dependence of ϕ on d and these invariants must be such that ϕ is unchanged under an inversion of the normal to $\mathcal{S}$, even though d, $\mathbf{g}$, $\mathbf{B}$, and $\mathbf{K}$ change sign.

In axishells, the above-mentioned tensors and vectors have the following simple representations:

$$\mathbf{E} = e_\sigma \mathbf{tt} + e_\theta \mathbf{e}_\theta \mathbf{e}_\theta \ , \quad \mathbf{K} = k_\sigma \mathbf{tt} + k_\theta \mathbf{e}_\theta \mathbf{e}_\theta \ , \quad \mathbf{B} = b_\sigma \mathbf{tt} + b_\theta \mathbf{e}_\theta \mathbf{e}_\theta \ , \tag{O.22}$$

$$\mathbf{g} = g\mathbf{t} \ , \quad \mathbf{k} = k_n \mathbf{e}_\theta \ , \tag{O.23}$$

where $\mathbf{tt}$, etc. denotes the direct (or tensor) product. With the aid of the Cayley-Hamilton Theorem, we find that those independent, *non-zero* invariants that involve the 2nd-order tensors only reduce to

$$\operatorname{tr}\mathbf{B} = b_\sigma + b_\theta \ , \quad \operatorname{tr}\mathbf{E} = e_\sigma + e_\theta \ , \quad \operatorname{tr}\mathbf{K} = k_\sigma + k_\theta \tag{O.24}$$

$$\operatorname{tr}\mathbf{B}^2 = b_\sigma^2 + b_\theta^2 \ , \quad \operatorname{tr}\mathbf{E}^2 = e_\sigma^2 + e_\theta^2 \ , \quad \operatorname{tr}\mathbf{K}^2 = k_\sigma^2 + k_\theta^2 \tag{O.25}$$

$$\operatorname{tr}\mathbf{B}\cdot\mathbf{E} = b_\sigma e_\sigma + b_\theta e_\theta \ , \ \operatorname{tr}\mathbf{B}\cdot\mathbf{K} = b_\sigma k_\sigma + b_\theta k_\theta \ , \ \operatorname{tr}\mathbf{E}\cdot\mathbf{K} = e_\sigma k_\sigma + e_\theta k_\theta \ , \tag{O.26}$$ [6]

where tr = trace. This subset of invariants (which comprises all the invariants in the classical theory of elastically isotropic shells) exhibits *interchangeability*, i.e., each invariant is unchanged if the subscripts σ and θ are interchanged.

The invariants involving $\mathbf{g}$ are

$$\mathbf{g}\cdot\mathbf{g} = g^2 \tag{O.27}$$

$$\mathbf{g}\cdot(\mathbf{B}, \mathbf{E}, \mathbf{K})\cdot\mathbf{g} = g^2(b_\sigma, e_\sigma, k_\sigma) \tag{O.28}$$

$$\mathbf{g}\cdot(\mathbf{B}\cdot\mathbf{E}, \mathbf{B}\cdot\mathbf{K}, \mathbf{E}\cdot\mathbf{K})\cdot\mathbf{g} = g^2(b_\sigma e_\sigma, b_\sigma k_\sigma, e_\sigma k_\sigma) \tag{O.29}$$

$$\mathbf{g}\cdot\mathbf{B}\cdot\mathbf{E}\cdot\mathbf{K}\cdot\mathbf{g} = g^2 b_\sigma e_\sigma k_\sigma \ . \tag{O.30}$$

These are all of the form

---

[6]The invariant $\operatorname{tr}\mathbf{B}\cdot\mathbf{E}\cdot\mathbf{K}$ can be expressed in terms of the invariants listed in (O.24)-(O.26)—see equation (2.11) of Simmonds (1985).

$$g^2(1\, or\, b_\sigma)\times(1\, or\, e_\sigma)\times(1\, or\, k_\sigma)\,, \tag{O.31}$$

where the "or" indicates that one or the other of the terms in the parentheses is to be taken.

The invariants involving **k** only are likewise of the form

$$k_n^2(1\, or\, b_\theta)\times(1\, or\, e_\theta)\times(1\, or\, k_\theta)\,. \tag{O.32}$$

For simplicity, we ignore invariants such as $d\mathbf{g}\cdot(\mathbf{te}_\theta - \mathbf{e}_\theta\mathbf{t})\cdot\mathbf{k}$ or $\mathbf{g}\cdot\mathbf{B}\cdot(\mathbf{te}_\theta - \mathbf{e}_\theta\mathbf{t})\cdot\mathbf{k}$, formed with the permutation tensor $\mathbf{te}_\theta - \mathbf{e}_\theta\mathbf{t}$.

Note that (O.31) and (O.32) do not exhibit interchangeability. However, it would be wrong to conclude, as the form of these expressions suggests, that there is a bias toward the σ-coordinate; by multiplying various of the invariants involving g or k_n by the invariants in (O.24) and adding the resulting expressions to other of the invariants involving g or k_n, we, ultimately, may obtain new (but not necessarily independent) invariants of the form

$$g^2(1\, or\, b_\sigma\, or\, b_\theta)\times(1\, or\, e_\sigma\, or\, e_\theta)\times(1\, or\, k_\sigma\, or\, k_\theta) \tag{O.33}$$

or

$$k_n^2(1\, or\, b_\sigma\, or\, b_\theta)\times(1\, or\, e_\sigma\, or\, e_\theta)\times(1\, or\, k_\sigma\, or\, k_\theta)\,. \tag{O.34}$$

For example, we get $g^2 b_\sigma e_\theta k_\theta$ from (O.24) and (O.31) by forming

$$g^2 b_\sigma[(e_\sigma + e_\theta)(k_\sigma + k_\theta) - e_\sigma(k_\sigma + k_\theta) - k_\sigma(e_\sigma + e_\theta) + e_\sigma k_\sigma]\,. \tag{O.35}$$

If the dimensionless strain-energy density, ϕ, is to be invariant under a reversal of the normal to $\mathcal{S}$, then those invariants that change sign must enter the functional form of ϕ either multiplied by one another or else multiplied by the base-reference deviation, d. Using, for conciseness, relations such as

$$2e_\sigma e_\theta = (e_\sigma + e_\theta)^2 - (e_\sigma{}^2 + e_\theta^2) \tag{O.36}$$

and

$$b_\sigma k_\theta + b_\theta k_\sigma = (b_\sigma + b_\theta)(k_\sigma + k_\theta) - (b_\sigma k_\sigma + b_\theta k_\theta)\,, \tag{O.37}$$

and introducing, as before, dimensionless arguments, we find that

$$\begin{aligned}\phi = \phi\{&e_\sigma + e_\theta, e_\sigma e_\theta, g^2(1, e_\sigma), d^2/h^2;\\ &d\,[k_\sigma + k_\theta, e_\sigma k_\sigma + e_\theta k_\theta, g^2(k_\sigma, e_\sigma k_\sigma)],\\ &h^2[k_\sigma^2 + k_\sigma^2, k_\sigma k_\theta, k_n^2(1, e_\sigma, dk_\sigma, de_\sigma k_\sigma)];\\ &h^2(b_\sigma k_\sigma + b_\theta k_\theta, b_\sigma k_\theta + b_\theta k_\sigma, b_\sigma^2 + b_\theta^2, b_\sigma b_\theta), d\,(b_\sigma + b_\theta),\\ &g^2 b_\sigma(d, de_\sigma, h^2 k_\sigma), h^2 k_n^2 b_\sigma(d, de_\sigma, h^2 k_\sigma)\}\,.\end{aligned} \tag{O.38}$$

It is also possible to arrive at the right side of this expression by considering a shell of revolution in an arbitrary state of strain, taking all possible combinations of strain and curvature components that satisfy interchangeability, and, finally, restricting the strain tensors and vectors to have the forms $(0.22)_{1,2}$ and (O.23).

**4. Approximations to the strain-energy density*

In this subsection, we exploit the conditions and results of subsection O.2 to obtain approximations to Φ with error estimates. We shall follow closely, and adapt to our ends, the analysis for beamshells in subsection IV.N.2, excluding regions near any apices. There, a more careful analysis is needed, especially if the axishell is cone-like (or, worse yet, cusp-like) because the principal curvature, b_θ, is infinite at such singular points.

Many of our approximations exploit *thinness,* which means that at every point of the reference surface, $\mathcal{S}$, the ratio of the thickness, h, or the deviation, d, (which is of the order of h, at most) to *any* radius of curvature of either $\mathcal{S}$ or $\mathfrak{Z}$ is small. That is,

$$|hb_\sigma, hb_\theta, db_\sigma, db_\theta, hk_\sigma, hk_\theta, hk_n, dk_\sigma, dk_\theta, dk_n| \ll 1 . \tag{O.39}$$

As we discussed in Chapter IV, it is possible to conceive of cases in which, say, hk_σ will be locally large, for example near a concentrated load at an apex of an axishell. Though our analysis will be invalid near such points, we have some reason to believe that *global* deflections, predicted by using the resulting simplified strain-energy density, will be quite good.

An extension (which we shall not elaborate upon) of the analysis in subsection IV.N.2 yields the following approximations to ϕ.

(a) Let $\tilde{\phi}$ be the strain-energy density obtained from (O.18) by setting $\mathcal{Z}=0$ (i.e., by setting $b_\sigma = b_\theta = 0$). Thus,

$$\tilde{\phi} \equiv \tilde{\phi}(\mathcal{X}; \mathcal{Y}) . \tag{O.40}$$

If $b \equiv \sqrt{b_\sigma^2 + b_\theta^2}$ denotes the curvature norm, then, following the same line of reasoning that led from (IV.N.26) to (IV.N.31) in a beamshell, we find that

$$|\phi - \tilde{\phi}| = O(hb\phi) \;, \quad \| e \| \le e_0 \,, \tag{O.41}$$

which implies that, *in replacing the strain-energy density of an axishell by that of a flat plate, we make a relative error of* O(hb).

(b) Assume that $\tilde{\phi}(\mathcal{X}; \mathcal{Y})$ has a finite Taylor expansion about $\mathcal{Y}=0$ of the (symbolic) form

$$\begin{aligned}\tilde{\phi}(\mathcal{X}; \mathcal{Y}) = \tilde{\phi}(\mathcal{X}; 0) + \tilde{\phi},_{\mathcal{Y}}(\mathcal{X}; 0)\mathcal{Y} + \frac{1}{2}\tilde{\phi},_{\mathcal{YY}}(\mathcal{X}; 0)\mathcal{Y}^2 + \frac{1}{3!}\tilde{\phi},_{\mathcal{YYY}}(\mathcal{X}; 0)\mathcal{Y}^3 \\ + \frac{1}{4!}\tilde{\phi},_{\mathcal{YYY}}(\mathcal{X}; \tau\mathcal{Y})\mathcal{Y}^4 ,\end{aligned} \tag{O.42}$$

where $0 < \tau < 1$. (Actually, $\mathcal{Y}$ and $\tilde{\phi},_{\mathcal{Y}}$ are vectors and $\tilde{\phi},_{\mathcal{YY}}$ is a 2nd order tensor, etc.) As in (IV.N.32), we let $\bar{\phi}$ equal the first 2 terms on the right side of (O.42) plus those parts of the next 2 terms that involve squares or cubes of bending strains. Then, following the same line of reasoning we took in going from (IV.N.33) to (IV.N.34), we conclude that

$$\phi = \bar{\phi}[1 + O(hb + h^2k^2)] \ , \ \| e \| \le e_0 \ , \tag{O.43}$$

where $k \equiv \sqrt{k_\sigma^2 + k_\theta^2 + k_n^2}$ is the norm of the bending strains. We note that if $d=0$ or if $d = O(h^2 b)$ (as with midsurface coordinates in transversely homogeneous shells), then the error estimate (O.43) continues to apply if we neglect terms that are cubic in the bending strains. On the other hand, if $d = O(h)$, then neglect of such terms in $\bar{\phi}$ leads to a larger relative error in (O.43) of $O(hb + hk)$. Equation (IV.N.32) makes these facts obvious.

(c) Again consider a finite Taylor expansion of $\tilde{\phi}$, but now with respect to the argument $\mathcal{A} \equiv g^2, \mathcal{Y}$ about $\mathcal{A}=0$. If we take the remainder term in this expansion to be $(1/4!)\phi,_{\mathcal{AAAA}}(\mathcal{X}_0; \tau\mathcal{A})$, where $\mathcal{X}_0 \equiv e_\sigma, e_\theta, 0, d^2/h^2$ and $0 < \tau < 1$, and take $\hat{\phi}$ to be the polynomial part of the expansion with powers of the bending strains higher than 3 cast out, then, just as (IV.N.37) follows from (IV.N.35), so

$$\phi = \hat{\phi}[1 + O(hb + h^2k^2 + g^2)] + O(g^4) \ , \ \| e \| \le e_0 \ . \tag{O.44}$$

This approximation is obviously useful in problems with small shearing strains.

The error estimates discussed in this subsection are useful not only for estimating the errors in approximating ϕ by simpler forms, but also for evaluating approximations in the equations of motion. If we wish, however, to obtain quantitative results, we must turn to experiments or to constitutive models for 3-dimensional bodies, such as we consider in the next subsection.

*5 *Strain-energy density by descent from 3-dimensions*

Simmonds (1986, 1987) has derived strain-energy densities for elastically isotropic axiplates whose 3-dimensional strain-energy density (strain energy per unit undeformed volume) is of the form

$$W = W(I_1, I_2, I_3) \ , \tag{O.45}$$

where the I_i are the standard, 3-dimensional strain invariants (Green & Zerna, 1968). According to (O.41), a strain-energy density for an axiplate may be used for an axishell of the same construction to within a relative error of $O(hb)$, where b is a measure of the curvature of the undeformed reference surface of the axishell.

For simplicity, we consider axiplates of constant density and constant, undeformed thickness, h, a typical particle of which has initial position

$$\mathbf{x} = r\mathbf{e}_r(\theta) + z\mathbf{e}_z \ , \ 0 \le r \le a \ , 0 \le \theta < 2\pi \ , -h/2 \le z \le h/2 \ . \tag{O.46}$$

[This equation is a special case of (A.2) with $R = \sigma = r$ and $Z = \zeta = z$.] In deriving constitutive relations in elasticity, it is sufficient to consider static deformations, so that, in the case at hand, the particle takes some new position

$$\bar{\mathbf{x}} = \bar{r}(r, z)\mathbf{e}_r(\theta) + \bar{z}(r, z)\mathbf{e}_z = \bar{\mathbf{y}}(r, \theta) + p(r, z)\bar{\mathbf{t}} + q(r, z)\bar{\mathbf{b}} \ , \tag{O.47}$$

where, from (B.14), $h\bar{\mathbf{y}} = \int_{-h/2}^{h/2} \bar{\mathbf{x}} dz$. (In a Kirchhoff motion, $p=0$ and $q=z$.)

The symmetric, 3-dimensional metric tensor **G** of the deformed axiplate is defined, implicitly, by

$$d\overline{\mathbf{x}}\cdot d\overline{\mathbf{x}} \equiv d\mathbf{x}\cdot\mathbf{G}\cdot d\mathbf{x} . \tag{O.48}$$

The matrix of the *physical components* of **G** in the coordinate system (r, θ, z) is denoted and defined by

$$\begin{bmatrix} G_r & 0 & 2\Gamma \\ 0 & G_\theta & 0 \\ 2\Gamma & 0 & G_z \end{bmatrix} \equiv \begin{bmatrix} \overline{\mathbf{x}}_{,r}\cdot\overline{\mathbf{x}}_{,r} & 0 & \overline{\mathbf{x}}_{,r}\cdot\overline{\mathbf{x}}_{,z} \\ 0 & r^{-2}\overline{\mathbf{x}}_{,\theta}\cdot\overline{\mathbf{x}}_{,\theta} & 0 \\ \overline{\mathbf{x}}_{,z}\cdot\overline{\mathbf{x}}_{,r} & 0 & \overline{\mathbf{x}}_{,z}\cdot\overline{\mathbf{x}}_{,z} \end{bmatrix} . \tag{O.49}$$

Thus,

$$I_1 = \mathrm{tr}\,\mathbf{G} = G_r + G_\theta + G_z \tag{O.50}$$

$$I_2 = ½(\mathrm{tr}\,\mathbf{G})^2 - ½\mathrm{tr}\,\mathbf{G}^2 = G_rG_\theta + G_\theta G_z + G_zG_r - 4\Gamma^2 \tag{O.51}$$

$$I_3 = \det\mathbf{G} = G_rG_\theta G_z - 4\Gamma^2 G_\theta , \tag{O.52}$$

where tr = trace.

If the material is *incompressible*, i.e., if $I_3 = 1$, so that W has the dimensionless, functional form

$$W = Ew(I_1, I_2) , \tag{O.53}$$

where E is a "Young's modulus", then Simmonds (1986) has shown (with the aid of a 2-term Taylor expansion in Γ^2) that

$$\phi = h^{-1}\int_{-h/2}^{h/2} w(K_1, K_2)dz + O(\overline{\Gamma}^2) , \tag{O.54}$$

where

$$K_1 \equiv G_r + G_\theta + \frac{1}{G_rG_\theta} , \quad K_2 \equiv G_rG_\theta + \frac{G_r + G_\theta}{G_rG_\theta} , \tag{O.55}$$

and

$$\overline{\Gamma}^2 \equiv h^{-1}\int_{-h/2}^{h/2} \Gamma^2 dz . \tag{O.56}$$

(We may identify $\overline{\Gamma}$ with the 1-dimensional transverse shear strain g.)

To proceed, we set

$$\mathcal{W}(G_r, G_\theta) \equiv w(K_1(G_r, G_\theta), K_2(G_r, G_\theta)) , \quad \delta \equiv ½h/L^* , \tag{O.57}$$ [7]

[7]The reader should have no difficulty in distinguishing the symbols $\mathcal{W}$ and δ, as used in this subsection, from their more general use to denote the external mechanical power and a variation, respectively.

where, as with beamshells in subsection (IV.N.3), L^* denotes the characteristic wavelength of the deformation on the midplane of the axiplate. Further, for conciseness, we introduce the following notation: if $f=f(X,Y)$, $X=X(x,y,\delta)$, and $Y=Y(x,y,\delta)$, then

$$\overset{0}{f} \equiv f \text{ at } \delta=0 \text{ and } f_{(m,n)} \equiv \frac{\partial^{m+n} f}{\partial X^m \partial Y^n}\,. \tag{O.58}$$

It follows from (O.47) and equations (19), (23), (25), and (26) of Simmonds (1986), that

$$\lambda_r^2 \equiv \bar{\mathbf{y}}_{,r}\cdot\bar{\mathbf{y}}_{,r} = \overset{0}{G}_r \;,\; \lambda_\theta^2 \equiv r^{-2}\bar{\mathbf{y}}_{,\theta}\cdot\bar{\mathbf{y}}_{,\theta} = \overset{0}{G}_\theta \tag{O.59}$$

and hence that

$$\overset{0}{\mathcal{W}}(G_r, G_\theta) = \mathcal{W}(\overset{0}{G}_r, \overset{0}{G}_\theta) \equiv \mathcal{W}(\lambda_r, \lambda_\theta)\,. \tag{O.60}$$

These relations, plus equations (35), (38), and (41) of Simmonds (1987), imply that

$$\begin{aligned}\phi = \mathcal{W} + (1/4h\textstyle\int_{-h/2}^{h/2} q_0^2 dz)[(\mathcal{W}_{(1,0)} + 2\lambda_r^2\mathcal{W}_{(2,0)})k_r^2 + 4\mathcal{W}_{(1,1)}\lambda_r\lambda_\theta k_r k_\theta \\ + (\mathcal{W}_{(0,1)} + 2\lambda_\theta^2\mathcal{W}_{(0,2)})k_\theta^2] \\ + O[(e_\sigma^2 + e_\theta^2)\delta^2 + \delta^3 + \bar{\Gamma}^{-2}]\,,\end{aligned} \tag{O.61}$$

where $\mathcal{W}$ is to be expressed as a function of λ_r and λ_θ and, from (IV.N.65) and (O.47), $q_0=(q)_{\delta=0}$. In an incompressible body,

$$q_0 \equiv z/\lambda_r\lambda_\theta\,. \tag{O.62}$$

As an example, consider the neo-Hookean material

$$w = (1/6)(I_1 - 3) = w(I_1, 0)\,. \tag{O.63}$$

It follows from (O.55) and (O.60) that

$$\mathcal{W}(\lambda_r, \lambda_\theta) = (1/6)(\lambda_r^2 + \lambda_\theta^2 + \lambda_r^{-2}\lambda_\theta^{-2} - 3) \tag{O.64}$$

so that, with the error terms omitted, (O.61) takes the form

$$\begin{aligned}\phi = \frac{1}{6}\Bigg\{ & \lambda_r^2 + \lambda_\theta^2 + \frac{1}{\lambda_r^2\lambda_\theta^2} - 3 \\ & + \frac{h^2}{12\lambda_r^6\lambda_\theta^6}[(3 + \lambda_r^4\lambda_\theta^2)\lambda_\theta^2 k_r^2 + 4\lambda_r\lambda_\theta k_r k_\theta + (3 + \lambda_r^2\lambda_\theta^4)\lambda_r^2 k_\theta^2]\Bigg\}\,.\end{aligned} \tag{O.65}$$

[This is equation (45) of Simmonds (1986) with his erroneous factor of 1/8 replaced by 1/6.]

To compute a strain-energy density for *compressible* axiplates, Simmonds (1987) assumes plane stress; that is, he assumes that $\partial W/\partial\Gamma_z = 0$, where Γ_z is the transverse normal component of the 3-dimensional Lagrangian strain. Because $\mathbf{e}_z \cdot \mathbf{G} \cdot \mathbf{e}_z = G_z = 1 + 2\Gamma_z$, the assumption of plane stress implies that

$$\partial W / \partial G_z = 0 . \tag{O.66}$$

But $W = W(I_1, I_2, I_3)$ and, by (O.50)-(O.52), I_1, I_2, and I_3 are functions of G_r, G_θ, and G_z. Hence, by the chain rule,

$$W_1 + (G_r + G_\theta) W_2 + G_r G_\theta W_3 = 0 , \tag{O.67}$$

where

$$W_i = \partial W / \partial I_i = W_i(G_r, G_\theta, G_z, \Gamma^2) \ , \ i = 1, 2, 3 . \tag{O.68}$$

Suppose that (O.67) can be solved for G_z in the form

$$G_z = \hat{G}_z(G_r, G_\theta, \Gamma^2) . \tag{O.69}$$

Substituting (O.69) into (O.50)-(O.52) and the resulting expressions into

$$W = C w(I_1, I_2, I_3) , \tag{O.70}$$

where C is any convenient material constant having the same dimensions as W, we obtain an expression for w of the form

$$w = \hat{w}(G_r, G_\theta, \hat{G}_z(G_r, G_\theta, \Gamma^2), \Gamma^2) . \tag{O.71}$$

In 1st-approximation shell theory, the effects of transverse shearing strain are ignored. To estimate the errors so incurred in the strain-energy density, we assume that $\hat{w}$ is a differentiable function of Γ^2 for all values of G_r and G_θ of interest. Then, by the mean-value theorem, (O.50)-(O.52), and (O.67),

$$\begin{aligned} w &= \hat{w}(G_r, G_\theta, \hat{G}_z(G_r, G_\theta, 0), 0) + (\partial \hat{w} / \partial \Gamma^2)_* \Gamma^2 \\ &\equiv \mathcal{W}(G_r, G_\theta) - 4(w_2 + G_\theta w_3)_* \Gamma^2 , \end{aligned} \tag{O.72}$$

where the asterisk (*) indicates that a function is to be evaluated at $(G_r, G_\theta, \Gamma_*^2)$, with Γ_* some value between 0 and Γ.

As explained in detail in Simmonds (1987), we may derive from (O.72) an expression for the dimensionless strain-energy density, ϕ, of an axiplate by following the same steps that led from (O.57) to (O.61), but *without invoking incompressibility*. The only difference is that, in place of (O.62), we must take

$$q_0 = z \sqrt{\hat{G}_z(\lambda_r^2, \lambda_\theta^2, 0)} , \tag{O.73}$$

$\hat{G}_z$ being the function in (O.69).

As an example, consider the following 3-dimensional strain-energy density proposed by Blatz & Ko (1962) for a compressible, polyurethane rubber:

$$W = \tfrac{1}{2}\mu \left[I_1 + \frac{1 - 2\nu}{\nu} I_3^{-\nu/(1-2\nu)} - \frac{1 + \nu}{\nu} \right] . \tag{O.74}$$

As the strains grow small, ν may be identified with Poisson's ratio and μ with the shear modulus, $\frac{1}{2}E/(1+\nu)$, where E is Young's modulus. Note that (O.74) approaches a neo-Hookean material as $\nu \to \frac{1}{2}$.

In view of (O.50)-(O.52), (O.67) takes the form

$$1 - G_r G_\theta (G_r G_\theta G_z - 4G_\theta \Gamma^2)^{-(1-\nu)/(1-2\nu)} = 0\,, \tag{O.75}$$

which yields

$$G_z = (G_r G_\theta)^{-\nu/(1-2\nu)} + 4G_r^{-1}\Gamma^2\,. \tag{O.76}$$

Hence, from (O.50), (O.52), (O.72), and (O.74), and with $C = E$ in (O.70),

$$\mathcal{W} = \frac{1}{4(1+\nu)}\left[G_r + G_\theta + \frac{1-\nu}{\nu}(G_r G_\theta)^{-\nu/(1-\nu)} - \frac{1+\nu}{\nu}\right]. \tag{O.77}$$

(Another obvious choice for C is μ.) Furthermore, from (O.73) and (O.76),

$$q_0 = z\,(\lambda_r \lambda_\theta)^{-\nu/(1-\nu)} \tag{O.78}$$

so that (O.61) for compressible materials, with the error terms deleted, becomes

$$\phi = \frac{1}{4(1+\nu)}\Bigg\{\left[\lambda_r^2 + \lambda_\theta^2 + \frac{1-\nu}{\nu}(\lambda_r\lambda_\theta)^{-2\nu/(1-\nu)} - \frac{1+\nu}{\nu}\right]$$
$$\frac{h^2}{12(\lambda_r\lambda_\theta)^{2(1+\nu)/(1-\nu)}}\Bigg[\left(\frac{1+\nu}{1-\nu} + \lambda_r^{2/(1-\nu)}\lambda_\theta^{2\nu/(1-\nu)}\right)\lambda_\theta^2 k_r^2 \tag{O.79}$$
$$+ \frac{4\nu}{1-\nu}\lambda_r\lambda_\theta k_r k_\theta + \left(\frac{1+\nu}{1-\nu} + \lambda_r^{2\nu/(1-\nu)}\lambda_\theta^{2/(1-\nu)}\right)\lambda_r^2 k_\theta^2\Bigg]\Bigg\}.$$

As $\nu \to \frac{1}{2}$, this expression approaches (O.65).

At the end of the book, Appendix B, compiled by Jim Fulton, lists some 3-dimensional strain-energy densities that have been proposed in the literature. Using (O.61), with q_0 given either by (O.62) or (O.73), one may compute the corresponding strain-energy densities for axiplates (and hence for an axishells).

P. Alternative Strains and Stresses

Use of $\{\mathbf{T}, \mathbf{B}\}$ as a reference frame in section L produced e_σ, e_θ, g, and β as basic kinematic variables, subject to the compatibility condition (M.2). If $\bar{\mathbf{y}}$ or the *displacement*

$$\mathbf{u} \equiv \bar{\mathbf{y}} - \mathbf{y} \equiv u\mathbf{e}_r + w\mathbf{e}_z \tag{P.1}$$

is needed, say for the acceleration term in (D.3) or in the boundary conditions, or if the external loads are functions of $\bar{\mathbf{y}}$, then we can always regard (L.1) as an additional field equation.

Alternatively, it might be more convenient in some cases to use $\mathbf{u}$ and g as the basic variables, thereby eliminating (L.1) and (M.2) as field equations. A problem here is that e_σ, k_σ, k_θ, and k_n turn out to be rather complicated, non-rational functions of $\mathbf{u}$ and g. This may be remedied by taking $\{\bar{\mathbf{y}}', \mathbf{m}\}$ as a new ortho*non*normal reference frame, which leads to new extensional and bending strains that are polynomials in u, w, and their derivatives. Thus, let

$$\bar{\mathbf{y}}' = \mathbf{t} + \mathbf{u}' \equiv \lambda_\sigma \bar{\mathbf{t}} \equiv \lambda_\sigma(\cos\bar{\alpha}\,\mathbf{e}_r + \sin\bar{\alpha}\,\mathbf{e}_z)\,, \tag{P.2}$$

where λ_σ is the *meridional stretch* and

$$\bar{\alpha} \equiv \alpha + \beta + \gamma\,. \tag{P.3}$$

This angular relation—see Fig. 4.2—defines a new shear strain, γ. From (P.2),

$$\lambda_\sigma^2 \equiv 1 + 2\gamma_\sigma\,, \tag{P.4}$$

where

$$\gamma_\sigma \equiv \mathbf{u}'\cdot\mathbf{t} + \tfrac{1}{2}\mathbf{u}'\cdot\mathbf{u}' = u'\cos\alpha + w'\sin\alpha + \tfrac{1}{2}(u'^2 + w'^2) \tag{P.5}$$

is the *Lagrangian meridional strain.*

We now express the kinematic variables $\mathbf{v}' - \omega\mathbf{m}$, $r^{-1}\mathbf{e}_r\cdot\mathbf{v}$, ω', and $r^{-1}\omega$, that appear in the deformation power, (H.4), as combinations of the time derivatives of the new strains. In carrying out these calculations, we note from (D.4) and (P.2) that

$$\mathbf{m} = \lambda_\sigma(-\sin\bar{\alpha}\,\mathbf{e}_r + \cos\bar{\alpha}\,\mathbf{e}_z) \equiv \lambda_\sigma\bar{\mathbf{b}}\,. \tag{P.6}$$

From (E.3) and (P.3),

$$\omega = \dot{\beta} = \dot{\bar{\alpha}} - \dot{\gamma}\,, \tag{P.7}$$

so that (D.4), (P.2), (P.4), and (P.6) yield

$$\mathbf{v}' - \omega\mathbf{m} = \lambda_\sigma^{-2}\dot{\gamma}_\sigma\bar{\mathbf{y}}' + \dot{\gamma}\mathbf{m}\,. \tag{P.8}$$

Hence, if we set

$$\mathbf{N}_\sigma \equiv \tilde{N}_\sigma\bar{\mathbf{y}}' + \tilde{Q}\mathbf{m}\,, \tag{P.9}$$

it follows from (P.8) that

$$\mathbf{N}_\sigma\cdot(\mathbf{v}' - \omega\mathbf{m}) = \tilde{N}_\sigma\dot{\gamma}_\sigma + \lambda_\sigma^2\tilde{Q}\dot{\gamma}\,. \tag{P.10}$$

Next, from (P.1),

$$r^{-1}\mathbf{e}_r\cdot\mathbf{v} = \lambda_\theta^{-1}\dot{\gamma}_\theta\,, \tag{P.11}$$

where

$$\lambda_\theta \equiv \bar{r}/r = 1 + u/r \tag{P.12}$$

is the *hoop stretch* and

$$\gamma_\theta \equiv u/r + \tfrac{1}{2}(u/r)^2 \tag{P.13}$$

is the *Lagrangian hoop strain.* Thus, if we define a *modified hoop stress resultant* by

$$N_\theta^* \equiv \lambda_\theta^{-1} N_\theta \,, \tag{P.14}$$

then

$$N_\theta r^{-1} \mathbf{e}_r \cdot \mathbf{v} = N_\theta^* \dot{\gamma}_\theta \,. \tag{P.15}$$

There seem to be at least two different sets of bending strains that might be useful. The first, that we shall denote by $\{\kappa_\sigma, \kappa_\theta, \kappa_n\}$, has the simplest expressions in terms of u and w.

To define κ_σ, we note that any meridional bending strain must be closely related to the difference between

$$\frac{\bar{\mathbf{y}}'' \cdot \bar{\mathbf{b}}}{\lambda_\sigma^{\,2}} = \frac{\bar{\mathbf{y}}'' \cdot \mathbf{m}}{\lambda_\sigma^{\,3}} \,, \tag{P.16}$$

the meridional curvature of $\mathcal{J}$, and $\mathbf{y}'' \cdot \mathbf{b} = \alpha'$, the meridional curvature of $\mathcal{S}$. But, by (E.1), (E.2), (P.1), and (P.2),

$$\bar{\mathbf{y}}'' = \alpha' \mathbf{b} + u'' \mathbf{e}_r + w'' \mathbf{e}_z \tag{P.17}$$

and

$$\mathbf{m} = \mathbf{b} - w' \mathbf{e}_r + u' \mathbf{e}_z \,. \tag{P.18}$$

Hence, by (P.2)-(P.4) and (P.6),

$$\bar{\mathbf{y}}'' \cdot \mathbf{m} = \lambda_\sigma^{\,2}(\alpha' + \beta' + \gamma') = \alpha' + \lambda_\sigma^{\,2}\gamma' + \lambda_\sigma^{\,2}\beta' + 2\gamma_\sigma \alpha' \tag{P.19}$$

is quadratic in u and w. This leads us to define

$$\kappa_\sigma \equiv \lambda_\sigma^{\,2}\beta' + 2\gamma_\sigma \alpha' \,, \tag{P.20}$$

so that, from (P.19),

$$\begin{aligned} \kappa_\sigma &= \bar{\mathbf{y}}'' \cdot \mathbf{m} - \alpha' - \lambda_\sigma^{\,2}\gamma' \\ &= (\cos\alpha + u')w'' - (\sin\alpha + w')u'' \\ &\qquad + (u'\cos\alpha + w'\sin\alpha)\alpha' - (1 + \underline{2\gamma_\sigma})\gamma' \,. \end{aligned} \tag{P.21}$$

As

$$\lambda_\sigma^{\,2}\omega' = \lambda_\sigma^{\,2}\dot{\beta}' = (\lambda_\sigma^{\,2}\beta')^{\cdot} - 2\dot{\gamma}_\sigma \beta' \,, \tag{P.22}$$

it follows from (L.3) and (P.20) that

$$\lambda_\sigma^{\,2}\omega' = \dot{\kappa}_\sigma - 2(\alpha + \beta)'\dot{\gamma}_\sigma \,. \tag{P.23}$$

We underlined the strain term in (P.21), which we could have expressed in terms of $\mathbf{u}$ via (P.5), to show what can be be neglected in a small-strain, large-rotation problem.

If we define a *modified meridional stress couple*

$$M_\sigma^* \equiv \lambda_\sigma^{-2} M_\sigma \,, \tag{P.24}$$

then, by (P.23),

$$M_\sigma \omega' = M_\sigma^* \dot{\kappa}_\sigma - 2(\alpha + \beta)' M_\sigma^* \dot{\gamma}_\sigma . \tag{P.25}$$

Finally, to express $\mathbf{M}_\theta \cdot \mathbf{e}_r r^{-1} \omega$ as an inner product of stress couples times strain rates, *linear* in u and w, we set

$$\mathbf{M}_\theta \equiv M_\theta^* \bar{\mathbf{y}}' + M_n^* \mathbf{m} \tag{P.26}$$

and note by (P.2) and (P.8) that

$$\begin{aligned} \bar{\mathbf{y}}' \cdot \mathbf{e}_r r^{-1} \omega &= r^{-1} \lambda_\sigma (\dot{\bar{\alpha}} - \dot{\gamma}) \cos \bar{\alpha} \\ &= \dot{\kappa}_\theta - \frac{\sin \bar{\alpha}}{r \lambda_\sigma} \dot{\gamma}_\sigma - \frac{\lambda_\sigma \cos \bar{\alpha}}{r} \dot{\gamma} , \end{aligned} \tag{P.27}$$

where

$$r \kappa_\theta \equiv \lambda_\sigma \sin \bar{\alpha} - \sin \alpha = w' . \tag{P.28}$$

[If $\gamma = 0$, (P.21) and (P.28) yield the bending strains introduced by Simmonds (1985).] In a similar way, we may write

$$\begin{aligned} \mathbf{m} \cdot \mathbf{e}_r r^{-1} \omega &= -r^{-1} \lambda_\sigma (\dot{\bar{\alpha}} - \dot{\gamma}) \sin \bar{\alpha} \\ &= \dot{\kappa}_n - \frac{\cos \bar{\alpha}}{r \lambda_\sigma} \dot{\gamma}_\sigma + \frac{\lambda_\sigma \sin \bar{\alpha}}{r} \dot{\gamma} , \end{aligned} \tag{P.29}$$

where

$$r \kappa_n \equiv \lambda_\sigma \cos \bar{\alpha} - \cos \alpha = u' . \tag{P.30}$$

Thus,

$$\begin{aligned} \mathbf{M}_\theta \cdot \mathbf{e}_r r^{-1} \omega = M_\theta^* \dot{\kappa}_\theta + M_n^* \dot{\kappa}_n - (r \lambda_\sigma)^{-1} (M_\theta^* \sin \bar{\alpha} + M_n^* \cos \bar{\alpha}) \dot{\gamma}_\sigma \\ - r^{-1} \lambda_\sigma (M_\theta^* \cos \bar{\alpha} - M_n^* \sin \bar{\alpha}) \dot{\gamma} . \end{aligned} \tag{P.31}$$

Altogether then, adding (P.10), (P.15), (P.25), and (P.31), we have

$$\tau : \dot{\varepsilon} = N_\sigma^* \dot{\gamma}_\sigma + N_\theta^* \dot{\gamma}_\theta + Q^* \dot{\gamma} + M_\sigma^* \dot{\kappa}_\sigma + M_\theta^* \dot{\kappa}_\theta + M_n^* \dot{\kappa}_n , \tag{P.32}$$

where

$$N_\sigma^* \equiv \tilde{N}_\sigma - 2(\alpha + \beta)' M_\sigma^* - (r \lambda_\sigma)^{-1} (M_\theta^* \sin \bar{\alpha} + M_n^* \cos \bar{\alpha}) \tag{P.33}$$

and

$$Q^* \equiv \lambda_\sigma^2 \tilde{Q} - r^{-1} \lambda_\sigma (M_\theta^* \cos \bar{\alpha} - M_n^* \sin \bar{\alpha}) \tag{P.34}$$

are *modified meridional and transverse shear stress resultants*, respectively.

Our second set of bending strains, that we shall denote by $\{K_\sigma, K_\theta, K_n\}$, are chosen so that, if $\gamma = 0$, then K_σ and K_θ agree with those of Budiansky (1968), specialized to axishells. Budiansky's bending strains are defined so that the associated field equations reduce, when linearized, to those of Sanders (1959) and Koiter (1960). The Sanders-Koiter field equations display a particularly concise form of the static-geometric duality (Budiansky & Sanders, 1963) and are characterized by bending strains that can be expressed solely in terms of linearized rotations. Thus, $\{K_\sigma, K_\theta, K_n\}$ are to be such that, if $\gamma = 0$, their linear parts will be given by

$$K_\sigma^L = \beta_L{}' \ , \ rK_\theta^L = \beta_L \cos\alpha \ , \ rK_n^L = -\beta_L \sin\alpha \ , \tag{P.35}$$

where

$$\beta_L \equiv u' \sin\alpha + w' \cos\alpha \tag{P.36}$$

is the linearized rotation-displacement expression.

From (P.21), (P.22), and (P.28), it is clear that the following expressions are polynomials in u and w and, when linearized, will reduce to (P.35):

$$\begin{aligned} K_\sigma &\equiv \lambda_\sigma{}^2 \beta' \\ &= \kappa_\sigma - 2\gamma_\sigma \alpha' \\ &= (\cos\alpha + u')w'' - (\sin\alpha + w')u'' \\ &\qquad + (u' \cos\alpha + w' \sin\alpha - \underline{2\gamma_\sigma})\alpha' - (1 + \underline{2\gamma_\sigma})\gamma' \end{aligned} \tag{P.37}$$

$$\begin{aligned} rK_\theta &\equiv \lambda_\sigma \sin\bar{\alpha} - (1 + \gamma_\sigma)\sin\alpha \\ &= r\kappa_\theta - \gamma_\sigma \sin\alpha \\ &= w' - \underline{\gamma_\sigma} \sin\alpha \end{aligned} \tag{P.38}$$

$$\begin{aligned} rK_n &\equiv \lambda_\sigma \cos\bar{\alpha} - (1 + \gamma_\sigma)\cos\alpha \\ &= r\kappa_n - \gamma_\sigma \cos\alpha \\ &= u' - \underline{\gamma_\sigma} \cos\alpha \, . \end{aligned} \tag{P.39}$$

As before, we could have used (P.5) to express the last lines of (P.37)-(P.39) in terms of displacements only.

To find the set of stresses conjugate to the set of strains, $\{\gamma_\sigma, \gamma_\theta, \gamma, K_\sigma, K_\theta, K_n\}$, we proceed as before. The result is that the deformation power density takes the form

$$\tau{:}\dot{\varepsilon} = \bar{N}_\sigma \dot{\gamma}_\sigma + N_\theta^* \dot{\gamma}_\theta + Q^* \dot{\gamma} + M_\sigma^* \dot{K}_\sigma + M_\theta^* \dot{K}_\theta + M_n^* \dot{K}_n \, , \tag{P.40}$$

where

$$\bar{N}_\sigma \equiv N_\sigma^* + 2\alpha' M_\sigma^* + r^{-1}(M_\theta^* \sin\alpha + M_n^* \cos\alpha) \, . \tag{P.41}$$

Q. Elastostatics

With equilibrium equations, compatibility conditions, and constitutive relations, we have a complete set of field equations with which to solve static boundary value problems. The form of these equations depends on the basis $\mathcal{B}$ we choose in the rz-plane and on whether we work with the strain-energy density, Φ, the mixed-energy density, Ψ, or (if it exists) the complementary stress-energy density, Φ_c.

There seem to be four reasonable choices for $\mathcal{B}$: the fixed basis $\{\mathbf{e}_r, \mathbf{e}_z\}$, the initial basis $\{\mathbf{t}, \mathbf{b}\}$, the spin basis $\{\mathbf{T}, \mathbf{B}\}$, or the deformed basis $\{\overline{\mathbf{y}}', \mathbf{m}\}$. As the last three bases reflect, in different ways, the two intrinsic, distinguished directions in a shell, tangential and normal, they might seem, at first glance, to be preferable to the 1st basis, $\{\mathbf{e}_r, \mathbf{e}_z\}$. However, in practice, it turns out that the latter leads to much simpler results because, in this basis, the force equilibrium equations can be satisfied identically, leaving a moment equilibrium equation that is closely analogous to the strain-rotation compatibility condition (M.2). (The analogy is exact in linear theory.) We shall therefore use $\{\mathbf{e}_r, \mathbf{e}_z\}$ almost exclusively in the next few sections.

Thus, let

$$\mathbf{N}_\sigma = H\mathbf{e}_r + V\mathbf{e}_z \ , \ \mathbf{p} = p_H\mathbf{e}_r + p_V\mathbf{e}_z \ , \tag{Q.1}$$

where H and V are mnemonics for "horizontal" and "vertical". Substituting these component representations into (D.3), we have

$$(rH)' - N_\theta + rp_H = 0 \tag{Q.2}$$

$$(rV)' + rp_V = 0 \ . \tag{Q.3}$$

The horizontal equilibrium equation is satisfied identically by setting

$$N_\theta = (rH)' + rp_H \ , \tag{Q.4}$$

while the vertical equilibrium equation can be integrated to yield

$$rV = r_0V_0 - \int_0^\sigma rp_V ds \ . \tag{Q.5}$$

In a few important cases, the above integral can be evaluated explicitly. For gravity loading, $p_V = -mg$, so that

$$rV = r_0V_0 + mg\tilde{A}(\sigma) \ , \tag{Q.6}$$

where $\tilde{A} = \int_0^\sigma r(s)ds$ is the area of the initial surface of revolution (per unit circumferential radian) between its base and the current parallel σ=constant; for a constant normal pressure, p_0,

$$p_V = -p_0(\overline{r}/r)\mathbf{m}\cdot\mathbf{e}_z = -p_0(\overline{r}/r)[(1+e_\sigma)\cos(\alpha+\beta) - g\sin(\alpha+\beta)] \ , \tag{Q.7}$$

which, in view of the compatibility condition (M.2), can be rewritten as

$$p_V = -p_0 \bar{r}\bar{r}'/r\,. \tag{Q.8}$$

Hence,

$$rV = r_0 V_0 + \tfrac{1}{2} p_0 (\bar{r}^2 - \bar{r}_0^2)\,. \tag{Q.9}$$

It remains to consider the equation of static moment equilibrium, which follows from (L.15) upon setting the inertia term to 0. As this equation involves N_σ and Q, we have, upon equating the right sides of $(L.4)_1$ and $(Q.1)_1$,

$$N_\sigma = H\cos(\alpha+\beta) + V\sin(\alpha+\beta) \tag{Q.10}$$

$$Q = -H\sin(\alpha+\beta) + V\cos(\alpha+\beta)\,, \tag{Q.11}$$

so that moment equilibrium takes the form

$$\begin{aligned}(rM_\sigma)' - M_\theta\cos(\alpha+\beta) + M_n\sin(\alpha+\beta)&\\ + r\{[(1+e_\sigma)\cos(\alpha+\beta) - g\sin(\alpha+\beta)]V&\\ - [(1+e_\sigma)\sin(\alpha+\beta) + g\cos(\alpha+\beta)]H + l\} &= 0\,.\end{aligned} \tag{Q.12}$$

1. General field equations using the modified strain-energy density $\Upsilon = \Phi - gQ$

Transverse shearing strain seems best handled by assuming that the constitutive relation, $(O.3)_3$, can be inverted in the form $g = \hat{g}(\Lambda)$, where $\Lambda = e_\sigma, e_\theta, Q, k_\sigma, k_\theta, k_n$ is an argument list. With the Legendre transformation,

$$\Upsilon \equiv \Phi(e_\sigma, e_\theta, g(\Lambda), k_\sigma, k_\theta, k_n) - g(\Lambda)Q = \Upsilon(\Lambda)\,, \tag{Q.13}$$

the constitutive relations, (O.3), take the form

$$\begin{aligned}N_\sigma &= \Upsilon_{,e_\sigma}\,,\quad N_\theta = \Upsilon_{,e_\theta}\,,\quad g = -\Upsilon_{,Q}\\ M_\sigma &= \Upsilon_{,k_\sigma}\,,\quad M_\theta = \Upsilon_{,k_\theta}\,,\quad M_n = \Upsilon_{,k_n}\,,\end{aligned} \tag{Q.14}$$

where we have set to 0 the unknown (and unknowable!) Lagrange multiplier in (O.3).

We now use (L.9) and (L.10) to express k_θ and k_n in terms of β, and use (Q.11) to express Q in terms of H, V, and β. With

$$\Xi \equiv e_\sigma, e_\theta, H, V, k_\sigma, \beta \tag{Q.15}$$

denoting a new argument list, we have the following set of 2 *algebraic* equations and 5, 1st-order differential equations for M_σ and the 6 unknowns composing Ξ:

From (Q.10) and $(Q.14)_1$,

$$H\cos(\alpha+\beta) + V\sin(\alpha+\beta) = \Upsilon_{,e_\sigma}(\Xi)\,. \tag{Q.16}$$

From $(Q.14)_4$,

$$M_\sigma = \Upsilon_{,k_\sigma}(\Xi)\,. \tag{Q.17}$$

From (M.2),

$$(re_\theta)' = (1 + e_\sigma)\cos(\alpha + \beta) - \cos\alpha + \Upsilon_{,Q}(\Xi)\sin(\alpha + \beta)\,. \tag{Q.18}$$

From (Q.12) and $(Q.14)_{5,6}$,

$$\begin{aligned}(rM_\sigma)' = {} & \Upsilon_{,k_\theta}(\Xi)\cos(\alpha + \beta) - \Upsilon_{,k_n}(\Xi)\sin(\alpha + \beta) \\ & + r\{[(1 + e_\sigma)\sin(\alpha + \beta) - \Upsilon_{,Q}(\Xi)\cos(\alpha + \beta)]H \\ & - [(1 + e_\sigma)\cos(\alpha + \beta) + \Upsilon_{,Q}(\Xi)\sin(\alpha + \beta)]V - l\}\,.\end{aligned} \tag{Q.19}$$

From (L.8),

$$\beta' = k_\sigma\,. \tag{Q.20}$$

From (Q.4) and $(Q.14)_2$,

$$(rH)' = \Upsilon_{,e_\theta}(\Xi) - rp_H\,. \tag{Q.21}$$

From (Q.3),

$$(rV)' = -rp_V\,. \tag{Q.22}$$

If, by use of (Q.22), V can be expressed as a function of $e_\sigma, e_\theta, H, k_\sigma$, and β, then the system of equations simplifies, because the number of equations and unknowns reduces by 1. If the deformed position $\bar{\mathbf{y}}$ is needed, we can complete the system of equations by adding (M.7). Other obvious simplifications occur if we make the classical assumption that Υ is independent of Q and k_n, i.e., $\Upsilon = \Phi(e_\sigma, e_\theta, k_\sigma, k_\theta)$, so that $g = M_n = 0$.

2. General field equations using a mixed-energy density Ψ

As we have seen with beamshells, the use of the mixed-energy density, Ψ, instead of Φ has a profound effect on the form of the governing field equations. In the theory of rubber-like, axisymmetric membranes, a strain-energy density of the form $\Phi = \hat{\Phi}(\lambda_\sigma, \lambda_\theta) = \tilde{\Phi}(e_\sigma, e_\theta)$ seems always to be used. Likewise, the work by Chernykh (1980), Keppel (1984), Taber (1985, 1987), and Simmonds (1985, 1986, 1987) on elastic axishells undergoing large strains has employed Φ. Nevertheless, we believe, in static problems at least, that there are compelling reasons to work with Ψ. Most papers on the nonlinear theory of axishells—there are hundreds—assume small strains which means that Φ and hence Ψ may be taken to be quadratic. In this special but important case, there is no question in our minds that the equations developed by Reissner (1950, 1963a, 1969, 1972), which involve Ψ and consist of two nonlinear, coupled ordinary differential equations for the rotation β and a stress function F, embody the best approach. First, although the equations for axishells are of 6th order,

Reissner's equations are, essentially, only of 4th order because integrals for V and $\bar{y}$ separate out. Second, much of the static-geometric duality, which his equations exhibit when linearized and which proves so useful there, persists in the full nonlinear equations. Third, Reissner's equations are numerically robust. And fourth, the equations for the the torsionless, axisymmetric deformation of plates or membranes fall out immediately. It seems profitable, therefore, to see what form the field equations take when $\Psi = \Psi(N_\sigma, N_\theta, Q, k_\sigma, k_\theta, k_n)$ is assumed to be known.

For conciseness, we use the notation

$$\Psi_{,1} \equiv \frac{\partial \Psi}{\partial N_\sigma} \ , \ \Psi_{,2} \equiv \frac{\partial \Psi}{\partial N_\theta} \ , \ \Psi_{,12} \equiv \frac{\partial^2 \Psi}{\partial N_\sigma \partial N_\theta} \ , \text{ etc.}, \tag{Q.23}$$

and, to see the large-strain generalization of the Reissner equations, set

$$rH = F \ , \ F <> -\beta \, , \tag{Q.24}$$

the 2nd expression indicating that in *linear* theory, F is the static dual of $-\beta$. Substituting the constitutive relations, (O.5), into the equation of moment equilibrium, (Q.12), and using (L.8)-(L.10), (Q.4), (Q.10), (Q.11), and $(Q.24)_1$ to express the arguments of Ψ in terms of β, F, and V, we obtain an equation of the form

$$\begin{aligned} r(\sigma)[\Psi_{,42}(\sigma, F, F', \beta, \beta')F'' + \Psi_{,44}(\sigma, F, F', \beta, \beta')\beta''] \\ + G(\sigma, F, F', \beta, \beta', rV, l, p_H, p_V, p_H') = 0 \, . \end{aligned} \tag{Q.25}$$

If, as in (Q.9), rV can be expressed explicitly in terms of strains and rotations and, ultimately, in terms of β and F, then rV need not be treated as an additional unknown; otherwise, we must augment (Q.25) by the differential equation of vertical equilibrium, $(rV)' = -rp_V$.

A 2nd equation relating F and β, which follows from the compatibility condition, (M.2), upon using the same substitutions that got us from (Q.12) to (Q.25), has the form

$$\begin{aligned} -r(\sigma)[\Psi_{,22}(\sigma, F, F', \beta, \beta')F'' + \Psi_{,24}(\sigma, F, F', \beta, \beta')\beta''] \\ + \tilde{G}(\sigma, F, F', \beta, \beta', rV, p_H, p_V, p_H') = 0 \, . \end{aligned} \tag{Q.26}$$

Again, the same remarks as we made following (Q.25) concerning rV as an unknown apply here.

In (Q.25) and (Q.26), we have a pair of coupled, 2nd-order *quasilinear* ordinary differential equations. If we make the classical assumption that Ψ does not depend on Q or k_n, then (Q.25) and (Q.26) simplify in their details but not in general form. (This is in contrast to problems in general shell theory where the unknowns depend on two spatial variables; there, dropping Q and k_n from Ψ *lowers* the order of the governing equations.) For numerical work, it is often useful to have a system of 1st-order differential equations. For this purpose, it may be best to work in the spin basis $\{\mathbf{T}, \mathbf{B}\}$ and abandon F and β as unknowns.

R. The Simplified Reissner Equations for Small Static Strains

Building on the work of his father, H. Reissner (1912), and Meissner (1913) in *linear* theory, E. Reissner, in a sequence of papers (1950, 1963a, 1969, 1972), has developed and refined the *nonlinear* theory of axishells suffering small, static strains. In this section, we shall derive 2 simplified versions of Reissner's equations, the first of which is due to Koiter (1980). The 2nd derivation illustrates that the "best", i.e., formally simplest, form of the equations of general shell theory need not be the best form for special classes of shells. Both derivations are based on the mixed, quadratic strain-energy density (O.11), repeated here for convenience:

$$\Psi = -\tfrac{1}{2}[A(N_\sigma{}^2 + N_\theta{}^2 - 2\nu_e N_\sigma N_\theta) + A_s Q^2] + \tfrac{1}{2}[D(k_\sigma^2 + k_\theta^2 + 2\nu_b k_\sigma k_\theta) + D_n k_n^2]\,, \tag{R.1}$$

where A_s is a transverse shear compliance and D_n is a normal bending stiffness. From (O.5) follows

$$e_\sigma = A(N_\sigma - \nu_e N_\theta)\ ,\ e_\theta = A(N_\theta - \nu_e N_\sigma)\ ,\ g = A_s Q \tag{R.2}$$

$$M_\sigma = D(k_\sigma + \nu_b k_\theta)\ ,\ M_\theta = D(k_\theta + \nu_b k_\sigma)\ ,\ M_n = D_n k_n\,, \tag{R.3}$$

where we have set the unknown Lagrange multiplier $\lambda = 0$.

Substituting (R.2) into the compatibility condition (M.3) and recalling that $r' = \cos\alpha$, we have

$$A\{r(N_\sigma + N_\theta)' - (1+\nu_e)[(rN_\sigma)' - N_\theta\cos\alpha]\} + (\sin\alpha + rk_\theta)A_s Q - [1 + A(N_\sigma - \nu_e N_\theta)]rk_n = 0\,. \tag{R.4}$$

Using (L.10) to write the static version of (L.12) in the form

$$(rN_\sigma)' - N_\theta\cos\alpha = rk_n N_\theta + r(\alpha' + k_\sigma)Q - rp_T \tag{R.5}$$

and replacing the terms in the first brackets in (R.4) by the right side of (R.5), we obtain

$$rA(N_\sigma + N_\theta)' + \underline{[A_s(\sin\alpha + rk_\theta) - (1+\nu_e)Ar(\alpha' + k_\sigma)]Q} - [1 + \underline{\underline{A(N_\sigma + N_\theta)}}]rk_n + (1+\nu_e)Arp_T = 0\,. \tag{R.6}$$

Shortly, we shall discuss under what conditions the singly and doubly underlined terms can be neglected.

Likewise, using (L.10) to write $M_\theta\cos(\alpha+\beta) = M_\theta\cos\alpha + rk_n M_\theta$ and substituting (R.2) and (R.3) into the static version of the moment equilibrium condition,(L.15), we have

$$D\{r(k_\sigma + k_\theta)' + (1 - \nu_b)[(-rk_\theta)' + k_\sigma \cos\alpha)]\}$$
$$+ [D_n \sin(\alpha+\beta) - rD(k_\theta + \nu_b k_\sigma)]k_n + [1 + (A - A_s)N_\sigma - \nu_e AN_\theta]rQ + rl = 0 . \quad \text{(R.7)}$$

Using (L.9) to replace $\sin(\alpha+\beta)$ by $\sin\alpha + rk_\theta$ and (M.11) to replace the terms in the first brackets in (R.7) by $-(\alpha' + k_\sigma)rk_n$, we have

$$\begin{aligned} rD(k_\sigma + k_\theta)' + \underline{\{D_n(\sin\alpha + rk_\theta) - Dr[k_\sigma + k_\theta + (1-\nu_b)\alpha']\}k_n} \\ + [1 + \underline{\underline{(A - A_s)N_\sigma - \nu_e AN_\theta}}]rQ + rl = 0 . \end{aligned} \quad \text{(R.8)}$$

In the spirit of Koiter's analysis (1980), we now make various simplifications in (R.6) and (R.8) by exploiting the relative errors we make in adopting the uncoupled, quadratic expression, (R.1), for Ψ.

First, if $A = (E_e h)^{-1}$ and A_s are of comparable size, then, as the linear stress-strain relations that follow from (R.1) contain relative errors of the order of the maximum extensional strain, and as these stress-strain relations have been introduced into (R.6) and (R.8), it is consistent to neglect the doubly underlined terms in these equations. Next, *if we make the single, standard assumption of classical 1st-approximation shell theory that the effects of the transverse shear stress resultant, Q, and the normal bending strain, k_n, are negligible in the expression for the mixed energy,* Ψ, i.e., if we neglect any term with a factor of A_s or D_n, then it is only *reasonable* to ignore the singly underlined expressions in (R.6) and (R.8) since the terms that comprise these expressions are clearly of the same type. Thus, upon using (L.8)-(L.10) to express $k_\sigma + k_\theta$ and k_n in terms of β, and (Q.4), (Q.10), (Q.11), and (Q.24) to express $N_\sigma + N_\theta$ and Q in terms of F, V, and p_H, we obtain the simplified, Reissner-Meissner-Reissner (RMR) equations[8]

$$\begin{aligned} Ar[F' + r^{-1}F\cos(\alpha+\beta)]' + \cos\alpha - \cos(\alpha+\beta) \\ + Ar\{[V\sin(\alpha+\beta) + rp_H]' + (1+\nu_e)p_T\} = 0 \end{aligned} \quad \text{(R.9)}^9$$

$$Dr\{\beta' + r^{-1}[\sin(\alpha+\beta) - \sin\alpha]\}' - F\sin(\alpha+\beta) + rV\cos(\alpha+\beta) + rl = 0 . \quad \text{(R.10)}$$

In the classical, *linear* theory of axishells, Simmonds (1975) has shown rigorously that the pointwise errors in the *solutions* of the RMR equations are of the same relative error as those inherent in the classical, uncoupled, quadratic form of Ψ. It would be of great interest to see if, and under what conditions, this analysis can be extended to the nonlinear RMR equations.

---

[8]So named by Koiter (1980) to honor the contributions of H. Reissner (1912), E. Meissner (1913), and E. Reissner (1949, 1950, 1963a, 1969, 1972).

[9]The term $(1+\nu_e)p_T$ was inadvertently omitted from the left side of equation (5.17) of Libai & Simmonds (1983). However, it *does* appear in equation (85) of Simmonds & Danielson (1972), of which the aforementioned equation (5.17) is a special case.

The significance of the RMR equations, as Koiter has emphasized in his derivation (which differs from ours), is that they also follow, after 2 integrations, from a set of canonical, intrinsic, tensorially invariant equations of general, nonlinear, 1st-approximation theory, specialized to axishells. Moreover, except for the the factor $(1+\nu_e)$ multiplying p_T and the term $1-\nu_b^2$ in the bending stiffness D, (R.9) and (R.10) are *completely free of Poisson ratios.*

The following auxiliary quantities, which may appear in boundary conditions, are obtained as follows: from (L.8), (L.9), and $(R.3)_1$,

$$M_\sigma = D\{\beta' + \nu_b r^{-1}[\sin(\alpha+\beta) - \sin\alpha]\} \; ; \qquad (R.11)$$

from (Q.4), (Q.11), and $(R.2)_2$,

$$e_\theta = A\{F' + rp_H - \nu_e[r^{-1}F\cos(\alpha+\beta) + V\sin(\alpha+\beta)]\} \; ; \qquad (R.12)$$

and finally, from (M.7) and (P.1),

$$\begin{aligned} w &= w(0) + \int_0^\sigma [\sin(\alpha+\beta) - \sin\alpha + e_\sigma \sin(\alpha+\beta) + g\cos(\alpha+\beta)]ds \\ &= w(0) + \int_0^\sigma [\cos\alpha\sin\beta - 2\sin\alpha\sin^2(\beta/2) + e_\sigma\sin\alpha \\ &\qquad + \underline{2e_\sigma\cos(\alpha+\beta/2)\sin(\beta/2) + g\cos(\alpha+\beta)}]ds \, . \end{aligned} \qquad (R.13)^{10}$$

An alternative, simplified form of Reissner's (1963a) equations may be obtained by eliminating $\nu_e(rN_\sigma)'$ instead of $(1+\nu_e)(rN_\sigma)'$ from the strain-rotation compatibility condition and by eliminating $\nu_b(rk_\theta)'$ instead of $(1-\nu_b)(rk_\theta)'$ from moment equilibrium. Thus, with $r' = \cos\alpha$ and the strain-stress relations (R.2), (M.2) may be given the form

$$\begin{aligned} &A\{(rN_\theta)' - N_\sigma\cos(\alpha+\beta) - \nu_e[(rN_\sigma)' - N_\theta\cos(\alpha+\beta)]\} \\ &\qquad + A_s Q\sin(\alpha+\beta) + \cos\alpha - \cos(\alpha+\beta) = 0 \, , \end{aligned} \qquad (R.14)$$

which, in view of the the static form of (L.12), reduces to

$$\begin{aligned} &A[(rN_\theta)' - N_\sigma\cos(\alpha+\beta)] + [A_s\sin(\alpha+\beta) - \nu_e Ar(\alpha' + k_\sigma)]Q \\ &\qquad + \cos\alpha - \cos(\alpha+\beta) + \nu_e Arp_T = 0 \, . \end{aligned} \qquad (R.15)$$

By (Q.10), (Q.11), and (Q.24),

$$N_\sigma\cos(\alpha+\beta) = r^{-1}F + Q\sin(\alpha+\beta) \, . \qquad (R.16)$$

Hence,

---

[10]In linear theory, the 1st and 3rd terms in the 2nd line are comparable whereas in moderate rotation theory (which we shall discuss presently) the 2nd term comes into play as well. It is consistent with the approximations introduced in the derivation of the RMR equations to neglect the underlined terms, as we shall do henceforth.

$$A\,[(rN_\theta)' - r^{-1}F] + \underline{[(A_s - A)\sin(\alpha+\beta) - \nu_e Ar(\alpha' + k_\sigma)]Q}$$
$$+ \cos\alpha - \cos(\alpha+\beta) + \nu_e Arp_T = 0\,. \tag{R.17}$$

As before, we now argue that if the effects of the transverse shear stress resultant are ignored *in the mixed-energy density* Ψ, it is only consistent to ignore the underlined term above. Finally, with $N_\theta = F' + rp_H$, we have the first of our 2 simplified equations:

$$A\,(rF'' + F'\cos\alpha - r^{-1}F) + \cos\alpha - \cos(\alpha+\beta) + A\,[(r^2p_H)' + \nu_e rp_T] = 0\,. \tag{R.18}$$

Proceeding in an analogous fashion to simplify the equation of moment equilibrium, we substitute the stress-strain relations, (R.2) and (R.3), into the static form of (L.15) to get

$$D\{(rk_\sigma)' - k_\theta\cos(\alpha+\beta) + \nu_b[(rk_\theta)' - k_\sigma\cos(\alpha+\beta)]\} + D_nk_n\sin(\alpha+\beta)$$
$$+ [1 + (A - A_s)N_\sigma - \nu_e AN_\theta]rQ + rl = 0\,, \tag{R.19}$$

which, in view of (L.10) and the compatibility condition, (M.12), may be written

$$D\,[(rk_\sigma)' - k_\theta\cos(\alpha+\beta)] + [D_n\sin(\alpha+\beta) + \nu_b Dr\alpha']k_n$$
$$+ [1 + (A - A_s)N_\sigma - \nu_e AN_\theta]rQ + rl = 0\,. \tag{R.20}$$

If we use (L.9) and (L.10) to set

$$k_\theta\cos(\alpha+\beta) = r^{-1}\sin\beta + k_n\sin(\alpha+\beta)\,, \tag{R.21}$$

we have

$$D\,[(rk_\sigma)' - r^{-1}\sin\beta] + \underline{[(D_n - D)\sin(\alpha+\beta) + \nu_b Dr\alpha']k_n}$$
$$+ [1 + \underline{\underline{(A - A_s)N_\sigma - \nu_e AN_\theta}}]rQ + rl = 0\,. \tag{R.22}$$

The arguments here for neglecting the singly and doubly underlined terms are the same as we used to neglect similar terms in (R.8). Thus, with $k_\sigma = \beta'$ and $rQ = -F\sin(\alpha+\beta) + rV\cos(\alpha+\beta)$, we have our 2nd simplified equation:

$$D\,(r\beta'' + \beta'\cos\alpha - r^{-1}\sin\beta) - F\sin(\alpha+\beta) + rV\cos(\alpha+\beta) + rl = 0\,. \tag{R.23}$$

We shall refer to (R.18) and (R.23) as the *simplified Reissner equations (for classical, nonlinear axishells)*. They are significantly simpler than Koiter's RMR equations and are virtually free of Poisson ratios of stretching and bending—ν_e rears its head only in the term $\nu_e Arp_T$ in (R.18). The auxiliary equations associated with (R.18) and (R.23) for computing H, M_σ, e_θ, and w remain $(Q.24)_1$ and (R.11)-(R.13).

The linearity of (R.18) in F and the near linearity in β of the coefficient of D in (R.23) suggests that, in specific boundary value problems, a reformulation as nonlinear integral equations should be profitable for both theoretical and numerical work.[11]

Reissner's 1963 equations include the effects of transverse shearing strain and variable thickness. If (in our notation) we take h to be constant and set $A_s/A=0$ in Reissner's equation (29), then his undifferentiated stress function term becomes $r^{-1}F[\cos^2\alpha+\nu_e r(\cos\alpha)']$, whereas the analogous term in (R.18) is $r^{-1}F$. Furthermore, in his equation (28) there appears the term

$$r^{-1}\{\cos(\alpha+\beta)[\sin(\alpha+\beta)-\sin\alpha]-\nu_e r\alpha'[\cos(\alpha+\beta)-\cos\alpha]\}\ , \tag{R.24}$$

whereas in (R.23) the analogous term is simply $r^{-1}\sin\beta$.

For a slightly different derivation and justification of the simplified Reissner equations, see Simmonds & Libai (1987a).

1. Membrane theory

If we set $D=l=0$, then (R.23) reduces to a transcendental equation having the solution

$$\sin(\alpha+\beta)=\frac{rV}{\sqrt{F^2+r^2V^2}}\ ,\quad \cos(\alpha+\beta)=\frac{F}{\sqrt{F^2+r^2V^2}}\ . \tag{R.25}$$

Substituting (R.25) into (R.18) and the auxiliary equations $(Q.24)_1$, (R.12) and (R.13), we get

$$A(rF''+F'\cos\alpha-r^{-1}F)+\cos\alpha-F(F^2+r^2V^2)^{-1/2} + A[(r^2p_H)'+\nu_e rp_T]=0 \tag{R.26}$$

$$H=r^{-1}F \tag{R.27}$$

$$e_\theta=A(F'+rp_H-\nu_e r^{-1}\sqrt{F^2+r^2V^2}) \tag{R.28}$$

$$w=w(0)+\int_0^\sigma\{rV(F^2+r^2V^2)^{-1/2}-\sin\alpha + A[r^{-1}(F^2+r^2V^2)-\nu_e(F'+rp_H)]\sin\alpha\}ds\ . \tag{R.29}$$

Because $D=0$, (R.11) yields $M_\sigma=0$.

Flügge & Riplog (1956) have used these equations to study cylindrical, conical, and spherical membranes under radial edge loads only and have shown that, near the edge(s), there must be a flat, narrow annular region whose width is determined by the condition that its inner edge be stress free. Clark & Narayanaswamy (1967) have studied this problem for arbitrary aximembranes and have shown that the governing differential equation in the flat region is linear

---

[11]Indeed, because $r^2F''+rr'F'-F=r(rF')'-F\equiv\mathcal{L}F$, the change of variable $s\equiv\int r^{-1}(\sigma)d\sigma$ reduces $\mathcal{L}$ to a constant coefficient operator. The associated Green's function is therefore a linear combination of exponentials of $s(\sigma)$.

and can be solved in closed form. They have also studied arbitrary aximembranes under axial edge loads only. Reissner & Wan (1965) have shown that a shallow, spinning membrane may exhibit a flattened region at a sufficiently high angular velocity. Simmonds (1961) has studied uniformly rotating membranes with arbitrary meridians. Other special cases of the small-strain, nonlinear membrane equations that have been studied will be mentioned in section S. The nonlinear theory of membranes undergoing large strains is studied in Section T.

2. Moderate rotation theory

This is a theory in which all trigonometric functions of β are expanded in powers of β and terms of $O(\beta^2)$ are neglected compared to 1. Thus, (R.18), (R.23), and the auxiliary equations $(Q.24)_1$ and (R.11)-(R.13) reduce to

$$A\,(rF''+F'\cos\alpha-r^{-1}F)+\tfrac{1}{2}\beta^2\cos\alpha+\beta\sin\alpha+A\,[(r^2p_H)'+\nu_e rp_T]=0 \quad (R.30)$$

$$D\,(r\beta''+\beta'\cos\alpha-r^{-1}\beta)-F\,(\sin\alpha+\beta\cos\alpha)+rV(\cos\alpha-\beta\sin\alpha)+rl=0 \quad (R.31)$$

$$H=r^{-1}F \quad (R.32)$$

$$M_\sigma=D\,[\beta'+\nu_b r^{-1}(\beta\cos\alpha-\tfrac{1}{2}\beta^2\sin\alpha)]\,. \quad (R.33)$$

$$e_\theta=A\{F'+rp_H-\nu_e[r^{-1}F\,(\cos\alpha-\beta\sin\alpha)+V(\sin\alpha+\beta\cos\alpha)]\} \quad (R.34)$$

$$w=w(0)+\int_0^\sigma(\beta\cos\alpha+e_\sigma\sin\alpha-\tfrac{1}{2}\beta^2\sin\alpha)ds\,. \quad (R.35)$$

We emphasize that the accuracy of the moderate rotation theory depends on the geometry of the reference meridian, the thickness of the shell, the boundary conditions, and the nature and magnitude of the external loads. Sometimes, in a specific problem, the error in the moderate rotation theory can be estimated from a scaling of the full, simplified Reissner equations, but, usually, an *a posteriori* analysis is required.

3. Nonlinear shallow shell theory

Shallow shell theory is a special case of moderate rotation theory in which α^2, $r\alpha\alpha'$, and $\alpha\beta$ are neglected compared to 1. Thus,

$$\cos\alpha=r'\approx 1\ ,\ \ \sin\alpha=z'\approx dz/dr \quad (R.36)$$

and

$$F'=\frac{dF}{dr}\cos\alpha\approx\frac{dF}{dr} \quad (R.37)$$

$$rF''+F'\cos\alpha=r\frac{d^2F}{dr^2}+\frac{dF}{dr}(\cos\alpha-\alpha'\sin\alpha)\approx r\frac{d^2F}{dr^2}+\frac{dF}{dr} \quad (R.38)$$

$$p_T = p_H \cos(\alpha+\beta) + p_V \sin(\alpha+\beta) \approx p_H + p_V(\alpha+\beta)\,, \tag{R.39}$$

so that (R.30)-(R.35) reduce to

$$A\left[r\frac{d^2F}{dr^2} + \frac{dF}{dr} - \frac{F}{r}\right] + \frac{dz}{dr}\beta + \tfrac{1}{2}\beta^2 + Ar\,[rp_H{}' + (2+\nu)p_H] = 0 \qquad (\text{R.40})^{12}$$

$$D\left[r\frac{d^2\beta}{dr^2} + \frac{d\beta}{dr} - \frac{\beta}{r}\right] - F\left[\frac{dz}{dr} + \beta\right] + rV + rl = 0 \tag{R.41}$$

$$H = r^{-1}F \tag{R.42}$$

$$M_\sigma = D\left[\frac{d\beta}{dr} + \nu_b\frac{\beta}{r}\right]\,. \tag{R.43}$$

$$e_\theta = A\left[\frac{dF}{dr} + rp_H - \nu_e\frac{F}{r}\right] \tag{R.44}$$

$$w = w(0) + \int_0^r \left[\beta + \frac{dz}{dr}e_\sigma\right] dr\,. \tag{R.45}$$

Weinitschke (1987a) has proved that if $H > 0$ everywhere, then solutions to (R.40), (R.41), and appropriate boundary conditions prescribing edge stresses, moments, and/or displacements for dome- or ring-like shallow axishells are unique.

S. Special Cases of the Simplified Reissner Equations

1. Cylindrical shells and membranes

Here, $r=R$, $\sigma=z$, and $\alpha=\pi/2$ so that the simplified Reissner equations, (R.18) and (R.23), and the auxiliary equations, $(\text{Q.24})_1$ and (R.11)-(R.13), take the form

$$AR\,(F'' - R^{-2}F) + \sin\beta + A\,(R^2 p_H{}' + \nu_e p_T) = 0 \tag{S.1}$$

$$DR\,(\beta'' - R^{-2}\sin\beta) - F\cos\beta - RV\sin\beta + Rl = 0\,. \tag{S.2}$$

$$H = R^{-1}F \tag{S.3}$$

[12]Rough order of magnitude arguments suggest that the relative error we make in ignoring the term $Ar\nu_e(dz/dr+\beta)p_V$ is of the same order of magnitude as the inherent relative error in shallow shell theory.

$$M_\sigma = D\,[\beta' - \nu_b R^{-1}(1-\cos\beta)] \tag{S.4}$$

$$e_\theta = A\,[F' + Rp_H + \nu_e(R^{-1}F\sin\beta - V\cos\beta)] \tag{S.5}$$

$$w = w(0) + \int_0^z [e_\sigma - 2\sin^2(\beta/2)]ds\,. \tag{S.6}$$

We shall not attempt an exhaustive discussion of the many practical problems to which these equations apply. Rather, we shall consider 2 idealized, but, we believe, illuminating examples. Except for a few cursory remarks, we leave the important question of stability to Section W.

(a) Cylindrical shell under axial edge forces only. If the only external load is a net axial force of magnitude $2\pi RP$ applied to the ends of the shell, which are free to move (save that rigid body movements are suppressed), then $P=V$, the vertical force per unit *undeformed* length along a parallel circle. By inspection, $F=\beta=0$ is a solution of (S.1) and (S.2). From (S.3), (Q.4), (Q.10), and (Q.11), $N_\theta = Q = 0$, $N_\sigma = P$, while (S.5) and (S.6) yield the radial displacement, $u = \mathrm{R}e_\theta = -\nu_e APR$, and the *overall end lengthening*,

$$\Delta \equiv w(L) - w(0) = APL\,. \tag{S.7}$$

Suppose that the ends of the cylindrical shell are simply-supported, i.e., suppose that

$$u(0) = u(L) = M_\sigma(0) = M_\sigma(L) = 0\,. \tag{S.8}$$

Then, the trivial solution $F=\beta=0$ fails to meet the above condition of zero radial displacement at the edges and must be modified.

It is instructive to see first what *membrane theory* predicts. From (R.26), we have

$$AR\,(F'' - R^{-2}F) - \frac{F}{\sqrt{F^2 + R^2P^2}} = 0\,. \tag{S.9}$$

The boundary condition $e_\theta = 0$ follows from (R.28) as

$$F' - \nu_e R^{-1}\sqrt{F^2 + R^2P^2} = 0\ ,\ \ z = 0, L\,, \tag{S.10}$$

while the boundary condition $M_\sigma = 0$ is satisfied trivially because $D=0$.

The form of (S.9) and (S.10) suggests that we nondimensionalize F by setting

$$F \equiv PR\varepsilon f\,, \tag{S.11}$$

where, for the moment, ε is an unspecified dimensionless parameter. It is not difficult to see that solutions of (S.9) will exhibit boundary layers of width $O(R\sqrt{AP})$ at $z=0$ and $z=L$. [Such boundary layers in elastic membranes suffering small strains were discovered by Bromberg & Stoker (1945).] Because f' will therefore be large compared with f/R at the edges of the shell, the boundary condition (S.10), with $F = PR\varepsilon f$, implies that $P^{-1}F' = O(\varepsilon f/\sqrt{AP}) = O(1)$. This suggests that we set

$$\varepsilon = \sqrt{AP} \ , \tag{S.12}$$

which is of the order of magnitude of the square root of the extensional strains. Thus, substituting (S.11), (S.12), and

$$z = R\varepsilon x \ , \ d(\ )/dx = (\ )^{\cdot} \tag{S.13}$$

into (S.9) and (S.10), we obtain the differential equation

$$\ddot{f} - \frac{f}{\sqrt{1+\varepsilon^2 f^2}} - \varepsilon^2 f = 0 \tag{S.14}$$

and boundary conditions

$$\dot{f} = \nu\sqrt{1+\varepsilon^2 f^2} \ , \ x = 0 \, , \, \varepsilon^{-1}L/R \ . \tag{S.15}$$

The nonlinear boundary value problem (S.14) and (S.15) has the formal solution

$$f = f_0(x) + \varepsilon^2 f_1(x) + \cdots \ , \tag{S.16}$$

where, to within transcendentally small terms,

$$f_0(x) = \nu[\exp(x - \varepsilon^{-1}L/R) - \exp(-x)] \ . \tag{S.17}$$

Clearly, $f_0'(x) > 0$ and it can be shown that $N_\theta = P\dot{f}(x) > 0$. As $N_\sigma = P\sqrt{1+\varepsilon^2 f^2}$ is also positive, the membrane solution is stable. If $P < 0$, the stress resultants are negative and the membrane solution is unstable. This instability is manifested mathematically by ε being imaginary so that $f_0(x)$ becomes oscillatory when expressed in terms of $z = R\varepsilon x$.

At the other end of the spectrum from membrane theory lies *inextensional bending theory*, obtained by setting $A = 0$ in (S.1) and (S.5). Aside from an eversion ($\beta = \pi$), the resulting simplified Reissner equations equations have only the trivial solution $F = \beta = 0$. Hence, $M_\sigma \equiv 0$ while $e_\theta \equiv 0$ because $A = 0$.

If A, $D \neq 0$, a solution lies somewhere between the extreme states of membrane stretching and inextensional bending and may be conveniently located in the stretching-bending spectrum by specifying *Reissner's (1950) transition number*

$$\rho \equiv PR\sqrt{A/D} = \frac{R\varepsilon^2}{\sqrt{AD}} = \left[\frac{P}{h\sqrt{E_e E_b}}\right]\left[\frac{R}{h}\right] \ , \tag{S.18}$$

the first factor on the right representing a typical level of strain in the shell and the 2nd factor measuring thinness. In a sense to be made precise, if $\rho \ll 1$, bending dominates; if $\rho \gg 1$ stretching dominates; and if $\rho = O(1)$, both are of equal importance.

In membrane theory, both $(PR)^{-1}F$ and β are $O(\varepsilon)$ and these quantities approach 0 as the shell is made stiffer. We shall therefore regard the simplified

Reissner equations equations as a singular perturbation of those of membrane theory. With $F = PR\varepsilon f$, $z = R\varepsilon x$, and $\varepsilon = \sqrt{AP}$, as in membrane theory, and

$$\beta = \varepsilon g \, , \tag{S.19}$$

(S.1) and (S.2) together with the boundary conditions $e_\theta = M_\sigma = 0$, which come from (S.4) and (S.5), take the form

$$\ddot{f} - \varepsilon^2 f + \varepsilon^{-1} \sin \varepsilon g = 0 \tag{S.20}$$

$$\rho^{-2}(\ddot{g} - \varepsilon \sin \varepsilon g) - f \cos \varepsilon g - \varepsilon^{-1} \sin \varepsilon g = 0 \tag{S.21}$$

$$\dot{f} = \nu_e(\cos \varepsilon g - \varepsilon \sin \varepsilon g) \ , \ x = 0, \varepsilon^{-1} L/R \tag{S.22}$$

$$\dot{g} = \nu_b(1 - \cos \varepsilon g) \ , \ x = 0, \varepsilon^{-1} L/R. \tag{S.23}$$

To within relative errors of $O(\varepsilon^2)$, these equations reduce to

$$\ddot{f} + g = 0 \tag{S.24}$$

$$\rho^{-2}\ddot{g} - f - g = 0 \tag{S.25}$$

$$\dot{f} = \nu_e \ , \ \dot{g} = 0 \ , \ x = 0, \varepsilon^{-1} L/R \, . \tag{S.26}$$

Using (S.24) to eliminate g from the remaining equations, we obtain

$$\rho^{-2} \ddot{\ddot{f}} - \ddot{f} + f = 0 \tag{S.27}$$

$$\dot{f} = \nu_e \ , \ \dddot{f} = 0 \ , \ x = 0, \varepsilon^{-1} L/R \, . \tag{S.28}$$

If $P > 0$, a simple perturbation analysis shows that, as $\rho \to \infty$, the solution of (S.27) and (S.28) is given, in a 1st approximation, by

$$f \sim -\nu_e(e^{-x} + \rho^{-3} e^{-\rho x}) \tag{S.29}$$

near $x = 0$. Thus, although membrane behavior is dominant for large ρ, there is a weak contribution from bending, confined to a very narrow boundary layer within the already narrow stretching boundary layer. On the other hand, as $\rho \to 0$ (i.e., if bending dominates),

$$f \sim (\nu_e/\sqrt{2\rho}) e^{-(\sqrt{\rho/2})x} [\sin(\sqrt{\rho/2})x - \cos(\sqrt{\rho/2})x] \, . \tag{S.30}$$

To discuss the solution of (S.27) and (S.28) for arbitrary values of the Reissner transition number, we look for solutions of the form

$$f = C \exp(\sqrt{\rho}\, w x) = C \exp(\lambda^{-1} z/R) \ , \ \lambda^4 \equiv AD/R^2 \, . \tag{S.31}$$

It follows from (S.27) that w must satisfy

$$w^4 - \rho w^2 + 1 = 0 \, . \tag{S.32}$$

The behavior of the 4 roots of this equation—call them w_k, $k = 1, 2, 3, 4$—as a function of ρ are most easily discussed with reference to Fig. 5.3, which contains the unit circle in the complex w-plane. Roots in the left-half plane correspond to boundary-layers at $z=0$; those in the right-half plane to boundary-layers at $z=L$. Starting with $\rho \gg 2$, we have 4 real, symmetrically located roots on the real w-axis, 2 outside the unit circle associated with bending, and 2 inside the unit circle associated with stretching. As $\rho \to 2$, there are double roots at ± 1, and the distinction between bending and stretching solutions is no longer tenable. As ρ moves from 2 to the *critical value* $\rho_{cr} = -2$, the roots assume symmetrically located positions on the unit circle, pass through the points $\pm e^{\pm i\pi/4}$ at $\rho = 0$, and become, again, double roots at $\pm i$. Finally, as $\rho \to -\infty$, one pair of roots on the imaginary w-axis approaches the origin while the other pair approaches infinity.

The significance of ρ_{cr} is that, at this value, f first becomes purely oscillatory, which we interpret as *potential buckling*. This means that if we substitute $\Sigma_1^4 C_k \exp(\sqrt{\rho}\, w_k x)$ into the boundary conditions (S.28), then, for each L/R, there is a value of $\rho \leq -2$ where the determinant of coefficients vanishes. The maximum value of ρ as L/R varies is -2 and the associated *net critical axial load* is

$$P_{cr} = -\frac{2h^2\sqrt{E_e E_b}}{R} = -\frac{Eh^2}{\sqrt{3(1-\nu^2)}R}, \tag{S.33}$$

where, in the last expression, we have given E_e, E_b, and ν_e their classical values. A more detailed analysis may be found on pp. 484-490 of the book by Flügge (1973). We particularly commend to the reader his Fig. 8.27 on p. 488, which shows successive stages of the radial deflection u as the axial load P approaches P_{cr}.

We warn the reader that the buckling of axially-loaded, simply-supported, circular cylindrical shells is much more complicated that the above analysis would suggest. First, thin shells are apt to buckle into an *unsymmetric* pattern, and second, slight geometric imperfections, which are always present in real

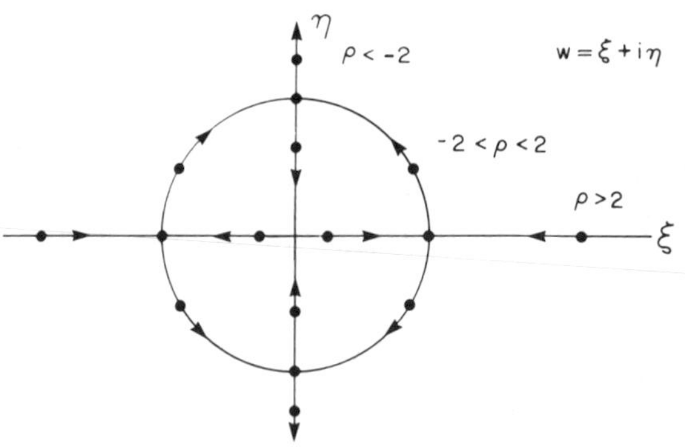

Fig. 5.3. Movement of the roots of $w^4 - \rho w^2 + 1 = 0$ as ρ decreases.

shells, tend to lower dramatically the theoretical buckling loads for perfect geometrical shapes. See the book by Yamaki (1984) for an extensive (though not exhaustive!) treatment of this problem.

(b) Cylindrical shell under axial and radial edge forces only. To focus on essentials, we consider a semi-infinite shell ($z > 0$) whose end $z = 0$, which is free to move radially (axial rigid body movement being suppressed), is acted upon by the stress resultant

$$\hat{\mathbf{N}}_\sigma = P(\sin B\,\mathbf{e}_r - \cos B\,\mathbf{e}_z)\ ,\ \ 0 \le B \le \pi/2\ . \tag{S.34}$$

In this case, $V = P\cos B$.

In the 1st example the governing equations, to a 1st approximation, turned out to be linear. Here, this cannot always be so because, in the membrane limit, the edge of the shell must turn through an angle B if the edge load is to be taken by membrane forces alone. Examining this membrane limit first, we have, from (R.26) and (R.27),

$$AR(F'' - R^{-2}F) - F(F^2 + R^2P^2\cos^2 B)^{-1/2} = 0 \tag{S.35}$$

$$R^{-1}F(0) = -P\sin B\ . \tag{S.36}$$

This last equation suggests the nondimensionalization

$$F = PRf\sin B\ . \tag{S.37}$$

Substituting (S.37) into (S.35) and setting $z = R\varepsilon x$, $(\ )^{\cdot} = d(\ )/dx$, we get

$$\ddot{f} - \frac{f}{\sqrt{f^2\sin^2 B + \cos^2 B}} - \varepsilon^2 f = 0 \tag{S.38}$$

$$f(0) = -1\ . \tag{S.39}$$

To a 1st approximation, the last term in (S.38) is negligible and we shall drop it.

Once f has been found, we may compute β from

$$\sin\beta = -\frac{f\sin B}{\sqrt{f^2\sin^2 B + \cos^2 B}}\ ,\ \ \cos\beta = \frac{\cos B}{\sqrt{f^2\sin^2 B + \cos^2 B}}\ , \tag{S.40}$$

which follow from (R.26) and (S.37) with $\alpha = \pi/2$ and $r = R$.

First, 2 special values of the deformed edge angle B deserve mention. If $B = 0$, (S.38) reduces to $\ddot{f} - f = 0$ which, in view of the boundary condition (S.39), has the solution $f = -e^{-x}$. However, note by (S.37) that $F \equiv 0$ so that $N_\sigma \equiv P$ and $N_\theta \equiv 0$, as should be.

If $B \to \pi/2$, (S.38) reduces to $\ddot{f} - \mathrm{sgn}\,f = 0, f \ne 0$. Near $x = 0$ where $f = -1$, this equation has the solution $f = -\tfrac{1}{2}x^2 + Cx - 1$, where C is an unknown constant and (S.40) yields $\beta = \pi/2$. If $C = \sqrt{2}$, $f(\sqrt{2}) = \dot{f}(\sqrt{2}) = 0$, which implies that, at $z = R\varepsilon\sqrt{2} = R\sqrt{2AP}$, $N_\sigma = N_\theta = 0$. Alternatively, another solution of (S.38) and (S.40) is the trivial one $f = \beta = 0$. The picture of the deformation supplied by Flügge & Riplog (1956) is that, in the small region $0 < z < R\sqrt{2AP}$, the membrane rotates through a right angle to form a flat flange, while the remaining portion of the membrane remains undeformed.

If $0 < B < \pi/2$, (S.38) (with the last term omitted) has the 1st integral

$$\tfrac{1}{2}\dot{f}^2 \sin^2 B - \sqrt{f^2 \sin^2 B + \cos^2 B} = -\cos B\ , \tag{S.41}$$

where a constant of integration has been set to 0 since f and $\dot{f} \to 0$ as $x \to \infty$. This becomes a *universal* equation, i.e., one free of B, that we may solve once for all if we take β as the dependent variable, via $\tan\beta = -f\tan B$, and introduce the new independent variable, $\zeta = \sqrt{2\sec B}\, x$. Upon taking the square root of the resulting equation, and noting that, as we move into the shell, β must decreases towards 0 from its positive edge value of B, we have

$$d\beta/d\zeta = -\cos^{3/2}\beta\sqrt{1-\cos\beta}\ . \tag{S.42}$$

Fig. 5.4 is a graph of $-d\beta/d\zeta$ *vs* $\cos\beta$. The actual solution corresponds to that portion of the curve between $\cos B$ and 1.

Because ζ does not enter the differential (S.42) explicitly, any solution will maintain its shape if ζ is replaced by ζ plus a constant. Thus, if we have a solution $\beta = U(\zeta)$ for a particular initial value $\beta_* = U(0)$, $0 < \beta_* < \pi/2$, the solution of (S.42) for any other initial value, say β_0, may be expressed as $\beta = U(\zeta - \zeta_0)$, where $\zeta_0 > 0$ if $\beta_0 > \beta_*$. We therefore call $U(\zeta)$ a *universal solution*. A convenient choice for β_* is $\cos^{-1}3/4$ because $-d\beta/d\zeta$ has a maximum there (of $\sqrt{3}/4$). Values of $U(\zeta)$, obtained by solving (S.42) numerically, are given in Table 5.2 and graphically in Fig. 5.5. The inverse of the function U, call it V, can be computed explicitly by separating variables in (S.42) and integrating from $\cos\beta_* = 3/4$ to $\cos\beta$ to obtain

$$\zeta = V(\beta)\ , \tag{S.43}$$

where

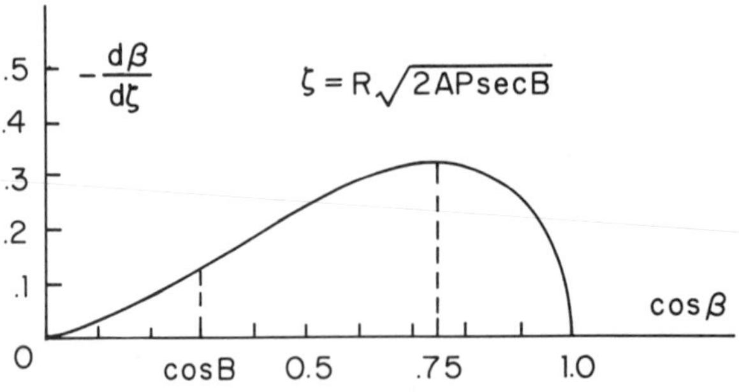

Fig. 5.4. Phase portrait of $d\beta/d\zeta = -\cos^{3/2}\beta\sqrt{1-\cos\beta}$.

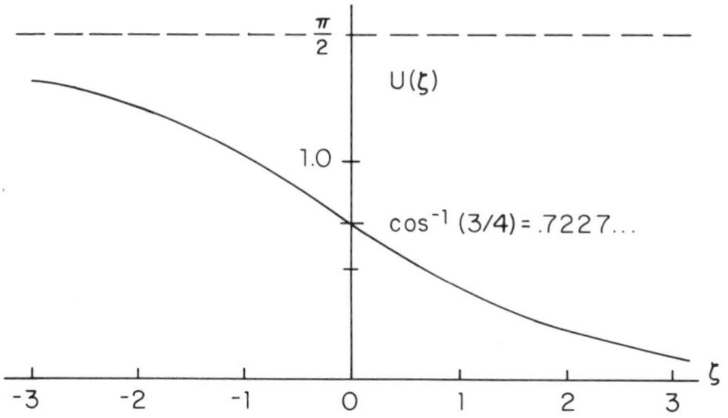

Fig. 5.5. The Universal function $U(\zeta)$.

$$
\begin{aligned}
V(\beta) &\equiv \int_{\beta}^{\beta_*} \frac{d\gamma}{\cos^{3/2}\gamma\sqrt{1-\cos\gamma}} \\
&= 2\int_{\sqrt{3}/2}^{\sqrt{\cos\beta}} \frac{du}{u^2(1-u^2)\sqrt{1+u^2}} \\
&= \int_{\sqrt{3}/2}^{\sqrt{\cos\beta}} \left[\frac{2}{u^2}+\frac{1}{1-u}+\frac{1}{1+u}\right]\frac{du}{\sqrt{1+u^2}} \qquad \text{(S.44)} \\
&= 2\left[\sqrt{\frac{7}{3}}-\sqrt{\frac{1+\cos\beta}{\cos\beta}}\right] \\
&\quad + \frac{1}{\sqrt{2}}\left[\sinh^{-1}\left(\frac{1+\sqrt{\cos\beta}}{1-\sqrt{\cos\beta}}\right)-\sinh^{-1}\left(\frac{1-\sqrt{\cos\beta}}{1+\sqrt{\cos\beta}}\right)+\sinh^{-1}\left(\frac{2-\sqrt{3}}{2+\sqrt{3}}\right)\right]\,.
\end{aligned}
$$

Returning to the full simplified Reissner equations, (S.1) and (S.2), retaining the dimensionless variables and parameters, $F = PRf\sin B$, $z = R\varepsilon x$, $\varepsilon = \sqrt{AR}$, and $\rho = PR\sqrt{A/D}$, we have

$$\ddot{f}\sin B + \sin\beta - \underline{\varepsilon^2 f} = 0 \qquad \text{(S.45)}$$

$$\rho^{-2}(\ddot{\beta} - \underline{\varepsilon^2\sin\beta}) - f\sin B\cos\beta - \cos B\sin\beta = 0\,. \qquad \text{(S.46)}$$

From (S.3) and (S.4), the boundary conditions at the end of the shell take the form

$$f(0) = -1\ ,\ \ \dot{\beta}(0) = \underline{\varepsilon\nu_b[1-\cos\beta(0)]}\,, \qquad \text{(S.47)}$$

where the underlined terms are negligible in a 1st approximation.

First, we consider $B = \pi/2$.

Table 5.2

Solution of $d\beta/d\zeta = -\cos^{3/2}\beta\sqrt{1-\cos\beta}$, $\beta(0) = \cos^{-1}(3/4)$

ζ	$U(\zeta)$	$U(-\zeta)$	ζ	$U(\zeta)$	$U(-\zeta)$
.00	.7227	.7227	2.60	.1453	1.3109
.10	.6903	.7552	2.70	.1356	1.3218
.20	.6580	.7875	2.80	.1264	1.3321
.30	.6260	.8195	2.90	.1179	1.3419
.40	.5945	.8510	3.00	.1099	1.3512
.50	.5636	.8819	3.10	.1025	1.3599
.60	.5334	.9121	3.20	.0955	1.3682
.70	.5041	.9415	3.30	.0891	1.3760
.80	.4757	.9700	3.40	.0830	1.3835
.90	.4482	.9975	3.50	.0774	1.3906
1.00	.4219	1.0240	3.60	.0721	1.3973
1.10	.3966	1.0495	3.70	.0672	1.4036
1.20	.3725	1.0738	3.80	.0626	1.4097
1.30	.3496	1.0971	3.90	.0584	1.4154
1.40	.3277	1.1193	4.00	.0544	1.4209
1.50	.3070	1.1404	4.10	.0507	1.4261
1.60	.2875	1.1604	4.20	.0472	1.4310
1.70	.2690	1.1795	4.30	.0440	1.4357
1.80	.2516	1.1975	4.40	.0410	1.4402
1.90	.2352	1.2146	4.50	.0382	1.4445
2.00	.2197	1.2308	4.60	.0356	1.4486
2.10	.2052	1.2461	4.70	.0332	1.4525
2.20	.1916	1.2605	4.80	.0309	1.4562
2.30	.1789	1.2742	4.90	.0288	1.4597
2.40	.1670	1.2871	5.00	.0268	1.4631
2.50	.1558	1.2993	∞	0	$\pi/2$

If we set

$$g = -\rho f \ , \ y = \sqrt{\rho}\, x \ , \tag{S.48}$$

then, with the underlined terms omitted, (S.45)-(S.47) reduce to

$$\frac{d^2 g}{dy^2} - \sin\beta = 0 \ , \ \frac{d^2\beta}{dy^2} + g\cos\beta = 0 \tag{S.49}$$

$$g(0) = \rho \ , \ \frac{d\beta}{dy}(0) = 0 \ . \tag{S.50}$$

Subtracting $d\beta/dy$ times $(S.49)_2$ from dg/dy times $(S.49)_1$, we obtain the 1st integral

$$\frac{1}{2}\left[\frac{dg}{dy}\right]^2 - g\sin\beta - \frac{1}{2}\left[\frac{d\beta}{dy}\right]^2 = 0\,, \tag{S.51}$$

where a constant of integration has been set to 0 since all terms on the left approach 0 as $y \to \infty$. A study of the solutions of the above equations is yet to be done. We merely note here that, as $\rho \to \infty$, the solutions must approach those of membrane theory, although the approach of β to β_M cannot be uniform in any interval containing the point $x = \sqrt{2}$ because β_M had a jump of $\pi/2$ there. We also note that, as $\rho \to 0$, the solutions must approach those of *linear bending theory*,

$$g \sim \rho e^{-y}(\cos y - \sin y)\ ,\ \beta \sim 2\rho e^{-y}(\cos y + \sin y)\,. \tag{S.52}$$

If $0 \leq B < \pi/2$, and if we make the change of variables and parameter,

$$q = -f\tan B\ ,\ \zeta = \sqrt{2\sec B}\,x\ ,\ \kappa = \rho\cos B\,, \tag{S.53}$$

then (S.45)-(S.47), with the underlined terms omitted, reduce to

$$\frac{d^2q}{d\zeta^2} - \sin\beta = 0\ ,\ \kappa^{-2}\frac{d^2\beta}{d\zeta^2} + q\cos\beta - \sin\beta = 0 \tag{S.54}$$

$$q(0) = \tan B\ ,\ \frac{d\beta}{d\zeta}(0) = 0\,. \tag{S.55}$$

The differential equations have a 1st integral which, when we impose decay conditions as $\zeta \to \infty$, becomes

$$\frac{1}{2}\left[\frac{dq}{d\zeta}\right]^2 - q\sin\beta - \cos\beta - \frac{1}{2}\kappa^{-2}\left[\frac{d\beta}{d\zeta}\right]^2 = -1\,. \tag{S.56}$$

The solutions of (S.54)-(S.56) are yet to be studied, but we expect these to approach the membrane solutions, $q_M = \tan U(\zeta)$ and $\beta_M = U(\zeta)$, uniformly on $0 < \zeta$, as $\rho \to \infty$, while if $\rho \to 0$, then the solutions should approach those of linear bending theory; i.e.,

$$\begin{aligned} q &\sim e^{-\sqrt{\kappa/2}\,\zeta}\tan B\,[\cos\sqrt{\kappa/2}\,\zeta + \sin\sqrt{\kappa/2}\,\zeta] \\ \beta &\sim -\kappa e^{-\sqrt{\kappa/2}\,\zeta}\tan B\,[\cos\sqrt{\kappa/2}\,\zeta - \sin\sqrt{\kappa/2}\,\zeta]\,. \end{aligned} \tag{S.57}$$

2. *Circular plates and membranes*

The simplest class of axishells after cylindrical shells is circular (or axi-) plates. Here, $\sigma = r$ and $\alpha = 0$ so that the simplified Reissner equations, (R.18) and (R.23), and the auxiliary equations, $(Q.24)_1$ and (R.11)-(R.13), reduce to

$$\begin{aligned} &A(rF'' + F' - r^{-1}F) + 1 - \cos\beta \\ &\qquad + A[r^2 p_H{}' + (2 + \nu_e\cos\beta) r p_H + \nu_e p_V \sin\beta] = 0 \end{aligned} \tag{S.58}$$

$$D(r\beta'' + \beta' - r^{-1}\sin\beta) - F\sin\beta + rV\cos\beta + rl = 0 \tag{S.59}$$

$$H = r^{-1}F \tag{S.60}$$

$$M_\sigma = D(\beta' + \nu_b r^{-1}\sin\beta) \tag{S.61}$$

$$e_\theta = A[F' + rp_H - \nu_e(r^{-1}F\cos\beta + V\sin\beta)] \tag{S.62}$$

$$w = w(0) + \int_0^r \sin\beta ds\ . \tag{S.63}$$

There is a large literature on the nonlinear theory of elastic axiplates, mostly based on the Föppl-Kármán equations in which β^2 is neglected compared to 1. This approximation reduces (S.58) and (S.59) to

$$A(rF'' + F' - r^{-1}F) + \tfrac{1}{2}\beta^2 + Ar[r^2 p_H{}' + (2+\nu_e)p_H] = 0 \qquad \text{(S.64)}^{13}$$

$$D(r\beta'' + \beta' - r^{-1}\beta) - F\beta + rV + rl = 0\ . \tag{S.65}$$

These equations also follow from (R.40) and (R.41) of nonlinear shallow shell theory.

The associated auxiliary conditions of the Föppl-Kármán theory are

$$H = r^{-1}F \tag{S.66}$$

$$M_\sigma = D(\beta' + \nu_b r^{-1}\beta) \tag{S.67}$$

$$e_\theta = A[F' + rp_H - \nu(r^{-1}F + V\beta)] \tag{S.68}$$

$$\bar{z} = \bar{z}(0) + \int_0^r \beta ds\ . \tag{S.69}$$

Solutions of various problems involving the Föppl-Kármán equations may be found in the first 5 sections of Chapter 13 of the book by Timoshenko & Woinowsky-Krieger (1959) and in the first 7 sections of Chapter 3 of the book by Chia (1980). The reader should also consult the massive bibliography (12,000+ entries!) of papers on all types of linear and nonlinear plate problems, compiled by Naruoka (1981). Dickey (1976) has proved the existence of a solution of the Föppl-Kármán equations for a clamped or simply-supported plate under variable normal pressure.

The following, almost trivial, example shows that the Föppl-Kármán *equations need not be a 1st-approximation to the simplified Reissner equations.* Consider an annular plate of inner radius, a, and outer radius, b, free of any external loads. The governing equations are (S.58) and (S.59) with $p_H = p_V = V = l = 0$, and the boundary conditions, $H(a) = H(b) = M_\sigma(a) = M_\sigma(b)$

[13] As in (R.40)— see footnote 12—we have neglected the term $Ar\nu_e p_V\beta$.

$=0$. Aside from the trivial solution, $F=\beta=0$, this boundary value problem admits an (unstable) everted solution in which $\beta(r)=\pi$ and

$$A(F''+F'-r^{-1}F)+2=0 \tag{S.70}$$

$$F(a)=F(b)=0\,, \tag{S.71}$$

i.e.,

$$F=-\frac{1}{A}\left[\frac{a^2}{b^2-a^2}\ln\left(\frac{a}{b}\right)\left(\frac{r}{b}-\frac{b}{r}\right)-\frac{r}{b}\ln\left(\frac{b}{r}\right)\right]. \tag{S.72}$$

Of course, the annular plate must be narrow $(b-a \ll b+a)$ for the solution to remain in the small-strain range. The Föppl-Kármán equations yield only the trivial solution, $F\equiv\beta\equiv 0$.

Other, more elaborate comparisons between the plate theories of Föppl-Kármán and Reissner have been made by Weinberg (1962), Hart & Evans (1964), and Reissner (1978). We highlight some of these.

Weinberg has studied the postbuckling behavior of an axiplate under a compressive, horizontal edge thrust, $-H_0$. In analyizing this problem using the Föppl-Kármán equations, Friedrichs & Stoker (1941, 1942) showed that if b is the radius of the plate, then there is a narrow boundary-layer of width $O(H_0 b^3/D)$ in which the horizontal stress resultant changes from $-H_0$ to its constant interior value, H_0. Arguing that the solutions of Reissner's equations must exhibit a similar boundary layer, Weinberg (in our notation) sets $\beta=\lambda G$, where $\lambda\equiv bH_0\sqrt{A/D}$, writes down the boundary-layer equations, and expands all trigonometric functions in powers of λ. The resulting equations are then solved by a combination of analytic and numeric methods. The end product is a set of series expansions in powers of λ^2 of various stresses and kinematic quantities at the edge and in the interior for simply-supported and clamped plates. In particular,

$$\textit{for simple-support}: N_\theta(b)=(H_0^{3/2}b/\sqrt{D})(-1.61428-0.07237\lambda^2+\cdots) \tag{S.73}$$

$$\textit{for clamping}: N_\theta(b)=(H_0^{3/2}b/\sqrt{D})(-0.35479+0.02601\lambda^2+\cdots)\,, \tag{S.74}$$

where N_θ is the hoop stress resultant. These results are independent of Poisson's ratio as a consequence of the boundary-layer approximation. As Weinberg notes, the errors in the Föppl-Kármán equations are small and most pronounced, as one would expect, for the simply-supported plate.

The compressive hoop stress suggests that, if the plate is sufficiently thin, it is liable to undergo secondary, *unsymmetric* buckling. This possibility, which has been analyized by Yanowitch (1956), is beyond the scope of this chapter. For further discussion of secondary buckling, see Cheo & Reiss (1974).

Hart & Evans (1964) consider an annular plate $(a\le r\le b)$ under vertical edge loads only $(rV=Pb,\ p_H=l=0)$. They denote the clamped boundary condition, $\beta=e_\theta=0$ at $r=a, b$, as case A and the partially clamped boundary condition, $\beta = N_r = 0$ at $r = a,\, b$, as case B. These cases are sketched in Fig. 5.6. First, Hart & Evans solve the Föppl-Kármán equations by singular perturbation techniques for large values of the dimensionless parameter

$$k^2 \equiv 12(1-\nu^2)\left[\frac{Pa}{EhR}\right]^{2/3}\left[\frac{R}{h}\right]^2 , \tag{S.75}$$

where $R \equiv \frac{1}{2}(a+b)$ is the mean radius of the plate. To a 1st approximation, the perturbation solutions for cases A and B consist of a sum of rapidly decaying boundary-layer solutions plus interior solutions that satisfy the Föppl membrane equations. The situation is complicated in case B because the membrane solution is singular at the edges.

Next, for a few moderate values of k, Hart & Evans solve numerically both the Föppl-Kármán and Reissner equations for several plate geometries and several thickness-to-mean-radius ratios. Results may be found in their last 11 figures and in their last 2 tables which we do not attempt to reproduce here. In conclusion, Hart & Evans state that

> It appears that the use of the von Karman equations in this problem gives results very close to those afforded by the Reissner equations. The maximum difference in solutions of 9 per cent in [the dimensionless radial stress] p and 4 per cent in [the dimensionless rotation function] q were recorded for a narrow ring at a very high radial edge stress ... well in excess of the usual elastic limit for steel. The von Karman equations appear to underestimate the small values of ... p corresponding to small values of k.

Reissner (1978) compares the solutions of his equations with those of Föppl-Kármán for relatively narrow annular plates made of a polar orthotropic material such that $e_r = k_r = 0$, $e_\theta = B_\theta N_\theta$, and $M_\theta = D_\theta k_\theta$. If the edges of the plate are free to move and acted upon by vertical forces only, Reissner finds that

> the use of small finite-deflection [i.e., moderate rotation] theory for this case amounts to an underestimation of the force necessary to produce an angular deflection of 45° by nearly 20 per cent.

A much simpler (though perhaps less realistic) picture of the differences in the solution of the Föppl-Kármán and simpified Reissner equations is gotten

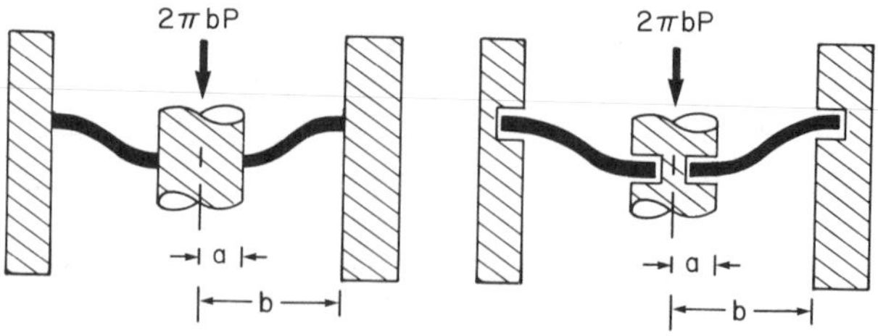

Fig. 5.6. Clamped and partially clamped annular plates under vertical edge loads.

if we consider an annular membrane with Poisson's ratio 1/3, free of surface loads ($p_H = p_V = 0$). The outer edge of the membrane is attached to a rigid hole of radius b while the inner edge is attached to a rigid but movable hub of radius a, which is acted upon by a downward vertical force of magnitude $2\pi bP$. This is a "worse case" analysis as the addition of bending stiffness can only reduce the angle through which radial fibers rotate.

With $rV = bP$ and the change of variables

$$r^2 = b^2 x \ , \ F = bPx^{-1/2} f \, , \tag{S.76}$$

(R.26) reduces to

$$4APx \frac{d^2 f}{dx^2} + 1 - \frac{f}{\sqrt{f^2 + x}} = 0 \, , \tag{S.77}$$

while (R.28), with $\nu = 1/3$, yields the boundary conditions

$$x \frac{df}{dx} = \frac{f}{2} + \frac{\sqrt{f^2 + x}}{6} \quad , \ x = \frac{a^2}{b^2}, 1 \, . \tag{S.78}$$

We now consider 2 different scalings of the variables in this nonlinear, 2-point boundary value problem.

First, we introduce the small, dimensionless parameter

$$\lambda \equiv (4AP)^{1/3} \tag{S.79}$$

and assume an expansion of the form

$$\tilde{f} \equiv \lambda f = \tilde{f}_0(x) + \lambda^2 \tilde{f}_1(x) + \cdots \, . \tag{S.80}$$

Substituting these expressions into (S.77) and (S.78), expanding the square roots, and equating to 0 coefficients of like powers of λ^2, we obtain an infinite sequence of boundary value problems, the first being

$$\frac{d^2 \tilde{f}_0}{dx^2} + \frac{1}{2\tilde{f}_0^2} = 0 \tag{S.81}$$

$$x \frac{d\tilde{f}_0}{dx} = \frac{2}{3} \tilde{f}_0 \ , \ x = \frac{a^2}{b^2}, 1 \, , \tag{S.82}$$

which are just the Föppl equations. Schwerin (1929) noted that (S.81) and (S.82) have the one-term solution

$$\tilde{f}_0 = (9/4)^{1/3} x^{2/3} \, . \tag{S.83}$$

A measure of the error in the Schwerin solution is the ratio of the 2nd term to the 1st in $\sqrt{\tilde{f}^2 + \lambda^2 x}$, i.e.,

$$\text{error} = \frac{\lambda^2 x}{\tilde{f}_0^2} = \left[\frac{4}{9}\right]^{2/3} \frac{\lambda^2}{x^{1/3}} \le \left[\frac{4b}{9a}\right]^{2/3} \lambda^2 . \tag{S.84}$$

Thus, the error in the Schwerin solution is small so long as $(a/b)^{2/3} \gg (b/a)^{1/3}\lambda^3 = O(\text{strains})$.

If $a/b = O(\lambda^3)$, we surmise that, to a 1st approximation, the solution of the Reissner membrane equations, (S.77) and (S.78), for $(a/b)^2 \le x \le 1$, will vary rapidly if $(a/b)^2 \le x \le \lambda^{3/2}$ and blend into the Schwerin solution on the interval $\lambda^{3/2} \le x \le 1$. To be more quantitative, we introduce a 2nd scaling of the variables in (S.77) by setting

$$f \equiv \lambda^3 y \ , \ x \equiv \lambda^6 z \ , \tag{S.85}$$

so obtaining

$$z\frac{d^2y}{dz^2} + 1 - \frac{y}{\sqrt{y^2+z}} = 0 . \tag{S.86}$$

In the special case $a = 0$, we have a circular membrane under a central, vertical point load and the solution of (S.86) has the form

$$y(z,\lambda) = -z\ln z + C(\lambda)z - (4/3)z^{3/2}\ln z + (4/3)[8/3 + C(\lambda)]z^{3/2} + \cdots \tag{S.87}$$

for small z, where C is an unknown function of λ.[14] Because $f = \lambda^{-1}\tilde{f} = \lambda^3 y$, y must blend into $\lambda^{-4}\tilde{f}$ as $z \to \infty$. This condition may be given mathematical form by introducing the *intermediate* variable

$$\eta \equiv \lambda^{-3/2} x = \lambda^{9/2} z , \tag{S.88}$$

expressing $\lambda^{-4}\tilde{f}(x,\lambda)$ and $y(z,\lambda)$ in terms of η, and then requiring, for *fixed* η, that these 2 solutions agree asymptotically as $\lambda \to 0$. That is,

$$y(\lambda^{-9/2}\eta, \lambda) \sim \lambda^{-4}\tilde{f}(\lambda^{3/2}\eta, \lambda) \quad \text{as } \lambda \to 0 \text{ , } \eta \text{ fixed} . \tag{S.89}$$

In view of (S.83), (S.89) yields, to lowest order,

$$y(\lambda^{-9/2}\eta, 0) \sim \lambda^{-3}(9/4)^{1/3}\eta \quad \text{as } \lambda \to 0 \text{ , } \eta \text{ fixed} , \tag{S.90}$$

or, in terms of z,

$$y(z, 0) \sim (9/4)^{1/3} z^{2/3} \quad \text{as } z \to \infty . \tag{S.91}$$

Fig. 5.7 displays the solution of (S.86) subject to the "boundary conditions" (S.87) and (S.91), together with Schwerin's 1-term solution of the Föppl equations for the limiting case $\lambda \to 0$. The corresponding graphs of

---

[14]Using several analytical/numerical methods and the symbol manipulating program MACSYMA, Rohn England has found that $C(0) = -1.8\ldots$

$$\beta = \cos^{-1}\left[\frac{y}{\sqrt{y^2+z}}\right] \tag{S.92}$$

are also shown in Fig. 5.7.

Other comparisons of solutions of the Föppl and Reissner membrane equations are given by Clark & Narayanaswamy (1967) and Weinitschke (1980). The first pair of authors consider an annular membrane, fixed at its outer edge, $r=b$, and acted upon along its inner edge, $r=a$, by a downward net vertical force of magnitude $2\pi P\,(b/a)$. Some interesting inequalities are derived and graphs are presented of stress, rotation, and displacement for $a/b=0.1$ and 0.5 and for several values of the large dimensionless parameter $2\pi bEh/P$.

Weinitschke (1980), using power series methods, has proved the existence of solutions of the Föppl equations for an annular membrane under normal pressure with either the radial displacement or the stress resultant, N_r, specified at the edges. This extends his earlier (1970) work on circular membranes. He presents many graphs of stress, rotation, and displacement for $a/b=0.1, 0.5$, and 0.9 and for different boundary conditions. Weinitschke (1980, 1987b), Grabmüller & Weinitschke (1986), and Grabmüller & Novak (1987a) have proved existence for these same problems in both the Föppl and Reissner theories, using alternative, integral equation methods. Various aspects of *uniqueness* of positive (tensile) solutions in circular and annular membranes in the Föppl and Reissner theories have been discussed by Grabmüller & Weinitschke (1986), Weinitschke (1987b), Grabmüller & Novak (1987b), and Grabmüller & Pirner (1987).

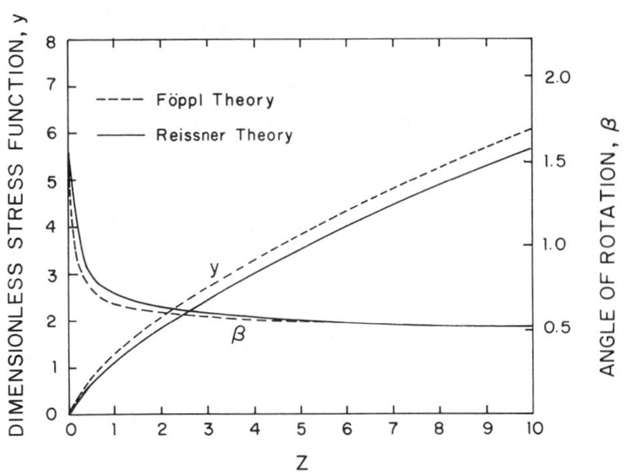

Fig. 5.7. Dimensionless stress function and angle of rotation of a point-loaded membrane as $\lambda=(4P/Eh)^{1/3}\to 0$ according to the Föppl and Reissner membrane theories.

3. Conical shells

Here, $\alpha=\alpha_0$ and $r=a+\sigma\cos\alpha_0$, where a is the minimum radius of a conical frustum. The simplified Reissner equations, (R.18) and (R.23), and the auxiliary conditions, (Q.24) and (R.11)-(R.13), specialize to

$$A\,[(a+\sigma\cos\alpha_0)F''+F'\cos\alpha_0-(a+\sigma\cos\alpha_0)^{-1}F\,]+\cos\alpha_0-\cos(\alpha_0+\beta)$$
$$+A\{[(a+\sigma\cos\alpha_0)^2p_H]'+\nu_e(a+\sigma\cos\alpha_0)p_T\}=0 \tag{S.93}$$

$$D\,[(a+\sigma\cos\alpha_0)\beta''+\beta'\cos\alpha_0-(a+\sigma\cos\alpha_0)^{-1}\sin\beta]-F\sin(\alpha_0+\beta)$$
$$+(a+\sigma\cos\alpha_0)[V\cos(\alpha_0+\beta)+l\,]=0 \tag{S.94}$$

$$H=(a+\sigma\cos\alpha_0)^{-1}F \tag{S.95}$$

$$M_\sigma=D\{\beta'+\nu_b(a+\sigma\cos\alpha_0)^{-1}[\sin(\alpha_0+\beta)-\sin\alpha_0]\} \tag{S.96}$$

$$e_\theta=A\{F'+(a+\sigma\cos\alpha_0)p_H$$
$$-\nu_e[(a+\sigma\cos\alpha_0)^{-1}F\cos(\alpha_0+\beta)+V\sin(\alpha_0+\beta)]\} \tag{S.97}$$

$$w=w(0)+\int_0^\sigma[\sin(\alpha_0+\beta)-\sin\alpha_0+e_\sigma\sin\alpha_0]ds\,. \tag{S.98}$$

We recover the equations for a cylindrical shell, (S.1)-(S.6), by setting $\alpha_0=\pi/2$, and those for a plate, (S.58)-(S.63), by setting $\alpha_0=0$.

The conical shell interpolates the cylindrical shell and the circular plate so that all the analytical challenges of the latter structures are compounded in the former. We shall discuss but one application of these equations, namely, to the study of the buckling of a family of axially loaded, simply-supported conical shells of equal area $\mathcal{A}$, ranging from cylindrical shells to circular plates. These equiareal shells have a maximum radius b, a slant height

$$L=[b-\sqrt{b^2-(\mathcal{A}/\pi)\cos\alpha_0}\,]\sec\alpha_0\,, \tag{S.99}$$

a minimum radius

$$a=\sqrt{b^2-(\mathcal{A}/\pi)\cos\alpha_0}\,, \tag{S.100}$$

and a net axial load of magnitude $2\pi bP$ so that $(a+\sigma\cos\alpha_0)V=bP$. The aim of the subsequent analysis is to produce curves of dimensionless axial load versus dimensionless end shortening, as shown in Fig. 5.8.

For a *steep conical shell*, we expect solutions that do not differ greatly from those for a cylindrical shell; that is, as the axial load is increased from 0, we expect the initially narrow edge-layers to extend further and further into the interior of the shell (where the linear membrane solution, $F_M=bP\cot\alpha_0$ predominates) until the layer from the wide edge engulfs the shell at the critical load, P_{cr}. This behavior and the structure of the governing simplified Reissner equations suggests that we introduce the scaled, dimensionless variables

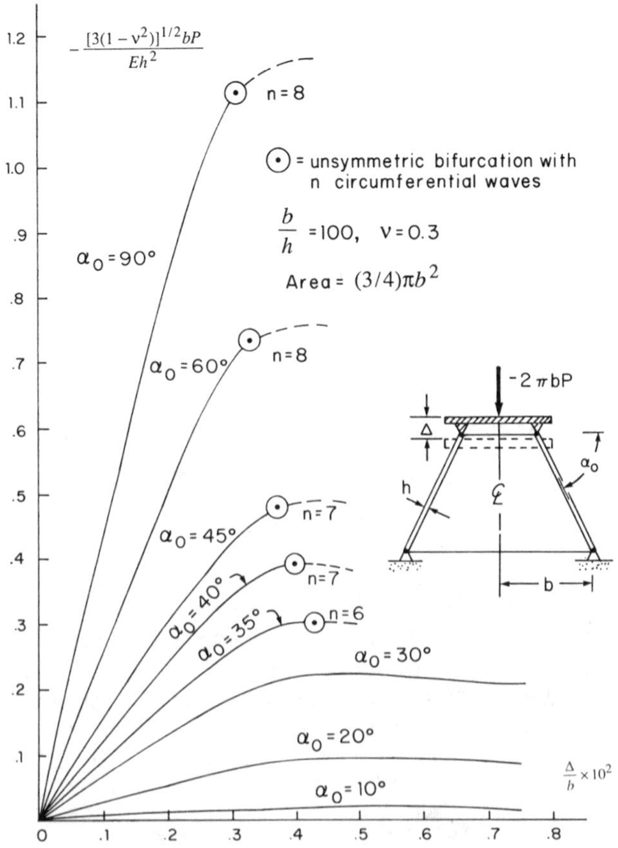

Fig. 5.8. Load versus end-shortening curves for several equiareal conical shells. (Calculations done by D. Bushnell using BOSOR5.)

$$F \equiv bP(\lambda f + \cot\alpha_0) \ , \ \beta \equiv \lambda\rho g \ , \ s \equiv (a/b)\sec\alpha_0 + \sigma/b \ , \qquad \text{(S.101)}$$

where

$$\lambda^4 \equiv \frac{DA}{b^2\sin^2\alpha_0} = \left[\frac{E_b}{E_e}\right]\left[\frac{h}{b\sin\alpha_0}\right]^2 , \ \rho \equiv bP\sqrt{A/D} \ . \qquad \text{(S.102)}$$

Substituting these expressions into (S.93), (S.94), and into the boundary conditions $M_\sigma = e_\theta = 0$ that follow from (S.96) and (S.97), we have

$$\lambda^2 \sin\alpha_0 \cos\alpha_0 \mathcal{L}f + s(\lambda\rho/2)^{-1}\sin(\lambda\rho g/2)\sin(\alpha_0 + \lambda\rho g/2) - \lambda = 0 \qquad \text{(S.103)}$$

$$\lambda^2 \sin\alpha_0 \cos\alpha_0 \{\mathcal{L}g + [g - (\lambda\rho)^{-1}\sin(\lambda\rho g)]\sec^2\alpha_0\} \\ - sf\sin(\alpha_0 + \lambda\rho g) - \lambda^{-1} s\sin(\lambda\rho g)\csc\alpha_0 = 0\,, \tag{S.104}$$

where

$$\mathcal{L}(\,) \equiv s^2(\,)_{,ss} + s(\,)_{,s} - (\,)\sec^2\alpha_0 \;,\; d(\,)/ds \equiv (\,)_{,s}\,, \tag{S.105}$$

is a linear, equidimensional differential operator and, at $s = (a/b,\ 1)\sec\alpha_0$,

$$\lambda s\sin\alpha_0\cos\alpha_0 f_{,s} - \nu_e[\lambda f\sin\alpha_0\cos(\alpha_0 + \lambda\rho g) + \cos(\lambda\rho g)] = 0 \tag{S.106}$$

$$s\cos\alpha_0 g_{,s} + \nu_b(\lambda\rho/2)^{-1}\sin(\lambda\rho g/2)\cos(\alpha_0 + \lambda\rho g/2) = 0\,. \tag{S.107}$$

With RO denoting *relative order of magnitude*, these differential equations take the form

$$\lambda^2\sin\alpha_0\cos\alpha_0\mathcal{L}f + sg\sin\alpha_0 - \lambda + \underline{\tfrac{1}{2}s\lambda\rho g^2\cos\alpha_0} = \mathrm{RO}(\lambda^2\rho^2 g^2) \tag{S.108}$$

$$\lambda^2\sin\alpha_0\cos\alpha_0\mathcal{L}g - sf\sin\alpha_0 - s\rho g\csc\alpha_0 - \underline{\tfrac{1}{2}s\lambda\rho fg\cos\alpha_0} = \mathrm{RO}(\lambda^2\rho^2 g^2)\,, \tag{S.109}$$

while the boundary conditions at $s = (a/b,\ 1)\sec\alpha_0$ become

$$\lambda s\sin\alpha_0\cos\alpha_0 f_{,s} - \nu_e[\lambda f\sin\alpha_0(\cos\alpha_0 - \underline{\lambda\rho g\sin\alpha_0}) + 1] = \mathrm{RO}(\lambda^2\rho^2 g^2) \tag{S.110}$$

$$s\cos\alpha_0 g_{,s} + \nu_b g(\cos\alpha_0 - \underline{\tfrac{1}{2}\lambda\rho g\sin\alpha_0}) = \mathrm{RO}(\lambda^2\rho^2 g^2)\,. \tag{S.111}$$

If $\lambda \ll 1$, i.e., if the shell is not too shallow, then neglect of the underlined and relative order terms leaves a set of linear equations with a relative error of $O(\lambda\rho g)$. Multiplying (S.109) by an unknown complex constant w^2, adding the resulting expression to (S.108), neglecting the underlined and RO-terms, and setting $y \equiv f + w^2 g$, we obtain the single, complex-valued, 2nd-order differential equation

$$\lambda^2\sin\alpha_0\cos\alpha_0\mathcal{L}y - sw^2 y\sin\alpha_0 - \lambda = 0\,, \tag{S.112}$$

provided we choose w to be any *one* of the 4 roots of

$$w^4 - \rho_* w^2 + 1 = 0 \;,\; \rho_* \equiv \rho\csc^2\alpha_0\,. \tag{S.113}$$

An equivalent but more complicated set of linear equations for the axisymmetric buckling of a conical shell was derived by Seide (1956), who took the components of displacement in the tangential and normal directions as his unknowns. However, Seide obtained *homogeneous* differential equations because he neglected (in our notation) the initial stress term $P/(s\sin\alpha_0\cos\alpha_0)$ on the left side of his 1st stress-strain relation. As we shall see, his conclusion that the net critical axial load, P_{cr}, of a cone

> is the bucking load of a cylinder having a thickness equal to the projection of the cone thickness on a plane perpendicular to the longitudinal axis

remains valid because the nonhomogeneous term in (S.112) cannot effect the results of a linear buckling analysis.

The solution of (S.112) may be expressed in terms of Bessel functions of order $2\cos\alpha_0$ and argument $2i\lambda^{-1}w\sqrt{s\sec\alpha_0}$. However, if the conical shell does not have an apex, i.e., if s is bounded away from 0, then standard methods of asymptotic integration of differential equations with a large parameter [Erdélyi (1956, Chapt. 4), Olver (1974, Chapt. 10)] yield essentially the same results in a simpler way. Thus, under the change of variables

$$y = s^{-1/4}Y \ , \ s = (t/2)^2 \ , \tag{S.114}$$

(S.112) reduces to the canonical form

$$Y_{,tt} + \left[\frac{w^2}{\lambda^2 \cos\alpha_0} + \frac{1}{t^2}\left(\frac{1}{4} - \frac{4}{\cos^2\alpha_0} \right) \right] Y + \frac{1}{\lambda (t/2)^{3/2} \sin\alpha_0 \cos\alpha_0} = 0 \,. \tag{S.115}$$

The ratio of the 2nd term in brackets to the 1st is

$$\frac{\lambda^2}{w^2}\left(\frac{\cos^2\alpha_0}{16} - 1 \right) \frac{b}{r} \,, \tag{S.116}$$

which is $O(\lambda^2)$, as long as $0 < a \le r$. It now follows from Erdélyi (1956, Section 4.3) that

$$\begin{aligned} f + w^2 g &= s^{-1/4} Y \\ &= s^{-1/4}[1 + O(\lambda)] \\ &\quad \times \{ C_1 \exp[-2w\lambda^{-1}(\sec\alpha_0 - \sqrt{s\sec\alpha_0})] \\ &\quad + C_2 \exp[-2w\lambda^{-1}(\sqrt{s\sec\alpha_0} - \sqrt{a/b}\sec\alpha_0)] - \lambda w^{-2} s^{-3/4} \csc\alpha_0 \} \end{aligned} \tag{S.117}$$

$$\begin{aligned} \lambda(f + w^2 g)_{,s} &= w s^{-3/4}\sqrt{\sec\alpha_0}\,[1 + O(\lambda)] \\ &\quad \times \{ C_1 \exp[-2w\lambda^{-1}(\sec\alpha_0 - \sqrt{s\sec\alpha_0})] \\ &\quad - C_2 \exp[-2w\lambda^{-1}(\sqrt{s\sec\alpha_0} - \sqrt{a/b}\sec\alpha_0] + O(\lambda^2) \} \,. \end{aligned} \tag{S.118}$$

If, for simplicity, we now set $\nu_e = \nu_b = \nu$, then the boundary conditions (S.110) and (S.111) imply, to lowest order, that $s\lambda(f + w^2 g)_{,s} \sin\alpha_0 = \nu$ at $s = (a/b,\ 1)\sec\alpha_0$. This yields 2 linear algebraic equations for C_1 and C_2, whose determinant reduces to

$$\exp[-4w\lambda^{-1}(1 - \sqrt{a/b})\sec\alpha_0] - 1 \tag{S.119}$$

and which vanishes if the argument of the exponential is a multiple of $2\pi i$. This can happen only if w is imaginary, i.e., with reference to (S.113) and Fig. 5.3 of subsection S.1, only if $\rho\csc^2\alpha_0 \le -2$. Since $a = b - L\cos\alpha_0$, it follows from (S.100), (S.113), and (S.119) that if w is imaginary, then

$$-\tfrac{1}{2}\rho\csc^2\alpha_0 = \tfrac{1}{2}(|w|^2 + |w|^{-2}) \ge 1 \,, \tag{S.120}$$

where

$$|w|^2 = (m\pi\lambda/2L)^2 b(\sqrt{b} + \sqrt{b - L\cos\alpha_0})^2 \ , \quad m = 1, 2, \cdots \ . \tag{S.121}$$

As equality holds in (S.120) if $w^2 = -1$, i.e., if

$$L = m\pi\lambda b\,[1 - (1/4)n\pi\lambda\cos\alpha_0]\ , \tag{S.122}$$

we define the critical value of the Reissner transition number to be $\rho_{cr} \equiv -2\sin^2\alpha_0$. By $(S.102)_2$, this yields the *net critical axial load*

$$P_{cr} = -\frac{2\sqrt{E_e E_b}\,h^2\sin^2\alpha_0}{b} = -\frac{Eh^2\sin^2\alpha_0}{\sqrt{3(1-\nu^2)}\,b}\ , \tag{S.123}$$

if we give E_e and E_b their classical values. A comparison of the last term in (S.123) with the last term in (S.33) for the cylindrical shell confirms Seide's statement quoted above.

Let us set $\mathcal{A}/\pi = (3/4)b^2$ so that, in the limiting case of an annular plate, $a/b = 0.5$. Then, from (S.99), $L = b\,[1 - \sqrt{1-(3/4)\cos\alpha_0}\,]\sec\alpha_0$ and (S.121) reduces to

$$|w|^2 = (m/\mu)^2\ , \tag{S.124}$$

where

$$\mu = \frac{2}{\pi}[12(1-\nu^2)]^{1/4}\left[\frac{b}{h}\right]^{1/2}\left\{\frac{1 - [1 - (3/4)\cos\alpha_0]^{1/4}}{\cos\alpha_0(\csc\alpha_0)^{1/2}}\right\}. \tag{S.125}$$

For a cylindrical shell, the factor in braces is 3/16; for a plate, it is 0.

In Fig. 5.9, we have graphed (S.120), with $|w|^2$ replaced by the right side of (S.124) and the minimum of the resulting expression taken over $m = 1, 2, \cdots$. Given the base radius, thickness, and elastic parameters of any member of the equiareal family of simply-supported conical shells, Fig. 5.9 allows us to compute the associated *axisymmetric, limit-point load* to within a relative error of $O(\lambda) = O(\sqrt{h/b})$.

As $\alpha_0 \to 0$, $\lambda \to \infty$ and the condition $\lambda \ll 1$, on which the preceding analysis was predicated, no longer obtains. In a shallow conical shell, more of the external vertical load is carried by the transverse shear stress resultant than in a steep conical shell of the same thickness. Moreover, the nonlinear terms in the strain-kinematic relations assume a greater importance.

To be more quantitative, we rewrite (S.93) and (S.94) and the boundary conditions $e_\theta = M_\sigma = 0$ [that follow from (S.97) and (S.96] in the form

$$A\mathcal{L}_r F + (\alpha_0 + \tfrac{1}{2}\beta)\beta = \mathrm{RO}(\alpha_0^2, \beta^2) \tag{S.126}$$

$$D\mathcal{L}_r\beta - F(\alpha_0 + \beta) + bP = \mathrm{RO}(\alpha_0^2, \beta^2) \tag{S.127}$$

$$rF_{,r} = \nu_e[F + bP(\alpha_0 + \beta)] + \mathrm{RO}(\alpha_0^2, \beta^2) \tag{S.128}$$

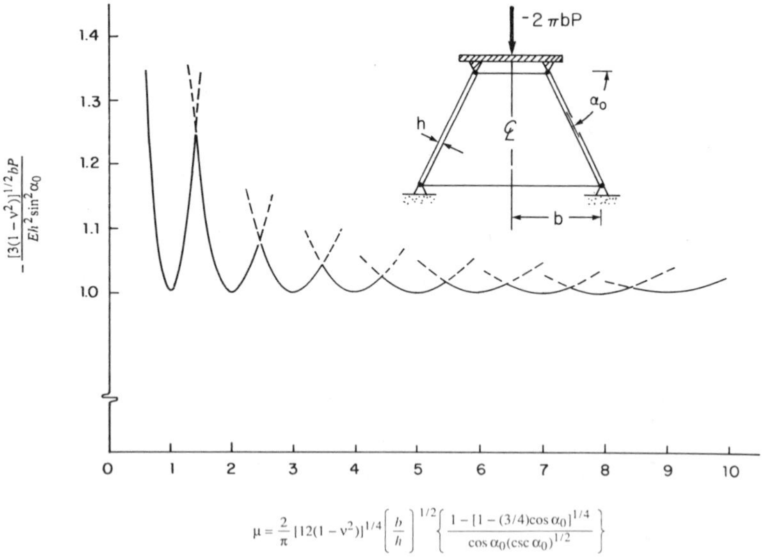

Fig. 5.9. Diagram for determining axisymmetric limit-point load of a conical shell [to within a relative error of $O(\sqrt{h/b})$].

$$r\beta_{,r} = -\nu_b\beta + RO(\alpha_0^2, \beta^2)\,, \tag{S.129}$$

where

$$\mathcal{L}_r(\,) \equiv r(\,)_{,rr} + (\,)_{,r} - r^{-1}(\,)\,. \tag{S.130}$$

With the relative error terms omitted, these are just the equations of nonlinear shallow shell theory.

An effective way of solving these equations numerically is to first recast them as *nonlinear integral equations.* To this end, we first introduce the dimensionless variables and parameters

$$F \equiv \sqrt{D/A}\,\alpha_0\hat{f}\ ,\ \ \beta \equiv \alpha_0\hat{g}\ ,\ \ r \equiv bx^2$$
$$x_* \equiv \sqrt{a/b}\ ,\ \ \hat{\lambda}^2 \equiv \frac{\sqrt{DA}}{b\alpha_0}\ ,\ \ \hat{\rho} \equiv \frac{bP\sqrt{A/D}}{\alpha_0^2}\,. \tag{S.131}$$

Substituting these expressions into (S.126)-(S.130) and, for simplicity, setting $\nu_e = \nu_b = \nu$, we obtain a pair of differential equations and boundary conditions that may be written in the following complex-valued form:

$$(\hat{\lambda}^2\mathcal{B} - i4x)w = -4x(i\hat{\rho} + \tfrac{1}{2}\hat{g}^2 - i\hat{f}\hat{g})\ ,\ \ x_* < x < 1 \tag{S.132}$$

$$xw_{,x} = 2\nu(\bar{w} + \alpha_0^2\hat{\rho})\ ,\ \ x = x_*, 1\,, \tag{S.133}$$

where

$$w \equiv \hat{f} + i\hat{g} \quad \text{and} \quad \mathcal{B}(\ ) \equiv x(\ )_{,xx} + (\ )_{,x} - 4x^{-1}(\ )\,, \tag{S.134}$$

$\mathcal{B}$ being a mnemonic for *Bessel operator*.

Let $G^{\pm}(x, t)$ denote the Green's functions that satisfy

$$(\hat{\lambda}^2 \mathcal{B} - i4x)G^{\pm}(x, t) = \delta(x - t) \ , \ x_* < x < 1 \tag{S.135}$$

and the homogeneous boundary conditions

$$xG_{,x}^{\pm}(x, t) = \pm 2\nu \overline{G}^{\pm}(x, t) \ , \ x = x_*, 1\,, \tag{S.136}$$

where $\delta(x - t)$ is the Dirac delta functional and an overbar denotes complex conjugation. Note that if $G^+ \equiv G(x, t; \nu)$, then $G^- = G(x, t; -\nu)$.

Starting from the identity

$$\int_{x_*}^{1} [w(x)(\hat{\lambda}^2 \mathcal{B} - i4x)G^{\pm}(x, t) - G^{\pm}(x, t)(\hat{\lambda}^2 \mathcal{B} - i4x)w(x)]dx$$

$$= \hat{\lambda}^2 \int_{x_*}^{1} \{w(x)[xG_{,x}^{\pm}(x, t)]_{,x} - G^{\pm}(x, t)[xw_{,x}(x)]_{,x}\}dx \tag{S.137}$$

$$= w(t) + 2\int_{x_*}^{1} [xG^{\pm}(x, t)(2i\hat{\rho} + \hat{g}^2(x) - 2i\hat{f}(x)\hat{g}(x)]dx\,,$$

we integrate on the left once by parts and introduce the boundary conditions (S.133) and (S.136) to obtain

$$\begin{aligned} w(t) = -2\{\hat{\rho}[\nu\hat{\lambda}^2 \alpha_0^2 G^{\pm}(x, t)|_{x_*}^{1} + 2i \int_{x_*}^{1} xG^{\pm}(x, t)dx] \\ + \int_{x_*}^{1} xG^{\pm}(x, t)[\hat{g}^2(x) - 2i\hat{f}(x)\hat{g}(x)]dx \\ + \nu\hat{\lambda}^2[G^{\pm}(x, t)\overline{w}(x) \mp \overline{G}^{\pm}(x, t)w(x)]_{x_*}^{1}\}\,. \end{aligned} \tag{S.138}$$

Note that

$$G^-\overline{w} + \overline{G}^- w = 2\mathcal{R}G^-\overline{w} \ , \ G^+\overline{w} - \overline{G}^+ w = i2\mathcal{I}G^+\overline{w}\,, \tag{S.139}$$

where $\mathcal{R}$ stands for "the real part of" and I stands for "the imaginary part of". Thus, if we take the real part of (S.138) with the *upper* sign and the imaginary part of (S.138) with the *lower* sign, we obtain equations of the form

$$\hat{f} = \hat{\rho}k^+ - \mathcal{K}^+(\hat{f}, \hat{g}) \ , \ \hat{g} = \hat{\rho}k^- - \mathcal{K}^-(\hat{f}, \hat{g})\,, \tag{S.140}$$

where

$$k^{\pm}(t) \equiv -2\nu\hat{\lambda}^2 \alpha_0^2 \,{}^{I}_{\mathcal{R}}G^{\pm}(x, t)|_{x_*}^{1} \pm 4\int_{x_*}^{1} [x \,{}^{I}_{\mathcal{R}}G^{\pm}(x, t)]dx \tag{S.141}$$

$$\mathcal{K}^{\pm}(t) \equiv 2\int_{x_*}^{1} [x \,{}^{\mathcal{R}}_{I}G^{\pm}(x, t)\hat{g}^2(x) \pm 2 \,{}^{I}_{\mathcal{R}}G^{\pm}(x, t)\hat{f}(x)\hat{g}(x)]dx\,. \tag{S.142}$$

Let

$$U(x) \equiv J_2(2\hat{\lambda}^{-1} i^{3/2} x) = \mathrm{ber}_2(2\hat{\lambda}^{-1} x) + i\,\mathrm{bei}_2(2\hat{\lambda}^{-1} x) \qquad \text{(S.143)}$$

$$V(x) \equiv Y_2(2\hat{\lambda}^{-1} i^{3/2} x) = \mathrm{ker}_2(2\hat{\lambda}^{-1} x) + i\,\mathrm{kei}_2(2\hat{\lambda}^{-1} x)\,. \qquad \text{(S.144)}$$

Then

$$G^+ = G(x, t; \nu) = \begin{cases} A_<(\nu)U(x) + B_<(\nu)V(x) \ , & x_* < x < t \\ A_>(\nu)U(x) + B_>(\nu)V(x) \ , & t < x < x_* \end{cases} . \qquad \text{(S.145)}$$

The 4 complex constants, $A_<(\nu)$, $B_<(\nu)$, $A_>(\nu)$, and $B_>(\nu)$, are determined from the set of 4 linear algebraic equations that follow from the standard conditions on a Green's function, namely,

$$x_* G,_x^+(x_*, t) - 2\nu \overline{G}^+(x_*, t) = 0 \ , \ G,_x^+(1, t) - 2\nu \overline{G}^+(1, t) = 0 \qquad \text{(S.146)}$$

$$G^+(t+, t) - G^+(t-, t) = 0 \ , \ G,_x^+(t+, t) - G,_x^+(t-, t) = 1\,. \qquad \text{(S.147)}$$

The Green's function G^- is obtained from (S.145) by replacing ν by $-\nu$. A numerical solution of (S.140) and, in particular, the calculation of the critical load parameter, $\hat{\rho}_{cr}$, has not been done as far as we know.

Wempner (1958) has noted that:

> The case in which both edges are restrained from radial displacement is of little importance, since this constitutes a very stiff spring, in which inelastic deformations accompany any appreciable deflection.

Thus, in the analysis of so-called Belleville springs, it is more realistic to assume that one or both of the edges are free to move horizontally; the reader should have no trouble in making the appropriate modifications of the boundary conditions (S.133).

Wempner (1964) has studied the axisymmetric deformations of a shallow conical shell whose edges are either (I) free of radial loads and couples or (II) have a rigid plug preventing radial displacement at the inner edge. Wempner takes the radial displacement, u, and the rotation, β, as his unknowns. In effect, he constructs a Green's function for the operator $\hat{\lambda}^2 \mathcal{B}$ instead of for the operator $\hat{\lambda}^2 \mathcal{B} - i4x$, as we do. His resulting integral equations have algebraic rather than transcendental kernels, but whereas our integral operators $\mathcal{K}^{\pm}$ in (S.140) act on only nonlinear terms, Wempner's act on linear and nonlinear terms, possibly slowing the rate of convergence of an iterative method.

Wempner solves his integral equations numerically and plots load-deflection curves using the following data:

$$a/b = 0.5 \ , \ h/b = 0.05 \ , \ \alpha_0 = 0.1 \ , \ \nu = 0.3\,. \qquad \text{(S.148)}$$

If both edges are free, Wempner obtained a monotonically increasing curve whereas if the inner edge is simply-supported and the outer edge free, he finds a critical (limit-point) load of

$$\frac{(1-\nu^2)P}{Eh} = 27.6\times 10^{-6} \qquad \text{(S.149)}$$

which converts, in our notation, to $\hat{\rho}_{cr} = -0.200$.

As we stressed earlier, a complete—or should we say reliable—buckling analysis should consider the possibility of unsymmetric deformation and the effect of slight imperfections, though such an analysis is beyond the scope of this book. Fig. 5.8 indicates when and why such considerations are necessary.

4. Spherical shells and membranes

Here, $r = R\sin\alpha$ and $\sigma = R\alpha$ so that (R.18), (R.23) and the associated auxiliary equations, (Q.24) and (R.11)-(R.13), reduce to

$$AR^{-1}(F''\sin\alpha + F'\cos\alpha - F\csc\alpha) + \cos\alpha - \cos(\alpha+\beta) + AR\,[(p_H\sin^2\alpha)' + \nu_e p_T \sin\alpha] = 0 \qquad \text{(S.150)}$$

$$DR^{-1}(\beta''\sin\alpha + \beta'\cos\alpha - \sin\beta\csc\alpha) - F\sin(\alpha+\beta) + R\,[V\cos(\alpha+\beta) + l\,]\sin\alpha = 0 \qquad \text{(S.151)}$$

$$H = R^{-1}F\csc\alpha \qquad \text{(S.152)}$$

$$M_\sigma = DR^{-1}\{\beta' + \nu_b[\csc\alpha\sin(\alpha+\beta) - 1]\} \qquad \text{(S.153)}$$

$$e_\theta = AR^{-1}\{F' + R^2 p_H \sin\alpha - \nu_e[F\csc\alpha\cos(\alpha+\beta) + VR\sin(\alpha+\beta)]\} \qquad \text{(S.154)}$$

$$w = w(0) + R\int_0^\alpha [2\sin(\beta/2)\cos(\hat{\alpha}+\beta/2) + e_\sigma \sin\hat{\alpha}]\,d\hat{\alpha}\,, \qquad \text{(S.155)}$$

where a prime now denotes differentiation with respect to α.

We shall focus on eversion, concentrated loading, and pressure loading, and merely mention other problems that have been considered in the literature.

(a) Eversion of a hemispherical shell. The solution of this problem is easy to realize physically by bisecting a thin rubber ball and turning one of the halves inside out. It will be observed that the everted shape is again hemispherical, except for a narrow, outward turning lip at the edge where one can literally get the feel of a boundary layer solution. Even though the ball is made of rubber and even though the intermediate stages of deformation may (but need not) involve large strains, the strains in the final equilibrium shape are small and axisymmetric. Thus, the simplified Reissner equations are applicable. If a uniform bending moment of the proper magnitude is then applied along the edge of the shell, the lip disappears and the shape is that of a complete spherical shell that has been everted (were this possible physically). Antman (1979) has analyized this latter problem using 3-dimensional elasticity theory.

By inspection, the *perfect eversion* $\beta=-2\alpha, F=0$ is a solution of (S.150) and (S.151), if there are no distributed loads. We expect this solution—which, by (S.153), predicts that $M_\sigma=-2(1+\nu_b)DR^{-1}$ everywhere in the shell—to be quite accurate except in a narrow boundary layer whose angular width, as in linear theory, should be $O(AD/R^2)^{1/4}=O(\sqrt{h/R})$. Thus, to obtain solutions that satisfy the stress-free boundary conditions $H(\pi/2)=M_\sigma(\pi/2)=0$, we set

$$\alpha=\pi/2-\varepsilon x \ , \ F=Df/\varepsilon R \ , \ \beta=-2\alpha+\varepsilon g \ , \ \varepsilon^2\equiv\sqrt{AD}/R \qquad \text{(S.156)}$$

and assume regular perturbation expansions of the form

$$f(x,\varepsilon)=f_0+\varepsilon f_1+\cdots \quad , \quad g(x,\varepsilon)=g_0+\varepsilon g_1+\cdots . \qquad \text{(S.157)}$$

Substituting (S.156) and (S.157) into the simplified Reissner equations, (S.150) and (S.151), and the stress free boundary conditions that follow from (S.152) and (S.153), and setting $p_H=p_T=l=0$, we find, to lowest order, that

$$\frac{d^2f_0}{dx^2}-g_0=0 \ , \ \frac{d^2g_0}{dx^2}+f_0=0 \qquad \text{(S.158)}$$

$$f_0(0)=0 \ , \ \frac{dg_0}{dx}(0)=-2(1+\nu_b) . \qquad \text{(S.159)}$$

The solution of these equations that decays as $x\to\infty$ is

$$f_0=-2\sqrt{2}(1+\nu_b)e^{-x/\sqrt{2}}\sin(x/\sqrt{2}) , g_0=2\sqrt{2}(1+\nu_b)e^{-x/\sqrt{2}}\cos(x/\sqrt{2}) . \qquad \text{(S.160)}$$

Note from (S.154) that the hoop strain at the edge is $e_\theta(\pi/2)=2\varepsilon^2(1+\nu_b)=O(h/R)$.

(b) Eversion of a shallow spherical cap. An everted spherical shell, if sufficiently shallow, may snap back to its original shape. The problem of determining the edge angle α_1 for which the inverted shape is neutrally stable is difficult to analyze because the smaller α_1, the further the boundary layer penetrates the interior, and because nonlinear effects are important. On the other hand, the governing equations, to a 1st approximation, are those of shallow shell theory as we shall now show by appropriately scaling the simplified Reissner equations and setting a certain small parameter to 0.

Let

$$\alpha\equiv\alpha_1 x \ , \ \beta=-\alpha_1 g \ , \ F\equiv-\alpha_1\sqrt{D/A}f \ , \ \lambda^2=\frac{\alpha_1^2 R}{\sqrt{AD}} . \qquad \text{(S.161)[15]}$$

Then, (S.150) and (S.151), with the external loads set to 0, take the dimensionless form

[15] λ^2, which is proportional to the rise-to-thickness ratio, is a standard parameter in shallow shell theory.

$$\frac{\sin(\alpha_1 x)}{\alpha_1}\frac{d^2 f}{dx^2}+\cos(\alpha_1 x)\frac{df}{dx}-\frac{\alpha_1 f}{\sin(\alpha_1 x)}$$
$$+\lambda^2\frac{\sin(\alpha_1 x)\sin(\alpha_1 g)-2\sin^2(\alpha_1 g/2)\cos(\alpha_1 x)}{\alpha_1^2}=0 \tag{S.162}$$

$$\frac{\sin(\alpha_1 x)}{\alpha_1}\frac{d^2 g}{dx^2}+\cos(\alpha_1 x)\frac{dg}{dx}-\frac{\sin(\alpha_1 g)}{\sin(\alpha_1 x)}-\lambda^2 f\frac{\sin\alpha_1(x-g)}{\alpha_1}=0\,. \tag{S.163}$$

The associated boundary conditions of no edge loads follow from (S.152), (S.153), and (S.161) as

$$f(1)=0\,,\ \frac{dg}{dx}(1)+\nu_b\left\{\frac{\cos\alpha_1\sin[\alpha_1 g(1)]+2\sin\alpha_1\sin^2[\alpha_1 g(1)/2]}{\sin\alpha_1}\right\}=0\,. \tag{S.164}$$

In addition, there are regularity conditions at $x=0$. To within a relative error of $O(\alpha_1^2)$, (S.162) and (S.163) reduce to the pair of coupled nonlinear ordinary differential equations

$$x\frac{d^2 f}{dx^2}+\frac{df}{dx}-\frac{f}{x}+\lambda^2 g(x-\tfrac{1}{2}g)=0 \tag{S.165}$$

$$x\frac{d^2 g}{dx^2}+\frac{dg}{dx}-\frac{g}{x}-\lambda^2 f(x-g)=0 \tag{S.166}$$

on the interval $0<x<1$. The associated boundary conditions at the free edge $x=1$ reduce to

$$f(1)=0\ ,\ \frac{dg}{dx}(1)+\nu_b g(1)=0\,. \tag{S.167}$$

The change of variables $\hat{f}=\lambda f$, $\hat{g}=\lambda g$, $\xi=\lambda x$ will remove the parameter λ from the above differential equations and put it in the boundary conditions.

We could attempt to find the value λ_{CR} at which spontaneous snap-back occurs by solving the above boundary value problem and finding when the second variation of the strain-energy vanishes or, equivalently, when the load apex-deflection curve has a horizontal inflection point. Instead, we look to the existing literature where the solution to this special problem may be gleaned from the solutions to various snap-buckling problems.

Chien & Hu (1957) give an approximate analytical solution for the axisymmetric snapping of a shallow spherical cap (1) under uniform pressure and free of edge moments and horizontal forces and (2) under a uniform edge moment, again with no horizontal edge force. Snapping pressures and moments as well as the (smaller) pressures and moments necessary to snap the shell back to its original shape are computed. From the lower dashed curve in Fig. 3 of that paper, computed for a Poisson's ratio of 1/3, we read off 8.1 as the value of twice the rise to thickness ratio at which *no pressure* is required to snap the shell back. This yields $\lambda_{CR}^2 \approx 4\sqrt{2/3}\,(8.1) \approx 26.5$. Unfortunately, the curve of the "backward snapping moment," m_b, versus λ, given by Chien & Hu in their Fig. 7 (where our λ is denoted by k), does not extend to $m_b = 0$. However, making a smooth extrapolation, we obtain $\lambda_{CR}^2 \approx (4.25)^2 \approx 18$. These 2 values of λ_{CR}^2 should agree. We have not attempted to determine if the discrepancy is consistent with, and due to, the approximations employed.

A third estimate of λ_{CR}^2 may be inferred from Mescall's (1965) calculations of the snapping of an unconstrained, shallow spherical shell under a concentrated load. From his Fig. 2, we conclude that $\lambda_{CR}^2 \approx 33$. We tend to favor Mescall's result, which is based on a *numerical* solution of the shallow shell equations, because his companion analysis of a clamped shallow spherical shell under a concentrated load has been corroborated by Fitch (1968).[16]

The last numerical results we cite may be read off from Fig. 4 of a paper by Brodland & Cohen (1986) who, using a Mooney-Rivlin strain-energy density, have computed load-deflection curves for shallow spherical caps, pointed-loaded at the apex and freely supported at the boundary. If the strains are small, the material is linear with a Poisson's ratio of ½. By a slight interpolation, one obtains a value of λ_{CR}^2 that agrees with Mescall's.

(c) Spherical shells under concentrated radial loads. Here, $p_H = p_T = l = 0$ and $V = -(P/2\pi R)\csc\alpha$, where P is the magnitude of an *inward* radial point load at the south pole. If the shell is complete, there is an equilibrating inward point load at the north pole; if the shell is a segment, there is an equilibrating edge force. The boundary conditions at the south pole are

$$(F, \beta) \to 0 \text{ as } \alpha \to 0 . \tag{S.168}$$

These conditions guarantee, respectively, that the extensional and bending strain energies will be finite in a neighborhood of the concentrated load.

Henceforth, we consider a complete spherical shell; by symmetry, its equator can move only in a fixed horizontal plane. In this case, if we take the origin of coordinates to coincide with the center of the undeformed shell, we can use (S.154) and (S.155) to express the radial and vertical displacements of a particle as

$$u = A\{F'\sin\alpha - \nu_e[F\cos(\alpha+\beta) - (P/2\pi)\sin(\alpha+\beta)]\} \tag{S.169}$$

$$w = -R\int_{\alpha}^{\pi/2} [2\sin(\beta/2)\cos(\hat{\alpha}+\beta/2) + e_\sigma \sin\hat{\alpha}]d\hat{\alpha} . \tag{S.170}$$

Remarkably, the *linear membrane theory* solution,

---

[16]One might speculate that the different results are due the use of limiting processes with different types of loads, rather than to calculational errors. If so, there might exist a process-independent maximum value of λ_{CR}^2. We leave this question unanswered.

$$F_{LM} = -(P/2\pi)\cot\alpha\ ,\ \ \beta_{LM} = 0\ , \tag{S.171}$$

for a shell under equal and opposite point loads, is also an *exact* solution of the full, simplified Reissner equations, (S.150) and (S.151). However, F_{LM} is too singular at the poles to meet the first of the boundary conditions (S.168). (While *nonlinear membrane theory*, by permitting a cusp at the poles, leads to zero values of F there, a consideration of equilibrium in vertical and normal directions shows that the meridional and hoop stresses have opposite signs throughout the shell. Thus, a true spherical membrane will always wrinkle under a radial point load. (For further discussion of wrinkling in spherical membranes, see Section V.T.) However, we *do* expect the linear membrane solution to be an "interior" solution of the full equations, in some sense. This suggests the change of variable

$$F \equiv -(P/2\pi)(\cot\alpha + \phi)\ , \tag{S.172}$$

which reduces (S.150) and (S.151) to the form

$$-\varepsilon^4 p\mathcal{L}\phi + \sin\alpha\sin\beta + 2\cos\alpha\sin^2(\beta/2) = 0 \tag{S.173}$$

$$\mathcal{L}\beta + (\beta - \sin\beta)\csc\alpha + p[\phi\sin(\alpha+\beta) + \csc\alpha\sin\beta] = 0\ , \tag{S.174}$$

where

$$\mathcal{L} \equiv \sin\alpha\frac{d^2}{d\alpha^2} + \cos\alpha\frac{d}{d\alpha} - \csc\alpha \tag{S.175}$$

and

$$\varepsilon^2 \equiv \frac{\sqrt{AD}}{R} = O\left[\frac{h}{R}\right]\ ,\ p \equiv \frac{PR}{2\pi D} = O\left[\frac{PR}{Eh^3}\right]\ . \tag{S.176}$$

The dimensionless load p can range through through all positive numbers, but ε is always small. The boundary conditions for these equations are

$$(\phi, \beta) \to (-\alpha^{-1}, 0)\ \text{ as }\ \alpha \to 0 \tag{S.177}$$

and, by symmetry,

$$\phi = \beta = 0\ \text{ at }\ \alpha = \pi/2\ . \tag{S.178}$$

A closed-form solution of (S.173) and (S.174) seems impossible. However, by exploiting the presence of the small parameter ε and considering, separately, cases when p is $O(\varepsilon)$, $O(1)$, or $O(\varepsilon^{-1})$, we can obtain simplified, accurate asymptotic approximations to these 2 equations.

(i) $p = O(\varepsilon)$

In this case, the loads are relatively small so we expect linear theory to emerge as a 1st-approximation. Thus, with the scaling

$$\beta \equiv -\varepsilon^2 p\gamma\ , \tag{S.179}$$

(S.173) and (S.174) take the linearized form

$$\varepsilon^2 \mathcal{L}\phi + \gamma \sin\alpha = 0 \tag{S.180}$$

$$\varepsilon^2 \mathcal{L}\gamma - \phi \sin\alpha + \underline{p\gamma \csc\alpha} = 0\,. \tag{S.181}$$

The linearized form of (S.170) is

$$w = \varepsilon^2 pR \int_{\alpha}^{\pi/2} [\gamma \cos\hat{\alpha} + \underline{O(\varepsilon^2)}] d\hat{\alpha}\,. \tag{S.182}$$

It will become apparent in the next subsection that the relative errors in (S.180)-(S.182) resulting from linearization are O(p); thus, it is consistent to ignore the underlined terms in (S.181) and (S.182). Doing so, we may combine (S.180) and (S.181) into the single, complex-valued equation

$$(\varepsilon^2 \mathcal{L} - i \sin\alpha)(\phi + i\gamma) = 0\,, \tag{S.183}$$

whose solution can be expressed in terms of the Legendre functions P_ν^1 and Q_ν^1, where $\nu(\nu+1) = -i\varepsilon^{-2}$.[17] For a complete sphere,

$$\phi + i\gamma = (\pi/2)\csc(\pi\nu)[P_\nu^1(-\cos\alpha) - P_\nu^1(\cos\alpha)] \tag{S.184}$$

or, with the aid of equations **3.4**(14) and **3.6**(6) of Erdélyi *et al* (1953),

$$\begin{aligned} \phi + i\gamma &= \frac{d}{d\alpha}[Q_\nu(\cos\alpha) - (\pi/2)\cot(\pi\nu/2) P_\nu(\cos\alpha)] \\ &= \frac{d}{d\alpha}\{Q_\nu(\cos\alpha) + i(\pi/2)[1 + O(e^{-\sqrt{|\nu|/2}})] P_\nu(\cos\alpha)\}\,, \quad 0 < \alpha < \pi/2\,. \end{aligned} \tag{S.185}$$

[A slightly different form of solution in terms of a stress function and the mid-surface radial displacement is given by Koiter (1963a).]

The Legendre functions are difficult to compute. However, because $|\nu| \gg 1$, the following expansion developed by Szegö (1934) is particularly useful:

$$\begin{aligned} &Q_\nu(\cos\alpha) + i(\pi/2)P_\nu(\cos\alpha) \\ &\quad = (\alpha \csc\alpha)\{K_0[-(\nu + ½)i\alpha] \\ &\qquad - i(1/8)(\cot\alpha - \alpha^{-1})(\nu + ½)^{-1} K_1[-(\nu + ½)i\alpha] + \cdots\}\,, \end{aligned} \tag{S.186}$$

where $K_0, \cdots$ are modified Bessel functions. To obtain a "clean" expansion, we modify (S.183) by adding the negligible term $(1/4)\sin\alpha$ to the operator on the left[18] so that we may set $\nu + ½ = \varepsilon^{-1} e^{3i\pi/4}$. Thus, from (S.185) and (S.186),

---

[17]The reader should have no difficulty in recognizing when ν stands for Poisson's ratio and when, as here, it stands for the degree of a Legendre function.

[18]Such terms are precisely of the type that Simmonds (1975) showed rigorously to be negligible in linear theory.

$$\phi+i\gamma=\frac{d}{d\alpha}\left\{\sqrt{\alpha\csc\alpha}\left[\ker\alpha/\varepsilon+\frac{\varepsilon}{\sqrt{2}}\left[\frac{\alpha\cos\alpha-\sin\alpha}{8\alpha\sin\alpha}\right](\mathrm{kei}_1\alpha/\varepsilon-\ker_1\alpha/\varepsilon)+O(\varepsilon^2)\right]\right.$$
$$\left.+i\left[\mathrm{kei}\,\alpha/\varepsilon-\frac{\varepsilon}{\sqrt{2}}\left[\frac{\alpha\cos\alpha-\sin\alpha}{8\alpha\sin\alpha}\right](\mathrm{kei}_1\alpha/\varepsilon+\ker_1\alpha/\varepsilon)+O(\varepsilon^2)\right]\right\}$$

$$\sim\frac{d}{d\alpha}(\ker\alpha/\varepsilon+i\,\mathrm{kei}\,\alpha/\varepsilon)\ \text{as}\ \varepsilon\to 0\ ,\ 0<\alpha<\pi/2\,. \tag{S.187}$$

This last line is Reissner's (1947) shallow shell solution. Alternatively, we could have obtained this result directly, as does Leckie (1961), by applying Langer's (1935) method to (S.183). The asymptotic forms of the ker and kei functions for large values of their arguments show that *the deformation is confined to a boundary layer of width $O(\sqrt{h/R})$, centered at the poles*. Moreover, from (S.182) and (S.187), the vertical deflection is

$$w=-\varepsilon^2 pR\,\mathrm{kei}\,\alpha/\varepsilon\ ,\ 0<\alpha<\pi/2\,, \tag{S.188}$$

to within a transcendentally small error. Setting $\varepsilon^2 R=\sqrt{AD}$ and noting that kei $(0)=-\pi/4$, we obtain the dimensionless *linear load-displacement relation*

$$p=\frac{4}{\pi}\frac{w(0)}{\sqrt{AD}}=O\left[\frac{w(0)}{h}\right]. \tag{S.189}$$

(ii) $p=O(1)$

The results of the preceding subsection show that if the dimensionless load p is small, then equal and opposite, radially inward point loads on a complete spherical shell produce a small inward dimple in an immediate neighborhood of the point loads. The picture that emerges from the experimental work of Evan-Iwanovski, Cheng, & Loo (1962), the numerical work of Bushnell (1967), the approximate analytic work of Ashwell (1960), the asymptotic/numerical work of Wan (1980a, 1980b, 1984), Wan & Ascher (1980), Wan & Yun (1985), Parker & Wan (1984), and Ranjan & Steele (1977) is that, unless there is asymmetric bifurcation, this dimple will grow as the loads increase. However, because the strains must remain small, the dimple will resemble an inverted spherical cap, except in a neighborhood of the point load or in a neighborhood of the "edge" of the inverted cap. (The radius of this edge must be determined as part of the solution.) Thus, as the load grows, linear theory gives way to nonlinear *shallow shell theory* which, in turn, gives way to deep shell theory.

Considering first the range where shallow shell theory comes into play, we introduce into the simplified Reissner equations, (S.150) and (S.151), the change of variables

$$F=-\varepsilon p\sqrt{D/A}\hat{f}\ ,\ \beta=-\varepsilon p\hat{g}\ ,\ \alpha=\varepsilon\xi\,, \tag{S.190}$$

so obtaining

$$\frac{\sin(\varepsilon\xi)}{\varepsilon}\frac{d^2\hat{f}}{d\xi^2}+\cos(\varepsilon\xi)\frac{d\hat{f}}{d\xi}-\frac{\varepsilon\hat{f}}{\sin(\varepsilon\xi)}+\frac{\sin(\varepsilon\xi)\sin(\varepsilon p\hat{g})-2\cos(\varepsilon\xi)\sin^2(\varepsilon p\hat{g}/2)}{p\varepsilon^2}=0 \tag{S.191}$$

$$\frac{\sin(\varepsilon\xi)}{\varepsilon}\frac{d^2\hat{g}}{d\xi^2}+\cos(\varepsilon\xi)\frac{d\hat{g}}{d\xi}-\frac{\sin(\varepsilon p\hat{g})}{p\sin(\varepsilon\xi)}-\hat{f}\,\frac{\sin\varepsilon(\xi-p\hat{g})}{\varepsilon}+\cos\varepsilon(\xi-p\hat{g})=0\,, \tag{S.192}$$

where ε and p are given by (S.176). If ξ, $\hat{g}$, and p are O(1), then, to within a relative error of $O(\varepsilon^2)$, our differential equations take the form

$$\xi\frac{d^2\hat{f}}{d\xi^2}+\frac{d\hat{f}}{d\xi}-\frac{\hat{f}}{\xi}+\hat{g}(\xi-\tfrac{1}{2}p\hat{g})=0 \tag{S.193}$$

$$\xi\frac{d^2\hat{g}}{d\xi^2}+\frac{d\hat{g}}{d\xi}-\frac{\hat{g}}{\xi}-\hat{f}(\xi-p\hat{g})+1=0\,. \tag{S.194}$$

The associated boundary conditions, from (S.168), (S.171), (S.176), and (S.190), are

$$(\hat{f},\hat{g})\sim 0,0)\ \text{as}\ \xi\to 0\ \text{and}\ (\hat{f},\hat{g})\sim(\xi^{-1},0)\ \text{as}\ \xi\to\infty\,. \tag{S.195}$$

These equations are difficult to solve if p is O(1) because then we can neither linearize nor use singular perturbation methods. In view of the singularity at $\xi=0$, the best approach in this case, both theoretically and numerically, is to convert the boundary value problem (S.193)-(S.195) into a *complex-valued, nonlinear integral equation*. To this end, we multiply (S.194) by the imaginary unit i and add the resulting equation to (S.193) to obtain

$$\mathcal{B}(\hat{f}+i\hat{g})=-i+p\hat{g}(\tfrac{1}{2}\hat{g}-i\hat{f})\equiv -i+pN(\hat{f},\hat{g})\,, \tag{S.196}$$

where

$$\mathcal{B}\equiv\frac{d}{d\xi}\left[\xi\frac{d}{d\xi}\right]-\frac{1}{\xi}-i\xi \tag{S.197}$$

is a Bessel operator.

Let $G(\xi,\tau)$ denote the *Green's function* that satisfies the differential equation

$$\mathcal{B}G=\delta(\xi-\tau)\ ,\ 0<\xi<\infty\,, \tag{S.198}$$

the homogeneous boundary conditions

$$G(\xi,\tau)\sim 0\ \text{as}\ \xi\to 0\ \text{and}\ G(\xi,\tau)\sim 0\ \text{as}\ \xi\to\infty\,, \tag{S.199}$$

and the jump conditions

$$\frac{dG}{d\xi}(\tau+,\tau)-\frac{dG}{d\xi}(\tau-,\tau)=\frac{1}{\tau}\ ,\ G(\tau+,\tau)-G(\tau-,\tau)=0 \qquad (S.200)$$

implied by the delta function in (S.198). It follows that $G(\xi,\tau)=G_<(\xi,\tau)$ if $\xi<\tau$ and $G(\xi,\tau)=G_<(\tau,\xi)$ if $\tau<\xi$, where, with $i^{1/2}=e^{i\pi/4}$,

$$\begin{aligned}G_<(\xi,\tau)&\equiv -K_1(i^{1/2}\tau)I_1(i^{1/2}\xi)\\ &=-[\mathrm{ker}_1\,\tau\,\mathrm{ber}_1\,\xi-\mathrm{kei}_1\,\tau\,\mathrm{bei}_1\,\xi+i(\mathrm{ker}_1\,\tau\,\mathrm{bei}_1\,\xi+\mathrm{kei}_1\,\tau\,\mathrm{ber}_1\,\xi)]\ ,\ \xi<\tau\,.\end{aligned} \qquad (S.201)$$

Thus, the solution(s) of (S.196) satisfy

$$\hat{f}(\xi;p)+i\hat{g}(\xi;p)=\xi^{-1}+(\mathrm{ker}\,\xi+i\mathrm{kei}\,\xi)'+p\int_0^\infty G(\tau,\xi)N(\hat{f}(\tau;p),\hat{g}(\tau;p))d\tau. \qquad (S.202)$$

The nonlinear shallow shell approximation to the net vertical deflection, relative to the fixed equatorial plane of the spherical shell, follows from (S.170) and (S.190) as

$$w=\varepsilon^2 pR\int_\xi^\infty \hat{g}(\hat{\xi};p)d\hat{\xi}\,. \qquad (S.203)$$

Fulton (1988) has solved the integral equation (S.202) numerically for a range of values of p. Fig. 5.10a displays graphs of $\hat{f}$ and $\hat{g}$ for $p=1$ and $p=5$; Fig. 5.10b displays the associated dimensionless vertical deflection computed from (S.203).

The asymptotic solution of (S.202) as $p\to\infty$ is of considerable interest because it represents an *elastic hinge* where there is a rapid transition from the nearly undeformed spherical shape to the inverted spherical shape. While this solution may be extracted from (S.202) using the asymptotic forms of K_1 and I_1, it is easiest to obtain (and to generalize when we come to deep dimples) if we return to the coupled differential equations, (S.193) and (S.194), introduce the scaling and change of variable

$$\hat{f}\equiv a\tilde{f}_0\ ,\ \hat{g}\equiv a(1-\tilde{h}_0)\ ,\ \xi\equiv ap+\eta\,, \qquad (S.204)^{19}$$

and then, in the resulting equations, let $p\to\infty$. If $\eta=O(1)$, i.e., in a neighborhood of the hinge and away from the point load, we obtain the *shallow shell hinge equations*

$$\frac{d^2\tilde{f}_0}{d\eta^2}+\tfrac{1}{2}(1-\tilde{h}_0^{\,2})=0 \qquad (S.205)$$

[19]From Ranjan & Steele (1977), we know that $a=O(1)$. Unfortunately, their value $a=\sqrt{2}/3=0.471\cdots$ is incorrect, being based on the assumption that β is small compared to α at the hinge; obviously, $\beta=-\alpha$ there.

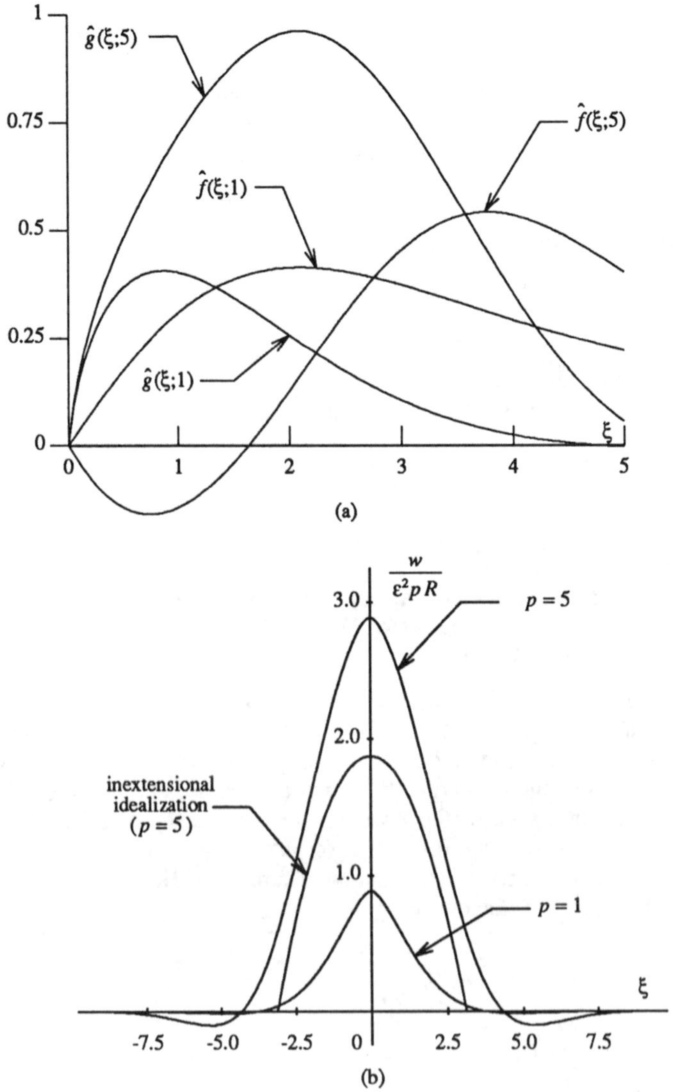

Fig. 5.10. (a) Shallow shell solutions; (b) vertical deflection.

$$\frac{d^2\tilde{h}_0}{d\eta^2} + \tilde{f}_0\tilde{h}_0 = 0 \,. \tag{S.206}$$

[Strictly speaking, (S.205) and (S.206) follow from (S.191), (S.192), and (S.204) as $p \to \infty$ *and* $\varepsilon p \to 0$.] The subscript 0, which distinguishes the shallow shell hinge functions from the deep shell hinge functions that we shall introduce in the next subsection, signifies that f_0 and h_0 are the limits of the latter as the hinge angle approaches 0.

On either side of the elastic hinge, the shell is, essentially, in a membrane state, the meridional stress resultant being tensile within the dimples and compressive outside. This observation and the abrupt change in deformed meridional angle across the hinge leads to the limiting boundary conditions

$$(\tilde{f}_0, \tilde{h}_0) \sim (0, \pm 1) \text{ as } \eta \to \pm\infty . \tag{S.207}$$

The differential equations have a 1st integral which, in view of (S.207), reads

$$\left[\frac{d\tilde{f}_0}{d\eta}\right]^2 - \left[\frac{d\tilde{h}_0}{d\eta}\right]^2 + \tilde{f}_0 - \tilde{f}_0 \tilde{h}_0^{\,2} = 0 . \tag{S.208}$$

Clearly, $\tilde{f}_0$ and $\tilde{h}_0$ are even and odd functions of η, respectively. Moreover, the 1st integral (and a little experimentation) implies that

$$\frac{d\tilde{h}_0}{d\eta}(0) = \sqrt{\tilde{f}_0(0)} . \tag{S.209}$$

Thus, in (S.205)-(S.207), we have a 2-point boundary-value problem on the semi-infinite interval $0 < \eta < \infty$ with but *one* unknown, $\tilde{f}_0(0)$, at the left end of the interval. Trial and error plus a standard numerical integration routine produces $\tilde{f}_0(0) = .896 \cdots$. Values of $\tilde{f}_0(\eta)$, $\tilde{f}_0{}'(\eta)$, $\tilde{h}_0(\eta)$, and $\tilde{h}_0{}'(\eta)$ are given in Table 5.3 and plotted in Fig. 5.11.

To find the unknown constant a in (S.204), it is necessary to consider terms of relative order p^{-1} in (S.191) and (S.192). Alternatively, we can follow Ranjan & Steele (1977) and determine a to a 1st approximation from the Principle of Stationary Potential Energy. This works because, in a variational principle, a gross quantity, such as the lowest natural frequency or a buckling load, can be determined to within a relative error that is of the order of the *square* of the relative pointwise error in the integrated functions. Hence, considering the southern hemisphere, we have

$$P\delta w(0) = \pi R^2 \delta \int_0^{\pi/2} [A(N_\sigma^2 + N_\theta^2 - \nu_e N_\sigma N_\theta) + D(k_\sigma^2 + k_\theta^2 + \nu_b k_\sigma k_\theta)] \sin\alpha \, d\alpha . \tag{S.210}$$

An order of magnitude analysis confirms the conclusion of Ranjan & Steele that, as $p \to \infty$, the strain energy is dominated by the contribution of the elastic hinge. Moreover, $w(0)$ may be computed by assuming that the deformation is inextensional[20], i.e., that $w(0) \sim a^2 p^2 \varepsilon^2 R$ as $p \to \infty$, so that (S.210)

Table 5.3
Shallow Shell Hinge Functions and Their Derivatives

η	$\tilde{f}_0(\eta)$	$\tilde{f}_0'(\eta)$	$\tilde{h}_0(\eta)$	$\tilde{h}_0'(\eta)$
0.000	0.896	0.000	0.000	0.946
0.250	0.879	-0.127	0.243	0.918
0.500	0.833	-0.236	0.464	0.842
0.750	0.763	-0.320	0.661	0.729
1.000	0.675	-0.375	0.827	0.595
1.250	0.578	-0.400	0.958	0.454
1.500	0.477	-0.397	1.055	0.321
1.750	0.381	-0.373	1.120	0.204
2.000	0.292	-0.335	1.159	0.109
2.250	0.213	-0.289	1.177	0.035
2.500	0.147	-0.240	1.179	-0.017
2.750	0.093	-0.193	1.169	-0.052
3.000	0.050	-0.149	1.153	-0.073
3.250	0.018	-0.110	1.134	-0.083
3.500	-0.004	-0.077	1.113	-0.084
3.750	-0.020	-0.050	1.092	-0.081
4.000	-0.030	-0.029	1.072	-0.074
4.250	-0.035	-0.012	1.055	-0.065
4.500	-0.037	-0.000	1.040	-0.055
4.750	-0.036	0.008	1.027	-0.045
5.000	-0.033	0.013	1.017	-0.036

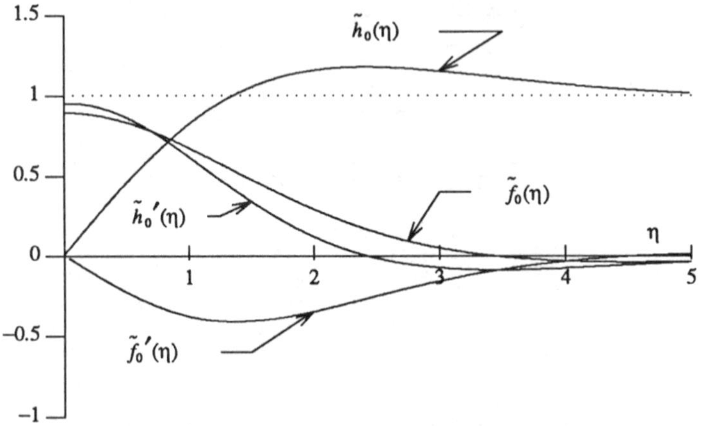

Fig. 5.11. Shallow shell hinge functions and their derivatives.

reduces to

$$\delta(a^2) \sim \int_0^\infty \left[\left(\frac{d\tilde{f}_0}{d\eta} \right)^2 + \left(\frac{d\tilde{h}_0}{d\eta} \right)^2 \right] d\eta \delta(a^3)$$

$$\equiv E_0 \delta(a^3) = (1.11 \cdots) \delta(a^3) . \tag{S.211}$$

Carring out the variation, we have $a = (2/3) E_0^{-1} = 0.600 \cdots$ and the dimensionless load-deflection curve in this range of p has the form

$$p \sim 1.66 \cdots) \sqrt{w(0)/\sqrt{AD}} . \tag{S.212}$$

(iii) $p = O(\varepsilon^{-1})$

Guided by the results of the preceding subsection, we now scale and change variables so that, as $\varepsilon \to 0$, we obtain the *(deep shell) hinge equations.* These equations contain the *hinge angle*, α_h, and, as $\alpha_h \to 0$, approach the shallow shell hinge equations, (S.205) and (S.206). Thus, with

$$F \equiv -\alpha_h \sqrt{D/A} \tilde{f} , \quad \beta = \alpha_h (\tilde{h} - 1) , \quad \alpha \equiv (\alpha_h + \varepsilon\eta) , \tag{S.213}$$

the simplified Reissner equations (S.150) and (S.151) reduce to

$$\frac{\sin(\alpha_h + \varepsilon\eta)}{\alpha_h} \frac{d^2 \tilde{f}}{d\eta^2} + \left[\frac{\varepsilon}{\alpha_h} \right] \cos(\alpha_h + \varepsilon\eta) \frac{d\tilde{f}}{d\eta} - \left[\frac{\varepsilon}{\alpha_h} \right]^2 \left[\frac{\alpha_h \tilde{f}}{\sin(\alpha_h + \varepsilon\eta)} \right]$$

$$+ \frac{\cos(\alpha_h \tilde{h} + \varepsilon\eta) - \cos(\alpha_h + \varepsilon\eta)}{\alpha_h^2} = 0 \tag{S.214}$$

$$\frac{\sin(\alpha_h + \varepsilon\eta)}{\alpha_h} \frac{d^2 \tilde{h}}{d\eta^2} + \left[\frac{\varepsilon}{\alpha_h} \right] \cos(\alpha_h + \varepsilon\eta) \frac{d\tilde{h}}{d\eta} - \left[\frac{\varepsilon}{\alpha_h} \right]^2 \frac{\sin[\alpha_h(1 - \tilde{h})]}{\sin(\alpha_h + \varepsilon\eta)}$$

$$+ \tilde{f} \frac{\sin(\alpha_h \tilde{h} + \varepsilon\eta)}{\alpha_h} - \left[\frac{\varepsilon}{\alpha_h} \right]^2 p \cos(\alpha_h \tilde{h} + \varepsilon\eta) = 0 . \tag{S.215}$$

Recalling that $O(\varepsilon p) = 1$, we obtain, in the limit as $\varepsilon \to 0$, the hinge equations

$$\frac{\sin \alpha_h}{\alpha_h} \frac{d^2 \tilde{f}}{d\eta^2} + \frac{\cos(\alpha_h \tilde{h}) - \cos \alpha_h}{\alpha_h^2} = 0 \tag{S.216}$$

---

[20]This inextensional approximation for $p = 5$ is shown in Fig. 5.10b.

$$\frac{\sin\alpha_h}{\alpha_h}\frac{d^2\tilde{h}}{d\eta^2}+\tilde{f}\frac{\sin(\alpha_h\tilde{h})}{\alpha_h}=0\,. \tag{S.217}$$

The associated boundary conditions are the same as those for the shallow shell hinge equations, namely (S.207). Moreover, the deep shell hinge equations have a 1st integral which, in view of (S.207), reduces to

$$\left[\frac{d\tilde{f}}{d\eta}\right]^2-\left[\frac{d\tilde{h}}{d\eta}\right]^2+2\tilde{f}\left[\frac{\cos(\alpha_h\tilde{h})-\cos\alpha_h}{\alpha_h\sin\alpha_h}\right]=0\,. \tag{S.218}$$

Like the shallow shell hinge functions $\tilde{f}_0$ and $\tilde{h}_0$, $\tilde{f}$ and $\tilde{h}$ are, respectively, even and odd functions of η so that their determination reduces to the solution of (S.216) and (S.217) on the interval $0<\eta<\infty$, subject to the boundary conditions

$$\tilde{f}'(0)=\tilde{h}(0)=0\ ,\ \ \tilde{h}'(0)=\sqrt{\tilde{f}(0)(2/\alpha_h)\tan(\alpha_h/2)}\ \text{ and }\ \tilde{h}(\infty)=1\,, \tag{S.219}$$

the third condition coming from the 1st integral (S.218).

To complete the asymptotic analysis of the deep dimple deformation, we use the inextensional approximation $w(0)\approx 4R\sin^2(\alpha_h/2)$ so that, with the aid of (S.213), the variational principle (S.210) takes the dimensionless, asymptotic form

$$4p\varepsilon\delta\sin^2(\alpha_h/2)\sim\delta\left\{\alpha_h^2\sin\alpha_h\int_0^\infty\left[\left[\frac{d\tilde{f}}{d\eta}\right]^2+\left[\frac{d\tilde{h}}{d\eta}\right]^2\right]d\eta\right\}. \tag{S.220}$$

To compute the hinge angle for small values of εp, we assume expansions of the hinge functions of the form

$$\tilde{f}=\tilde{f}_0+\alpha_h^2\tilde{f}_1+\cdots\ ,\ \ \tilde{h}=\tilde{h}_0+\alpha_h^2\tilde{h}_1+\cdots\,. \tag{S.221}$$

Then, (S.220) yields

$$\begin{aligned}\alpha_h&=\frac{2\varepsilon p}{3E_0}\left[1+\frac{4}{81E_0{}^2}\left[1-\frac{15E_1}{E_0}\right](\varepsilon p)^2+O(\varepsilon p)^4\right]\\&=(0.600\cdots)(\varepsilon p)[1+(0.0154\cdots)(\varepsilon p)^2+O(\varepsilon p)^4]\,,\end{aligned} \tag{S.222}$$

where E_0 is defined in (S.211) and

$$E_1\equiv 2\int_0^\infty\left[\left[\frac{d\tilde{f}_0}{d\eta}\right]\left[\frac{d\tilde{f}_1}{d\eta}\right]+\left[\frac{d\tilde{h}_0}{d\eta}\right]\left[\frac{d\tilde{h}_1}{d\eta}\right]\right]d\eta=0.0454\cdots\,. \tag{S.223}$$

The value of E_1 comes from Fulton (1988) where further details may be found. Fig. 5.12, taken from that dissertation, summarizes the load-deflection relation for the 3 ranges $p=O(\varepsilon^{-1})$, $p=O(1)$, and $p=O(\varepsilon^{-1})$.

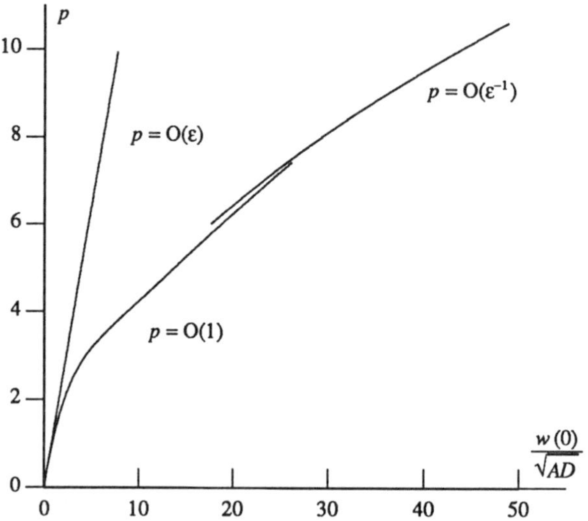

Fig. 5.12. Load-(apex) deflection curve for the 3 ranges $p = O(\varepsilon)$, $p = O(1)$, and $p = O(\varepsilon^{-1})$.

The stability of the axisymmetric solutions of the nonlinear equations for a complete spherical shell under equal and opposite inward radial point loads has been studied numerically by Bushnell (1967). His Fig. 3 shows that, as $R/h \to \infty$ (i.e., as $\varepsilon \to 0$), the shell will bifurcate into a 3-lobe pattern at $PR/Eh^3 \approx 6.2$. The value of $w(0)/h$ there is about 15. Bushnell uses $\nu = 0.3$ in his calculations so that, with $E_e = E_b = E$, we obtain a dimensionless bifurcation load of $p = PR/2\pi D \approx 10.8$ and a corresponding dimensionless vertical deflection of $w(0)/\sqrt{AD} \approx 49.6$.

Mescall (1964, 1965), Bushnell (1967), and Fitch (1968), among others, have solved the nonlinear equations for a shallow spherical shell under an inward radial point load. With $r = R\sin\alpha \approx R\alpha$ and the change of variables (S.190), (R.40) and (R.41) take the form of (S.193) and (S.194). The boundary conditions at the south pole are $(S.195)_1$, i.e., $(\hat{f}, \hat{g}) \to (0, 0)$ as $\xi \to 0$, and, at an unrestrained edge—imagine the cap resting on a rigid, greased horizontal table—

$$\hat{f}(\lambda) = 0 \ , \ \frac{d\hat{g}}{d\xi}(\lambda) + \nu_b \frac{\hat{g}(\lambda)}{\lambda} = 0 \tag{S.224}$$

or, at a clamped edge,

$$\hat{g}(\lambda) = 0 \ , \ \frac{d\hat{f}}{d\xi}(\lambda) - \nu_e \frac{\hat{f}(\lambda)}{\lambda} = 0 \, . \tag{S.225}$$

Here, λ is the standard shallow shell parameter defined in (S.161).

Fig. 1 of Mescall (1965), which is based on a numerical solution of these equations, shows, for an unrestrained shell , that the load-apex deflection curve is monotonic increasing if $\lambda^2 < 12.23$; for larger values of λ, snap buckling occurs. For a clamped shell, Mescall's Fig. 4 shows that, somewhere in the range $64 < \lambda^2 < 81$, the load-deflection curve ceases to be monotonic. (These results depend on λ *and* ν; unfortunately, Mescall fails to mention the value of ν used in his calculations.) Fitch (1968) has made a more detailed numerical study of the buckling of a clamped, shallow spherical cap under a concentrated load. As the load increases, he finds, with a Poisson's ratio of 1/3, that if $\lambda^2 < 61.8$, the shell is stable; if $61.8 < \lambda^2 < 84.6$, there is axisymmetric snap buckling; and if $\lambda^2 > 84.6$, axisymmetric bifurcation buckling occurs before axisymmetric snap-buckling.[21] At $\lambda^2 \approx 61.8$, the dimensionless critical buckling load is $p = PR/2\pi D \approx 14$. As $\lambda^2 \to \infty$, the dimensionless critical load for (bifurcation) buckling drops to $p = PR/2\pi D \approx 10.8$, which agrees with Bushnell's result for a complete spherical shell. Moreover, to quote from the discussion section of Fitch's paper,

> For [λ^2] greater than [84.6] the deflection of a clamped spherical cap under concentrated load will become asymmetric at a load below the critical load for axisymmetric snap-buckling. Depending on the value of λ the initial appearance of asymmetry will be characterized by a deflection pattern having three, four, or five circumferential waves. On a plot of load vs. apex deflection the bifurcation equilibrium path has a positive initial slope which is at most 30% less than the slope of the axisymmetric path at the critical load. A spherical cap with a small initial geometric imperfection would be expected to behave in qualitatively the same manner as the perfect cap. In particular the presence of the imperfection would not be expected to cause snap-buckling at a load below the critical load for the perfect structure.

A prolonged discussion of unsymmetrical behavior is beyond the scope of this book, but, to clear our conscience, we have warned the reader of the limits of axisymmetric theory. Further discussion on the stability of axishells may be found in Section W; the buckling of point-loaded spherical caps is discussed in subsection W.5.b.

(d) Clamped, internally pressurized spherical shells. If there is a positive internal pressure, p_0, then $\mathbf{p} = -p_0 \mathbf{m}(1+e_\theta)$. Hence, by (E.4), (E.5), and (L.2),

$$p_H = p_0(1+e_\theta)[(1+e_\sigma)\sin(\alpha+\beta) + g\cos(\alpha+\beta)] \;,\; p_T = p_0(1+e_\theta)g \,, \quad \text{(S.226)}$$

and from (Q.9),

$$V\sin\alpha = V\sin\alpha_0 + \tfrac{1}{2}p_0 R\,[(1+e_\theta)^2\sin^2\alpha - \sin^2\alpha_0]\,. \quad \text{(S.227)}$$

As we are neglecting the effects of transverse shearing strain and restricting attention in this subsection to dome-like shells under pressure loads only, we have

[21]For the sake of comparison, we have expressed all of Fitch's results in terms of λ^2 rather than in terms of λ, as he does.

$$p_H = p_0(1+e_\theta)(1+e_\sigma)\sin(\alpha+\beta) \ , \ p_T = 0 \tag{S.228}$$

$$V = \tfrac{1}{2} p_0 R (1+\varepsilon_\theta)^2 \sin\alpha \ , \ l = 0 \, . \tag{S.229}$$

It is worth noting that for a *complete*, internally pressurized spherical shell, the linear membrane solution,

$$F_{LM} = \tfrac{1}{2} p_0 R^2 (1+e)^2 \sin\alpha\cos\alpha \, , \ \beta_{LM} = 0 \, , \ e \equiv e_\sigma = e_\theta = \text{constant} \, , \tag{S.230}$$

is also an exact solution of the simplified Reissner equations, (S.150) and (S.151).[22] Often, it represents an interior solution of the governing equations, depending on how far we are from an edge, how high the pressure, how thin the shell, and how shallow. (This statement will be made more quantitative presently.)

Returning to the the spherical cap, we may simplify (S.228) and (S.229) by neglecting strains compared to 1, which is consistent with the approximations we made in deriving the simplified Reissner equations. Introducing the dimensionless variable and parameters,

$$\Phi = \frac{AF}{R} \ , \ \varepsilon^2 = \frac{\sqrt{AD}}{R} \ , \ p = p_0 \sqrt{A/D}\, R^2 \, , \tag{S.231}$$

we can write the field equations (S.150) and (S.151) as

$$\mathcal{L}\Phi + \varepsilon^2 p\,[\sin^2\alpha \sin(\alpha+\beta)]' + \sin\alpha\sin\beta + 2\cos\alpha\sin^2(\beta/2) = 0 \tag{S.232}$$

$$\varepsilon^4[\mathcal{L}\beta + (\beta - \sin\beta)\csc\alpha] - \Phi\sin(\alpha+\beta) + \tfrac{1}{2}\varepsilon^2 p \sin^2\alpha\cos(\alpha+\beta) = 0 \, , \tag{S.233}$$

where $\mathcal{L}$ is given by (S.175). The boundary conditions at a clamped edge are that the rotation and hoop strain vanish. With the aid of (S.154), (S.228), (S.229), and (S.231), we have

$$\beta(\alpha_1) = 0 \ , \ \Phi'(\alpha_1) - \nu_e \Phi(\alpha_1)\cot\alpha_1 + \varepsilon^2 p (1 - \tfrac{1}{2}\nu_e)\sin^2\alpha_1 = 0 \, . \tag{S.234}$$

Symmetry and (Q.10) imply that, at the center of the cap,

$$\beta(0) = \Phi(0) = 0 \, . \tag{S.235}$$

For some purposes, it is useful to subtract off the linear membrane solution in $(\text{S.230})_1$ by setting

$$\Phi = \varepsilon^2 p\,(\tfrac{1}{2}\sin\alpha\cos\alpha + f) \, , \tag{S.236}$$

in which case our dimensionless field equations and boundary conditions can be given the form

---

[22]Elementary considerations of static equilibrium and symmetry show that (S.230) holds regardless of the constitutive relations, so long as they are isotropic.

$$\varepsilon^2 p\{\mathcal{L}f + 2[\sin^2\alpha \sin(\beta/2)\cos(\alpha+\beta/2)]'\} + \sin\alpha\sin\beta + 2\cos\alpha\sin^2(\beta/2) = 0 \tag{S.237}$$

$$\varepsilon^2[\mathcal{L}\beta + (\beta - \sin\beta)\csc\alpha] - p\,[f\sin(\alpha+\beta) + \tfrac{1}{2}\sin\alpha\sin\beta] = 0 \tag{S.238}$$

$$\beta(\alpha_1) = 0 \ , \ f'(\alpha_1) - \nu_e f(\alpha_1)\cot\alpha_1 + \tfrac{1}{2}(1-\nu_e) = 0 \tag{S.239}$$

$$\beta(0) = f(0) = 0 \,. \tag{S.240}$$

We note that we can solve for f in terms of β by introducing the Green's function $G(\alpha, \gamma; \nu_e)$ which satisfies the differential equation

$$(G_\alpha \sin\alpha)_\alpha - G\csc\alpha = \delta(\alpha - \gamma) \tag{S.241}$$

and boundary conditions

$$G(0, \gamma; \nu_e) = 0 \ , \ G_\alpha(\alpha_1, \gamma; \nu_e) = \nu_e G(\alpha_1, \gamma; \nu_e) \,. \tag{S.242}$$

Explicitly, $G(\alpha,\gamma;\nu_e) = G_<(\alpha,\gamma;\nu_e)$ if $\alpha<\gamma$ and $G(\alpha,\gamma;\nu_e) = G_<(\gamma,\alpha;\nu_e)$ if $\gamma<\alpha$, where

$$G_<(\alpha,\gamma;\nu_e) \equiv -\frac{1}{2}\left[\cot(\gamma/2) + \left(\frac{1+\nu_e\cos\alpha_1}{1-\nu_e\cos\alpha_1}\right)\tan(\gamma/2)\cot^2(\alpha_1/2)\right] \times \tan(\alpha/2) \,. \tag{S.243}$$

With the aid of G, the solution of (S.237), subject to the boundary conditions $(\text{S.239})_2$ and $(\text{S.240})_2$, takes the form

$$\begin{aligned} f(\alpha) = &-\left(\frac{1-\nu_e}{1-\nu_e\cos\alpha_1}\right)\cos^2(\alpha_1/2)\tan(\alpha/2) \\ &- 2\int_0^{\alpha_1} G_\gamma(\gamma,\alpha;\nu_e)\sin^2\gamma\sin(\beta/2)\cos(\gamma+\beta/2)\,d\gamma \\ &- \varepsilon^{-2}p^{-1}\int_0^{\alpha_1} G(\gamma,\alpha;\nu_e)[\sin\gamma\sin\beta + 2\cos\gamma\sin^2(\beta/2)]\,d\gamma \,. \end{aligned} \tag{S.244}$$

This equation, substituted into (S.238), gives us a single integro-differential equation for β which may have some advantages from the viewpoint of functional analysis. [E.g., see Berger's (1967) work on the von Kármán equations.] It in turn can be reduced to the integral equation

$$\beta(\alpha) = \int_0^{\alpha_1} G(\gamma,\alpha;\infty)\{(\sin\beta-\beta) + \varepsilon^{-2}p[f\sin(\gamma+\beta) + \tfrac{1}{2}\sin\gamma\sin\beta]\}\,d\gamma \,, \tag{S.245}$$

where f is given by (S.244). We do not pursue this integral formulation further. Alternative integral formulations will be mentioned later.

To obtain detailed analytic results in any but the simplest problems, we must study the limiting forms of the boundary value problem (S.232)-(S.235) or, equivalently, (S.237)-(S.240), as $\varepsilon \to 0$. In doing so, is important to realize that, in addition to ε and p, the solutions of the simplified Reissner equations depend on the shallowness parameter

$$\lambda^2 \equiv \frac{\alpha_1^2 R}{\sqrt{AD}} , \tag{S.246}$$

which appears if we set $\alpha = \alpha_1 x = \varepsilon\lambda x$ so as make our differential equations apply on the unit interval $0 < x < 1$. Note that λ may range from 0 to ε^{-1} while p may range from 0 to ∞ subject, of course, to the restriction that the strains remain small.

Simmonds & Libai (1987b) set

$$\lambda = \Lambda\varepsilon^{-r} \ , \ p = P\varepsilon^{-s} , \tag{S.247}$$

where $\Lambda, P = O(1)$, and then looked for various scalings of the dependent and independent variables such that "meaningful" boundary-value problems emerged as $\varepsilon \to 0$. They found the 17 cases summarized in Fig. 5.13 and Table 5.4. In Fig. 5.13, the 17 cases correspond to various points, line segments, and regions in a portion of the *rs*-plane, bounded above by a small-strain boundary, which is *not* part of any of the cases, and on the right by the vertical line, $r = 1$ (corresponding to α_1 being a finite number, independent of ε), which *is* a part of some of the cases.

As shown in the table, the limiting equations found by Simmonds & Libai are of 3 types, namely, those of

(1) The linearized theory of deep shells, prestressed by pressure. The equations of the 3 cases in this group can all be solved by simple boundary layer techniques. In case A, the equations separate into those of membrane theory, first solved by Bromberg & Stoker (1945), plus those for a narrow bending boundary layer within the membrane boundary layer, first elucidated by Reissner (1950). In cases B and C, the membrane and bending boundary layers merge into a common layer of angular width $O(\varepsilon)$.

(2) The nonlinear shallow shell equations and various special cases thereof. The 5 cases in this group range from the theory of flat membranes of Föppl (1905) to the full-blown, nonlinear shallow shell theory of Marguerre (1938).

(3) The linear theory of plates. The 9 cases in this group come from the various forms taken on by the (auxiliary) compatibility condition for the stress function.

It is interesting to note that in no case does the meridional angle of rotation β become $O(1)$.

(e) Other problems. Lin & Wan (1985) have studied the asymptotic forms of the equations for a uniformly spinning, normally loaded, shallow spherical

Table 5.4
Summary of Changes of Variables Leading to Limiting Forms of the Simplified Reissner Equations for a Spherical Cap under Outward Pressure

$$\varepsilon^2 = \sqrt{AD}/R \ , \ \lambda^2 = \alpha_1^2 R/\sqrt{AD} \ , \ p = p_0\sqrt{A/D}R^2 \ , \ \lambda = \Lambda\varepsilon^{-r} \ , \ p = P\varepsilon^{-s}$$

DEEP SHELLS: $\Phi = \varepsilon^2 p(\frac{1}{2}\sin\alpha\cos\alpha + f)$

Case	α	f	β
A	$\alpha_1 - \varepsilon p^{1/2}\xi$	$\varepsilon p^{1/2}\hat{f}(\xi)$	$\varepsilon p^{1/2}\hat{\gamma}(\xi)$
	$\alpha_1 - \varepsilon p^{-1/2}\eta$	$\varepsilon p^{1/2}[\hat{f}(0) + p^{-1}\phi(\eta)]$	$\varepsilon p^{1/2}\gamma(\eta)$
B	$\alpha_1 - \varepsilon\zeta$	$\varepsilon\tilde{f}(\zeta)$	$\varepsilon p\tilde{\gamma}(\zeta)$
AB	$\alpha_1 - \varepsilon\zeta$	$\varepsilon\tilde{f}(\zeta)$	$\varepsilon P\gamma(\zeta)$

SHALLOW SHELLS: $\alpha = \alpha_1 x$

Case	Φ	β
C	$\varepsilon^3\lambda^{5/3}p^{2/3}\phi(x)$	$\varepsilon\lambda^{1/3}p^{1/3}\gamma(x)$
AC	$\varepsilon^3\lambda^3\tilde{\phi}(x)$	$\varepsilon\lambda\tilde{\gamma}(x)$
CD	$\varepsilon^3\lambda^{-1}\tilde{f}(x)$	$\varepsilon\lambda^{-1}\tilde{\gamma}(x)$
BE	$\Lambda^5\varepsilon^3 p\hat{\phi}(x)$	$\Lambda\varepsilon p\hat{\gamma}(x)$
ABCDE	$\Lambda^3\varepsilon^3\hat{\Phi}(x)$	$\Lambda\varepsilon\hat{B}(x)$

PLATES: $\alpha = \alpha_1 x$

Case	Φ	β
D	$\varepsilon^3\lambda^7 p^2 f(x)$	$\varepsilon\lambda^3 pg(x)$
E	$\varepsilon^3\lambda^5 p\bar{\phi}(x)$	$\varepsilon\lambda^3 pg(x)$
DE	$P\varepsilon^3\lambda^3 f(x)$	$P\varepsilon\lambda g(x)$
F	$\varepsilon^5\lambda^5 p^2\phi(x)$	$\varepsilon\lambda^3 pg(x)$
DF	$\Lambda^5\varepsilon^{10}p^2 f(x)$	$\Lambda^3\varepsilon^4 pg(x)$
G	$\varepsilon^5\lambda^3 p\tilde{f}(x)$	$\varepsilon\lambda^3 pg(x)$
EG	$\Lambda^3\varepsilon^8 p\phi(x)$	$\Lambda^3\varepsilon^4 pg(x)$
FG	$P\varepsilon^5\lambda\phi(x)$	$P\varepsilon\lambda g(x)$
DEFG	$\Lambda^3 P\varepsilon^6 f(x)$	$\Lambda^3 P\varepsilon^2 g(x)$

shell attached to a rigid hub, complementing earlier work by Simmonds (1962) and Reissner & Wan (1965).

With an eye to establishing a mechanical theory of tonometry, Updike & Kalnins (1970, 1972a, 1972b) have studied hemispherical shells pressed between 2 rigid plates. The same motivation has led to a series of papers by Taber (1982, 1983, 1984) dealing with fluid-filled spherical shells pressed upon by point loads or rigid indenters.

The analytic computation of nonlinear influence coefficients by perturbation methods for deep, edge-loaded, smooth axishells has been studied by Reissner (1959); by VanDyke (1966) (who corrected some numerical errors in Reissner's paper) for edge- and pressure-loaded spherical shells; and by and Ranjan & Steele (1980) for deep, edge- and pressure-loaded, smooth axishells. *Linear* influence coefficients for spherical caps of all shapes, prestressed by pressurization, have been computed by Cline (1963), although, in view of the analysis in Simmonds & Libai (1987b), we question the validity of Cline's results in the shallow shell range for the relatively high values of pressure he uses.

$$\epsilon^2 = \left[12(1-\nu^2)\right]^{-1/2} h/R$$

$$\lambda = \left[12(1-\nu^2)\right]^{1/4} (R/h)^{1/2} \alpha_1$$

$$= \Lambda \epsilon^{-r}$$

$$p = p_0\left[12(1-\nu^2)\right]^{1/2} R^2/Eh^2$$

$$= P\epsilon^{-s}$$

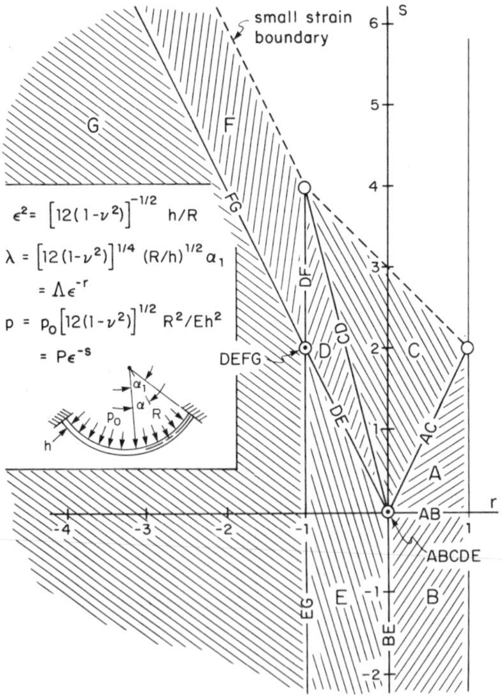

Fig. 5.13. Subsets of the *rs*-plane where the simplified Reissner equations have different forms.

Weinitschke & Lange (1987) use the method of matched asymptotic expansions to find the stress concentration factors for the circumferential stress and moment near a small, traction-free central hole in a pressurized, shallow spherical cap subject to various boundary conditions at its outer edge.

The literature on the buckling of spherical shells under *inward* point loads and pressure, uniform or variable, is vast and growing; several basic studies are discussed in an overview in Section W of the buckling of axishells.

5. Toroidal shells and membranes of general cross section

Toroidal shells find manifold applications—end caps on pressure vessels, transition sections, expansion joints, tires, inner tubes, fuel tanks on rockets, jet engine inlets, turbine shrouds, ... The first satisfactory treatment of the *linear* theory of toroidal axishells of *circular* cross section is due to Clark (1950) who used Langer's (1931) method of asymptotic integration for differential equations containing a large parameter and a transition point. Textbooks that either reproduce Clark's analysis or variants and extensions thereof include those of Axelrad (1983, pp. 110-118), Kraus (1967, pp. 278-282), Novozhilov (1970, pp. 338-356), Seide (1975, pp. 284-289), and Tsui (1968, pp. 142-156). In a recent note, Jenssen (1986) has improved the accuracy of Clark's particular solution of the governing equations.

Our interest lies in the nonlinear analysis of the special but challenging case of a closed, toroidal shell with a smooth but otherwise arbitrary cross section, under a uniform internal pressure p_0. Most of the analytic work in the literature has been confined to toroidal shells with circular undeformed meridians.

To set the stage, let us mention the history of toroidal shells, which is brief but instructive. In vol. 3 of the 9th edition of his famous treatise, Föppl (1922, p. 331), using linear theory, computed the meridional and hoop stress resultants in a toroidal membrane of *circular* cross section. However, as Dean (1939) first observed, the stress-strain and strain-displacement relations of linear membrane theory imply that the meridional derivative of the vertical displacement is infinite at the crowns of the membrane. The addition of flexural terms, he noted, would cure the problem; indeed, this procedure is recommended but not carried out by Flügge (1960, p. 100).

Dean's suggested resolution of the linear membrane paradox stood for over 20 years until Jordan (1962) remarked that:

> If [Dean's suggestion] were correct, as has been previously assumed, this would not only invalidate the physical concept of an ideal membrane but it would also lead to a more complicated ... analysis than appears to be warranted by the problem.... In the present paper the concept of an ideal membrane is revalidated by showing that an adequate non-linear membrane solution exists.

Jordan's problem was reworked in a more conventional and systematic way by Sanders & Liepins (1963) who, like Jordan, ended up with a *linear* differential equation containing a large parameter and a double turning point to which they applied standard asymptotic integration techniques. Dickey (1973) later discussed the existence, uniqueness, and asymptotic behavior of a set of nonlinear membrane equations essentially equivalent to those used by Jordan and Sanders & Liepins.

About the time of Sanders & Liepins' paper, Reissner (1963b) analyzed the inflation of a toroidal shell of circular cross section using a version of his moderate rotation theory of axishells. In terms of the *transition parameter*

$$\rho \equiv p_0 \left[\frac{a^2 b^5 A}{D^2} \right]^{1/3}, \tag{S.248}$$

where a is the distance to the center of the circular cross section and b is its radius, Reissner showed that, if $\rho \ll 1$, linear bending theory holds whereas, if $\rho \gg 1$, nonlinear membrane theory dominates; $\rho = O(1)$ is a transition region where elements of both theories mingle. Shortly thereafter, Reissner's 2 coupled differential equations for the transition region, which involve a large parameter and a simple turning point, were solved asymptotically by Rossettos & Sanders (1965). (Note that our definition of ρ is is that of Rossettos & Sanders which differs from that of Reissner.)

Several papers have treated thick and thin, internally pressurized toroidal shells made of an elastically isotropic, incompressible material. The strains have been allowed to be large[23] and sometimes the shell has been rotating steadily. Thus, Kydoniefs & Spencer (1965) have considered an internally pressurized body that, in its *deformed* state, occupies a region generated by rotating 2 concentric circles about an axis in their own plane. After an exact formulation, it is assumed

> that the radius of the larger of the circles which generates the torus is small compared to the overall radius of the torus, and ... the square of the ratio of these radii [is neglected] compared to unity. It is then found that the problem becomes, in effect, a perturbation of the problem of extension and inflation of a circular tube, the solution to which is known. A number of results can thus be obtained without imposing any restriction on the strain-energy function.
>
> To solve the equations for the 1st-order perturbation, it is further assumed that the strain-energy function has the neo-Hookean form. The equations of equilibrium are then solvable in closed form, and the complete configuration of the undeformed body which corresponds to the given deformed body is determined to this approximation. It is found that the undeformed body is generated by rotating *eccentric* circles about an axis in their plane.

Kydoniefs (1966) has added a uniform centrifugal load to the above problem and, for the special case of no internal pressure, has carried out a perturbation solution for a general strain-energy density. Hill (1980, 1982) has extended the

---

[23]Thus, some of the following comments, strictly speaking, do not fall within the confines of this subsection which assumes small strains. Nevertheless, it seems important to mention here these large-strain papers, for continuity, completeness, and because, whenever the small-strain equations predict high stresses, one is naturally curious about what happens in rubber-like materials.

results of these large-strain papers in 2 ways: he has worked with a general, elastically isotropic, incompressible material, and he has considered inflated, rotating, toroidal bodies whose *undeformed* cross sections are bounded by concentric circles.

Using the theory developed by Adkins & Rivlin (1952) for arbitrary, elastically isotropic membranes admitting large strains, Kydoniefs & Spencer (1967) have studied the finite inflation of toroidal membranes of circular *undeformed* cross section. (At the end of the first paragraph of their paper, Kydoniefs & Spencer state that

> it seems that if membrane theory is to be used, the assumption of a *circular deformed* cross section is incompatible with a continuous deformation of the membrane and uniform internal pressure, though it gives a reasonable stress distribution.

but give no evidence for this provocative statement!) As in earlier papers, a solution of the governing equations is sought in the form of a perturbation series in powers of the ratio of the minor to the major radius of the toroidal reference surface. Finally, numerical results are presented for the Mooney material, $W=C_1[I_1-3+\Gamma(I_2-3)]$. As Kydoniefs & Spencer emphasize,

> the deformation is very sensitive to changes of Γ, especially near $\Gamma=0$.

[The often singular nature of the neo-Hookean material ($\Gamma=0$) has also been noted by other writers.] Kydoniefs (1967) has extended the above analysis to toroidal aximembranes with undeformed, convex cross sections that are symmetric with respect to $z=0$. Kydoniefs uses the same perturbation expansion as before, but assumes a Mooney strain-energy density. As a numerical example, he solves the perturbation equations for an initially flat cross section and finds

> that the shape of the deformed cross section is circular to the first-order approximation and retains its central symmetry up to the second order.

(This result makes all the more mysterious the earlier statement of Kydoniefs & Spencer that implied that membrane theory is incompatible with the assumption of a deformed circular cross section.)

We have mentioned that, in pressurized toroidal shells with undeformed circular cross section, linear membrane theory delivers stress resultants that are finite everywhere (although, as Jordan and Sanders & Liepins have shown, nonlinear membrane theory predicts, in typical cases, that, near the crowns, the hoop stress resultant may be nearly 20% higher.) What, then, happens in a very thin, pressurized toroidal shell whose undeformed cross section does *not* have points of horizontal tangency lying on the same vertical line? As remarked by Flügge (1960, p. 33)

> a membrane stress system with finite values... is not possible in this shell under this load.

For pure membrane theory to work, the points of horizontal tangency in the *deformed* configuration must be in vertical opposition, but unless the undeformed cross section is a near circle, this is not possible if the extensional strains are to remain small and *axisymmetric*. Moreover, even if we invoke bending theory [as does Steele (1965) to obtain equilibrating shear stress resultants at the crowns], in what sense is membrane theory an "outer" solution of the full shell equations? Remarkably, no one to our knowledge has tackled this obvious question. We conjecture that wrinkling must occur. [A theory of wrinkled aximembranes that are inextensional, linearly elastic, or nonlinearly elastic, has been developed by Wu (1974a, 1978)—see subsection T.6.]

Returning to what is known in the small-strain, large-rotation theory of axishells, let us recast the analyses of Sanders & Liepins (1963), Reissner (1963b), and Rossettos & Sanders (1965) in terms of the simplified Reissner equations, (R.18) and (R.23). For a internally pressurized toroidal axishell with a circular reference meridian,

$$r = b(\tau + \sin\alpha)\ ,\ z = -b\cos\alpha\ ,\ 0 \le \alpha < 2\pi\ ,\ \tau \equiv a/b > 1\ , \qquad (S.249)$$

and, from (Q.9), we have, upon neglecting strains compared to 1,

$$rV = b\tau V(0) + \tfrac{1}{2}p_0 b^2(2\tau\sin\alpha + \sin^2\alpha)\ , \qquad (S.250)$$

where $V(0)$ is an unknown to be determined in the course of the analysis. Moreover, because $\mathbf{p} = -p_0(1+e_\theta)\mathbf{m} = p_0(1+e_\theta)[g\mathbf{T} - (1+e_\sigma)\mathbf{B}]$, we have, upon again neglecting strains compared to 1,

$$p_H = p_0 \sin(\alpha+\beta)\ ,\ p_T = 0\ . \qquad (S.251)$$

Thus, with a prime now denoting differentiation with respect to α, the simplified Reissner equations, (R.18) and (R.23), take the form

$$(A/b)\{\mathcal{L}F + p_0 b^2[(\tau + \sin\alpha)^2 \sin(\alpha+\beta)]'\} + \cos\alpha - \cos(\alpha+\beta) = 0 \qquad (S.252)$$

$$(D/b)[\mathcal{L}\beta + (\tau+\sin\alpha)^{-1}(\beta - \sin\beta)] - F\sin(\alpha+\beta) + [b\tau V(0) + \tfrac{1}{2}p_0 b^2(2\tau\sin\alpha + \sin^2\alpha)]\cos(\alpha+\beta) = 0\ , \qquad (S.253)$$

where

$$\mathcal{L} = (\tau + \sin\alpha)\frac{d^2}{d\alpha^2} + \cos\alpha\frac{d}{d\alpha} - \frac{1}{\tau + \sin\alpha}\ . \qquad (S.254)$$

By symmetry, particles do not move out of the equatorial plane $z = 0$. Thus, from (R.13), the vertical displacement w will be continuous at the crown $\alpha = 0$ (and hence, by symmetry, at the crown $\alpha = \pi$) if

$$\int_{-\frac{1}{2}\pi}^{\frac{1}{2}\pi} [\cos\alpha\sin\beta - 2\sin\alpha\sin^2(\beta/2) + e_\sigma\sin(\alpha+\beta) + g\cos(\alpha+\beta)]d\alpha = 0\ . \qquad (S.255)$$

Even though the simplified Reissner equations are predicated on the neglect of the transverse shearing strain g in the strain-energy density, it is consistent to retain g in (S.255). This is because, except in regions where shell theory is not a reasonable approximation, the transverse shear stress resultant Q is predicted accurately to a 1st-approximation and, therefore, so is $g = A_s Q$. (The transverse shear strain term can assume special importance in pressurized toroidal shells if the cross section of the undeformed reference surface does not have points of horizontal tangency on the same vertical line, because Q at such points may be unusually large.)

Symmetry further allows us to restrict the differential equations (S.252) and (S.253) to the domain $-\frac{1}{2}\pi < \alpha < \frac{1}{2}\pi$ and to impose the boundary conditions

$$F(\pm\tfrac{1}{2}\pi) = \beta(\pm\tfrac{1}{2}\pi) = 0 \,. \tag{S.256}$$

The 5th auxiliary condition (S.255) determines $V(0)$.

The linear membrane solution alluded to earlier is

$$F_{LM} = p_0 b^2(\tfrac{1}{2}\sin\alpha + \tau)\cos\alpha \ , \ \beta_{LM} = \frac{Ap_0 b\tau\cos\alpha}{2(\tau + \sin\alpha)\sin\alpha} \,. \tag{S.257}$$

Note that β has simple poles at the crowns of the reference surface. However, F_{LM} is regular everywhere which suggests that we set

$$F = p_0 b^2[(\tfrac{1}{2}\sin\alpha + \tau)\cos\alpha + f\,] \,. \tag{S.258}$$

This step enables us to reduce (S.252) and (S.253) to the form

$$\begin{aligned} &\varepsilon^2 p\{\mathcal{L}f + 2[(\tau + \sin\alpha)^2 \sin(\beta/2)\cos(\alpha + \beta/2)]'\} \\ &\qquad + \sin\alpha\sin\beta + 2\cos\alpha\sin^2(\beta/2) = \tfrac{1}{2}\varepsilon^2 p\tau(\tau + \sin\alpha)^{-1}\cos\alpha \end{aligned} \tag{S.259}$$

$$\begin{aligned} &\varepsilon^2[\mathcal{L}\beta + (\tau + \sin\alpha)^{-1}(\beta - \sin\beta)] \\ &\qquad - p[f\sin(\alpha + \beta) + (\tau + \tfrac{1}{2}\sin\alpha)\sin\beta - k\cos(\alpha + \beta)] = 0 \,, \end{aligned} \tag{S.260}$$

where

$$\varepsilon^2 \equiv \frac{\sqrt{AD}}{b} \ , \ p \equiv p_0\sqrt{A/D}\,b^2 = (\varepsilon/\tau)^{2/3}\rho \ , \ k \equiv \frac{\tau V(0)}{p_0 b} \,. \tag{S.261}$$

If $\tau = 0$, we have the equations for a spherical cap under outward pressure, (S.237) and (S.238).

Equations (S.259), (S.260), and the associated boundary conditions

$$f(\pm\tfrac{1}{2}\pi) = \beta(\pm\tfrac{1}{2}\pi) = 0 \tag{S.262}$$

depend on τ, p, and ε. As with the spherical cap, we could set $\tau = T\varepsilon^{-r}$, $p = P\varepsilon^{-s}$, and scale the variables so as to obtain "meaningful" boundary value problems in the limit as $\varepsilon \to 0$. Instead, following Reissner (1963b) and Rossettos & Sanders (1965), we shall assume that there are interior layers at the crowns and derive simplified equations containing the transition parameter ρ defined by (S.248). Thus, focusing on a neighborhood of the lower crown ($\alpha = 0$), we set

$$\alpha = \varepsilon^{2/3}\tau^{1/3}\eta \ , \ f = \varepsilon^{4/3}\tau^{-1/3}\phi \ , \ \beta = \varepsilon^{4/3}\tau^{-1/3}p\gamma \,. \tag{S.263}$$

Then, (S.259) and (S.260) may be given the form

$$\frac{d^2\phi}{d\eta^2} + \eta\gamma - \tfrac{1}{2} = O[(\varepsilon/\tau)^{2/3}, (\varepsilon/\tau)^{4/3}(\rho, \tau^2\rho)] \qquad \text{(S.264)}$$

$$\frac{d^2\gamma}{d\eta^2} - \eta\phi - \rho\gamma + c = O[(\varepsilon/\tau)^{2/3}, (\varepsilon/\tau)^{4/3}\rho] \,, \qquad \text{(S.265)}$$

where $c \equiv \varepsilon^{-2}k$. As usual, the order estimates assume that all variables are O(1) and that differentiation with respect to η does not increase orders of magnitude.

With the O-terms neglected, we have the 1st-approximation equations of Reissner. However, $(\varepsilon/\tau)^{2/3} = O(hb/a^2)^{1/3}$ may not be particularly small in practice and some of the terms on the right sides of (S.264) and (S.265) can be retained profitably. To prepare for a perturbation expansions in powers of $\varepsilon^{2/3}$, we follow Rossettos & Sanders (1965) and introduce the Liouville transformation [Erdélyi (1956, Sect. 4.5)]

$$x = \varepsilon^{-2/3}\tau^{-1/3}s(\alpha) \;,\; f = \varepsilon^{4/3}\tau^{-1/3}t(\alpha)\phi(x) \;,\; \beta = \varepsilon^{4/3}\tau^{-1/3}pt(\alpha)\gamma(x) \,, \qquad \text{(S.266)}$$

where

$$\begin{aligned} s(\alpha) &= \left[\frac{3}{2}\int_0^{\alpha} \left[\frac{\tau\,|\sin\hat{\alpha}|}{\tau + \sin\hat{\alpha}} \right]^{1/2} d\hat{\alpha} \right]^{2/3} \operatorname{sgn}\alpha \\ &= \alpha\left[1 - \frac{\alpha}{5\tau} + \frac{(102 - 25\tau^2)\alpha^2}{1050\tau^2} + O(\alpha^3) \right] \end{aligned} \qquad \text{(S.267)}$$

and

$$\begin{aligned} t(\alpha) &= \left[\frac{\tau s(\alpha)}{(\tau + \sin\alpha)\sin\alpha} \right]^{1/4} \\ &= 1 - \frac{3\alpha}{10\tau} + \frac{(53 + 10\tau^2)\alpha^2}{280\tau^2} + O(\alpha^3) \,. \end{aligned} \qquad \text{(S.268)}$$

The effect of this transformation is the following: if in the differential equation

$$\varepsilon^2 L\psi + \psi\sin\alpha = \varepsilon^2\theta(\alpha) \qquad \text{(S.269)}$$

we make the change of variable

$$\psi = \varepsilon^{4/3}\tau^{-1/3}t(\alpha)z(x) \,, \qquad \text{(S.270)}$$

then, assuming $\tau = O(1)$, we obtain

$$\frac{d^2z}{dx^2} + xz = \frac{s\theta}{t\sin\alpha} + O(\varepsilon^{4/3}) \,. \qquad \text{(S.271)}$$

But (S.267) and (S.268) imply that

$$\alpha = s + \frac{s^2}{5\tau} + O(s^3)\ ,\ t = 1 - \frac{3s}{10} + O(s^2)\,. \tag{S.272}$$

Hence, with $s = \varepsilon^{2/3}\tau^{1/3}x$ [from $(S.266)_1$] and $\theta(\alpha) = \theta(0) + \theta'(0)\alpha + O(\alpha^2)$, the right side of (S.271) takes the form

$$\theta(0) + (\varepsilon/\tau)^{2/3}\left[\tau\theta'(0) + \frac{\theta'(0)}{10}\right] + O(\varepsilon^{4/3})\,. \tag{S.273}$$

Applying this result to (S.259) and (S.260), we have

$$\frac{d^2\phi}{dx^2} + x\gamma - \tfrac{1}{2} + (9/20)(\varepsilon/\tau)^{2/3}x = O(\varepsilon^{4/3}) \tag{S.274}$$

$$\frac{d^2\gamma}{dx^2} - x\phi - \rho\gamma + c - (3/10)(\varepsilon/\tau)^{2/3}x\rho\gamma = O(\varepsilon^{4/3})\,. \tag{S.275}$$

As we move away from the lower crown, the solutions of these equations must approach those of linear membrane theory; equivalently, the second derivative terms in the above equations must approach zero. Thus, we have the boundary conditions

$$\phi \sim \frac{c}{x} + \frac{3\rho}{10x}(\varepsilon/\tau)^{2/3} + O(\varepsilon^{4/3})\ ,\ \gamma \sim \frac{1}{2x} - \frac{9}{20}(\varepsilon/\tau)^{2/3} + O(\varepsilon^{4/3}) \tag{S.276}$$

as $|x| \to \infty$. Assuming expansions of the form

$$\phi(x,\varepsilon) = \phi_0(x) + \varepsilon^{2/3}\phi_1(x) + \cdots\ ,\ \gamma(x,\varepsilon) = \gamma_0(x) + \varepsilon^{2/3}\gamma_1(x) + \cdots\,, \tag{S.277}$$

we obtain equations and boundary conditions for ϕ_0, γ_0, ϕ_1, and γ_1 of precisely the same form as equations (54) and (55) of Rossettos & Sanders (1965). These authors express the solutions of these equations in terms a canonical set of eight functions of x and ρ, $\{f_A, \cdots, f_D, g_A, \cdots, g_D\}$, half of which are even in x and half of which are odd. Graphs of $f_A, \cdots, g_D$ and their derivatives with respect to x, which were computed numerically for $\rho = .5$, 1.0, 5, and 15, are displayed in Figs. 3 and 4 of their paper.

The integral condition (S.255), that determines the constant c, has contributions from without and within the interior layer at the lower crown. We may systematically account for these by introducing the *intermediate variable*

$$\zeta \equiv \varepsilon^{-1/3}\alpha\,. \tag{S.278}$$

Then, with the aid of $(S.266)_1$, the integral condition may be given the form

$$\int_{-\frac{1}{2}\pi}^{-\varepsilon^{1/3}\zeta}[\cdots]d\alpha + \int_{\varepsilon^{1/3}\zeta}^{\frac{1}{2}\pi}[\cdots]d\alpha$$
$$+\,\varepsilon^{2/3}\tau^{1/3}\int_{-\varepsilon^{-2/3}\tau^{-1/3}s(\varepsilon^{1/3}\zeta)}^{\varepsilon^{-2/3}\tau^{-1/3}s(\varepsilon^{1/3}\zeta)}[\cdots](d\alpha/ds)dx \equiv \varepsilon^2 p\,(I_o + I_i) = 0\,. \tag{S.279}$$

Into the first 2 integrals (which compose I_o) we insert the "outer" solutions of

(S.259) and (S.260) which, in view of (S.257), have formal expansions of the form

$$f(\alpha, p, \varepsilon) = \varepsilon^2[f_0(\alpha, p) + O(\varepsilon^2)] \,,\; \beta(\alpha, p, \varepsilon) = \varepsilon^2 p\,[g_0(\alpha, p) + O(\varepsilon^2)] \,, \quad \text{(S.280)}$$

where

$$f_0 = \frac{c\cos\alpha - p\,(\tau + \tfrac{1}{2}\sin\alpha)g_0}{\sin\alpha} \,,\quad g_0 = \frac{\tau\cos\alpha}{2(\tau + \sin\alpha)\sin\alpha} \,. \quad \text{(S.281)}$$

Into the last integral in (S.279) we insert the expansions in (S.277).

By the Kirchhoff hypothesis, the direct and maximum bending stresses are given in terms of the stress resultants (N) and stress couples (M) by

$$\sigma_D = \frac{N}{h} \,,\quad \max\sigma_B = \frac{6M}{h^2} \,. \quad \text{(S.282)}$$

Rossettos & Sanders have shown that $\sigma_B = O(\varepsilon^{2/3}\sigma_D)$ and, moreover, that the most significant correction to the predictions of linear membrane theory are those to N_θ, which are of relative order $\varepsilon^{2/3}$. Thus, if we assume that the unknown constant $c = c_0 + \varepsilon^{2/3}c_1 + \cdots$, it is only necessary to compute c_0 to find a 1st-approximation to M_σ and M_θ and a 2nd-approximation to N_θ in the interior layer. This we shall do; the reader may refer to Rossettos & Sanders (1965) for the calculation of higher order corrections.

Turning to the integrands in (S.279) which come from (S.255), we first use $(R.2)_3$ and (R.23) to write

$$g = A_s Q = (A_s D/b^2)(\tau + \sin\alpha)^{-1}[\mathcal{L}\beta + (\tau + \sin\alpha)^{-1}(\beta - \sin\beta)] \,, \quad \text{(S.283)}$$

and use (Q.4), (Q.10), $(R.2)_1$, (S.250), (S.251), and (S.258) to write

$$\begin{aligned} e_\sigma &= A\left[\frac{F\cos(\alpha+\beta) + V\sin(\alpha+\beta)}{b\,(\tau + \sin\alpha)} - \nu_e(F' + rp_H)\right] \\ &= \varepsilon^2 p\left\{\frac{(\tau + \tfrac{1}{2}\sin\alpha)\cos\beta + f\cos(\alpha+\beta) + k\sin(\alpha+\beta)}{\tau + \sin\alpha}\right. \quad \text{(S.284)} \\ &\qquad \left. - \nu_e[\tfrac{1}{2} + f' + 2(\tau + \sin\alpha)\cos(\alpha + \beta/2)\sin(\beta/2)]\right\}. \end{aligned}$$

Thus, in the outer region where $\varepsilon^{1/3}\,|\zeta| \le |\alpha| \le \tfrac{1}{2}\pi$, we have

$$I_o = \left(\int_{-\frac{1}{2}\pi}^{-\varepsilon^{1/3}\zeta}, \int_{\varepsilon^{1/3}\zeta}^{\frac{1}{2}\pi}\right)\left[g_0\cos\alpha + \left(\frac{\tau + \frac{1}{2}\sin\alpha}{\tau + \sin\alpha} - \frac{\nu_e}{2}\right)\sin\alpha + O(\varepsilon^2 p)\right]d\alpha$$

$$= -\tau\int_{\varepsilon^{1/3}\zeta}^{\frac{1}{2}\pi}\frac{d\alpha}{\tau^2 - \sin^2\alpha} + O(\varepsilon^2 p) \tag{S.285}$$

$$= -\frac{\pi}{2\sqrt{\tau^2 - 1}} + \frac{\varepsilon^{1/3}\zeta}{\tau} + O(\varepsilon^{2/3}) .$$

In the inner region, where $x = O(1)$, the dominant contribution to I_i comes from the $\cos\alpha\sin\beta$-term in the integrand. Recalling that $\beta = \varepsilon^{4/3}\tau^{-1/3}pt(\alpha)\gamma(x)$, $s = \varepsilon^{2/3}\tau^{1/3}x$, and noting (S.272), we have

$$t(\alpha)(d\alpha/ds) = 1 - (\varepsilon/\tau)^{2/3}(x/10) + O(\varepsilon^{4/3}) . \tag{S.286}$$

Hence, making liberal use of the large x behavior listed by Rossettos & Sanders for the eight canonical functions $f_A, \cdots, g_D$, we find that

$$I_i = \int_{-x^*}^{x^*}\{[\gamma_0(x) + \varepsilon^{2/3}\gamma_1(x)][1 - (\varepsilon/\tau)^{2/3}(x/10)]\}dx + O(\varepsilon^{4/3}) , \tag{S.287}$$

where $x^* \equiv \varepsilon^{-2/3}\tau^{-1/3}s(\varepsilon^{1/3}\zeta)$. Using equations (61) and (63) of Rossettos & Sanders, we have

$$\int_{-x^*}^{x^*}\gamma_0(x)dx = (2c/\rho)\int_0^{x^*}[f_B(x) + 1]dx = (2c/\rho)I(\rho) + O(1/x^{*3}) \tag{S.288}$$

$$\begin{aligned}\varepsilon^{2/3}\int_{-x^*}^{x^*}\gamma_1(x)dx &= (3/10)(\varepsilon/\tau)^{2/3}\int_0^{x^*}[3f_B(x) + \rho f_C(x)]dx \\ &= -(9/10)(\varepsilon/\tau)^{2/3}x^* + O(\varepsilon^{2/3})\end{aligned} \tag{S.289}$$

$$\begin{aligned}-(1/10)(\varepsilon/\tau)^{2/3}\int_{-x^*}^{x^*}x\gamma_0(x)dx &= -(1/10)(\varepsilon/\tau)^{2/3}\int_0^{x^*}xf_A(x)dx \\ &= -(1/10)(\varepsilon/\tau)^{2/3}x^* + O(\varepsilon^{2/3}) ,\end{aligned} \tag{S.290}$$

where

$$I(\rho) \equiv \int_0^{\infty}[f_B(x) + 1]dx . \tag{S.291}$$

Thus,

$$I_i = (2c/\rho)I(\rho) - (\varepsilon^{1/3}/\tau)\zeta + O(\varepsilon^{2/3}) . \tag{S.292}$$

This expression, (S.285), and $I_o + I_i = 0$ imply that

$$c_0 = \frac{\pi\rho}{(4\sqrt{\tau^2 - 1})I(\rho)} . \tag{S.293}$$

[For c_1, see equation (72) of Rossettos & Sanders(1965).] Fig. 5 of Rossettos &

Sanders displays graphs of $I_1(x, \rho) \equiv \int_0^x [f_B(\hat{x})+1]d\hat{x}$ *vs.* x for $\rho = .5, 1, 5, 15$. As these authors note, $I_1(6, \rho) \approx I(\rho)$, so we read-off $I(.5) = .5\cdots$, $I(1) = 1.6\cdots$, $I(5) = 1.5\cdots$, and $I(15) = 2.1\cdots$.

Finally, note from (S.266), (S.284), and (S.286) that

$$\frac{N_\theta}{p_0 b} = \tfrac{1}{2} + f' + 2(\tau + \sin\alpha)\cos(\alpha + \beta/2)\sin(\beta/2)$$
$$= \tfrac{1}{2} + (\varepsilon/\tau)^{2/3}\phi'(x) + O(\varepsilon^{4/3})\,. \qquad \text{(S.294)}$$

Figures 5.14 and 5.15, which are reproduced from Rossettos & Sanders (1965) (with slight changes to conform with our notation), show, respectively, 2 deformed meridians for $\rho = 0.3$ and $\rho = 15$ and graphs of $N_\theta / p_0 b$ *vs.* α for $\rho = 0.3$, 0.5, 1, and 15, computed for $\tau = 1.5$ and $\varepsilon^2 = 0.0288$.

T. Nonlinear, Large-Strain Membrane Theory (Including Wrinkling)

1. What is a membrane?

The field equations of the membrane (or momentless) theory of shells—one of the simpler and more important subtheories of shells—are derived, formally, by deleting the stress couples from the equations of motion. Despite its limitations, membrane theory has been used extensively because it often delivers good, practical results, relatively easily.

We shall restrict our discussion in this section to the *elastostatics* of aximembranes. For *elastodynamic* examples, see subsection U.5. If we set $l = \omega = 0$ in (D.5), it follows that $\mathbf{N}_\sigma \cdot \mathbf{m} = 0$; hence the transverse shear stress

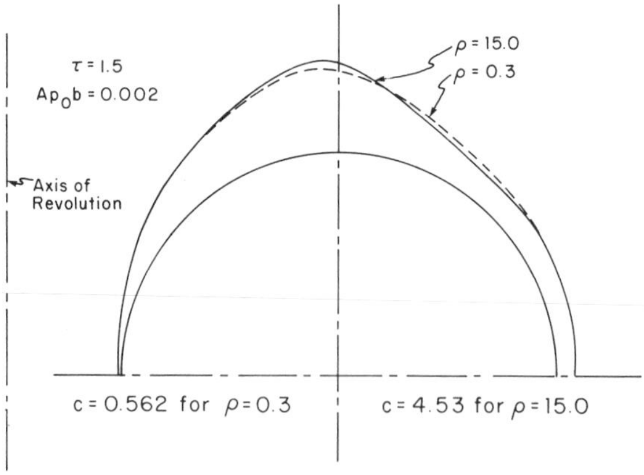

Fig. 5.14. Deformed meridian (magnified) (from Rossettos & Sanders, 1965).

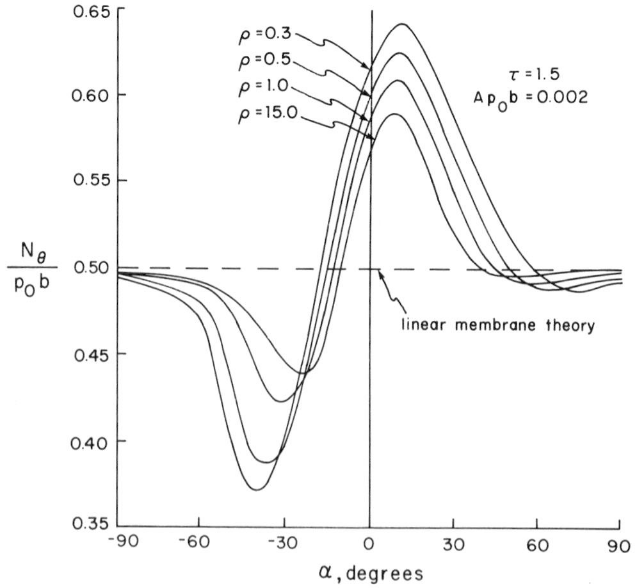

Fig. 5.15. Hoop stress resultant N_θ (from Rossettos & Sanders, 1965).

resultant is also zero, leaving the stress resultants N_σ and N_θ as the only static variables.[24] We emphasize that the vanishing of Q is a *consequence* of the membrane assumption and not a part of it. Nevertheless, it is convenient to group $Q=0$ with the membrane assumptions and to delete the transverse shearing energy from the strain-energy density Φ and the mixed-energy density Ψ.

The membrane assumption is justified in 2 important, distinct cases: (a) as an "interior" approximation to the full shell equations and (b) as a model for a shell that *cannot* support bending stresses. In case (a), there is extensive experimental and theoretical evidence showing that if we are away from boundaries and discontinuities, if the external loads vary smoothly, and that if neither short-wavelength vibrations nor buckling is of concern, then neglect of the stress couples in the equilibrium equations yields a good description of the response of the shell to external loads.[25] Indeed, a qualitative examination of the equations for axishells, such as (R.9) and (R.10), reveals the existence of a small

[24]It is understood, implicitly, that the transverse shearing strain g vanishes together with Q, so that $\bar{\alpha}=\alpha+\beta$, $\cos(\alpha+\beta)==\bar{\mathbf{t}}\cdot\mathbf{e}_r$, $\lambda_\sigma=1+e_\sigma$, and $\lambda_\theta=1+e_\theta$.

[25]In general, we must add the proviso that the shell must be constrained against inextensional bending. However, in axishells this is an automatic consequence of the permitted deformations.

parameter, $\varepsilon^2 \equiv \sqrt{AD}/R$, multiplying the *higher derivatives*. (Here, R is a typical dimension of the reference surface, such as a radius of curvature.) If the scale of differentiation is O(R), as in case (a), then the ε^2-terms are relatively small and may be suppressed. This gives the equations of membrane theory. However, in boundary or interior layers, the scale of differentiation, as we have seen in various examples in Section V.S, may be O(εR), and the bending terms may not be neglected. Boundary layers, of course, are the well-known mechanism which permits satisfaction of all the boundary conditions at an edge of a shell while interior layers (as we saw in subsection V.S.4 with dimpled spherical shells and in subsection V.S.5 with pressurized toroidal shells) prevent the formation of discontinuities in slope or displacement. Furthermore, within boundary or interior layers, the dominant bending stresses are usually of the same order of magnitude as the dominant membrane stresses. In summary, most axishell solutions can be closely approximated by an interior membrane state plus one or more boundary or interior layers (or both) in which bending, nonlinear membrane, and 3-dimensional effects intermix, often on different length scales.

Case (b) concerns shells (or, more precisely, models of shells) that cannot support stress couples—i.e., *true membranes*. Here, we have in mind biologic membranes, cloth, and rubber balloons or other inflatables. Roughly speaking, the membrane model is valid if the geometry, external loading, and boundary conditions are such that $\sqrt{M_\sigma{}^2 + M_\theta{}^2} \ll h\sqrt{N_\sigma{}^2 + N_\theta{}^2}$, *everywhere.*

In many cases, the limiting behavior of a shell as $h/R \to 0$ can be modeled by that of a true membrane, providing the load is sufficiently high. We cite as an example the asymptotic behavior of a pressurized spherical cap discussed in subsection S.4. (See also Fig. 5.13). Taking there $r = 1$ (to make α_1 independent of h), we find that, if the shell is deep, nonlinear membrane behavior emerges for pressures such that $s > 0$; if the shell is shallow or flat, nonlinear membrane behavior emerges if $s \geq 2$. We observe that $s = 0$ or $s = 2$ corresponds to pressures such that $p_0/E = O(h^2/R^2)$ or $O(h/R)$, respectively. The case $s = 4$, where $p_0/E = O(1)$, is a large-strain problem, lying deep in the "membrane region" of Fig. 5.13. In this region, appropriate membrane behavior can be expected, even if the thickness is relatively large as in rubber-like or biologic shells.

Concerning the theory of true membranes, we note that:

(i) Boundary conditions or other constraints on rotation, that normally are realized by bending stiffness, may produce angular discontinuities in the deformed reference surface. The possibility of such discontinuities, especially at boundaries, should always be considered in any membrane problem. For a more extensive discussion of boundary conditions in membrane theory, see Section V.

(ii) To accommodate external loads, true membranes change shape, sometimes considerably, as we have seen in some of the examples in Section S. To cite one, a tubular *shell* (of finite thickness) responds to a radial edge load by forming a narrow bending layer at the edge, while a tubular *membrane* responds with an edge rotation of ninety degrees.

(iii) An attempt to impose compressive stresses on membranes results in wrinkling, to be discussed later in this section.

The question: Is there a solution of the membrane problem?, i.e., Given the undeformed metric, loading, and boundary data, is there a deformed shape? is not easy to answer, except in the trivial case of applied edge moments or transverse edge shears when it is "No". For example, what can one say about the pressurized toroidal aximembrane of tilted cross section discussed in subsection S.5 in which a net axial force has to be carried by the membrane across the deformed crowns where $\bar{\alpha}=0, \pi$? Nevertheless, solutions do exist for many geometries, loads, materials, and boundary conditions, so that seeking a membrane solution to a shell problem is worthwhile, even if an existence theorem is lacking.

The predicted behavior of a membrane depends strongly on the degree of nonlinearity we incorporate into our theory—we can ignore nonlinear effects (i.e., linearize), we can include nonlinear geometrical effects only (nonlinear, small-strain theory), or we can include both geometrical and material nonlinearities (large-strain theory). Moreover, some authors have introduced material nonlinearities into geometrically linear theories.

Linear membrane theory is not shape-adjustable. It leads to a 2nd-order differential system which can accommodate only 2 types of boundary conditions, namely, a specified axial load at one edge plus the axial position at the other[26] or, specified axial position at both ends (an apex of an axishell should be regarded as an edge in this regard). Any boundary conditions beyond these can be satisfied only by invoking bending boundary layers, as discussed before.

The nonlinear, small-strain theory of membranes includes terms that are nonlinear in the rotations and is shape-adjustable. The governing differential equations are of 4th order which means that 2 boundary conditions can be specified at each edge. As a consequence, whereas linear shell theory would predict boundary layers of width $O(\sqrt{hR})$, the nonlinear, small-strain theory predicts boundary layers of width $O(\sqrt{e}R)$, where e is a typical extensional strain. Away from edges, the predictions of this and linear theory should be close. More details appear later and in Section V. Several examples of small-strain nonlinear membrane problems were given in Section S.

In large-strain problems, $e=O(1)$ and small-strain membrane boundary layers are obliterated. Moreover, the formulation of the governing equations and the solution techniques differ essentially from those of the small-strain case. We shall give a few examples presently.

---

[26]The latter boundary condition is a rigid body displacement and, as such, can be specified at any convenient location along the axishell. We can specify, instead, the tangential displacement $\mathbf{u}\cdot\mathbf{t}$.

Fig. 5.16a depicts a long cylindrical shell subjected to an inclined edge load. In subsection S.1 we determined the stresses and deformation using nonlinear, small-strain theory. Here, we supplement this analysis with a qualitative exposition. A linear analysis, using membrane theory alone, cannot accommodate the inclined load—only the evocation of a boundary layer from the full linear shell equations, as shown in Fig 5.16b, can save the day. Fig. 5.16c shows the deformation predicted by nonlinear, small-strain membrane theory with its characteristic boundary layer which turns the axial fibers at the edge into a direction parallel to the edge load. Fig. 5.16d shows the prediction of large-strain membrane theory in which the boundary layer of the preceding figure has widened considerably. In Fig. 5.16e, the angle B of the edge load is negative. Fig. 5.16f shows a deformation that is impossible in a membrane because no axial force can be transmitted through a point where the angle of the deformed meridian, $\alpha+\beta$, is zero The true solution must involve an eversion.

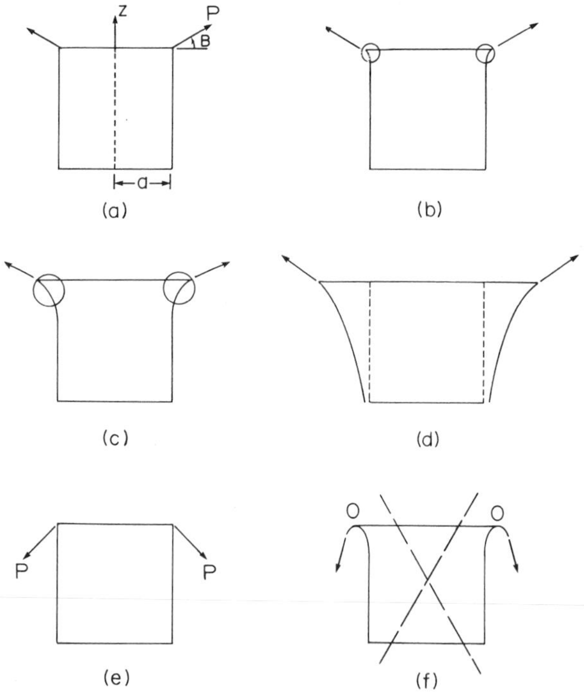

Fig. 5.16. Cylindrical shell with an inclined edge load: (a) basic configuration; (b) membrane solution with bending edge zone; (c) nonlinear, small-strain response; (d) nonlinear, large-strain response; (e) $B<0$ in basic configuration; (f) unacceptable membrane response. (Note that an axial force has to be transmitted through the point O where $\bar{\alpha}=0$.)

2. Constitutive relations

In discussing constitutive relations in large-strain, aximembrane problems, it is common and advantageous to introduce, as arguments of the strain-energy density Φ, the *meridional and hoop stretches*

$$\lambda_\sigma \equiv 1 + e_\sigma \ , \ \lambda_\theta \equiv 1 + e_\theta = \bar{r}/r \tag{T.1}$$

instead of the strains. Often, Φ is derived by descent from a 3-dimensional strain-energy density, W, assuming incompressibility. In compressible materials, one may assume plane stress, as did Simmonds (1987), who derived Φ from a class of 3-dimensional strain-energy densities proposed by Blatz & Ko (1962).

Some strongly deformed materials are elastically isotropic while others can be adequately assumed to be. In these cases, W is a function only of the 3 standard strain invariants I_1, I_2, and I_3. (Of course, W depends also on temperature under non-isothermal conditions and on undeformed position if the material is inhomogeneous.) In terms of the principal *3-dimensional stretches* Λ_1, Λ_2, and Λ_3,

$$I_1 = \Lambda_3^2 + \Lambda_2^2 + \Lambda_3^2 \ , \ I_2 = \Lambda_1^2\Lambda_2^2 + \Lambda_2^2\Lambda_3^2 + \Lambda_3^2\Lambda_1^2 \ , \ I_3 = (\Lambda_1\Lambda_2\Lambda_3)^2 \ . \tag{T.2}$$

In incompressible bodies, $\Lambda_3 = (\Lambda_1\Lambda_2)^{-1}$, so that

$$I_1 = \Lambda_1^2 + \Lambda_2^2 + (\Lambda_1\Lambda_2)^2 \ , \ I_2 = \Lambda_1^2\Lambda_2^2 + \Lambda_1^{-2} + \Lambda_2^{-2} \ , \ I_3 = 1 \ . \tag{T.3}$$

(Other forms of the basic strain invariants are sometimes used.)

Green & Adkins (1970), who assume that the 3-dimensional stresses do not vary through the thickness of a true membrane, give detailed expressions for the stress resultants in terms of the strain-energy density for both compressible and incompressible materials. See their equations (4.9.24)-(4.9.28) and (4.11.7). With $\Phi = hW$, $\lambda_i = \Lambda_i$, and $\lambda_3 = \bar{h}/h$, their results for an *incompressible* isotropic material, in our notation,[27] read

$$N_\sigma = \frac{\partial \Phi}{\partial \lambda_\sigma} = 2h\lambda_\sigma^{-1}(\lambda_\sigma^2 - \lambda_3^2)\left[\frac{\partial W}{\partial I_1} + \lambda_\theta^2 \frac{\partial W}{\partial I_2}\right] \tag{T.4}$$

$$N_\theta = \frac{\partial \Phi}{\partial \lambda_\theta} = 2h\lambda_\theta^{-1}(\lambda_\theta^2 - \lambda_3^2)\left[\frac{\partial W}{\partial I_1} + \lambda_\sigma^2 \frac{\partial W}{\partial I_2}\right] \ . \tag{T.5}$$

Some of the more widely used forms of W are listed in Appendix B of this book. The assumption $\Phi \approx hW$ is very common in large-strain membrane analyses. Its validity has been argued for thin shells of constant thickness, in the limit as $h \to 0$. The approximation has also been used for aximembranes of nonuniform thickness $h = h_0 f(\sigma)$, in the limit as $h_0 \to 0$.

---

[27]Green & Adkins use h_0 for *half* of the undeformed thickness and their stress resultants are measured per unit *deformed* length. For the compressible case, see also Foster (1967), equation (25).

Recent years have seen widespread attempts to model skin, arteries, blood cells, bladders, and heart valves (among other animal parts) as membranes. Though nonisotropy and time-dependent, inelastic behavior are always present to some degree, strain-energy densities have been often assigned to these biologic structures on the basis of crude experiments. Nonlinear elastic membrane theory is usually valid under restricted conditions only (such as cyclic loading), but, by permitting quantification, can help in the understanding of complex biomechanical phenomena. The book by Evans & Skalak (1980) studies biologic membranes in detail. Fung (1980) discusses strain-energy densities for different types of tissue, the most prominent forms being polynomial, exponential, or some combination of the 2. For artery walls and lung parenchyma, he advocates the form

$$\rho_0 W = \tfrac{1}{2} C \exp(a_1 E_\sigma^2 + a_2 E_\theta^2 + 2 a_4 E_\sigma E_\theta) \equiv \tfrac{1}{2} C \exp(Q) \,. \qquad \text{(T.6)}^{28}$$

Here, E_σ and E_θ are Lagrangian strains, but the engineering strains e_σ and e_θ can be used to the same degree of accuracy. The initial mass density ρ_0 can obviously be absorbed into C.

There are many advantages in working with stress resultants rather than with strains or stretches. For this purpose we need the form of the stress-energy density $\Psi(N_\sigma, N_\theta)$, yet little work has been done on reasonable forms of Ψ for strongly stretched membranes. Attempts to invert (T.4) and (T.5) often lead to very cumbersome results, or, at best, are left implicit. See Fung (1979, 1980) who presents methods for obtaining Ψ for quadratic and mixed exponential-quadratic forms. For example, inversion of (T.6) yields

$$\rho_0 W_c = \tfrac{1}{2} C \left[(Q - 1) \exp(Q) + 1 \right] , \qquad \text{(T.7)}$$

where W_c is the complementary strain-energy density such that $\Psi = h W_c$. In (T.7), Q is expressed in terms of the 3-dimensional stresses $S_\sigma = N_\sigma / h$ and $S_\theta = N_\theta / h$ through the equation

$$P = Q \exp(2Q) \, , \qquad \text{(T.8)}$$

where

$$P = (C^2 \Delta)^{-1} (a_2 S_\sigma^2 + a_1 S_\theta^2 - 2 a_4 S_\sigma S_\theta) \;\; , \;\; \Delta \equiv a_1 a_2 - a_4^2 \, . \qquad \text{(T.9)}$$

Fung calls $Q(P)$, which is determined by (T.8), the *universal function* because it does not involve material properties.

The wealth of proposed strain-energy densities—many of which are listed in Appendix B of this book—makes it possible to model many strongly elastic or pseudo-elastic materials by a proper fit of constants to experimental data. This, however, creates the risk of carrying such relations beyond their intended range of validity. Thus, care should be taken not to interpret a *constitutive defect* as a physical reality. This is illustrated by Jafari *et al* (1984) who show,

---

[28]The reader will have no trouble in distinguishing the Q in (T.6) from the symbol for the transverse shear stress resultant.

in tubes and spherical membranes made of *harmonic* materials, that finite inflation pressures can produce infinite stresses as a consequence of the Jacobian of the stress-strain relations vanishing.

3. Field equations and boundary conditions

Using the membrane assumption, we set $Q=0$ in (Q.11) which, together with (Q.3), (Q.4), and (Q.10), implies

$$H=N_\sigma \cos\bar{\alpha}\ ,\quad V=N_\sigma \sin\bar{\alpha} \tag{T.10}$$

$$N_\sigma=(r\sin\bar{\alpha})^{-1}(r_0V_0-\int_0^\sigma rp_V ds) \tag{T.11}$$

$$N_\theta=(rN_\sigma \cos\bar{\alpha})'+rp_H\ , \tag{T.12}$$

where $\bar{\alpha}=\alpha+\beta$. In *linear theory*, we replace $\bar{\alpha}$ by α so that

$$N_\sigma^L=(r\sin\alpha)^{-1}(r_0V_0-\int_0^\alpha rp_V ds) \tag{T.13}$$

$$N_\theta^L=(rN_\sigma \cos\alpha)'+rp_H\ . \tag{T.14}$$

Furthermore, the deformation in this case can be found by linearizing (M.2) and (M.6) with the result that

$$\beta_L \sin\alpha=e_\sigma^L\cos\alpha-(re_\theta^L)' \tag{T.15}$$

$$\bar{r}-r=re_\theta^L \tag{T.16}$$

$$w_L-w_0=\int_0^\sigma(\beta_L\cos\alpha+e_\sigma^L\sin\alpha)ds\ . \tag{T.17}$$

In the above, the constitutive relations are used to express e_σ^L and e_θ^L in terms of N_σ^L and N_θ^L. Linear constitutive relations are normally used, but *material nonlinearity* has occasionally been introduced at this stage by some authors. The boundary conditions which can be imposed on the solutions of (T.13)-(T.17) agree with the results of our previous qualitative discussion on the form and boundary conditions of linear membrane theory.

If the rotations are not small, we cannot replace $\bar{\alpha}$ by α and the stress resultants in (T.11) and (T.12) will no longer be independent of the deformation. Moreover, the exact expression, (M.2), rather than its linearized form, (T.15), must be used to compute β. This leads to a 2nd-order differential equation in β, which increases the order of the whole differential system to 4.

To obtain this 4th-order system, we introduce the modified stress-energy density

$$\Omega(N_\sigma,N_\theta)=\Psi(N_\sigma,N_\theta)-N_\sigma-N_\theta\ . \tag{T.18}$$

Then,

$$\lambda_\sigma = -\Omega,_{N_\sigma} \ , \ \lambda_\theta = -\Omega,_{N_\theta} \ , \tag{T.19}$$

and the compatibility condition (M.2) reduces to

$$(r\Omega,_{N_\theta})' - \Omega,_{N_\sigma} \cos\bar{\alpha} = 0 \ . \tag{T.20}$$

Substitution of (T.11) and (T.12) into (T.20) yields a 2nd-order, quasilinear differential equation for $\bar{\alpha}$. To obtain the deformed position, we use

$$\bar{r} = r\lambda_\theta \ , \ \bar{z} = \bar{z}_0 + \int_0^\sigma \lambda_\sigma \sin\bar{\alpha}\, ds \ . \tag{T.21}$$

This completes the system of equations.

In some applications, it is useful to work with the stress function $F = rH$ as the basic field variable rather than $\bar{\alpha}$. From (T.10) follows

$$\sin\bar{\alpha} = (F^2 + r^2V^2)^{-1/2} rV \ , \ \cos\bar{\alpha} = (F^2 + r^2V^2)^{-1/2} F \ , \tag{T.22}$$

where V, given by (Q.5), is assumed to be a known function. Substitution of (T.22) into (T.11) and (T.12) and the resulting expressions into (T.20) leads to the a single differential equation for F. See also subsection R.1.

Classical boundary conditions for the nonlinear theory of elastic aximembranes are derived in Section V from a variational principle. For convenience, we summarize the main results here:

> On ∂U, kinematical conditions are prescribed, i.e., the components $(\bar{r}, \bar{z})$ of the deformed position. The deformed angle $\bar{\alpha}$ cannot be prescribed on ∂U, but must come out of the solution. On ∂N, loads are prescribed, i.e., V and $\bar{\alpha}$ or, alternatively, V and H. (Thus, while it is acceptable to prescribe $\bar{\alpha}$ on ∂N, its prescription on ∂U will, in general, result in a discontinuity in slope on ∂U.) Force boundary conditions are subject to overall axial equilibrium that follows from (Q.5) as
>
> $$r_1 V_1 = r_0 V_0 - \int_0^L r p_V d\sigma \ . \tag{T.23}$$
>
> Thus, V cannot be specified on both edges simultaneously. The boundary conditions we have cited are consistent with the Principle of Virtual Work and are not restricted to potential edge loads.

The formulation of nonlinear membrane theory presented above, while complete, is not always convenient for applications, mainly because Ω is not always available and because, in numerical work, a system of 1st-order differential equations (plus, possibly, algebraic equations) is simpler to use than a single, quasilinear differential equation. Moreover, for special geometries and loadings, direct application of the unreduced equations is sometimes simpler. Nevertheless, if the strains are small, our single equation for F reduces to the more manageable (R.26), which has been applied to the solution of several problems. See Sections R and S for details.

To formulate the field equations of the nonlinear theory of elastic membranes in terms of λ_σ, λ_θ, and $\bar{\alpha}$, we use (T.11), (T.12), (T.20), and (T.21) to obtain

$$\left[r \frac{\partial \Phi}{\partial \lambda_\sigma} \sin \bar{\alpha} \right]' = - r p_V \tag{T.24}$$

$$\left[r \frac{\partial \Phi}{\partial \lambda_\sigma} \cos \bar{\alpha} \right]' = \frac{\partial \Phi}{\partial \lambda_\theta} - r p_H \tag{T.25}$$

$$(r \lambda_\theta)' = \lambda_\sigma \cos \bar{\alpha} \tag{T.26}$$

$$\bar{z}' = \lambda_\sigma \sin \bar{\alpha} \ , \ \bar{r} = \lambda_\theta r \ . \tag{T.27}$$

This system is convenient for extracting approximate solutions via power series or finite differences. In any iterative method, we need, at each stage, to express λ_σ in terms of λ_θ and $\partial\Phi/\partial\lambda_\sigma$. While this can be done analytically for many forms of Φ, a numerical method might be preferable if the extraction is cumbersome.

In (T.24) and (T.25), p_V and p_H can be arbitrary functions of σ, λ_σ, λ_θ, or $\bar{\alpha}$. For normal pressure, we have, noting (Q.8),

$$r p_V = - p \bar{r} \bar{r}' \ , \ r p_H = p \bar{r} \bar{z}' \ , \tag{T.28}$$

where the internal pressure p is force per unit *deformed area*. For a given $p(\bar{r})$, (T.24) can be integrated readily. Membrane caps subject to gravity or pressure normal to the *undeformed* shape are treated by Dickey (1987), who formulates his equations in terms of a stress function and radial and axial displacements.

Some authors refer their equations to the deformed configuration which, in some cases, reveals a formal simplicity. For example, the field equations of a pressure loaded aximembrane, so written exhibit the following static-geometric duality:

$$\frac{d}{d\bar{r}} (\bar{r} T_\sigma) = T_\theta \ , \ \frac{d}{d\bar{r}} (\bar{r} \bar{b}_\theta) = \bar{b}_\sigma \tag{T.29}$$

$$T_\sigma \bar{b}_\sigma + T_\theta \bar{b}_\theta = p \ , \tag{T.30}$$

where $T_\sigma = \lambda_\theta^{-1} N_\sigma$ and $T_\theta = \lambda_\sigma^{-1} N_\theta$ are forces per *unit deformed length* and $\bar{b}_\sigma$ and $\bar{b}_\theta$ are the principal curvatures of the deformed reference surface. To complete the set of field equations, one must relate the curvatures to the stretches [as, for example, in (U.37) and (U.38)] and add constitutive relations. For additional literature, see Foster (1967), Green & Adkins (1970), Libai & Simmonds (1983, Chapter VI.C), Wu (1979), and Yang & Feng (1970), among others.

4. Some special large-strain problems

In Section S, we solved several membrane problems using the nonlinear, small-strain theory. Here, we consider a few large-strain problems.

(a) Cylindrical membranes. The differential equations of pressure-loaded, cylindrical membranes can be reduced to quadratures by using the 1st integral of Pipkin (1968). To derive this 1st integral, we start with $(T.29)_1$, written in the form

$$\frac{N_\theta}{\lambda_\sigma} = \frac{d}{d\bar{r}}\left[\bar{r}\,\frac{N_\sigma}{\lambda_\theta}\right] = R\,\frac{dN_\sigma}{d\bar{r}}\,, \tag{T.31}$$

where R is the radius of the undeformed cylindrical reference surface and $\lambda_\theta = \bar{r}/R$. Expressing N_σ and N_θ as derivatives of the strain-energy density and rearranging terms, we obtain

$$\frac{\partial\Phi}{\partial\lambda_\theta}\,\frac{d\lambda_\theta}{d\bar{r}} + \frac{\partial\Phi}{\partial\lambda_\sigma}\,\frac{d\lambda_\sigma}{d\bar{r}} = \frac{d}{d\bar{r}}\left[\lambda_\sigma\,\frac{\partial\Phi}{\partial\lambda_\sigma}\right]. \tag{T.32}$$

If Φ is a function of λ_σ and λ_θ only (and thus does not depend explicitly on the coordinate), then the left side of (T.32) is a total derivative and we have

$$\Phi - \lambda_\sigma\,\frac{\partial\Phi}{\partial\lambda_\sigma} = C\,, \tag{T.33}$$

where C is a constant of integration. In Section V, we show that (T.33) is a special case of a more general integral for cylindrical shells that is equivalent to the Euler-Lagrange equations of a variational principle.

Following Wu (1970b), we assume that (T.33) can be solved for the meridional stretch in the form

$$\lambda_\sigma = \hat{\lambda}_\sigma(\bar{r}, C)\,. \tag{T.34}$$

If (T.34) is introduced into the integrated form of (T.24), we find that

$$\frac{\partial\Phi}{\partial\lambda_\sigma}\sin\bar{\alpha} = \frac{1}{R}\int_{\bar{r}_0}^{\bar{r}} p(\sigma)\sigma d\sigma + V_0\,. \tag{T.35}$$

Thus, we obtain $\bar{\alpha}$ in terms of $\bar{r}$ and various constants. To express $\bar{r}$ in terms of z, i.e., to reintroduce material coordinates, we must solve

$$\frac{dz}{d\bar{r}} = \frac{1}{\lambda_\sigma \cos\bar{\alpha}}\,, \tag{T.36}$$

where λ_σ and $\bar{\alpha}$ are to be expressed in terms of $\bar{r}$ via (T.34) and (T.35). The deformed axial coordinate may be obtained by solving

$$\frac{d\bar{z}}{d\bar{r}} = \tan\bar{\alpha}\,. \tag{T.37}$$

The solution procedure we have outlined is not devoid of numerical difficulties, the chief being the satisfaction of boundary conditions which may lead to transcendental equations. Still, Wu (1970a,b) used it successfully to solve several problems involving Neo-Hookean and Mooney materials, including the stretching of a tube by a radial edge force until it flattens into an annulus. With a neo-Hookean material, (T.36) for this latter problem reduces to

$$\lambda_\sigma = \frac{d\bar{r}}{dz} = \frac{1}{\sqrt{2}}\{f(\bar{r}) + [f^2(\bar{r}) + 12(R/\bar{r})^2]^{1/2}\}^{1/2}, \qquad \text{(T.38)}$$

where b is the (stress-free) inner radius of the annulus, and

$$f \equiv (\bar{r}/R)^2 - (b/R)^2 - 2(R/b). \qquad \text{(T.39)}$$

The stress resultants are found easily in terms of λ_σ and $\lambda_\theta = \bar{r}/R$. Here, we have the large-strain solution to the problem of Flügge & Riplog (1956) that we discussed in subsection S.1 within the framework of small-strain theory.

Several authors, assuming a Mooney-Rivlin material, have considered the axial stretching of a cylindrical membrane attached to rigid boundary rings. See Fig. 5.17. Stoker (1963) obtained an approximate solution using finite differences while Wu (1970b) discussed the reduction of the system of equations to quadratures but did not solve any specific cases. Yang & Feng (1970) integrated the governing differential equations numerically using a shooting method together with a Runge-Kutta integration process. Their results are presented in graphical form. Finally, Wu (1974b) showed that if the stretching is large enough, the deformed shape is a uniformly stretched catenoid.

If an aximembrane is inflated slowly, then, for some geometries and materials, there may be a critical pressure, p_M, where the graph of pressure *vs.* radial stretch has a maximum. Kydoniefs & Spencer (1969) studied this problem for a closed cylindrical membrane under internal pressure only. Benedict *et al* (1979) extended the analysis to simultaneous extension and inflation, using an exponential strain-energy density. In both studies, a p_M was found; in the latter it was further shown that p_M decreased as the axial elongation increased.

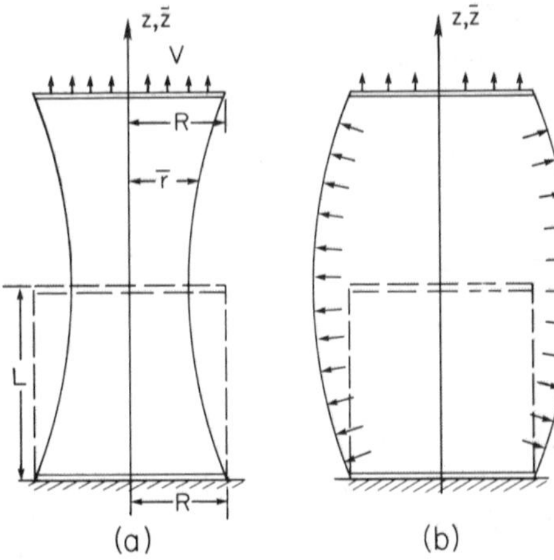

Fig. 5.17. Axial extension of a cylindrical membrane: (a) unpressurized; (b) pressurized.

(b) Spherical membranes. The simplest problem is to determine the deformed radius, $\bar{R}$, of a closed spherical membrane (i.e., a spherical balloon) as a function of its undeformed radius, R, its undeformed thickness, h, and the internal pressure, p, given the strain-energy density. Conversely, given experimental data, one should be able to infer something about the strain-energy density. Of special interest is the possible existence of a maximum pressure, p_M, beyond which the radius will increase with decreasing pressure. Such a phenomenon leads to a limit-point instability but, in contrast to small-strain buckling, the causes are mainly material nonlinearities. If, beyond the limit point, the deformation is *volume controlled*, then the pressure will decrease with increasing volume until, possibility, some minimum pressure, p_m, is reached beyond which the pressure again builds as the volume increases.

Another possibility in slowly inflated spherical membranes is bifurcation into a nonspherical aximembrane, followed by a return to sphericity at higher pressures. This has been observed in experiments on rubber balloons.

If there is spherical symmetry, then

$$\lambda \equiv \lambda_\sigma = \lambda_\theta = \bar{R}/R \ , \ \bar{b}_\sigma = \bar{b}_\theta = \bar{R}^{-1} = (\lambda R)^{-1} \ , \ T \equiv T_\sigma = T_\theta \ . \qquad \text{(T.40)}$$

From (T.30) we obtain

$$T = \tfrac{1}{2} p\bar{R} = \tfrac{1}{2}\lambda pR \ \text{ or } \ N \equiv N_\sigma = N_\theta = \tfrac{1}{2}\lambda^2 pR \ . \qquad \text{(T.41)}$$

Introduction of $N = \partial\Phi/\partial\lambda$ into (T.41) yields the required p–λ relation. For incompressible materials—see (T.4)—we have

$$p = \frac{4h}{\lambda R}(1-\lambda^{-6})\left[\frac{\partial W}{\partial I_1} + \lambda^2 \frac{\partial W}{\partial I_2}\right] \equiv \hat{p}(\lambda) \ . \qquad \text{(T.42)}$$

For specific forms of W, $\hat{p}(\lambda)$ may have a maximum and, possibly, a subsequent minimum. These stationary values occur at critical points, λ_{cr}, where $d\hat{p}/d\lambda = 0$. For a Mooney material,

$$\frac{d\hat{p}}{d\lambda} = \frac{4h}{R}\left[\left[\frac{7}{\lambda^8} - \frac{1}{\lambda^2}\right]C_1 + \left[1 + \frac{5}{\lambda^6}\right]C_2\right] = 0 \ . \qquad \text{(T.43)}$$

This equation has been studied by Green & Adkins (1970) who show that, if $0 < C_2/C_1 < 0.21$, then there is a maximum pressure as well as a minimum pressure at a larger value of radial stretch. If $C_2/C_1 > 0.21$, the p-λ curve increases monotonously. In rubbers, $C_2/C_1 < 0.21$ ordinarily. In a neo-Hookean material, $C_2 = 0$ and $p_M = 24(7)^{-7/6}(h/R)C$ at $\lambda_{cr}^6 = 7$; there is no p_m. Using an incremental elastic material model, Durban & Baruch (1974) found that $p_M = (4/3e)E\,(h/R)$ at $\lambda_{cr} = e^{1/3}$, where e is the base of the natural logarithms. And Needleman (1977), working with the 3-dimensional strain-energy density

$$W = \sum_r (\mu_r/\alpha_r)(\lambda_\sigma^{\alpha_r} + \lambda_\theta^{\alpha_r} + \lambda_3^{\alpha_r} - 3) \qquad \text{(T.44)}$$

proposed by Ogden (1972), obtained values of p_M and p_m for 2 "real" materials. In (T.44), the μ_r's and α_r's are experimentally determined constants. The deformed thickness can be found from $\bar{h} \approx \lambda_3 h$.

The bifurcation of an inflated spherical membrane into an aximembrane of more general shape has been considered by Needleman (1977), who quotes earlier work by Feodosev (1968), Shield (1972), and Needleman (1976), and who used a Ritz-Galerkin procedure to solve, approximately, the nonlinear equations for an *imperfect* spherical membrane. In each of his 2 material models, he found a loss of sphericity at values of λ greater than the lowest λ_{cr}. He also found that there was a trend back to sphericity as λ increased. In earlier work, the bifurcation point(s) were calculated by superimposing small nonspherical deformations on a large spherical deformation. The eigenfunctions were Legendre polynomials, with $n = 1$ yielding the smallest λ_{cr}. (See Section W for the corresponding energy method.)

Air-supported structures and biologic cells are examples of structures that can be modeled as inflated spherical membranes to which additional loads are applied. Feng and Yang (1973), using numerical integration, studied a complete, gas-filled spherical membrane of Mooney material compressed between 2 rigid plates, as shown in Fig. 5.18a. Lardner & Pujara (1980) extended the

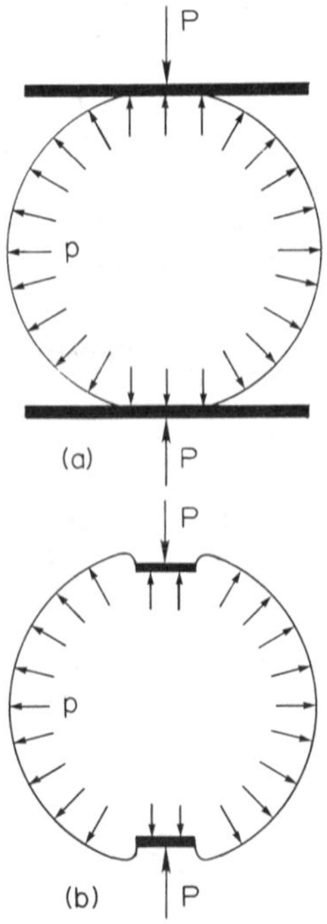

Fig. 5.18. Inflated spherical membranes under superimposed loads:
(a) compressed between rigid plates; (b) indented by rigid disks.

analysis to liquid-filled membranes (which are common in biological problems) and used the following strain-energy density which appears to be suitable for red blood cells:

$$W = (1/4)B\,(\tfrac{1}{2}\bar{I}_1^2 + \bar{I}_1 - \bar{I}_2) + (1/8)\bar{I}_2^2 \,, \tag{T.45}$$

where

$$\bar{I}_1 \equiv \lambda_\sigma{}^2 + \lambda_\theta^2 - 2 \;,\; \bar{I}_2 \equiv \lambda_\sigma{}^2\lambda_\theta^2 - 1 \,. \tag{T.46}$$

Using the Rayleigh-Ritz method in conjunction with the Principle of Stationary Potential Energy, Taber (1984) computed load-deflection curves for a liquid-filled spherical membrane indented by rigid disks, as shown in Fig. 5.18b. This study has application to tonometry, the measurement of inner ocular pressure in the eyeball.

(c) Plane circular membranes. Here, the classical, large-strain problem is to compute the inflated shape of an initially flat, circular membrane of radius b, with or without initial extension, fixed at its edge. If there is no stretching before inflation, λ_θ is initially 1 everywhere and remains equal to 1 at the edge. At the center of the membrane, axisymmetry requires that $\bar{\alpha} = \bar{r} = 0$, $\lambda_\sigma = \lambda_\theta$, and $N_\sigma = N_\theta$.

The field equations can be taken to be (T.24)-(T.26) with $\alpha = 0$, $\sigma \equiv r$, and $\bar{\alpha} = \beta$. Although many investigators prefer other forms and use various numerical integration schemes, all start from the center of the membrane and integrate until the assigned edge value of λ_θ is reached; solutions are adjusted to the initial radius by scaling. These investigators have compared their results with experiments of Treloar (1944) on rubber to evaluate various proposed strain-energy densities. For details, see Green & Adkins (1970), Klingbeil & Shield (1964), Hart-Smith & Crisp (1967), Yang & Feng (1970), and Pujara & Lardner (1978).

Instead of numerical integration, one can use the Theorem of Stationary Potential together with a Rayleigh-Ritz procedure. In this way Tielking & Feng (1974) obtained the deformed profile of an inflated circular membrane with a rigid inclusion.

Exact solutions for inflated circular membranes can be obtained for *the inverse problem:* To find the the thickness distribution and pressure that produce a *given* final shape. Thus, Hart-Smith & Crisp (1967) solved the problem of inflating an initially flat circular membrane into a preassigned spherical cap. Their method uses the fact that, if the deformed shape is spherical, then $\lambda_\sigma = \lambda_\theta = \lambda$ everywhere and, from (T.40) and (T.41), $T = T_\sigma = T_\theta = \tfrac{1}{2}p\bar{R}$, where $\bar{R}$ is the radius of the sphere . However, λ is not constant because, for the deformed spherical membrane,

$$\lambda_\sigma = \bar{R}\,\frac{d\bar{\alpha}}{dr} \;,\; \lambda_\theta = \frac{\bar{r}}{r} = \frac{\bar{R}\sin\bar{\alpha}}{r} \,. \tag{T.47}$$

Since $\lambda_\sigma = \lambda_\theta$, we have

$$\frac{dr}{r} = \frac{d\bar{\alpha}}{\sin\bar{\alpha}}, \tag{T.48}$$

which implies that $r = C\tan(\bar{\alpha}/2)$; C is determined by the boundary condition $\bar{\alpha}(b) = \hat{\alpha}$, where $(\hat{\ })$ denotes, as usual, a prescribed quantity. Thus,

$$r = b\,\frac{\tan(\bar{\alpha}/2)}{\tan(\hat{\alpha}/2)}, \tag{T.49}$$

which yields

$$\bar{\alpha} = 2\tan^{-1}[(r/b)\tan(\hat{\alpha}/2)] \text{ and } \lambda = \frac{\cos^2(\bar{\alpha}/2)}{\cos^2(\hat{\alpha}/2)}. \tag{T.50}$$

Substitution of (T.50) and the value of T into the constitutive relation determines h in terms of $\bar{R}$, $\hat{\alpha}$, p, and the material constants.

The more elaborate problem of finding the initial thickness distribution that leads to an inflated paraboloidal membrane was solved by Vaughan (1980).

(d) Annular membranes. As in Fig. 5.19, consider an annular membrane fixed to a rigid wall at its outer radius, b; its inner edge is fixed to a freely moving rigid hub of radius a to which a downward vertical force of magnitude P is applied. As $a/b \to 0$, we approach a complete membrane under a concentrated or point load at its center. The small-strain, moderate-rotation theory of Föppl (1905), discussed at some length in subsection S.2, predicts a finite deflection Δ under such a loading. In particular, if $\nu = 1/3$, there is a simple, closed-form solution of the governing equations discovered by Schwerin (1929) which yields the load-deflection relation

$$P = \frac{1}{3}\pi E h b\left[\frac{\Delta}{b}\right]^3. \tag{T.51}$$

As Schwerin's solution predicts arbitrarily large stresses and strains in a neighborhood of the concentrated load, the assumptions underlying Föppl's theory are violated and the validity of (T.51) is questionable.

To see what an exact analysis predicts, Fulton & Simmonds (1986) applied large-strain theory to membranes with various strain-energy densities and hub radii and solved numerically (T.24)-(T.27) (in a slightly modified form). Their major conclusion is that as $a/b \to 0$, quadratic, neo-Hookean, or Mooney forms of W lead to infinite deflections, but the Rivlin-Saunders form, $W = C_1[(I_1 - 3) + \Gamma f(I_2 - 3)]$, with $f(x) = x^p$, does not, if $p > 2$. Typical deformed profiles for this latter material for a fixed (dimensionless) load and various hub to outer radius ratios are shown in Fig. 5.19.

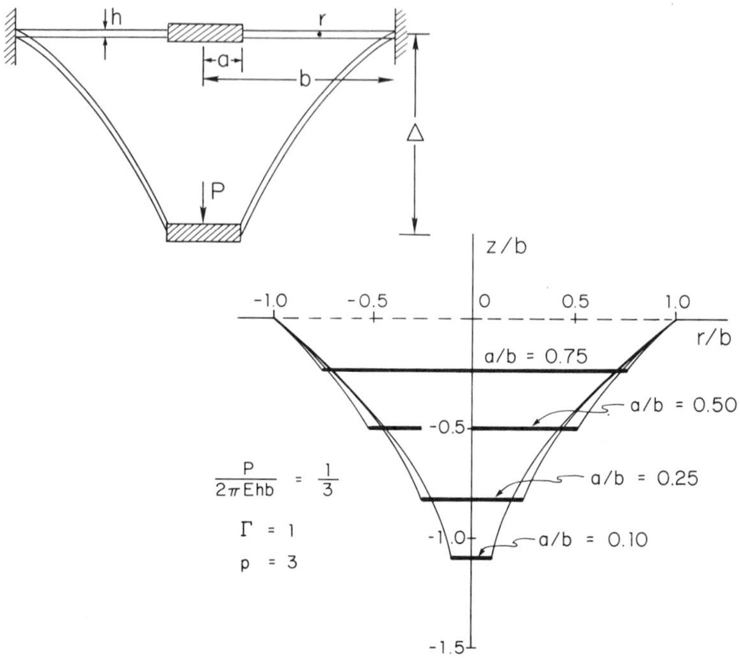

Fig. 5.19. Deformed profiles for a Rivlin-Saunders material for various hub ratios at a fixed vertical load.

5. Asymptotic approximations

If the strains in a membrane are large everywhere, the strain-energy density can be simplified by dropping some of the less important terms. This suggests the possibility of obtaining closed-form *asymptotic* solutions. Extending the work of Isaacson (1965), Wu (1979) has shown that if

$$W \sim C_1[I^2-(2+C_2)J]\,,\ I \equiv \lambda_\sigma+\lambda_\theta\,,\ J \equiv \lambda_\sigma\lambda_\theta\,,\ C_1, C_2 \text{ constant} \qquad \text{(T.52)}$$

as the stretches approach infinity, then the equilibrium equations for any inflated aximembrane imply that the deformed shape approaches a sphere as the pressure grows. See also the work of Foster (1967) who used a neo-Hookean material. Antman & Calderer (1985a) extended this work by analyizing the restrictions that must be placed on the strain-energy density of a spheroidal *shell* if, under increasing internal pressure, it is to approach a spherical shape. For more details, the reader is referred to Antman's original paper as well as to an accompanying paper by Antman & Calderer (1985b) on the asymptotic state of inflated, noncircular, closed cylindrical *beamshells*.

6. Wrinkled membranes

Very thin membranes will wrinkle rather than support compressive stresses. Often, this drastic alteration of shape results in the formation of wrinkled *tension fields* over portions of the deformed surface. These fields, in which the principal stress resultants are everywhere nonnegative, provide a mechanism for carrying the imposed loads. Within a tension field, the crests and troughs of the wrinkles are parallel to the direction of principal tension, the stress resultant in the normal direction being zero.

Wrinkling of natural objects and artifacts is not uncommon. In some cases, wrinkling can be avoided by pressurization, the addition of tensile edge loads, or by a change in initial shape, i.e., by a change in design; in other cases, it can be tolerated so long as it does not lead to catastrophic failure.

Apparently, Wagner (1929) was the first to apply tension field theory to the analysis of the stable, postbuckling behavior of thin, stiffened, rectangular plates under edge shear stresses. His pioneering work has been widely used in aeronautical and civil engineering designs. Other studies include that of Reissner (1938) on the wrinkling of an annular membrane in shear; those of Stein & Hedgepath (1961), Comer & Levy (1963), and Koga (1972) on wrinkling in bent, beam-like inflated tubes; those of Mansfield (1969, 1970) on the wrinkling of plane membranes; those of Wu (1974a, 1978) on the wrinkling of membranes of revolution; and that of Zak (1982) on the wrinkling of films of arbitrary shape.

The basic idea in the theory of wrinkled membranes is to avoid studying the wrinkled region in detail by replacing it with a smoothed-out *pseudo-surface*. Not only must this pseudo-surface be in equilibrium, but, on it, the minimum principal stress resultant must be zero. Obviously, the stretch of the pseudo-surface in the direction of the zero minimum principal stress resultant will not be equal to that of the actual wrinkled surface—otherwise there would be no wrinkles! To model the behavior of the pseudo-surface, one may attempt to introduce pseudo-constitutive relations. Thus, Reissner (1938) used a special curvilinear, orthotropic material law and Stein & Hedgepath (1961) introduced a variable Poisson's ratio.

We shall follow Wu (1978) who introduced the *wrinkle strain*, $\bar{\lambda}_\theta - \lambda_\theta$, as an additional variable in the analysis of meridionally wrinkled membranes of revolution. Here, λ_θ is the actual circumferential stretch and $\bar{\lambda}_\theta$ is the stretch of the pseudo-surface. The meridional stretch of the membrane and the pseudo-surface are taken to have a common value, λ_σ. The need for an additional variable is obvious since the requirement that one of the stress resultants vanish—N_θ in this case—introduces a field equation in addition to the standard ones of equilibrium, compatibility, and constitution.

For the pseudo-surface, the field equations (T.24)-(T.27) retain their general form, except that $\partial\Phi/\partial\lambda_\theta$ is set to zero in (T.25)[29] and λ_θ is replaced by $\bar{\lambda}_\theta$ in (T.26) and $(T.27)_2$. The actual stretches, λ_σ and λ_θ, should still be used in the constitutive relations, (T.4) or (T.5)[30] (or in their equivalents), but, by setting $N_\theta = 0$ in (T.5), we get an equation that expresses λ_θ in terms of λ_σ. The substitution of this expression into (T.4) allows us to solve for $\partial\Phi/\partial\lambda_\sigma$ $(=N_\sigma)$ in terms of λ_σ only. These results, substituted into (T.24)-(T.27), lead to a complete set of field equations that can be solved by quadratures to find $\lambda_\sigma, \bar{\lambda}_\theta, \bar{\alpha}, \bar{r}$, and $\bar{z}$.

The size of the wrinkled region may be determined from the requirement that its boundary be continuous with the regular part of the deformed surface. Moreover, at such a boundary we must have $N_\theta = 0$ and $\lambda_\theta = \bar{\lambda}_\theta$. Locating this boundary is not always simple.

Using the approach outlined in the last two paragraphs, Wu (1978) analyized the in-plane stretching of an annular Mooney membrane. The reader should consult his paper for details.

Zak (1982) considered the form of the equilibrium equations for the pseudo-surface in a coordinate system aligned with the directions of the principal stress resultants, N_I and N_{II}. Using the fact that $N_{II} = 0$, Zak showed, in the absence of external surface loads, that the crests of the wrinkles form a family of straight lines, i.e., the pseudo-surface is *ruled*. For a hemispherical membrane held at its base and pulled by a concentrated radial force at its pole, it follows that the pseudo-surface must be a cone. If the force is sufficiently small (so that $\lambda_\sigma \approx 1$) or if the membrane is inextensional, then the height of the cone becomes

$$H = \sqrt{(\pi/2)^2 - 1}\,R = (1.211 \cdots)R\,, \qquad \text{(T.53)}$$

where R is the radius of the hemispherical membrane. See Fig. 5.20.

A related practical phenomenon occurs if an air-supported (i.e., pressurized) membrane, in the initial shape of a spherical cap, is pulled down at its apex by a central weight. Here, because there is a surface load, the wrinkles emanating from the point of application of the load form curved lines. In practice, water and/or snow can accumulate in this depression which, ultimately, can lead to collapse.[31] This mode of failure, known as *ponding collapse*, has been analyized by Malcolm & Glockner (1981) who used the fact [from (T.29)] that $T^* = \bar{r}\,\bar{T}_\sigma =$ constant in the wrinkled region, where $\bar{T}_\sigma$ is the meridional force per unit of circumferential distance on the pseudo-surface. Assuming that wrinkling

[29] It is more convenient to use the equivalent equation (T.29) which can be integrated immediately to yield $\bar{r}\,\bar{T}_\sigma = rN_\sigma =$ constant, where $\bar{T}_\sigma$ is measured per unit length of the pseudo-surface.

[30] This is an assumption which is taken in some average sense. The *detailed* stress distribution is, of course, more complicated.

[31] An apex load is not necessary for collapse: an initial depression of the apex might serve as the triggering mechanism.

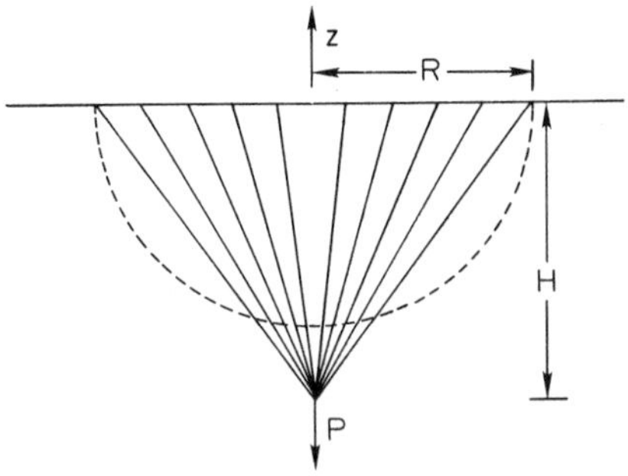

Fig. 5.20. A hemispherical membrane pulled into a cone.

starts at $\bar{r}=\bar{r}_A$ and that the ponding substance (with specific weight ρ) accumulates up to the crest, where, as in Fig. 5.21, $\bar{r}=\bar{r}_B$, $\bar{z}=\bar{z}_B$, we have the following equilibrium equations:

$$T^* \cos\bar{\alpha} = \tfrac{1}{2} p(\bar{r}^2 - \bar{r}_B^2) \ , \ \bar{r}_B < \bar{r} < \bar{r}_A \ (\bar{\alpha} < \pi/2) \qquad \text{(T.54)}$$

$$T^* \frac{d}{d\bar{r}}(\cos\bar{\alpha}) = T^* \bar{\alpha}' = [p - \rho(\bar{z}_B - \bar{z})]\bar{r} \equiv p_n \bar{r} \ , \ 0 < \bar{r} < \bar{r}_B \ (\bar{\alpha} > \pi/2) \qquad \text{(T.55)}$$

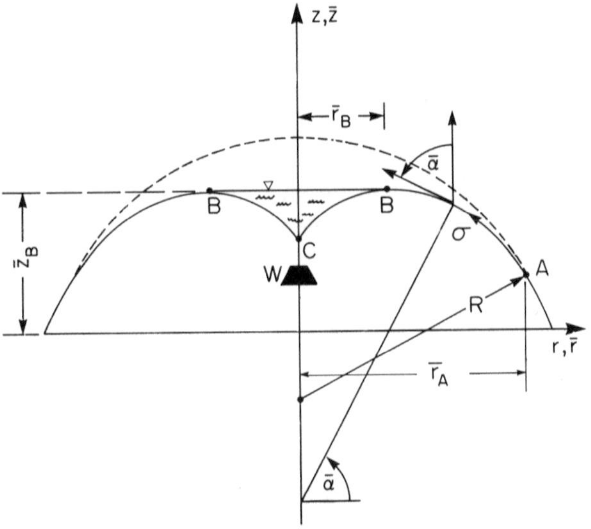

Fig. 5.21. Deflection of an air-supported spherical membrane.

$$T^* \cos \bar{\alpha} = -W/2\pi \ , \ \bar{r} = 0 \, . \tag{T.56}$$

Here, W is the central weight and $\bar{\alpha}$ is the angle between the positive $\bar{z}$-direction and a tangent to the deformed meridian. At $\bar{r} = \bar{r}_A$, $\cos \bar{\alpha} = \bar{r}_A / R$. We must also satisfy the inextensionality condition $L_{AB} + L_{BC} = R \sin^{-1}(\bar{r}_A / R)$ or an equivalent relation for extensible membranes. Here, L_{AB} and L_{BC} denote distances between corresponding points along the meridian.

Malcolm & Glockner solved this system numerically for W in terms of the remaining parameters. Their results, compared with experiments, were too high, but they found that the addition of extra slack (as an imperfection) gave better agreement. No doubt the addition of extensionality could improve the results.

For further studies of this and related problems, see Ahmadi & Glockner (1983), Lukasiewicz & Glockner (1983), and Szyszkowski & Glockner (1984a,b).

Tension field theory is also useful in the analysis of a circular plate subjected to high pressure, provided that the outer edges are *elastically restrained* in their plane. Inward movement of the outside supports may lead to circumferential buckling of the outer region of the plate, which can be modeled as a wrinkled membrane; in the inner region, the Föppl-Kármán equations apply. Croll (1985) analyized a simply-supported plate with an in-plane elastic restraint, such that $N_r = -k_r u$ at $r = a$, where k_r is a constant. He also found the radius, b, of the transition circle between the two regions in terms of the dimensionless parameter $K_r \equiv k_r a / Eh$. Sample results are: $K_r = 1, b/a = 0.8$; $K_r = 0.5$, $b/a = 0.66$.

U. Elastodynamics

Because axishells, by definition, cannot deform inextensionally (except, trivially, as rigid bodies), many interesting phenomena, such as large amplitude, low frequency flexural vibrations (found in beamshells), are absent and the literature on their dynamics is not large. A few topics will be mentioned at the end of this section, but most of our discussion will be devoted to formulations of the field equations.

There are a variety of component forms of the equations of motion of elastic axishells, depending on the basis in the rz-plane and the dependent variables. We make no attempt to list all possibilities. Rather, we concentrate on displacement, rotation-stress resultant, system, and intrinsic formulations, with and without transverse shearing strain.

1. Displacement-shear strain form

In the basis $\{\mathbf{e}_r, \mathbf{e}_z\}$, the equations of balance of linear momentum, which follow from (D.3), (P.1), and (Q.1), are simple:

$$(rH)' - N_\theta + rp_H = mr\ddot{u} \tag{U.1}$$

$$(rV)' + rp_V = mr\ddot{w}\ . \tag{U.2}$$

The equation of balance of rotational momentum is (L.15). Introducing the modified transverse shear stress resultant defined by (P.34) and noting from (P.1) and (P.2) that

$$\bar{\alpha} \equiv \alpha + \beta + \gamma = \tan^{-1}\left[\frac{\sin\alpha + w'}{\cos\alpha + u'}\right] = \bar{\alpha}(u', w'; \alpha)\,, \tag{U.3}$$

we have

$$(rM_\sigma)' + rQ^* + rl = Ir\ddot{\beta}\,, \tag{U.4}$$

where

$$\dot{\beta} = \bar{\alpha}_{,u'}(u', w'; \alpha)\dot{u}' + \bar{\alpha}_{,w'}(u', w'; \alpha)\dot{w}' - \dot{\gamma}\,. \tag{U.5}$$

Here,

$$-\lambda_\sigma^2\bar{\alpha}_{,u'} = \sin\alpha + w'\ \ ,\ \ \lambda_\sigma^2\bar{\alpha}_{,w'} = \cos\alpha + u'\,, \tag{U.6}$$

and

$$\lambda_\sigma^2 = 1 + 2(u'\cos\alpha + w'\sin\alpha) + u'^2 + w'^2\,. \tag{U.7}$$

We assume that the strain-energy density, Φ, is a function of σ and the six strain measures, $\gamma_\sigma, e_\theta, \gamma, \kappa_\sigma, \kappa_\theta, \kappa_n$, introduced in Section P. Using the strain-displacement relations (L.7), (P.5), (P.21), (P.28), and (P.30), we have

$$\begin{aligned}\Phi &= \tilde{\Phi}(\gamma_\sigma(u', w'), e_\theta(u), \gamma, \kappa_\sigma(u', u'', w', w'', \gamma'), \kappa_\theta(w'), \kappa(u')) \\ &= \bar{\Phi}(u, u', u'', w', w'', \gamma, \gamma')\,.\end{aligned} \tag{U.8}$$

The simplest way to express the equations of motion and boundary conditions in terms of u, w, and γ is to start with the following alternative form of the Principle of Virtual Work:

$$[r(\hat{\mathbf{N}}_\sigma\cdot\delta\mathbf{u} + \hat{M}_\sigma\delta\beta)]_0^L + \int_0^L r[(\mathbf{p} - m\ddot{\mathbf{u}})\cdot\delta\mathbf{u} + (l - I\dot{\beta})\delta\beta]d\sigma = \int_0^L r\delta\bar{\Phi}d\sigma\,, \tag{U.9}$$

where $\mathbf{u} = u\mathbf{e}_r + w\mathbf{e}_z$, $\mathbf{N}_\sigma = H\mathbf{e}_r + V\mathbf{e}_\theta$, and $\mathbf{p} = p_H\mathbf{e}_r + p_V\mathbf{e}_z$. In view of (U.8),

$$\delta\bar{\Phi} = \bar{\Phi}_{,u}\delta u + \cdots + \bar{\Phi}_{,\gamma'}\delta\gamma'\,. \tag{U.10}$$

Introducing this expression together with the component representations of $\mathbf{u}$, $\mathbf{N}_\sigma$, and $\mathbf{p}$ into (U.9), setting $\delta\beta = \bar{\alpha}_u{}'\delta u' + \bar{\alpha}_w{}'\delta w' - \delta\gamma$ *in the integrand* on the left of (U.9), and integrating by parts to remove all spatial derivatives of variations, we obtain

$$\{r[(\hat{H}-H)\delta u+(\hat{V}-V)\delta w+(\hat{M}_\sigma-M_\sigma)\delta\beta]\}_0^L$$
$$+\int_0^L\{[(rH)'-N_\theta+rp_H-mr\ddot{u}]\delta u+[(rV)'+rp_V-mr\ddot{w}]\delta w \qquad \text{(U.11)}$$
$$+[(r\bar{\Phi},_{\gamma'})'-r\bar{\Phi},_\gamma+I\dot{\beta}]\delta\gamma\}d\sigma=0\,,$$

where

$$H=\frac{\partial\bar{\Phi}}{\partial u'}-\frac{1}{r}\left[r\frac{\partial\bar{\Phi}}{\partial u''}\right]'+\underline{(I\dot{\beta}-l)\bar{\alpha},_{u'}}$$
$$V=\frac{\partial\bar{\Phi}}{\partial w'}-\frac{1}{r}\left[\frac{\partial\bar{\Phi}}{\partial w''}\right]'+\underline{(I\dot{\beta}-l)\bar{\alpha},_{w'}} \qquad \text{(U.12)}$$

$$N_\theta=r\frac{\partial\bar{\Phi}}{\partial u}\ ,\ M_\sigma=\lambda_\sigma^2\tilde{\Phi},_{\kappa_\sigma}(\gamma_\sigma(u',w'),\ \cdots,\kappa(u'))\,. \qquad \text{(U.13)}$$

The coefficients of the independent variations δu, δw, and $\delta\gamma$ in the integrand in (U.11) must vanish, yielding 3 field equations for u, w, and γ. Boundary conditions may be read-off from the integrated term in (U.11), provided we note that, while edge values of δu and δw may be prescribed independently (if there are no geometric constraints), $\delta u'$, $\delta w'$, and $\delta\gamma'$ may not, but only the combination that we identify as $\delta\beta$.

In 1st-approximation shell theory, one sets $l=I=\gamma=0$ and ignores the underlined terms in (U.12). This yields the equations proposed by Simmonds (1985).

2. Rotation-stress resultant form

Here, we take $\beta, N_\sigma, N_\theta$, and Q as the unknowns and use the mixed-energy density, $\Psi=\Psi(N_\sigma, N_\theta, Q, k_\sigma, k_\theta, k_n)$, the latter 3 arguments being expressible in terms of β via (L.8)-(L.10). Our approach is based on the observation that the spatial derivative of the acceleration, $\dot{\mathbf{v}}$, can be expressed in terms of e_σ, g, β, and their time derivatives. Thus, from (L.1),

$$\mathbf{v}'=\dot{\mathbf{y}}'=(\dot{e}_\sigma-g\dot{\beta})\mathbf{T}+[(1+e_\sigma)\dot{\beta}+\dot{g}]\mathbf{B}\,. \qquad \text{(U.14)}$$

$$\dot{\mathbf{v}}'=[\ddot{e}_\sigma-g\ddot{\beta}-2\dot{g}\dot{\beta}-(1+e_\sigma)\dot{\beta}^2]\mathbf{T}+[(1+e_\sigma)\ddot{\beta}+\ddot{g}+2\dot{e}_\sigma\dot{\beta}-g\ddot{\beta}]\mathbf{B}\,. \qquad \text{(U.15)}$$

Dividing the equation of balance of linear momentum, (D.3), by r, differentiating with respect to σ, setting $\mathbf{N}_\sigma=N_\sigma\mathbf{T}+Q\mathbf{B}$, and noting from (E.4) and (E.5) that

$$\mathbf{T}'=(\alpha+\beta)'\mathbf{B}\ ,\ \mathbf{B}'=-(\alpha+\beta)'\mathbf{T}\,, \qquad \text{(U.16)}$$

we obtain the 2 scalar equations

$$N_\sigma'' + (r'/r)N_\sigma' + [(r'/r)' - (\alpha+\beta)'^2]N_\sigma - (N_\theta/r)'\cos(\alpha+\beta) - 2(\alpha+\beta)'Q' - [(r'/r)(\alpha+\beta)' + (\alpha+\beta)'']Q + p_T' - (\alpha+\beta)'p_B = m[\ddot{e}_\sigma - g\ddot{\beta} - 2\dot{g}\dot{\beta} - (1+e_\sigma)\dot{\beta}^2] \tag{U.17}$$

$$Q'' + (r'/r)Q' + [(r'/r)' - (\alpha+\beta)'^2]Q + 2(\alpha+\beta)'N_\sigma' + (N_\theta/r)'\sin(\alpha+\beta) + [(r'/r)(\alpha+\beta)' + (\alpha+\beta)'']N_\sigma + p_B' + (\alpha+\beta)'p_T = m[(1+e_\sigma)\ddot{\beta} + \ddot{g} + 2\dot{e}_\sigma\dot{\beta} - g\dot{\beta}^2]\,. \tag{U.18}$$

To these 2 equations, we adjoin the equation of balance of rotational momentum, (L.15), and the strain-rotation compatibility condition, (M.2). These are the 4 basic field equations. If the bending strain-rotation relations (L.8)-(L.10) are substituted into the constitutive relations, (O.5), and if these, in turn, are substituted into the field equations, we obtain a system of 4, 2nd-order, quasilinear partial differential equations for the 4 unknowns $\beta, N_\sigma, N_\theta$, and Q. We note that if N_θ and k_σ are uncoupled in Ψ, which is usual, then (M.2) reduces to a 1st-order equation.

Classical boundary conditions may be inferred from the integrated term in (U.11). In view of (L.7), (M.7), (O.5), (Q.10), and (Q.11), these conditions, plus the fact that (U.17) and (U.18) are *differentiated* forms of the equations of balance of linear momentum, require that we prescribe at $\sigma = 0, L$,

$$H = N_\sigma\cos(\alpha+\beta) - Q\sin(\alpha+\beta) \quad \text{or} \quad u = re_\theta = -r\Psi,_{N_\theta} \tag{U.19}$$

$$V = N_\sigma\sin(\alpha+\beta) + Q\cos(\alpha+\beta) \quad \text{or} \quad (rV)' = mr\ddot{\bar{z}} - rp_V \qquad \text{(U.20)[32]}$$

$$M_\sigma \quad \text{or} \quad \beta\,. \tag{U.21}$$

Even though our system of field equations consists of 4, 2nd-order partial differential equations, it is *not* of 8th order—in fact, because there are 3 boundary conditions at each edge of the axishell, the system is only of 6th order.[33]

In 1st-approximation theory, we take $\Psi = \Psi(N_\sigma, N_\theta, k_\sigma, k_\theta)$ and set $I = 0$ in (L.15), which yields

---

[32]Consider the differential equation $y' = f(x)$, subject to the auxiliary condition $y(a) = A$. If we differentiate to obtain the new equation, $\eta' = f'(x)$, where $\eta \equiv y'$, then the auxiliary condition satisfied by η is *not* $y(a) = A$, but rather the original differential equation, evaluated at some fixed point, say $x = a$, i.e., $\eta(a) = f(a)$. The analogues of this auxiliary condition are the boundary conditions (U.20); the analogues of $y(a) = A$ are the conventional classical boundary conditions

$$V = N_\sigma\sin(\alpha+\beta) + Q\cos(\alpha+\beta) \quad \text{or} \quad w = \bar{z}(0,t) + \int_0^\sigma [(1 - \Psi,_{N_\theta})\sin(\alpha+\beta) - \Psi,_Q\cos(\alpha+\beta)]ds\,.$$

For a similar transformation in a simpler setting, see (III.J.10).

[33]See Courant & Hilbert (1962, vol. 2, p. 13) where it is shown that a general, 2nd-order partial differential equation in 2 independent variables is equivalent to a system of 3, 1st-order partial differential equations.

$$rQ = (1 - \Psi,_{N_\sigma})^{-1}[\Psi,_{k_\theta}\cos(\alpha+\beta) - (r\Psi,_{k_\sigma})'] . \tag{U.22}$$

With Q expressed as above (and thus, ultimately, in terms of β, N_σ, and N_θ) and with

$$e_\sigma = -\Psi,_{N_\sigma} \ , \ e_\theta = -\Psi,_{N_\theta} \ , \ g = M = 0 \ , \tag{U.23}$$

the compatibility condition, (M.2), and the 2 differentiated equations of balance of linear momentum, (U.17) and (U.18), reduce to 3 equations for β, N_σ, and N_θ, but the system is still of 6th order.

3. A system of 1st-order equations

Here, the procedure is virtually the same as we used for beamshells in subsection IV.P.3, the object being to express the field equations as a system of 1st-order partial differential equations.

Four of the field equations consist of the equations of balance of linear momentum, (L.12) and (L.13), the equation of balance of rotational momentum, (L.15)—with $\ddot{\beta}$ replaced by ω—and the compatibility condition, (M.2). We use the constitutive relations, (O.5), to express $e_\sigma, \cdots, M_n$ in terms of $N_\sigma, \cdots, k_n$. With the abbreviations

$$\frac{\partial\Psi}{\partial N_\sigma} \equiv \Psi,_1 \ , \cdots , \ \frac{\partial\Psi}{\partial k_n} \equiv \Psi,_6 \ , \ \frac{\partial^2\Psi}{\partial N_\sigma^2} \equiv \Psi,_{11} \ , \ \cdots , \tag{U.24}$$

it follows that

$$M_\sigma{}' = \Psi,_{41}N_\sigma{}' + \Psi,_{42}N_\theta{}' + \cdots + \Psi,_{46}k_n{}' \tag{U.25}$$

$$e_\theta{}' = -\Psi,_{21}N_\sigma{}' - \cdots - \Psi,_{26}k_n{}' . \tag{U.26}$$

But from (L.8)-(L.10),

$$(rk_\theta)' = rk_\theta{}' + k_\theta\cos\alpha = (\alpha' + k_\sigma)\cos(\alpha+\beta) + (\sin\alpha)' \tag{U.27}$$

$$(rk_n)' = rk_n{}' + k_n\cos\alpha = -(\alpha' + k_\sigma)\sin(\alpha+\beta) - (\cos\alpha)' . \tag{U.28}$$

Thus, k_θ and k_n as well as $k_\theta{}'$ and $k_n{}'$ can be expressed as functions of k_σ, β, and σ only, and our 4 field equations are seen to be linear in the spatial and time derivatives of the 8-component vector, $\underline{u} \equiv [N_\sigma, N_\theta, Q, k_\sigma, \beta, \omega, v_T, v_B]^T$.

Two of the 4 additional field equations we need are

$$\dot{\beta} = \omega \tag{U.29}$$

and

$$\beta' = k_\sigma . \tag{U.30}$$

Two more equations follow from the component form of the compatibility condition, $\mathbf{v}' = (\overline{\mathbf{y}}')^{\bullet}$, which, with the aid of (L.1) and (L.14)$_2$, takes the component form

$$v_T{}' - (\alpha' + k_\sigma) v_B = \dot{e}_\sigma - \omega g \tag{U.31}$$

$$v_B{}' + (\alpha' + k_\sigma) v_T = \dot{g} + \omega(1 + e_\sigma) . \tag{U.32}$$

But,

$$\dot{e}_\sigma = -\Psi_{,11}\dot{N}_\sigma - \cdots - \Psi_{,16}\dot{k}_n \; , \; \dot{g} = -\Psi_{,31}\dot{N}_\sigma - \cdots - \Psi_{,36}\dot{k}_n \tag{U.33}$$

and, using (L.8)-(L.10) and their time derivatives, we conclude that $\dot{k}_\theta$ and $\dot{k}_n$ may be expressed as functions of ω, β, and σ.

Thus, with the substitutions described above, the 8 equations (L.12), (L.13), (L.15), (M.2), and (U.29)-(U.32) reduce to a system of 1st-order, quasilinear partial differential equations of the form

$$A(\underline{u})\underline{u}' + B(\underline{u})\dot{\underline{u}} + C(\underline{u}) = 0 , \tag{U.34}$$

where A, B, and C are 8×8 matrices. The order of the system, for reasons explained earlier, is 6.

In the 1st-approximation theory of axishells, we set $I = 0$ and take

$$\Psi = \Psi(N_\sigma, N_\theta, k_\sigma, k_\theta(\beta)) = \tilde{\Psi}(N_\sigma, N_\theta, k_\sigma, \beta) . \tag{U.35}$$

The matrices A, B, and C simplify somewhat, but the order of the differential system (U.34) remains unchanged.

4. *Intrinsic form*

Our aim here is to take the *deformed metric coefficients* λ_σ and $\bar{r}$ as the basic unknowns, thereby excluding any explicit reference to displacements, rotations, or curvatures. For simplicity, we consider 1st-approximation shell theory only in which case $\lambda_\sigma = 1 + e_\sigma$ and $\bar{r} = r(1 + e_\theta)$.[34]

The reader should compare the results in this subsection with the corresponding results for birods in subsection III.J.2, which exemplify the intrinsic approach in the simplest possible setting. The intrinsic equations for axishells provide the simplest example of the intrinsic equations developed by Libai (1981,1983) for general shells, the main simplifications for axishells being (a) the vanishing of the curvature-generating function and (b) the possibility of expressing the slopes and curvatures of the deformed meridian explicitly in terms of the deformed metric coefficients so that time-rate relationships, such as equation (14) of Libai (1981), are not needed.

As a first step, we eliminate β from the field equations by using the relation

---

[34]In shearable shells the variables would be $\lambda_\sigma, \bar{r}$, and g. Note that in standard tensor notation $\lambda_\sigma = (\bar{a}_{11})^{1/2}$ and $\bar{r} = (\bar{a}_{22})^{1/2}$, where $\bar{a}_{\alpha\beta}$, $\alpha, \beta = 1, 2$ are the covariant components of the metric tensor of the deformed reference surface $\bar{\mathfrak{Z}}$.

$$\bar{r}_{,s} = \cos(\alpha + \beta) = \lambda_\sigma^{-1}\bar{r}' , \tag{U.36}$$

where $ds = \lambda_\sigma d\sigma$ is differential arc length along the deformed meridian, $\bar{\mathcal{M}}$. The curvatures of $\bar{\mathfrak{S}}$ and the bending strains may then be expressed as

$$\bar{r}\bar{b}_\theta = \sin(\alpha + \beta) = (1 - \bar{r}_{,s}^2)^{1/2} \tag{U.37}$$

$$\lambda_\sigma \bar{b}_\sigma = (\alpha + \beta)' = -(\bar{r}_{,s})'(1 - \bar{r}_{,s}^2)^{-1/2} \tag{U.38}$$

$$k_\sigma = \lambda_\sigma \bar{b}_\sigma - \alpha' \tag{U.39}$$

$$k_\theta = r^{-1}(\bar{r}\bar{b}_\theta - \sin\alpha) . \tag{U.40}$$

Henceforth, the curvatures and bending strains will be taken as *defined* by (U.37)-(U.40) in terms of the basic variables, λ_σ and $\bar{r}$.

To proceed further, it is convenient to take the acceleration in the component form

$$\ddot{\mathbf{y}} = a_\sigma \bar{\mathbf{t}} + a_b \bar{\mathbf{b}} . \tag{U.41}$$

Using (L.12), (L.13), and eliminating Q via (L.15), we express the acceleration components in terms of the stress resultants, stress couples, and loading as

$$mra_\sigma = (rN_\sigma)' - \bar{r}_{,s}(N_\theta + \bar{b}_\sigma M_\theta) + \bar{b}_\sigma (rM_\sigma)' + rp_T \tag{U.42}$$

$$mra_b = \{\lambda_\sigma^{-1}[\bar{r}_{,s}M_\theta - (rM_\sigma)']\}' + r\lambda_\sigma \bar{b}_\sigma N_\sigma + \bar{r}\bar{b}_\theta N_\theta + rp_B . \tag{U.43}$$

Because (**T**, **B**) and ($\bar{\mathbf{t}}$, $\bar{\mathbf{b}}$) coincide in 1st-approximation shell theory, p_T and p_B are loads (per unit undeformed area) in the $\bar{\mathbf{t}}$ and $\bar{\mathbf{b}}$ directions, respectively.

We also need an expression for the spin $\dot{\beta}$ in terms of λ_σ and $\bar{r}$. This follows from (U.36) and $\lambda_\sigma \bar{r}_{,s} = \bar{r}'$ as

$$\dot{\beta} = \lambda_\sigma^{-1}(\bar{r}_{,s}\dot{\lambda}_\sigma - \dot{\bar{r}}')(1 - \bar{r}_{,s}^2)^{-1/2} . \tag{U.44}$$

Our final step is to express $\ddot{\lambda}_\sigma$ and $\ddot{\bar{r}}$ in terms of the acceleration components a_σ and a_b. With the aid of (U.36) and (U.37), the radial component of (U.41) takes the form

$$\ddot{\bar{r}} = \bar{r}_{,s}a_\sigma - (1 - \bar{r}_{,s}^2)^{1/2}a_b . \tag{U.45}$$

The second expression we seek follows upon noting that $\dot{\mathbf{y}}' = (\lambda_\sigma \bar{\mathbf{t}})^\bullet = \dot{\lambda}_\sigma \bar{\mathbf{t}} + \lambda_\sigma \dot{\bar{\mathbf{t}}}$. Hence, with the aid of (U.38) and (U.44),

$$\begin{aligned} \ddot{\lambda}_\sigma &= (\dot{\mathbf{y}}' \cdot \bar{\mathbf{t}})^\bullet \\ &= (a_\sigma \bar{\mathbf{t}} + a_b \bar{\mathbf{b}})' \cdot \mathbf{t} + (\lambda_\sigma \bar{\mathbf{t}})^\bullet \cdot \dot{\mathbf{t}} \qquad \text{(U.46)} \\ &= a_\sigma' + (\bar{r}_{,s})'(1 - \bar{r}_{,s}^2)^{-1/2}a_b + [\lambda_\sigma(1 - \bar{r}_{,s}^2)]^{-1}(\dot{\bar{r}}' - \bar{r}_{,s}\dot{\lambda}_\sigma)^2 . \end{aligned}$$

Equations (U.45) and (U.46) are the two field equations of intrinsic axishell dynamics, where a_σ and a_b are given by the right sides of (U.42) and (U.43), with the stress resultants and couples expressed in terms of $\lambda_\sigma, \bar{r}, k_\sigma$, and k_θ via the constitutive relations, (O.3). The latter two quantities and the deformed curvatures, $\bar{b}_\sigma$ and $\bar{b}_\theta$, that appear on the right sides of (U.42) and (U.43) are to be expressed in terms of λ_σ and $\bar{r}$ via (U.37)-(U.40).

In a true intrinsic formulation, kinematic boundary conditions should be expressed in terms of the basic unknowns. Conditions on β or $\dot{\beta}$ can be written in terms of λ_σ and $\bar{r}$ with the aid of (U.36) and (U.44). Regarding a boundary condition of position—which, upon differentiation with respect to time, becomes a boundary condition of velocity—we assume that

$$\dot{\bar{\mathbf{y}}} = \hat{\mathbf{v}}(t) = \hat{v}_r(t)\mathbf{e}_r + \hat{v}_z(t)\mathbf{e}_z \tag{U.47}$$

is prescribed on ∂U. Equivalent conditions are

$$\dot{\bar{r}} = \hat{v}_r \ , \ \ a_\sigma = \bar{r}_{,s}\dot{\hat{v}}_r + (1 - \bar{r}_{,s}^{\,2})^{1/2}\dot{\hat{v}}_z \ \text{ on } \partial U \ , \tag{U.48}$$

where $(U.48)_2$ was obtained by differentiating (U.47) with respect to time. Note that when we replace a_σ by $(mr)^{-1}$ times the right side of (U.42), we are replacing a kinematic boundary condition by one involving derivatives of stresses. For a similar transformation in birods or general shells, see, respectively, (III.J.10) or equation (22) of Libai (1983).

As in birods and general shells, uniform axial loads (such as gravity) of the form $m^{-1}\mathbf{p} = C\mathbf{e}_z$, C = constant, drop out of the field equations (U.45) and (U.46) and reappear only if the kinematical boundary condition, $(U.48)_2$, is prescribed. Also, the intrinsic equations are unaffected by a superimposed uniform axial velocity.[35] These last two properties make the intrinsic equations convenient for approximation schemes in which the left sides of (U.45) and (U.46) are integrated stepwise in time, while the right sides are spatially discretized using finite difference or element methods.

If required, the axial velocity and position can be obtained from a time integration of the axial component of (U.41). However, if the motion of, say, the lower edge, $\bar{z}_0(t)$, is prescribed or calculated, then it may be simpler to apply the equation

$$\bar{z}(\sigma, t) = \bar{z}_0(t) + \int_0^\sigma \lambda_\sigma (1 - \bar{r}_{,s}^{\,2})^{1/2} d\bar{\sigma} \, , \tag{U.49}$$

which follows from $\bar{z}_{,s} = \sin(\alpha + \beta)$ and (U.37).

[35]The discussion regarding kinematic boundary conditions and uniform loads and velocities applies, essentially, also to the rotation-stress resultant formulation of subsection U.2, which shares many of the features of the intrinsic approach.

5. Some special topics

Finite axisymmetric oscillations of circular and annular plates are of practical interest and have been studied by several authors. Because of nonlinear terms in the extensional strain-displacement relations, transverse displacements w of the order of the thickness of the plate induce, via stress-strain laws, stress resultants that act through the curvature of the deformed plate to produce restoring forces of the same order of magnitude as those coming from bending stresses. The net result is that a finitely deformed plate acts not unlike a shallow shell in which there are tensile membrane forces whose stiffening effect tends to raise the natural frequencies of an infinitesimally deformed plate.

Thus, Yamaki (1961) and Wah (1963), using the von Kármán plate equations with an inertial term, $m\ddot{w}$, added in accordance with D'Alembert's Principle, apply Galerkin procedures in the spatial domain to obtain approximate nonlinear ordinary differential equations in time which they solved in terms of elliptic integrals. Their results are quoted by Leissa (1969). In his recent book, Niordson (1985) studies the oscillations of a clamped circular plate. Assuming, as an approximation, that w is sinusoidal in time, he reduces the axisymmetric von Kármán equations to two coupled ordinary differential equations for the spatial variation of w and a stress function. These equations are then solved by expanding the unknowns in perturbation series in powers of an amplitude parameter. Niordson's results for the lowest frequency are within a few percent of those of Yamaki and Wah—remarkable agreement in view of the different approximations used, especially as Wah uses Berger's (1955) *non-rational* approximation to the von Kármán equations.

The "shell effect" is also important in computing the influence of radial compressive preloads on the natural frequencies of circular or annular plates. Here we see an apparent paradox because the frequencies increase, rather than decrease, with the loads, as might be expected.[36] For a theoretical study, see Massonet (1948); for experimental results, see Rosen & Libai (1975). The vibrations of arbitrary, preloaded shells is treated by Budiansky (1968).

Another topic of obvious interest is the large, purely radial oscillations of cylindrical and spherical shells. The simple geometry allows one to first use finite elasticity theory and only afterwards to invoke thinness (if necessary!) to obtain explicit solutions.

Knowles (1960, 1962) has studied problems with cylindrical geometry. Among other things, he gives asymptotic solutions for the period of oscillation of a tube of Mooney-Rivlin material. Shahinpoor & Nowinsky (1971) study tubes of this material in more detail and consider the displacement field. Using a result of Burt & Reid (1976) relating to solutions of a nonlinear Klein-Gordon equation, Roger & Baker (1980) have determined the radial oscillations of tubes made of a 4-parameter incompressible material (that includes the Mooney-

[36]This is attributed, in part, to initial imperfections that are amplified by the compressive loading and thus enhance the "shell effect."

Rivlin material as a special case).

Wang (1965) has treated the large, purely radial oscillations of thin, spherical shells made of an arbitrary, incompressible material. He expressed the period of oscillation as an integral, but does not treat specific cases. This has been done by Mukherjee & Chakraborty (1985) who, using the aforementioned results of Burt & Reid (1976), compute explicit expressions for displacements and periods for a neo-Hookean material.

V. Variational Principles for Axishells

The derivation of variational principles for axishells follows the same patterns as set in Chapters III and IV for birods and beamshells. To avoid repetition, we refer the reader to these chapters for motivation and certain details.

1. Hamilton's Principle

We start from the weak form of the equations of motion, (G.1), where we now set

$$\mathbf{V} = \delta\bar{\mathbf{y}} \ , \ \Omega = \delta\beta \ , \ \mathbf{v} = \dot{\bar{\mathbf{y}}} \ , \ \omega = \dot{\beta} \, . \tag{V.1}$$

Using the results of Sections L and O, we have

$$\delta\bar{\mathbf{y}}' = \mathbf{T}\delta e_\sigma + \mathbf{B}\delta g + \mathbf{m}\delta\beta \tag{V.2}$$

$$\begin{aligned} &\mathbf{N}_\sigma \cdot \delta\bar{\mathbf{y}}' + r^{-1} N_\theta \mathbf{e}_r \cdot \delta\bar{\mathbf{y}} + M_\sigma \delta\beta' + (r^{-1}\mathbf{M}_\theta \cdot \mathbf{e}_r - \mathbf{N}_\sigma \cdot \mathbf{m})\delta\beta \\ &\quad = N_\sigma \delta e_\sigma + N_\theta \delta e_\theta + Q\delta g + M_\sigma \delta k_\sigma + M_\theta \delta k_\theta + M_n \delta k_n \\ &\quad = \delta\Phi(e_\sigma, e_\theta, g, k_\sigma, k_\theta, k_n) \, , \end{aligned} \tag{V.3}$$

the last line holding if and only if the axishell is elastic. Introducing these expressions into (G.1), we obtain

$$\begin{aligned} &\int_{t_1}^{t_2} [(r\mathbf{N}_\sigma \cdot \delta\bar{\mathbf{y}} + rM_\sigma \delta\beta)_a^b + \int_a^b (\mathbf{p} \cdot \delta\bar{\mathbf{y}} + l\delta\beta) r d\sigma + \delta(\mathcal{K} - \mathcal{E})] dt \\ &\qquad = \int_a^b r\,(m\mathbf{v} \cdot \delta\bar{\mathbf{y}} + I\omega\delta\beta) \Big|_{t_1}^{t_2} d\sigma \, , \end{aligned} \tag{V.4}$$

where $\mathcal{K}$ as in (H.3), and

$$\mathcal{E} \equiv \int_a^b \Phi r d\sigma \tag{V.5}$$

are, respectively, the kinetic and strain energies of the axishell (per circumferential radian). For inelastic axishells, replace $\delta\Phi$ by the middle line of (V.3). This latter form of the Principle of Virtual Work has many uses.

We extend (V.4) to the entire axishell by setting $a = 0$ and $b = L$. On $\sigma = 0, L$, the edge kinematic vector y defined in (I.1) satisfies the holonomic constraints

$$\mathcal{R}(y) = 0\,, \tag{V.6}$$

so that

$$\delta\mathcal{R}(y) \equiv A(y)\delta y = 0\,, \tag{V.7}$$

where the variation of y is given by

$$\delta y \equiv [\delta\bar{\mathrm{y}}_0, \delta\beta_0, \delta\bar{\mathrm{y}}_L, \delta\beta_L]^T\,. \tag{V.8}$$

The virtual work of the edge loads (per radian in the circumferential direction) is then, as in (I.4) and (I.5),

$$[r(\mathbf{N}_\sigma\cdot\delta\bar{\mathrm{y}} + M_\sigma\delta\beta)]_0^L = \mathcal{N}^T\delta y = \sum_1^{6-m}(\mathcal{N}^T\mathcal{M}_k)\delta q_k + \mathcal{N}^T\mathcal{H}\,. \tag{V.9}$$

In the classical case, where some of the components of y are specified, this expression reduces to

$$\mathcal{N}^T\delta y = \hat{\mathcal{N}}_*^T\delta y_*\,. \tag{V.10}$$

In the above, $\hat{\mathcal{N}}_*^T$ is the *prescribed* part of the edge load vector and y_* is the *unspecified* part of the edge kinematic vector.

We now assume, as in Section K, that a load potential $\mathcal{P}$ exists such that

$$\hat{\mathcal{N}}^T\delta y + \int_0^L(\mathbf{p}\cdot\delta\bar{\mathrm{y}} + l\delta\beta)r\,d\sigma = -\delta\mathcal{P}\,. \tag{V.11}$$

Under this condition, (V.4) reduces to

$$\delta\int_{t_1}^{t_2}(\mathcal{K} - \mathcal{E} - \mathcal{P})dt = \int_0^L(m\mathbf{v}\cdot\delta\bar{\mathrm{y}} + I\omega\delta\beta)\Big|_{t_1}^{t_2} r\,d\sigma\,. \tag{V.12}$$

Finally, if conditions are such that the right side of (V.12) vanishes, then we obtain *Hamilton's Principle for axishells*,

$$\delta\int_{t_1}^{t_2}(\mathcal{K} - \mathcal{E} - \mathcal{P})dt = 0\,. \tag{V.13}$$

This equation has the same form as (III.K.9) and (IV.Q.11), but the functionals $\mathcal{K}$, $\mathcal{E}$, and $\mathcal{P}$, defined by (H.3), (V.5), and (V.11), are those for an axishell. For further discussion of the principle, see Section III.K.

2. *Variational principles for elastostatics*

(a) Principle of stationary total potential energy. In static problems, $\mathcal{K}$, $\mathbf{v}$, and ω are zero, and (V.12) reduces to the *Principle of Stationary Total Potential Energy for Axishells*,

$$\delta(\mathcal{E} + \mathcal{P}) = 0\,. \tag{V.14}$$

This principle states that among all admissible deformed positions $\bar{\mathrm{y}}$ and rotations β, sufficiently smooth on $0 < \sigma < L$ and satisfying (V.6) on the boundaries, only those that satisfy the equations of static equilibrium (including any prescribed edge loads) render the functional $\mathcal{E} + \mathcal{P}$ stationary. The functional

$\mathcal{E}+\mathcal{P}$ is to be expressed in terms of $\bar{\mathbf{y}}$ and β through the results of Sections K and L. In particular, see (K.1) for a criterion for $\mathcal{P}$ to exist and see (L.1), (L.2), and (L.7)-(L.10) for expressions relating the strains to $\bar{\mathbf{y}}$ and β.

(b) Extended variational principles. We follow the pattern of the previous chapters and start with the *Hu-Washizu Variational Principle for Axishells*,

$$\delta\Pi_1[e_\sigma, e_\theta, g, k_\sigma, k_\theta, k_n, N_\sigma, N_\theta, Q, M_\sigma, M_\theta, M_n, \bar{\mathbf{y}}, \beta, \mathcal{L}] = 0\,, \quad \text{(V.15)}$$

where

$$\Pi_1 \equiv \mathcal{E}+\mathcal{P}-\int_0^L [\lambda_T N_\sigma + \lambda_\theta N_\theta + gQ$$
$$+ (k_\theta + r^{-1}\sin\alpha)M_\theta + (k_n + r^{-1}\cos\alpha)M_n + (k_\sigma - \beta')M_\sigma \quad \text{(V.16)}$$
$$- (\mathbf{N}_\sigma\cdot\bar{\mathbf{y}}' + r^{-1}\mathbf{N}_\theta\cdot\bar{\mathbf{y}}_{,\theta} + r^{-1}\mathbf{M}_\theta\cdot\mathbf{e}_z)]r\,d\sigma - \mathcal{L}^T\mathcal{R}(y)\,.$$

In the above, $\lambda_T \equiv 1+e_\sigma$, $\lambda_\theta \equiv 1+e_\theta$, and $\mathcal{L}^T$ is a constant Lagrange multiplier vector with the same number of columns (m) as there are kinematical constraints. [If shearing deformations are ignored, then $\lambda_T = \lambda_\sigma$. The stretches λ_σ and λ_θ were introduced in Section P—see (P.2) and (P.12).]

If classical boundary conditions are prescribed, then

$$\mathcal{L}^T\mathcal{R}(y) = \{r[\tilde{\mathbf{N}}_\sigma\cdot(\bar{\mathbf{y}}-\hat{\bar{\mathbf{y}}}) + \tilde{M}_\sigma(\beta-\hat{\beta})]\}_{\partial U}\,. \quad \text{(V.17)}$$

Here, $\tilde{\mathbf{N}}_\sigma$ and $\tilde{M}_\sigma$ are Lagrange multipliers and $\hat{\bar{\mathbf{y}}}$ and $\hat{\beta}$ are specified values of $\bar{\mathbf{y}}$ and β on the ∂U parts of the edges. For future use, we note that, in tensor notation,

$$\mathbf{N}_\sigma\cdot\bar{\mathbf{y}}' + r^{-1}\mathbf{N}_\theta\cdot\bar{\mathbf{y}}_{,\theta} = \mathbf{N}^\alpha\cdot\bar{\mathbf{y}}_{,\alpha}\,, \quad \text{(V.18)}$$

where $\mathbf{N}^\alpha$ are the contravariant vector components of the stress resultant tensor $\mathbf{N}$. The reader should compare (V.16) and (V.17) with (IV.Q.17) and (IV.Q.18), the analogous equations for beamshells. The differences between these expressions arise from differences in geometry and the increased number of unknowns in axishells. As with beamshells, the field variables in Π_1 are subject to no constraints, save for smoothness conditions necessary for the integral in (V.16) to exist.

If we use the alternative formulation with $\overline{M}_\theta$ and k_θ, as in (L.16)-(L.18), and (O.7), we must replace the 3 terms involving M_θ, M_n, and $\mathbf{M}_\theta\cdot\mathbf{e}_z$ in Π_1 by the single term $r^{-1}[rk_\theta + \sin\alpha - \sin(\alpha+\beta)]\overline{M}_\theta$.

We now prove that those unknowns that make Π_1 stationary, and only those, satisfy the field equations of axishell theory. To this end, we compute the variation of Π_1, using (IV.Q.19), and integrate by parts to obtain

$$\delta\Pi_1 = \int_0^L \{[(\Phi,_{e_\sigma} - N_\sigma)\delta e_\sigma + (\Phi,_{e_\theta} - N_\theta)\delta e_\theta + (\Phi,_g - Q)\delta g$$
$$+ (\Phi,_{k_\sigma} - M_\sigma)\delta k_\sigma + (\Phi,_{k_\theta} - M_\theta)\delta k_\theta + (\Phi,_{k_n} - M_n)\delta k_n]$$
$$- (e_\sigma - \mathbf{T}\cdot\bar{\mathbf{y}}' + 1)\delta N_\sigma - (e_\theta - r^{-1}\bar{r} + 1)\delta N_\theta - (g - \mathbf{N}\cdot\bar{\mathbf{y}}')\delta Q$$
$$- (k_\sigma - \beta')\delta M_\sigma - [k_\theta - r^{-1}(\sin\bar{\alpha}_T - \sin\alpha)]\delta M_\theta - [k_n - r^{-1}(\cos\bar{\alpha}_T - \cos\alpha)]\delta M_n \tag{V.19}$$
$$- r^{-1}[(r\mathbf{N}_\sigma)' - N_\theta\mathbf{e}_r + r\mathbf{p}]\cdot\delta\bar{\mathbf{y}} - r^{-1}[(rM_\sigma)' - \mathbf{M}_\theta\cdot\mathbf{e}_r + r\mathbf{N}_\sigma\cdot\mathbf{m} + rl]\delta\beta\}\, r\,d\sigma$$
$$+ [\mathcal{N}^T - \hat{\mathcal{N}}^T - \mathcal{L}^T A(y)]\delta y - \delta\mathcal{L}^T\mathcal{R}(y) = 0 .$$

In this expression, $\bar{\alpha}_T = \alpha + \beta$; $\mathcal{N}^T$ is the stress vector at the boundary, as in (I.4); and $\hat{\mathcal{N}}^T$ is the *prescribed* edge loading. The boundary terms are the same as those which appear in the Principle of Virtual Work. The reader should compare (J.4) and (J.7) with $\mathcal{R}(y) = 0$ in place of $A\dot{y} = B$.

For classical boundary conditions, the boundary terms reduce to

$$[\mathcal{N}^T - \hat{\mathcal{N}}^T - \mathcal{L}^T A(y)]\delta y + \delta\mathcal{L}^T\mathcal{R}(y) = \{r[(\mathbf{N}_\sigma - \tilde{\mathbf{N}}_\sigma)\cdot\delta\bar{\mathbf{y}} + (M_\sigma - \tilde{M}_\sigma)\delta\beta$$
$$- (\bar{\mathbf{y}} - \hat{\bar{\mathbf{y}}})\cdot\delta\tilde{\mathbf{N}}_\sigma - (\beta - \hat{\beta})\delta\tilde{M}_\sigma]\}_{\partial U} \tag{V.20}$$
$$+ \{r[(\mathbf{N}_\sigma - \hat{\mathbf{N}}_\sigma)\cdot\delta\bar{\mathbf{y}} + (M_\sigma - \hat{M}_\sigma)\delta\beta]\}_{\partial N} ,$$

which is the analog of (IV.Q.21). Here, as before, the tilde (˜) denotes a Lagrange multiplier, a hat (ˆ) denotes a prescribed edge value, and ∂U and ∂N denote those parts of the boundary where loads or kinematical constraints are prescribed. See subsection IV.H.2 for a more precise definition and discussion.

As the variations in (V.19) are independent, a necessary condition for Π_1 to be stationary is that all the expressions in parentheses be zero. This yields the constitutive relations, the strain-kinematic relations, the equilibrium equations, all the boundary conditions, and expressions for the Lagrange multipliers in terms of $\mathcal{N}^T$. In the classical case, the Lagrange multiplier $\tilde{\mathcal{N}}^T$ must equal the unprescribed part of $\mathcal{N}^T$.

The process leading to (V.19) can be reversed, insuring the equivalence of the conditions stated above with the stationarity of Π_1.

Proceeding as in previous chapters, we now derive variational principles in terms of the mixed-energy density Ψ, defined in (O.4), and its modified form, Ω, defined by $\Omega \equiv \Psi - N_\sigma - N_\theta = \Phi - \lambda_T N_\sigma - \lambda_\theta N_\theta - gQ$. See also (T.18). Use of Ω facilitates the elimination of e_σ, e_θ, and g from the variational functional, resulting in

$$\delta\Pi_2[N_\sigma, N_\theta, Q, M_\sigma, M_\theta, M_n, k_\sigma, k_\theta, k_n, \bar{\mathbf{y}}, \beta, \mathcal{L}] = 0 , \tag{V.21}$$

where

$$\Pi_2 \equiv \mathcal{P} + \int_0^L [\Omega - (k_\theta + r^{-1}\sin\alpha)M_\theta - (k_n + r^{-1}\cos\alpha)M_n - (k_\sigma - \beta')M_\sigma + (\mathbf{N}_\sigma\cdot\bar{\mathbf{y}}' + r^{-1}\bar{r}N_\theta + r^{-1}\mathbf{M}_\theta\cdot\mathbf{e}_z)]rd\sigma - \mathcal{L}^T\mathcal{R}(y)\,. \tag{V.22}$$

In the stationary conditions, the stretches λ_T and λ_θ and strain g are replaced by $-\Omega,_{N_\sigma}$, $-\Omega,_{N_\theta}$, and $-\Omega,_Q$ respectively.

In the next variational principle, the bending strains and stress couples are eliminated completely from the functional. This is accomplished by expressing the bending strains, *a priori*, in terms of rotations, via the right sides of (L.8)-(L.10). Also, the stress couples are taken, *a priori*, to be given by the constitutive relations (O.3). The resulting new principle is

$$\delta\Pi_3[N_\sigma, N_\theta, Q, \bar{\mathbf{y}}, \beta, \mathcal{L}] = 0\,, \tag{V.23}$$

where

$$\Pi_3 \equiv \mathcal{P} + \int_0^L (\Omega + \mathbf{N}_\sigma\cdot\bar{\mathbf{y}}' + r^{-1}\bar{r}N_\theta)rd\sigma - \mathcal{L}^T\mathcal{R}(y)\,. \tag{V.24}$$

Except for obvious smoothness requirements, the arguments of Π_3 are subject to no constraints.

To find the stationary conditions, we calculate $\delta\Pi_3$ explicitly, so obtaining

$$\begin{aligned}\delta\Pi_3 = \int_0^L \{&(\Omega,_{N_\sigma} + \mathbf{T}\cdot\bar{\mathbf{y}}')\delta N_\sigma + (\Omega,_{N_\theta} + r^{-1}\bar{r})\delta N_\theta + (\Omega,_Q + \mathbf{B}\cdot\bar{\mathbf{y}}')\delta Q \\ &- r^{-1}[(r\mathbf{N}_\sigma)' - N_\theta\mathbf{e}_r + r\mathbf{p}]\cdot\delta\bar{\mathbf{y}} \\ &- r^{-1}[(rM_\sigma)' - \mathbf{M}_\theta\cdot\mathbf{e}_r + \mathbf{m}\cdot\mathbf{N}_\sigma]\delta\beta\}rd\sigma \\ &+ [\mathcal{N}^T - \hat{\mathcal{N}}^T - \mathcal{L}^T A(y)]\delta y - \delta\mathcal{L}^T\mathcal{R}(y)\,.\end{aligned} \tag{V.25}$$

Hence, the conditions for Π_3 to be stationary are the strain-kinematic relations for the extensional and shearing strains, the equations of equilibrium in the interior and on the boundary, the kinematic boundary conditions, and an expression for the Lagrange multiplier vector $\mathcal{L}$ in terms of the load vector $\mathcal{N}$. For classical boundary conditions, see (V.20).

Next, by eliminating the deformed position $\bar{\mathbf{y}}$ from the functional Π_3, we derive a variational principle in the stress resultants and rotations only. The penalty for this success is some restrictions on the arguments of the resulting functional.

We assume, as in Section IV.Q, that the surface force $\mathbf{p}$ is a function of $\bar{\mathbf{y}}$ that can be inverted to yield $\bar{\mathbf{y}} = \bar{\mathbf{y}}(\mathbf{p})$. We also assume that, on the edges, the stress resultant $\hat{\mathbf{N}}_\sigma$ is a function of $\bar{\mathbf{y}}$ with an inverse, $\bar{\mathbf{y}}(\mathbf{N}_\sigma)$. If the above conditions are satisfied, we can construct the *mixed* load potential

$$\mathcal{P}[\hat{\mathbf{N}}_\sigma, \beta, \mathbf{p}] \equiv \mathcal{P} + \int_0^L \mathbf{p}\cdot\bar{\mathbf{y}}(\mathbf{p}) r d\sigma + r\hat{\mathbf{N}}_\sigma\cdot\bar{\mathbf{y}}(\hat{\mathbf{N}}_\sigma)|_{\partial N} \tag{V.26}$$

such that

$$\delta\mathcal{P} = \int_0^L (\bar{\mathbf{y}}\cdot\delta\mathbf{p} - l\delta\beta) r d\sigma + [r(\bar{\mathbf{y}}\cdot\delta\hat{\mathbf{N}}_\sigma - \hat{M}_\sigma\delta\beta)]_{\partial N} . \tag{V.27}$$ [37]

Although more general boundary conditions can be accommodated, we shall assume, for simplicity, that these are classical, so that

$$\mathcal{L}^T \mathcal{R}(y) = [r(\tilde{\mathbf{N}}_\sigma\cdot(\bar{\mathbf{y}} - \hat{\bar{\mathbf{y}}}) + r\tilde{M}_\sigma(\beta - \hat{\beta})]_{\partial U} , \tag{V.28}$$

where $\tilde{\mathbf{N}}_\sigma$ and $\tilde{M}_\sigma$ are Lagrange multipliers.

We now integrate (V.24) by parts, require $\mathbf{N}_\sigma$ and N_θ to satisfy the equilibrium conditions

$$(r\mathbf{N}_\sigma)' - N_\theta \mathbf{e}_r + r\mathbf{p} = \mathbf{0} \text{ on } 0 < \sigma < L \tag{V.29}$$

$$\mathbf{N}_\sigma = \hat{\mathbf{N}}_\sigma \text{ on } \partial N , \tag{V.30}$$

and equate the Lagrange multiplier $\tilde{\mathbf{N}}_\sigma$ to $\mathbf{N}_\sigma$ on ∂U. The result of these operations is the new functional

$$\Pi_4 \equiv \mathcal{P} + \int_0^L \Omega r d\sigma + [r\mathbf{N}_\sigma\cdot\hat{\bar{\mathbf{y}}} - r\tilde{M}_\sigma(\beta - \hat{\beta})]_{\partial U} , \tag{V.31}$$

and the variational principle is

$$\delta\Pi_4[N_\sigma, N_\theta, Q, \beta, \tilde{M}_\sigma, \hat{N}_\sigma, \mathbf{p}] = 0 . \tag{V.32}$$

The arguments in (V.32) are constrained to satisfy the equilibrium equations (V.29) and (V.30).

In the last principle to be derived, we constrain β to satisfy $\beta = \hat{\beta}$ on ∂U. Then, the last term in Π_4 vanishes and we obtain

$$\Pi_5 \equiv \mathcal{P} + \int_0^L \Omega r d\sigma + (r\mathbf{N}_\sigma\cdot\hat{\bar{\mathbf{y}}})_{\partial U} , \tag{V.33}$$

with the associated principle

$$\delta\Pi_5[N_\sigma, N_\theta, Q, \beta, \hat{N}_\sigma, \mathbf{p}] = 0 . \tag{V.34}$$

As in (V.32), the arguments of Π_5 are also constrained to satisfy the equilibrium equations (V.29) and (V.30).

The stationary conditions common to both Π_4 and Π_5 are the force-kinematic equations, which connect $\mathbf{N}_\sigma$ and N_θ to $\bar{\mathbf{y}}$ and β through the constitutive relations and the strain-kinematic relations; moment equilibrium; any boundary conditions on M_σ; and any boundary conditions on $\bar{\mathbf{y}}$. In addition, in

[37] In the important special case of "dead loads", in which $\mathbf{p}$ and $\hat{\mathbf{N}}_\sigma$ are independent of $\bar{\mathbf{y}}$, (V.26) reduces to $\mathcal{P}[\beta] \equiv \mathcal{P} + \int_0^L \mathbf{p}\cdot\bar{\mathbf{y}} r d\sigma + r\hat{\mathbf{N}}_\sigma\cdot\bar{\mathbf{y}}|_{\partial N}$. Comparison with (V.9) and (V.11) shows that the $\mathbf{p}$ and $\hat{\mathbf{N}}_\sigma$ terms *cancel out* of $\mathcal{P}$. Also, (V.27) reduces to $\delta\mathcal{P} = -\int_0^L l\delta\beta r d\sigma - r\hat{M}_\sigma\delta\beta|_{\partial N}$.

(V.32) only, $M_\sigma = \tilde{M}_\sigma$ on ∂U. This *condition* gets replaced in (V.34) by the *constraint*, $\beta = \hat{\beta}$ on ∂U.

It should be noted that although the constitutive relations are incorporated in Ω and therefore are not needed in solving for the unknowns in Π_4 and Π_5, they may be needed in auxiliary equations to compute secondary unknowns.

Finally, we observe that the Π_5 variational functionals are of exactly the same form for birods, (III.K.25), beamshells, (IV.Q.35), and axishells, (V.33). This may be part of a more general pattern.

3. Remarks

(a) Compatibility conditions. An extensional-shear strain field (e_σ, e_θ, g) is derivable from a *kinematic field* $(\bar{y}, \beta)$ if and only if the former satisfies the compatibility condition (M.2). A bending strain field $(k_\sigma, k_\theta, k_n)$ is derivable from a kinematic field if and only if the former satisfies (M.12) and (M.13). The kinematic field itself is said to be *edge-compatible* if it satisfies any kinematic constraints at the boundary. If all of the above conditions are met, then the deformation field (or the stress field, via constitutive relations) is said to satisfy the equations of *edge-interior compatibility*.

In some extended variational principles, the strain-kinematic relations and the kinematic boundary constraints are stationary conditions, i.e., the edge-interior compatibility conditions are the stationary conditions. Thus, we can rephrase the variational principle for the mixed functional Π_5 as follows: Of all stress resultant fields (N_σ, N_θ, Q) that satisfy *force equilibrium* and of all bending fields $(k_\sigma, k_\theta, k_n)$ that satisfy *bending compatibility*, those that satisfy the equations of *extensional-shear force compatibility* and *moment equilibrium* render Π_5 stationary. The duality of equilibrium and compatibility in the variational principles as well as in other parts of shell theory should be apparent to the reader.

(b) Use of stress functions. Stress functions are introduced to satisfy identically some or all of the equilibrium equations and boundary conditions. For example, if we use (Q.4), (Q.10), and (Q.11) to express N_θ, N_σ, and Q in terms of the stress function $F = rH$, then force equilibrium is satisfied identically. The advantage of using F is that the number of fields in Π_4 and Π_5 is reduced, so that the variational principles (V.32) and (V.34), if V is known, reduce to

$$\delta\Pi_4[F, \beta, \tilde{M}_\sigma, \hat{N}_\sigma, \mathbf{p}] = 0 \tag{V.35}$$

$$\delta\Pi_5[F, \beta, \hat{N}_\sigma, \mathbf{p}] = 0\,, \tag{V.36}$$

where Π_4 and Π_5 may also depend on σ, either through the surface load $\mathbf{p}$ or through $r(\sigma)$, the r-coordinate of the reference meridian. The situation is more complicated, but manageable, if $\bar{r}$ is introduced as an additional unknown via $\mathbf{p}$, as in centrifugal loading. However, by using the constitutive relations, we may write

$$\bar{r} = r[1 + e_\theta(N_\theta, N_\sigma, Q, \beta)] \; . \qquad \text{(V.37)}$$

Substitution into (Q.4), (Q.10), and (Q.11) leads to three nonlinear algebraic equations which define, implicitly or explicitly, N_σ, N_θ, and Q in terms of F and β.

A mixed principle involving a mixed-energy density, with β and F as arguments, was used by Reissner (1963a) to study simplified versions of the field equations.

(c) Membrane axishells. Here, Ψ does not depend on the bending strains, k_σ, k_θ, and k_n. Further, moment equilibrium, (L.15), with no distributed surface couple, yields $Q=0$. Although $Q=0$ is not a constitutive assumption, it is convenient to introduce this *result* into the expression for Ψ and into (Q.4), (Q.10), and (Q.11). As arguments in the variational principle, we can use either F and V, as in (Q.26), or $(\alpha+\beta)$ and V, where now $\bar{\alpha}=\alpha+\beta$. Choosing the latter, we have

$$N_\sigma = V\csc\bar{\alpha} \;\; , \;\; N_\theta = (rV\cot\bar{\alpha})' + rp_H \; . \qquad \text{(V.38)}$$

The functionals (V.31) and (V.33) now have the identical form

$$\Pi_4[\bar{\alpha}, V, \hat{\mathbf{N}}_\sigma, \mathbf{p}] = \Pi_5[\bar{\alpha}, V, \hat{\mathbf{N}}_\sigma, \mathbf{p}]$$
$$= \mathcal{P} + \int_0^L \Omega(N_\sigma, N_\theta) r d\sigma + (r\mathbf{N}_\sigma \cdot \hat{\bar{\mathbf{y}}})_{\partial U} \; , \qquad \text{(V.39)}$$

where, by (V.38), N_σ and N_θ are functions of $\bar{\alpha}$, V, and possibly σ, *only.*

To derive differential equations and boundary conditions for membrane axishells, we take the variation of (V.39), noting (Q.3) and (V.38). There results

$$\int_0^L \{rV[(1+e_\sigma)\cos\bar{\alpha} - (r + re_\theta)']\csc^2\bar{\alpha}\delta\bar{\alpha}$$
$$+ r[(r + re_\theta)'\cot\bar{\alpha} - (1+e_\sigma)\csc\bar{\alpha} + \bar{z}']\delta V$$
$$+ r[\bar{r} - r(1+e_\theta)]\delta p_H\} d\sigma \qquad \text{(V.40)}$$
$$+ [r(\bar{r}\cot\bar{\alpha} + \bar{z})\delta(\hat{V} - V) - r\bar{r}V\csc^2\bar{\alpha}\delta(\hat{\bar{\alpha}} - \bar{\alpha})]_{\partial N}$$
$$+ \{[(\hat{\bar{r}} - r - re_\theta)\cot\bar{\alpha} + (\hat{\bar{z}} - \bar{z})]r\delta V - V[\hat{\bar{r}} - r(1+e_\theta)]\csc^2\bar{\alpha}\delta\bar{\alpha}\}_{\partial U} = 0 \; .$$

In (V.40), the strains are *defined* by $e_\sigma \equiv -\Psi_{,N_\sigma}$ and $e_\theta = -\Psi_{,N_\theta}$, and are expressed, through N_σ and N_θ, in terms of $\bar{\alpha}$ and σ. The stationary conditions of (V.40), on $0 < z < L$, are the differential equation

$$[r(1+e_\theta)]' = (1+e_\sigma)\cos\bar{\alpha} \qquad \text{(V.41)}$$

and the auxiliary conditions for calculating the deformed position,

$$\bar{r} = r(1+e_\theta) \;\; , \;\; \bar{z}' = (1+e_\sigma)\sin\bar{\alpha} \; . \qquad \text{(V.42)}$$

In addition, the admissibility condition $(rV)' + rp_V = 0$ must be satisfied.

On ∂U, the stationary conditions are

$$r(1+e_\theta) = \hat{\bar{r}} \;,\; \bar{z} = \hat{\bar{z}} \,, \tag{V.43}$$

i.e., the deformed position must assume its assigned value at an edge. In contrast, $\bar{\alpha}$ satisfies no condition on ∂U and, in fact, may be discontinuous there. This may be seen from (V.31): since $M_\sigma = 0$, the condition $\beta = \hat{\beta}$ on ∂U drops out! It follows from (V.40) that the elements of the *admissible set* must satisfy $V = \hat{V}$, $\bar{\alpha} = \hat{\bar{\alpha}}$ on ∂N. Note that the requirement that $\bar{\alpha} = \hat{\bar{\alpha}}$ on ∂N is a *force* condition, not a *kinematic* one. This is because, if edge loads are prescribed, $\bar{\alpha}$ must be such that $\mathbf{m} \cdot \mathbf{N}_\sigma = 0$. Thus, the requirement of force equilibrium at the edges pulls in a kinematic condition "through the back door". In any event, the number of conditions per edge is *two* compared with linear membrane theory where only *one* condition per edge may be prescribed.

(d) Moderate rotations and shallowness. If moderate rotations are assumed, then the equilibrium equations, (Q.10) and (Q.11), which define the *admissible set* for Π_4 and Π_5, will simplify, as will the expressions for the bending strains in terms of β, (L.8)-(L.10), which form a part of Ψ. Because terms of $O(\beta^2)$ are neglected compared with 1, we set, in the above,

$$\cos(\alpha+\beta) \approx \cos\alpha - \beta\sin\alpha \;,\; \sin(\alpha+\beta) \approx \sin\alpha + \beta\cos\alpha \tag{V.44}$$

$$\cos\alpha - \cos(\alpha+\beta) \approx \beta\sin\alpha + \tfrac{1}{2}\beta^2\cos\alpha \,,\; \sin(\alpha+\beta) \approx \beta\cos\alpha - \tfrac{1}{2}\beta^2\sin\alpha \,. \tag{V.45}$$

[Care must be taken to preserve the β^2-terms when differences in the trigonometric functions appear in the equations. This includes the $H\cos(\alpha+\beta)$-term in (Q.10).] Shallow shell theory makes the additional simplification that α is small and neglects $O(\alpha^2)$- and $O(\alpha\beta)$-terms compared with 1. In this case, we set

$$\cos(\alpha+\beta) \approx 1 \;,\; \sin(\alpha+\beta) \approx \alpha+\beta \tag{V.46}$$

$$\cos\alpha - \cos(\alpha+\beta) \approx \alpha\beta + \tfrac{1}{2}\beta^2 \;,\; \sin(\alpha+\beta) - \sin\alpha \approx \beta \tag{V.47}$$

in (Q.10), (Q.11), (L.9), and (L.10). Also, because $dr = \cos\alpha d\sigma \approx d\sigma$, we can replace the meridional variable σ by r. With these simplifications, we obtain the Π_4 and Π_5 variational principles for moderate rotations and shallow shells.

4. Examples

We now give several examples related to the Π_5-variational principle.

(a). In the first example, we show how the stress function approach leads to the field equations of axishells as stationary conditions. Starting from (V.33), we have for the variation of Π_5,

$$\delta\Pi_5 = \delta\mathcal{P} + \int_0^L (-\lambda_T \delta N_\sigma - \lambda_\theta \delta N_\theta - g\delta Q$$
$$+ M_\sigma \delta k_\sigma + M_\theta \delta k_\theta + M_n \delta k_n) r d\sigma + (r\delta \mathbf{N}_\sigma \cdot \hat{\mathbf{y}})_{\partial U} \,, \tag{V.48}$$

where $\delta\mathcal{P}$ is given in (V.27). In the above, the strains and moments are *defined* as in (O.5) and the bending strains are expressed in terms of β, as in (L.8)-(L.10).

Expressing N_σ, N_θ, and Q in terms of $F = rH$ and V, as in (Q.4), (Q.10), and (Q.11), and performing the variations, we have

$$\delta N_\theta = \delta F' + r\delta p_H \tag{V.49}$$

$$\begin{aligned} r\delta N_\sigma &= \delta F \cos\bar{\alpha}_T + r\delta V \sin\bar{\alpha}_T + rQ\,\delta\beta \\ r\delta Q &= -\delta F \sin\bar{\alpha}_T + r\delta V \cos\bar{\alpha}_T - rN_\sigma \delta\beta \end{aligned} \tag{V.50}$$

$$r(M_\sigma \delta k_\sigma + M_\theta \delta k_\theta + M_n \delta k_n) = rM_\sigma \delta\beta' + (M_\theta \cos\bar{\alpha}_T - M_n \sin\bar{\alpha}_T)\delta\beta \,, \tag{V.51}$$

where, as before, $\bar{\alpha}_T = \alpha + \beta$. Inserting these relations along with (V.27) for $\delta\mathcal{P}$ into the right side of (V.48), using (Q.3) to eliminate δp_V, integrating by parts where indicated, and setting F and V on ∂N to their prescribed values there, we obtain, finally,

$$\delta\Pi_5 = \int_0^L \{[(r\lambda_\theta)' - \lambda_T \cos\bar{\alpha}_T + g\sin\bar{\alpha}_T]\delta F$$
$$- [(rM_\sigma)' - M_\theta \cos\bar{\alpha}_T + M_n \sin\bar{\alpha}_T + \lambda_T rQ - grN_\sigma + rl]\delta\beta$$
$$+ (\bar{z}' - \lambda_T \sin\bar{\alpha}_T - g\cos\bar{\alpha}_T) r\delta V + (\bar{r} - r\,\lambda_\theta) r\delta p_H\} d\sigma \tag{V.52}$$
$$+ r(\hat{\bar{\mathbf{y}}} - \bar{\mathbf{y}}) \cdot \delta \mathbf{N}_\sigma |_{\partial U} + r(M_\sigma - \hat{M}_\sigma)\delta\beta |_{\partial N} \,.$$

The requirement that $\delta\Pi_5 = 0$ for all variations of F, V, p_H, and β establishes the stationary conditions. The interior conditions are recognized to be strain-rotation compatibility, (M.2), moment equilibrium, (L.15), and the auxiliary conditions, (M.6), needed for calculation the deformed position $\bar{\mathbf{y}}$ in terms of the strains. On the boundary, the stationary conditions are $\bar{\mathbf{y}} = \hat{\bar{\mathbf{y}}}$ on ∂U and $M_\sigma = \hat{M}_\sigma$ on ∂N, in accordance with our previous remarks on the Π_5 principle.

Note the important role played by δp_H and δV in obtaining the auxiliary stationary conditions. The external loads are to be treated as variables in the process and cannot be deleted *a priori*, even if they are constant in a specific class of problems. This somewhat unexpected turn of events stems from using a stress function, which reduces the number of stress-resultant variations; the corresponding load variations must be kept to avoid the inadvertent suppression of auxiliary conditions. If V is known and $\delta V \equiv 0$, the equation for $\bar{z}'$ disappears from (V.52), as it should, because in this case it determines merely a rigid body displacement and thus is not an integral part of the axishell equations.

(b). As a second example, we derive a first integral of the field equations for a cylindrical axishell of initial radius R, free of surface loads and subject to *arbitrary boundary conditions* (consistent, of course, with overall static equilibrium). Under these conditions, (V.33) reduces to boundary terms plus the integral

$$\int_0^L \Omega R d\sigma \equiv \int_0^L G(F, \beta, F', \beta') d\sigma . \qquad (V.53)$$

The functional form of G is determined via (L.8)-(L.10), (Q.4), (Q.10), and (Q.11). In particular, the dependence of G on its last two arguments comes from $N_\theta = F'$ and $k_\sigma = \beta'$.

As G does not depend explicitly on σ, it follows that a first integral of the associated Euler-Lagrange equations is H=constant, where H is the *Hamiltonian* of G [Gelfand & Fomin (1963), p. 70]. That is,

$$H = F' \frac{\partial G}{\partial F'} + \beta' \frac{\partial G}{\partial \beta'} - G = \text{constant} . \qquad (V.54)$$

With

$$\frac{\partial G}{\partial F'} = R \frac{\partial \Omega}{\partial N_\theta} = -R\lambda_\theta \ , \quad \frac{\partial G}{\partial \beta'} = R \frac{\partial \Omega}{\partial k_\sigma} = RM_\sigma \, , \qquad (V.55)$$

we obtain

$$H = \beta' M_\sigma - \lambda_\theta N_\theta - \Omega = \text{constant} , \qquad (V.56)$$

which may also be written in terms of the strain-energy density Φ as

$$H = \beta' M_\sigma + (1 + e_\sigma) N_\sigma + Qg - \Phi = \text{constant} . \qquad (V.57)$$

This expression, surprisingly enough, is identical in form to the energy integral of beamshell theory, (IV.O.35), the additional internal stress resultants and couples in the axishell theory being reflected only through Φ.

In a *membrane*, $M_\sigma = Q = 0$, and (V.57), with $\lambda_\sigma = (1 + e_\sigma)$, reduces to

$$\lambda_\sigma \frac{\partial \Phi}{\partial \lambda_\sigma} - \Phi = \text{constant} . \qquad (V.58)$$

This is Pipkin's (1968) 1st integral for a cylindrical aximembrane. It can also be derived directly from (V.34) by taking $\Omega = \Omega(N_\sigma, N_\theta)$.

Returning to (V.56) and taking, as an example, Ψ in the classical, quadratic form (O.11), with $A_s = D_n = 0$, we have, to a first approximation,

$$\tfrac{1}{2} D (k_\sigma^2 - k_\theta^2) + \underline{N_\sigma} + \tfrac{1}{2} A (N_\sigma{}^2 - N_\theta{}^2) = \text{constant} . \qquad (V.59)$$

It is interesting to note that the terms involving the Poisson ratios of extension, ν_e, and bending, ν_b, except for that involved in D, drop out. The small-strain approximation, implied by taking Ψ to be quadratic, permits us to delete the underlined term in (V.59) with the result that

$$\tfrac{1}{2}D\,(k_\sigma^2 - k_\theta^2) + N_\sigma - \tfrac{1}{2}AN_\theta{}^2 = \text{constant}\,, \qquad \text{(V.60)}$$

or, in terms of F and β,

$$\tfrac{1}{2}D\,[\beta'^2 - \underline{R^{-2}(1-\cos\beta)^2}] - R^{-1}F\sin\beta + V\cos\beta - \tfrac{1}{2}AF'^2 = \text{constant}\,. \qquad \text{(V.61)}$$

In many applications involving boundary layers, the underlined term is small compared with β'^2, and we have, to a 1st approximation,

$$\tfrac{1}{2}D\beta'^2 - R^{-1}F\sin\beta + V\cos\beta - \tfrac{1}{2}AF'^2 = \text{constant}\,, \qquad \text{(V.62)}$$

which is simply the dimensional form of (S.56). Note, however, that in very short cylindrical shells, we can have $\beta' \approx 0$—an extreme case being a cylindrical shell flattened into a washer—and the underlined term in (V.61) cannot be neglected.

(c). As a third example, we explore the possibility of associating a variational principle with the simplified Reissner equations of Section V.R by adopting, *ab initio*, a particular form of Ψ. Recall that in deriving the simplified Reissner equations, we often neglected the shearing strain, g, or the normal moment, M_n, compared with the other strain or moment terms in the governing equations. [See, in particular, the discussion in Section V.R concerning (R.6) and (R.8).] With this observation, we take

$$\Psi = -\tfrac{1}{2}A\,(N_\sigma^2 + N_\theta^2 + Q^2 - 2\nu N_\sigma N_\theta) + \tfrac{1}{2}D\,(k_\sigma^2 + k_\theta^2 + k_n^2 + 2\nu k_\sigma k_\theta)\,, \qquad \text{(V.63)}$$

where the coefficients of Q^2 and k_n^2 have been modified. We shall use the Π_5 variational principle defined by (V.33)-(V.34), with Ψ given by (V.63).

In deriving field equations, we use the stress-function approach (described in our first example) which led to (V.52). Our modified mixed-energy density (V.63) implies that

$$e_\sigma = A\,(N_\sigma - \nu N_\theta)\;,\quad e_\theta = A\,(N_\theta - \nu N_\sigma)\;,\quad g = AQ \qquad \text{(V.64)}$$

$$M_\sigma = D\,(k_\sigma + \nu k_\theta)\;,\quad M_\theta = D\,(k_\theta + \nu k_\sigma)\;,\quad M_n = Dk_n\,, \qquad \text{(V.65)}$$

and we insert these relations into (V.52). The field equations are obtained by setting to 0 the coefficients of δF and $\delta\beta$ in the resulting integral. After some calculations, we find that

$$\begin{aligned} &A\,(rF'' + F'\cos\alpha - r^{-1}F) + \cos\alpha - \cos(\alpha+\beta) \\ &\qquad + A\,[(r^2p_H)' + \nu r p_T] - \underline{\nu r g\,(\alpha+\beta)'} = 0 \end{aligned} \qquad \text{(V.66)}$$

$$\begin{aligned} &D\,(r\beta'' + \beta'\cos\alpha - r^{-1}\sin\beta) - F\sin(\alpha+\beta) + rV\cos(\alpha+\beta) \\ &\qquad + \underline{\nu r\,(M_n\alpha' + gN_\theta)} = 0\,. \end{aligned} \qquad \text{(V.67)}$$

These equations differ from the simplified Reissner equations by the underlined ν-terms, which involve shearing deformations and nonclassical moments—terms of little importance that are omitted in classical shell theory. Thus, we conclude that, but for a few minor terms, the variational principle for Π_5, with Ψ as in (V.63), is equivalent to the simplified Reissner equations.

W. The Mechanical Theory of Stability of Axishells

The underlying physical and mathematical ideas of structural stability are the same for all elastic bodies; hence the terminology, criteria, and concepts introduced in Section IV.R on the stability of beamshells apply here and need not be reviewed. Yet there are significant differences from a beamshell in the way a "true" shell resists external loads and buckles. For example, the critical (bifurcation) stress of a straight, edge-loaded, simply-supported beamshell, acting as a column, is

$$\frac{N_{cr}}{h} = \left[\frac{\pi^2}{12}\right]\left[\frac{E}{1-\nu^2}\right]\left[\frac{h}{L}\right]^2, \qquad \text{(W.1)}$$

and the critical stress for an edge-loaded, simply-supported square plate compressed along 2 opposite edges is

$$\frac{N_{xcr}}{h} = \left[\frac{\pi^2}{3}\right]\left[\frac{E}{1-\nu^2}\right]\left[\frac{h}{L}\right]^2. \qquad \text{(W.2)}$$

On the other hand, the critical bifurcation-buckling stress of an axially-compressed, circular cylindrical shell is, approximately,

$$\frac{N_{\sigma cr}}{h} = \left[\frac{1}{\sqrt{3}}\right]\left[\frac{E}{\sqrt{1-\nu^2}}\right]\left[\frac{h}{R}\right]. \qquad \text{(W.3)}^{38}$$

We note that the critical stress of a cylindrical shell of comparable dimensions to a beamshell or plate is much higher and is *linear* in h/R, whereas the critical stress of the beamshell or plate is *quadratic* in h/L. Moreover, the stable post-buckling behavior of the column (though close to being neutral) and that of the plate contrast with the strongly unstable behavior of the cylindrical shell at its critical point and its imperfection sensitivity. Indeed, experiments on very thin, unavoidably-imperfect, circular cylindrical shells, show large scatter and snap loads as low as 20% of the theoretical bifurcation load (Brush & Almroth, 1975, p. 186).

A second set of contrasting behaviors is provided by a beamshell tube, which has a critical, bifurcating pressure,

$$p_{cr} = \frac{1}{4}\left[\frac{E}{1-\nu^2}\right]\left[\frac{h}{R}\right]^3, \qquad \text{(W.4)}$$

and a closed spherical shell, where

$$p_{cr} = \left[\frac{2}{\sqrt{3}}\right]\left[\frac{E}{\sqrt{1-\nu^2}}\right]\left[\frac{h}{R}\right]^2. \qquad \text{(W.5)}$$

For the tube, the bifurcation pressure is proportional to $(h/R)^3$ whereas for the

[38] See (S.33).

spherical shell it is proportional to $(h/R)^2$—an order of magnitude larger. Further, as discussed in Chapter IV, a tube under fluid pressure is imperfection *insensitive*, but a spherical shell snaps at pressures much below the bifurcation and, in experiments, large scatter is observed. (See Kollar, 1982, Figs. 1 and 2; Kaplan, 1974; and Hoff, 1969, pp.1064-1068). Exceptions to this mode of behavior are found mostly in *shallow* cylindrical panels, spherical caps, and stiffened shells (whose behavior, in this sense, is similar to that of a thick shell).

Based on the above examples and many others, a general trend in the behavior of shells near bifurcation points may be seen. As noted by Koiter (1967a)

> all recent work tends to confirm earlier conjectures that the significant increase in the classical critical load of curved shell structures, in comparison with similar flat structures, has always to be paid for by increased imperfection sensitivity.

The reason for this increased resistance to buckling, with its consequent potential for catastrophe, lies in the response of a *true shell* to imposed changes in geometry and loading. By definition, a true shell can be *properly constrained* so that it cannot undergo a pure inextensional deformation—any imposed admissible displacement field must increase the extensional energy of the shell; moreover, a true shell can be *properly supported* so that any sufficiently smooth set of loads can be equilibrated primarily by membrane action. This dual behavior (which is a nonlinear extension of the *static-geometric duality* of linear theory) is not manifest in curved beamshells, which often deform inextensionally and can support only certain loads by membrane (direct) forces alone. If we think of a beamshell as a special case of a cylindrical shell, then its unavoidable flexibility becomes clear: there are no boundaries on planes normal to the z-direction on which geometrical constraints can be applied. In this respect, a beamshell acts like a very wide beam and not like a true shell, which is much stiffer in its response to loading. For further discussion, see Koiter (1982, Sect. 3).

Most books, chapters of books, or survey articles, that discuss buckling of shells of revolution, deal with specific geometries: cylinders, spheres, cones, tori, and planes.[39] Classical material on the buckling of circular plates and cylindrical, conical, and spherical shells may be found in the book by Timoshenko & Gere (1961). Flügge's book (1973) has a chapter on the buckling of cylindrical shells and Dym's (1974) treats the buckling and postbuckling of plates. Brush & Almroth (1975) devote a chapter of their book to cylindrical shells and another chapter to circular plates, conical and spherical shells, and toroidal segments. Yamaki's book (1984) is devoted exclusively to the stability, postbuckling, and imperfection sensitivity of circular cylindrical shells and includes many numerical and experimental results. The book by Kollar & Dulacska (1984) presents a practical approach to shell buckling with many

---

[39]Of course, there exist many numerical analyses of special purpose, one-of-a-kind, industrial shells.

design recommendations.

In addition to the survey articles on shell buckling already cited in Section IV.R, we mention Koiter's (1967a) article on buckling and postbuckling equations, Hoff's (1969) review summarizing activities in the '60's on cylindrical and spherical shells, Kollar's (1982) review on spherical shells, Bushnell's (1981) survey—now available in expanded, book form (1985)—with emphasis on computerized solutions, and Singer's review (1982) on experimental investigations.

Before proceeding with derivations, we emphasize, once again, that *our analysis is restricted to axisymmetric, fundamental equilibrium paths and to axisymmetric bifurcations.* This restricts the usefulness of our equations as *nonaxisymmetric* bifurcation is the rule, not the exception. Nevertheless, our development will be of value where axisymmetric deformation prevails and will serve as a prologue for nonaxisymmetric studies.

1. Buckling equations

We calculate, first, the incremental potential energy of an axishell due to the addition of an admissible deformation field, $(\mathbf{u}, \beta^{(1)})$. We adopt the variables and notation of Section IV.R on the stability of beamshells, but add a subscript σ to meridional quantities and a subscript θ to circumferential quantities. With this in mind, we can use (IV.R.5)-(IV.R.10) with (e, k) replaced by (e_σ, k_σ). To these 6 equations, we add the following expressions for the incremental circumferential and normal strains.

$$e_\theta^{(1)} = r^{-1}\mathbf{e}_r\cdot(\overline{\mathbf{y}} - \overline{\mathbf{y}}^{(0)}) = r^{-1}\mathbf{e}_r\cdot\mathbf{u} \tag{W.6}$$

$$k_\theta^{(1)} = r^{-1}[\sin(\overline{\alpha}_T^{(0)} + \beta^{(1)}) - \sin\overline{\alpha}_T^{(0)}] \tag{W.7}$$

$$k_n^{(1)} = r^{-1}[\cos(\overline{\alpha}_T^{(0)} + \beta^{(1)}) - \cos\overline{\alpha}_T^{(0)}]\,, \tag{W.8}$$

where, as before, $\overline{\alpha}_T = \alpha + \beta$. We express $k_\theta^{(1)}$ and $k_n^{(1)}$ in ascending degrees of nonlinearity in the incremental variables as

$$k_\theta^{(1)} = \sum_i k_{\theta i}^{(1)}\ ,\quad k_n^{(1)} = \sum_i k_{ni}^{(1)}\,, \tag{W.9}$$

where

$$k_{\theta 1}^{(1)} = r^{-1}\beta^{(1)}\cos\overline{\alpha}_T^{(0)} \tag*{(W.10)$_1$}$$

$$k_{\theta 2}^{(1)} = -\tfrac{1}{2}r^{-1}(\beta^{(1)})^2\sin\overline{\alpha}_T^{(0)} \tag*{(W.10)$_2$}$$

$$k_{\theta 3}^{(1)} = -\frac{1}{6r}(\beta^{(1)})^3\cos\overline{\alpha}_T^{(0)} \tag*{(W.10)$_3$}$$

$$k_{\theta 4}^{(1)} = \frac{1}{24r}(\beta^{(1)})^4\sin\overline{\alpha}_T^{(0)} \tag*{(W.10)$_4$}$$

$$k_{n1}^{(1)} = -r^{-1}\beta^{(1)}\sin\bar{\alpha}_T^{(0)} \tag{W.11$_1$}$$

$$k_{n2}^{(1)} = -\tfrac{1}{2}r^{-1}(\beta^{(1)})^2\cos\bar{\alpha}_T^{(0)} \tag{W.11$_2$}$$

$$k_{n3}^{(1)} = \frac{1}{6r}(\beta^{(1)})^3\sin\bar{\alpha}_T^{(0)} \tag{W.11$_3$}$$

$$k_{n4}^{(1)} = \frac{1}{24r}(\beta^{(1)})^4\cos\bar{\alpha}_T^{(0)}\,. \tag{W.11$_4$}$$

There are no higher degree terms in $e_\theta^{(1)}$.

Using a Taylor series expansion, we have for the increment of strain energy *per unit circumferential radian* due to the incremental deformations $\mathbf{u}$ and $\beta^{(1)}$,

$$\Delta\mathcal{E} = \int_0^L \left[\sum_{i=1}^{6} \left[\frac{\partial\phi}{\partial\varepsilon_i} \right]^{(0)} \varepsilon_i^{(1)} + \tfrac{1}{2} \sum_{i=1}^{6} \sum_{j=1}^{6} \left[\frac{\partial^2\phi}{\partial\varepsilon_i\partial\varepsilon_j} \right]^{(0)} \varepsilon_i^{(1)}\varepsilon_j^{(1)} + R \right] r\,d\sigma\,, \tag{W.12}$$

where

$$\varepsilon_i \equiv (e_\sigma, e_\theta, g, k_\sigma, k_\theta, k_n) \tag{W.13}$$

is a strain vector and R denotes terms involving third derivatives of the strain-energy density with respect to ε_i, evaluated at some unknown point between 0 and ε_i. The superscript (0) indicates that the strain-energy density derivatives are evaluated at the fundamental state.

For the moment, we consider deformation-independent (i.e., *dead)* loads only. Then, the quadratic increment of total potential energy takes the form

$$\begin{aligned}\Delta(\mathcal{E}+\mathcal{P}) = \int_0^L [&N_\sigma^{(0)}e_{\sigma2}^{(1)} + Q^{(0)}g_2^{(1)} + M_\theta^{(0)}k_{\theta2}^{(1)} + M_n^{(0)}k_{n2}^{(1)} \\ &+ \tfrac{1}{2}(N_\sigma^{(1)}e_{\sigma1}^{(1)} + N_\theta^{(1)}e_\theta^{(1)} + Q^{(1)}g_1^{(1)} \\ &+ M_\sigma^{(1)}k_\sigma^{(1)} + M_\theta^{(1)}k_{\theta1}^{(1)} + M_n^{(1)}k_{n1}^{(1)})]r\,d\sigma\,,\end{aligned} \tag{W.14}$$

where the linearized incremental stress resultants and couples are *defined* by[40]

$$\tau_i^{(1)} \equiv \sum_{j=1}^{6} \left[\frac{\partial^2\Phi}{\partial\varepsilon_i\partial\varepsilon_j} \right]^{(0)} \varepsilon_{j1}^{(1)}\,, \tag{W.15}$$

with

$$(\tau_1^{(1)}, \cdots, \tau_6^{(1)}) = (N_\sigma^{(1)}, N_\theta^{(1)}, Q^{(1)}, M_\sigma^{(1)}, M_\theta^{(1)}, M_n^{(1)})\,. \tag{W.16}$$

The quadratic form of Φ is given by (O.8). In this case the $\partial^2\Phi/\partial\varepsilon_i\partial\varepsilon_j$ are the elastic coefficients.

---

[40]See (IV.R.16)-(IV.R.18) for detailed examples.

As discussed in Section IV.R, the buckling equations are the Euler-Lagrange equations of the variational principle $\delta\Delta(\mathcal{E}+\mathcal{P})=0$. Using results from Chapter IV, in particular (IV.R.20), we have

$$\begin{aligned}\delta\Delta(\mathcal{E}+\mathcal{P})=-\int_0^L <&[(rM_\sigma^{(1)})'-(M_\theta^{(1)}-M_n^{(0)}\beta^{(1)})\cos\bar{\alpha}_T^{(0)}+(M_n^{(1)}+M_\theta^{(0)}\beta^{(1)})\sin\bar{\alpha}_T^{(0)}\\&+r(1+e_\sigma^{(0)})(Q^{(1)}+N_\sigma^{(0)}\beta^{(1)})-rg^{(0)}(N_\sigma^{(1)}-Q^{(0)}\beta^{(1)})-r(\mathbf{N}_\sigma^{(0)}\times\mathbf{e}_\theta)\cdot\mathbf{u}']\delta\beta^{(1)}\\&+\{[r(Q^{(1)}+N_\sigma^{(0)}\beta^{(1)})\mathbf{B}^{(0)}+r(N_\sigma^{(1)}-Q^{(0)}\beta^{(1)})\mathbf{T}^{(0)}]'-N_\theta^{(1)}\mathbf{e}_r\}\cdot\delta\mathbf{u}>d\sigma\\&+\{r[(Q^{(1)}+N_\sigma^{(0)}\beta^{(1)})\mathbf{B}^{(0)}+(N_\sigma^{(1)}-Q^{(0)}\beta^{(1)})\mathbf{T}^{(0)}]\cdot\delta\mathbf{u}+rM_\sigma^{(1)}\delta\beta^{(1)}\}_0^L=0.\end{aligned}\tag{W.17}$$

From the above, we obtain, by standard arguments, the following differential equations and boundary conditions for the buckling of an axishell subject to deformation-independent loads:

in the interior: $0<\sigma<L$

$$\begin{aligned}(rM_\sigma^{(1)})'-(M_\theta^{(1)}-M_n^{(0)}\beta^{(1)})\cos\bar{\alpha}_T^{(0)}+(M_n^{(1)}+M_\theta^{(0)}\beta^{(1)})\sin\bar{\alpha}_T^{(0)}&\\+r(1+e_\sigma^{(0)})(Q^{(1)}+N_\sigma^{(0)}\beta^{(1)})-rg^{(0)}(N_\sigma^{(1)}-Q^{(0)}\beta^{(1)})-r(\mathbf{N}_\sigma^{(0)}\times\mathbf{e}_\theta)\cdot\mathbf{u}'&=0\end{aligned}\tag{W.18}$$

$$[r(Q^{(1)}+N_\sigma^{(0)}\beta^{(1)})\mathbf{B}^{(0)}+r(N_\sigma^{(1)}-Q^{(0)}\beta^{(1)})\mathbf{T}^{(0)}]'-N_\theta^{(1)}\mathbf{e}_r=\mathbf{0}\,. \tag{W.19}$$

on the edges: $\sigma=0,L$

$$Q^{(1)}+N_\sigma^{(0)}\beta^{(1)}=0 \text{ or } \delta(\mathbf{B}^{(0)}\cdot\mathbf{u})=0 \tag{W.20}$$

$$N_\sigma^{(1)}-Q^{(0)}\beta^{(1)}=0 \text{ or } \delta(\mathbf{T}^{(0)}\cdot\mathbf{u})=0 \tag{W.21}$$

$$M^{(1)}=0 \text{ or } \delta\beta^{(1)}=0\,. \tag{W.22}$$

We have expressed the vector boundary terms in (W.17) in terms of the basis $\{\mathbf{T}^{(0)},\mathbf{B}^{(0)}\}$, but other choices, such as the fixed basis $\{\mathbf{e}_r,\mathbf{e}_z\}$, may be advantageous.

A useful and concise way of writing the above buckling equations is in terms of linearized vector increments. Thus, with $\mathbf{T}^{(1)}\equiv\beta^{(1)}\mathbf{B}^{(0)}$, $\mathbf{B}^{(1)}\equiv-\beta^{(1)}\mathbf{T}^{(0)}$, and $\mathbf{m}^{(1)}\equiv\mathbf{u}'\times\mathbf{e}_\theta$, we find for the increments of $\mathbf{N}_\sigma$ and $\mathbf{M}_\theta$,

$$\mathbf{N}_\sigma^{(1)}=(N_\sigma^{(1)}-Q^{(0)}\beta^{(1)})\mathbf{T}^{(0)}+(Q^{(1)}+N_\sigma^{(0)}\beta^{(1)})\mathbf{B}^{(0)} \tag{W.23}$$

$$\mathbf{M}_\theta^{(1)}=(M_\theta^{(1)}-M_n^{(0)}\beta^{(1)})\mathbf{T}^{(0)}+(M_n^{(1)}+M_\theta^{(0)}\beta^{(1)})\mathbf{B}^{(0)}\,. \tag{W.24}$$

With these expressions, we may write the variational equations, the buckling equations, and the boundary conditions, respectively, as

$$\begin{aligned}\delta\Delta(\mathcal{E}+\mathcal{P})=-\int_0^L \{&[(rM_\sigma^{(1)})'-\mathbf{M}_\theta^{(1)}\cdot\mathbf{e}_r+r(\mathbf{N}_\sigma^{(1)}\cdot\mathbf{m}^{(0)}+\mathbf{N}_\sigma^{(0)}\cdot\mathbf{m}^{(1)})]\delta\beta^{(1)}\\&+[(r\mathbf{N}_\sigma^{(1)})'-N_\theta^{(1)}\mathbf{e}_r]\cdot\delta\mathbf{u}\}d\sigma+[r(\mathbf{N}_\sigma^{(1)}\cdot\delta\mathbf{u}+M_\sigma^{(1)}\delta\beta^{(1)})]_0^L=0\,.\end{aligned}\tag{W.25}$$

in the interior: $0<\sigma<L$

$$(rM_\sigma^{(1)})' - \mathbf{M}_\theta^{(1)}\cdot\mathbf{e}_r + r(\mathbf{N}_\sigma^{(1)}\cdot\mathbf{m}^{(0)} + \mathbf{N}_\sigma^{(0)}\cdot\mathbf{m}^{(1)}) = 0 \tag{W.26}$$

$$(r\mathbf{N}_\sigma^{(1)})' - N_\theta^{(1)}\mathbf{e}_r = \mathbf{0}\,. \tag{W.27}$$

on the edge: $\sigma=0, L$

$$\mathbf{N}_\sigma^{(1)} = \mathbf{0} \text{ or } \delta\mathbf{u} = \mathbf{0}\ ,\ M_\sigma^{(1)} = 0 \text{ or } \delta\beta^{(1)} = 0\,. \qquad \text{(W.28)}^{41}$$

A simple, formal perturbation of the vector equilibrium equations (D.3) and (D.5) (with the inertia terms set to zero) also produces (W.26) and (W.27). Nevertheless, the energy method offers the important advantages of emphasizing the physical basis of stability analysis[42] and incorporating variational principles, such as (IV.R.1) or (IV.R.2). Approximate methods of stability analysis exploit this last feature.

To obtain a complete set of equations, we introduce the constitutive relations, (W.15), and the strain-kinematic relations, (IV.R.10)-(IV.R.13) and (W.6)-(W.11), into the equilibrium conditions, (W.18) and (W.19). This yields a 6th order system of homogeneous ordinary differential equations in $\mathbf{u}$ and $\beta^{(1)}$.

The results suggest that it is useful to express the buckling equations in the basis $\{\mathbf{e}_r, \mathbf{e}_z\}$. In doing so, we may eliminate $\mathbf{u}'$ in (W.18). With the aid of the relations

$$\mathbf{m}^{(1)} = \mathbf{u}'\times\mathbf{e}_\theta = e_{\sigma 1}^{(1)}\mathbf{B}^{(0)} - g_1^{(1)}\mathbf{T}^{(0)} - (\bar{\gamma}^{(0)})'\beta^{(1)} \tag{W.29}$$

$$N_\sigma^{(1)} = H^{(1)}\cos\bar{\alpha}_T^{(0)} + V^{(1)}\sin\bar{\alpha}_T^{(0)} + Q^{(0)}\beta^{(1)} \tag{W.30}$$

$$Q^{(1)} = -H^{(1)}\sin\bar{\alpha}_T^{(0)} + V^{(1)}\cos\bar{\alpha}_T^{(0)} - N_\sigma^{(0)}\beta^{(1)}\,, \tag{W.31}$$

where $V^{(1)} \equiv \mathbf{N}_\sigma^{(1)}\cdot\mathbf{e}_z$ and $H^{(1)} \equiv \mathbf{N}_\sigma^{(1)}\cdot\mathbf{e}_r$, we can rewrite (W.18) and (W.27) as follows:

$$\begin{aligned}(rM_\sigma^{(1)})' &- (M_\theta^{(1)} - M_n^{(0)}\beta^{(1)})\cos\bar{\alpha}_T^{(0)} + (M_n^{(1)} + M_\theta^{(0)}\beta^{(1)})\sin\bar{\alpha}_T^{(0)} \\ &+ r[(1+e_\sigma^{(0)})Q^{(1)} + e_{\sigma 1}^{(1)}Q^{(0)} - g^{(0)}N_\sigma^{(1)} - g_1^{(1)}N_\sigma^{(0)}] = 0\end{aligned} \tag{W.32}$$

$$(rH^{(1)})' = N_\theta^{(1)} \tag{W.33}$$

$$(rV^{(1)})' = 0\,. \tag{W.34}$$

The above equations are also the basis for the following stress function-rotation formulation of buckling. From (W.34)

---

[41]Mixed component conditions are also possible, as in (W.20).

[42]The perturbation (or loss-of-uniqueness approach) can determine the availability of a second equilibrium path but cannot explain, by itself, *why or how* a structure switches branches at a bifurcation point.

$$rV^{(1)} = r(0)V^{(1)}(0) = r(L)V^{(1)}(L) = C, \text{ a constant}, \tag{W.35}$$

and from (W.33),

$$N_\theta^{(1)} = (F^{(1)})', \quad F^{(1)} \equiv rH^{(1)}, \tag{W.36}$$

where $F^{(1)}$ is the *incremental stress function*. Introduction of (W.35) and (W.36) into (W.32) yields

$$\begin{aligned}(rM_\sigma^{(1)})' - M_\theta^{(1)} \cos\overline{\alpha}_T^{(0)} + M_n^{(1)} \sin\overline{\alpha}_T^{(0)} + [\mathbf{M}_\theta^{(0)}\cdot\mathbf{e}_z - r\mathbf{N}_\sigma^{(0)}\cdot(\overline{\mathbf{y}}^{(0)})']\beta^{(1)} \\ - \lambda_T^{(0)} F^{(1)} \sin\overline{\alpha}_T^{(0)} + C\lambda_T^{(0)} \cos\overline{\alpha}_T^{(0)} + \underline{r(Q^{(0)}e_{\sigma 1}^{(1)} - N_\sigma^{(0)} g_1^{(1)})} = 0,\end{aligned} \tag{W.37}$$

where $\lambda_T^{(0)} = 1 + e_\sigma^0$ and $\overline{\alpha}_T^{(0)} = \alpha + \beta^{(0)}$. Substitution of the constitutive relations (W.15) into (W.37) yields a 2nd-order differential equation in $\beta^{(1)}$ and $F^{(1)}$.[43]

A second relation between $\beta^{(1)}$ and $F^{(1)}$ comes from the kinematic expression

$$\mathbf{u}' = e_{\sigma 1}^{(1)} \mathbf{T}^{(0)} + g_1^{(1)} \mathbf{B}^{(0)} + \beta^{(1)} \mathbf{m}^{(0)}, \tag{W.38}$$

which is derived with the aid of (IV.R.12) and (IV.R.13). In a beamshell, the displacements are unconstrained, but in an axishell, the radial displacement is related to the hoop strain by (W.6). Using this relation and taking the dot product of (W.38) with $\mathbf{e}_r$, we obtain the *compatibility condition*

$$(re_\theta^{(1)})' = e_{\sigma 1}^{(1)} \cos\overline{\alpha}_T^{(0)} - g_1^{(1)} \sin\overline{\alpha}_T^{(0)} - \lambda_T^{(0)} \beta^{(1)} \sin\overline{\alpha}_T^{(0)}. \tag{W.39}$$

Incrementing (M.2) yields the same result. Finally, introduction of the constitutive and stress function relations (W.15), (W.30), (W.31), (W.35), and (W.36), reduces (W.39) to a 2nd-order differential equation for $F^{(1)}$ and $\beta^{(1)}$. The axial displacement follows from (W.38) as

$$(\mathbf{u}\cdot\mathbf{e}_z)' = e_{\sigma 1}^{(1)} \sin\overline{\alpha}_T^{(0)} + g_1^{(1)} \cos\overline{\alpha}_T^{(0)} + \lambda_T^{(0)} \beta^{(1)} \cos\overline{\alpha}_T^{(0)}. \tag{W.40}$$

Altogether, we have a homogeneous, variable coefficient system of 6th order consisting of 2, coupled 2nd-order differential equations for $F^{(1)}$ and $\beta^{(1)}$, namely (W.37) and (W.39), plus 2 quadratures, namely (W.35) and (W.40).

2. *Effects of other loads*

The buckling equations developed above are valid only for distributed or edge loads that are deformation-independent (i.e., "dead"). The treatment of other typical edge loads that have potentials, such as torsional or directional springs, presents no difficulties. However, as evidenced by our discussion of ring problems in Section IV.R, we must be careful in the evaluation of potentials of distributed loads.

---

[43]In many applications the underlined term in (W.37), which represents the effects of shear deformation, is negligible and only the constitutive relations for the incremental stress couples are needed. However, should the underlined terms be kept, it is easy enough to express $e_{\sigma 1}^{(1)}$ and $g_1^{(1)}$ in terms of $F^{(1)}$ and C via constitutive relations.

The commonest deformation-dependent distributed load is constant fluid pressure, $\mathbf{p}=p_0\mathbf{m}$. The associated potential for an arbitrary *closed* shell has been computed by Koiter [1967a, Eq. (3.4), p. 198], in terms of displacements, strains, and rotations. For axishells, we give a more direct derivation based on the formula

$$\hat{V}=(2\pi/3)\oint_{\mathcal{L}}\bar{r}(\mathbf{q}\times\mathbf{e}_\theta)\cdot d\mathbf{q} \tag{W.41}$$

for the volume contained within the toroidal surface generated by rotating a closed, planar, piecewise smooth curve $\mathcal{L}$ in the $\overline{rz}$-plane about the $\bar{z}$-axis. Here, $\mathbf{q}$ is the position vector of an generic point A on $\mathcal{L}$. For simplicity, we assume that each edge of the axishell is either held during buckling or else is an apex; this is the most common case. Then, taking $\hat{V}$ as the volume between the fundamental surface, with deformed position $\bar{\mathbf{y}}^{(0)}$, and the adjacent admissible surface, with deformed position $\bar{\mathbf{y}}$, we obtain for the increment of potential (per circumferential radian),

$$\Delta\mathcal{P}=(1/3)p_0\int_0^L[\bar{r}(\bar{\mathbf{y}}'\times\bar{\mathbf{y}})-\bar{r}^{(0)}(\bar{\mathbf{y}}^{(0)\prime}\times\bar{\mathbf{y}}^{(0)})]\cdot\mathbf{e}_\theta d\sigma\,. \tag{W.42}$$

To express $\Delta\mathcal{P}$ in terms of displacements, we substitute $\bar{r}=\bar{r}^{(0)}+u$ and $\bar{\mathbf{y}}=\bar{\mathbf{y}}^{(0)}+\mathbf{u}$ into (W.42), so obtaining

$$\Delta\mathcal{P}=-p_0\int_0^L[\bar{r}^{(0)}\mathbf{m}^{(0)}\cdot\mathbf{u}+u(\bar{\mathbf{y}}^{(0)}+\tfrac{1}{3}\mathbf{u})\times\mathbf{u}'\cdot\mathbf{e}_\theta]d\sigma\,. \tag{W.43}$$

Only the quadratic middle term in (W.43) appears in the final buckling equations, which the reader will have no difficulty in inferring from the results of the preceding subsection. If shallow buckling modes are anticipated, the fluid pressure potential may be approximated by the potential for a deformation-independent, constant normal load, as discussed by Koiter (1967a, especially p. 198).

3. Simplification of the buckling equations

To be concrete, we shall introduce various approximations into a "basic" set of equations which consist of the equilibrium conditions, (W.19) and (W.32), plus the constitutive and kinematic relations (W.6)-(W.8), (W.10), (W.11), (W.15), and (IV.R.10)-(IV.R.13). In parallel, we shall also simplify the stress function-rotation set of buckling equations consisting of (W.37) and (W.39) plus the auxiliary conditions, (W.30), (W.31), and (W.40).

(a) Neglect of the deformations of the fundamental state. These deformations enter the equations through the strains, $e_\sigma^{(0)}$ and $g^{(0)}$, and the rotation, $\beta^{(0)}=\alpha_T^{(0)}-\alpha$. The strains are seldom of significance, but the rotation may be important if there are concentrated loads, if there is significant buckling near an edge, or if the shell is shallow. In such cases, there may be snapping or unsymmetric bifurcation and a nonlinear prebuckling analysis may be necessary. With the availability of good computer codes, nonlinear prebuckling analyses are now routine in *numerical work*, but in most *classical* applications the geometry of the fundamental state is replaced by that of the undeformed shell. With this approximation and with the neglect of the strains in the fundamental state, the basic equations reduce to

$$(rM_\sigma^{(1)})' - (M_\theta^{(1)} - M_n^{(0)}\beta^{(1)})\cos\alpha + (M_n^{(1)} + M_\theta^{(0)}\beta^{(1)})\sin\alpha + r(Q^{(1)} + e_{\sigma 1}^{(1)}Q^{(0)} - g_1^{(1)}N_\sigma^{(0)}) = 0 \tag{W.44}$$

$$[r(Q^{(1)} + N_\sigma^{(0)}\beta^{(1)})\mathbf{b} + r(N_\sigma^{(1)} - Q^{(0)}\beta^{(1)})\mathbf{t}]' - N_\theta^{(1)}\mathbf{e}_r = \mathbf{0} \tag{W.45}$$

$$e_{\sigma 1}^{(1)} = \mathbf{t}\cdot\mathbf{u}' \ , \ g_1^{(1)} = \mathbf{b}\cdot\mathbf{u}' - \beta^{(1)} \ , \ e_\theta^{(1)} = r^{-1}\mathbf{e}_r\cdot\mathbf{u} \tag{W.46}$$

$$k_\sigma^{(1)} = (\beta^{(1)})' \ , \ k_{\theta 1}^{(1)} = r^{-1}\beta^{(1)}\cos\alpha \ , \ k_{n1}^{(1)} = -r^{-1}\beta^{(1)}\sin\alpha \, . \tag{W.47}$$

In the constitutive relations, $\partial^2\Phi/\partial\varepsilon_i\partial\varepsilon_j$ is to be evaluated at $\varepsilon_i = 0$ rather than at $\varepsilon_i = \varepsilon_i^{(0)}$.

For the simplified stress function-rotation buckling equations, we obtain

$$(rM_\sigma^{(1)})' - M_\theta^{(1)}\cos\alpha + M_n^{(1)}\sin\alpha + \mathbf{M}_\theta^{(0)}\cdot\mathbf{e}_z - rN_\sigma^{(0)})\beta^{(1)} - F^{(1)}\sin\alpha + C\cos\alpha + r(e_{\sigma 1}^{(1)}Q^{(0)} - g_1^{(1)}N_\sigma^{(0)}) = 0 \tag{W.48}$$

$$(re_\theta^{(1)})' = e_{\sigma 1}^{(1)}\cos\alpha - (g_1^{(1)} + \beta^{(1)})\sin\alpha \tag{W.49}$$

$$rN_\sigma^{(1)} = F^{(1)}\cos\alpha + C\sin\alpha + rQ^{(0)}\beta^{(1)} \ , \ N_\theta^{(1)} = (F^{(1)})' \tag{W.50}$$

$$rQ^{(1)} = -F^{(1)}\sin\alpha + C\cos\alpha - rN_\sigma^{(0)}\beta^{(1)} \tag{W.51}$$

$$(\mathbf{u}\cdot\mathbf{e}_z)' = e_{\sigma 1}^{(1)}\sin\alpha + (g_1^{(1)} + \beta^{(1)})\cos\alpha \, . \tag{W.52}$$

(b) Neglect of the stress couples and transverse shear stress resultant of the fundamental state in the buckling equations. In many applications, the fundamental state is almost purely membrane, with bending effects, if any, confined to narrow edge zones, If we ignore $M_\theta^{(0)}, M_n^{(0)}$, or $Q^{(0)}$ in the basic buckling equations, (W.44) and (W.45), we find that

$$(rM_\sigma^{(1)})' - M_\theta^{(1)}\cos\alpha + M_n^{(1)}\sin\alpha + r(Q^{(1)} - g_1^{(1)}N_\sigma^{(0)}) = 0 \tag{W.53}$$

$$[r(Q^{(1)} + N_\sigma^{(0)}\beta^{(1)})\mathbf{b} + rN_\sigma^{(1)}\mathbf{t}]' - N_\theta^{(1)}\mathbf{e}_r = \mathbf{0} \, . \tag{W.54}$$

The parallel set of stress function-rotation buckling equations, (W.48), (W.50), and (W.51), reduces to

$$(rM_\sigma^{(1)})' - M_\theta^{(1)}\cos\alpha + M_n^{(1)}\sin\alpha - (\beta^{(1)} + g_1^{(1)})rN_\sigma^{(0)} - F^{(1)}\sin\alpha + C\cos\alpha = 0 \tag{W.55}$$

$$rN_\sigma^{(1)} = F^{(1)}\cos\alpha + C\sin\alpha \ , \ N_\theta^{(1)} = (F^{(1)})' \tag{W.56}$$

$$rQ^{(1)} = -F^{(1)}\sin\alpha + C\cos\alpha - rN_\sigma^{(0)}\beta^{(1)} \ . \tag{W.57}$$

Of course, it might be argued that, in the Computer Age, if a numerical calculation is called for, one might as well retain all prebuckling terms.

(c) Constitutive assumptions. Many shells remain elastic only for sufficiently small strains; therefore, it is reasonable to assume that Φ is quadratic. At the same time, we drop from Φ the non-classical k_n-term whose effect is rarely incorporated in practice. The resulting basic set of equations for *shearable shells* is

$$(rM_\sigma^{(1)})' - M_\theta^{(1)}r' + r(1 - A_sN_\sigma^{(0)})Q^{(1)} = 0 \tag{W.58}$$

$$[r(Q^{(1)} + N_\sigma^{(0)}\beta^{(1)})\mathbf{b} + rN_\sigma^{(1)}\mathbf{t}]' - N_\theta^{(1)}\mathbf{e}_r = \mathbf{0} \tag{W.59}$$

$$M_\sigma^{(1)} = D[(\beta^{(1)})' + r^{-1}\nu\beta^{(1)}\cos\alpha] \ , M_\theta^{(1)} = D[r^{-1}\beta^{(1)}\cos\alpha + \nu(\beta^{(1)})'] \tag{W.60}$$

$$A_eN_\sigma^{(1)} = \mathbf{t}\cdot\mathbf{u}' + r^{-1}\nu\mathbf{e}_r\cdot\mathbf{u} \ , \ A_eN_\theta^{(1)} = r^{-1}\mathbf{e}_r\cdot\mathbf{u} + \nu\mathbf{t}\cdot\mathbf{u}' \tag{W.61}$$

$$A_sQ^{(1)} = \mathbf{b}\cdot\mathbf{u}' - \beta^{(1)} \ . \tag{W.62}$$

Substitution of (W.60)-(W.62) into (W.58) and (W.59) yields a scalar and a vector equation for $\beta^{(1)}$ and $\mathbf{u}$.

The stress function-rotation buckling equations for shearable axishells are

$$(rM_\sigma^{(1)})' - M_\theta^{(1)}r' - (rN_\sigma^{(0)}\beta^{(1)} + F^{(1)}\sin\alpha - C\cos\alpha)(1 - A_sN_\sigma^{(0)}) = 0 \tag{W.63}$$

$$\begin{aligned}(re_\theta^{(1)})' - e_{\sigma 1}^{(1)}r' + (1 - A_sN_\sigma^{(0)})\beta^{(1)}\sin\alpha \\ - A_sr^{-1}(F^{(1)}\sin\alpha - C\cos\alpha)\sin\alpha = 0\end{aligned} \tag{W.64}$$

$$M_\sigma^{(1)} = D[(\beta^{(1)})' + r^{-1}\nu\beta^{(1)}\cos\alpha] \ , \ M_\theta^{(1)} = D[r^{-1}\beta^{(1)}\cos\alpha + \nu(\beta^{(1)})'] \tag{W.65}$$

$$\begin{aligned} re_{\sigma 1}^{(1)} &= A_e[F^{(1)}\cos\alpha + C\sin\alpha - \nu r(F^{(1)})'] \\ re_\theta^{(1)} &= A_e[r(F^{(1)})' - \nu(F^{(1)}\cos\alpha + C\sin\alpha)]\end{aligned} \tag{W.66}$$

$$rg^{(1)} = rA_sQ^{(1)} = A_s(-F^{(1)}\sin\alpha + C\cos\alpha - rN_\sigma^{(0)}\beta^{(1)}) \ . \tag{W.67}$$

Substitution of (W.65) and (W.66) into (W.63) and (W.64) yields 2 equations in $\beta^{(1)}$ and $F^{(1)}$. Subsequent calculation of $\mathbf{u}$ uses (W.46)$_3$ and (W.52).

A further constitutive assumption with significant consequences is the Love-Kirchhoff hypothesis which, in our equations, is equivalent to setting $A_s = 0$. From (W.62), this implies that $\beta^{(1)} = \mathbf{b}\cdot\mathbf{u}'$ while $Q^{(1)}$ is *constitutively undefined*, remaining a parameter in (W.58) and (W.59). Upon elimination of $Q^{(1)}$, the latter equation reduces to a vector differential equation for $\mathbf{u}$; (W.63) and (W.64) remain 2 differential equations for $\beta^{(1)}$ and $F^{(1)}$, but are now considerably simplified.

(d) Shallow buckling modes: DMV theory. Frequently, the *buckling modes* of well-supported shells are shallow in the sense that the characteristic length l of the mode satisfies the inequality $|l\bar{\alpha}'_{max}| \ll 1$. Then, we may use the Donnell-Mushtari-Vlasov (DMV) theory of quasi-shallow shells[44] which leads to simplified buckling equations. Indeed, until the advent of large, finite-element computer codes, most buckling analyses used these simplified equations. Further discussion of the DMV theory may be found in Section IV.R and in the references cited there. In particular, the reader should consult chapters 5 and 6 of Brush & Almroth (1975) where the DMV equations are developed and applied to some classical buckling problems.

The essence of the DMV approximation is the deletion of the underlined terms with an α'-factor in the following expressions, the 1st for rotation in terms of displacements and the 2nd for force equilibrium in the **t**-direction:

$$\beta^{(1)} = (\mathbf{b}\cdot\mathbf{u})' + \underline{\alpha'(\mathbf{t}\cdot\mathbf{u})} \tag{W.68}$$

$$(rN_\sigma^{(1)})' - \underline{r\alpha'(Q^{(1)} + N_\sigma^{(0)}\beta^{(1)})} = N_\theta^{(1)}r' \,. \tag{W.69}$$

(In the variational formulation of buckling, the second approximation is a consequence of the first). We note that the expression for $\beta^{(1)}$ is the same as that for a flat plate. The other buckling equations retain their form (within the Love-Kirchhoff hypothesis).

The DVM approximation has also been used to study postbuckling, where the consequent equations are frequently called the *nonlinear Donnell shell equations*. Applying approximations (W.68) and (W.69) directly to the nonlinear equilibrium equations (L.12), (L.13), and (L.15), and setting $\cos(\alpha+\beta) \approx \cos\alpha$ and $\sin(\alpha+\beta) \approx \sin\alpha + \beta\cos\alpha$ where appropriate, we obtain for the axishell equilibrium equations

$$(rN_\sigma)' - N_\theta \cos\alpha + rp_T = 0 \tag{W.70}$$

$$(rQ)' + r(\alpha+\beta)'N_\sigma + N_\theta(\sin\alpha + \beta\cos\alpha) + p_B = 0 \tag{W.71}$$

$$(rM_\sigma)' - r'M_\theta + rQ = 0 \,, \tag{W.72}$$

and for the kinematics,

$$e_\sigma = \mathbf{u}'\cdot\mathbf{t} + \tfrac{1}{2}\beta^2 \;\;,\;\; e_\theta = r^{-1}(\mathbf{u}\cdot\mathbf{e}_r) \tag{W.73}$$

[44]The theory of quasi-shallow shells was actually developed, independently, by Libai (1962) and Koiter (1966), who each pointed out that if $Kl^2 \ll 1$, where K is the Gaussian curvature of the shell reference surface, then the geometrically shallow DMV theory, interpreted properly, can be applied to non-shallow shells.

$$\beta = (\mathbf{u}\cdot\mathbf{b})' \ , \ k_\sigma = \beta' \ , \ k_\theta = r^{-1}\beta\cos\alpha \, . \qquad \text{(W.74)}$$

The constitutive relations are (O.8) or (O.11) in their classical forms. Note that nonlinear effects are confined to the $\beta' rN_\sigma$- and $\beta N_\theta \cos\alpha$-terms in (W.71) and to the β^2-term in $(W.73)_1$.[45]

(e) Shallow axishells. Axishells that are shallow (in the sense of Marguerre) with respect to the *xy*-plane have received much attention because of the relative simplicity of their governing equations. The underlying assumptions and modes of behavior are similar to those of shallow arches that we discussed in Section IV.R. The basic geometric restriction is that $\alpha^2, \beta^2, r\alpha\alpha' \ll 1$[46] In addition, we assume a quadratic, mixed-energy density. An outline of the derivation for transversely loaded, shearable axishells is as follows.

From the shallowness approximation, we have

$$k_\theta = r^{-1}[\sin\alpha(\cos\beta - 1) + \cos\alpha\sin\beta] \approx r^{-1}\beta \, . \qquad \text{(W.75)}$$

Substitution into (L.15) and use of the constitutive relations yields

$$D\,[(r\beta')' - r^{-1}\beta] + r\,[1 + A_e(N_\sigma - \nu N_\theta) - A_s N_\sigma]Q = 0 \, . \qquad \text{(W.76)}$$

Following the same reasoning as in the shallow arch problem and using the stress function formulation as in (Q.4), (Q.5),(Q.11), and (Q.24), we have

$$rN_\sigma = F \ , \ N_\theta = F' \ , \ rV = r(0)V(0) - \int_0^r rp_V dr = rQ + (\alpha + \beta)F \, . \quad \text{(W.77)}$$

We shall also replace differentiation with respect to σ by differentiation with respect to r, the justification being that $dr = \cos\alpha d\sigma \approx d\sigma$.

A second basic equation follows from the compatibility condition (M.2) as

$$(re_\theta)' - e_\sigma + \tfrac{1}{2}\beta^2 + \alpha\beta + A_s Q(\alpha + \beta) = 0 \, . \qquad \text{(W.78)}$$

Introducing (W.77) into (W.76) and (W.78), we arrive at

$$D\,[(r\beta')' - r^{-1}\beta] + [1 + \underline{r^{-1}(A_e - A_s)F - \nu A_e F'}][rV - (\alpha + \beta)F] = 0 \quad \text{(W.79)}$$

$$A_e[(rF')' - r^{-1}F] + (\alpha + \tfrac{1}{2}\beta)\beta + \underline{A_s(\alpha + \beta)V} = 0 \, . \qquad \text{(W.80)}$$

If the extensional strains are small and shear effects negligible, the underlined terms may be dropped and we have the classical form of the shallow axishell equations, (R.40) and (R.41).

The displacement in the *z*-direction is determined from

---

[45] Within the Donnell theory, $(W.73)_1$ is an approximation to the *exact* expression $e_\sigma\cos\beta - g\sin\beta = 1 - \cos\beta + \mathbf{u}'\cdot\mathbf{t}$.

[46] The restriction on $\alpha\alpha'$ stems from the term $\nu[(rk_\theta)' - k_\sigma\cos\sigma]$ which, by (W.74), is $O(\alpha\alpha'\beta)$.

$$(\mathbf{u}\cdot\mathbf{e}_z)' = (1 + e_\sigma)\beta + (\underline{g} + \alpha e_\sigma)\,. \tag{W.81}$$

Note that shear effects are manifested on the right side of (W.81) to a lower order than in the field equations (W.79) and (W.80). Indeed, if $V/H = O(\alpha)$—see the discussion on shallow arches—then, for small strains and isotropic materials, all the shear terms in (W.79) and (W.80) are negligible and the only significant shear effect is the g-term in (W.81). This point is demonstrated in the example to follow.

The results of this subsection should be compared with those of Section R, which are based on the simplified Reissner equations. There, shear and nonlinear strain effects have been neglected, but loadings omitted in the present section were considered.

4. Buckling of axisymmetric plates

As an application, we shall use (W.63)-(W.66) to derive equations for the axisymmetric buckling of a shearable circular plate of constant properties, under a radial compressive stress resultant, P, at its outer edge, $r = b$. With $\alpha = 0$, $N_\sigma^{(0)} = -P$, and $' = d/dr$, we have

$$(rM_\sigma^{(1)})' - M_\theta^{(1)} + (rP\beta^{(1)} + C)(1 + PA_s) = 0 \tag{W.82}$$

$$(re_\theta^{(1)})' - e_{\sigma 1}^{(1)} = 0 \tag{W.83}$$

$$M_\sigma^{(1)} = D\,(\beta^{(1)\prime} + r^{-1}\nu\beta^{(1)})\ ,\ \ M_\theta^{(1)} = D\,(r^{-1}\beta^{(1)} + \nu\beta^{(1)\prime}) \tag{W.84}$$

$$re_{\sigma 1}^{(1)} = A_e(F^{(1)} - \nu rF^{(1)\prime})\ ,\ \ re_\theta^{(1)} = A_e(rF^{(1)\prime} - \nu F^{(1)}) \tag{W.85}$$

$$rg^{(1)} = A_s(C + rP\beta^{(1)}) \tag{W.86}$$

$$u = \mathbf{u}\cdot\mathbf{e}_r = re_\theta^{(1)} \tag{W.87}$$

$$w' = (\mathbf{u}\cdot\mathbf{e}_z)' = g^{(1)} + \beta^{(1)} = (1 + PA_s)\beta^{(1)} + r^{-1}CA_s\,. \tag{W.88}$$

These equations decouple into the set {(W.82), (W.84),(W.86), (W.88)}, which contain $\beta^{(1)}$ and w as unknowns, and the rest, which contain $F^{(1)}$ and u as unknowns. The latter have no effect on the buckling load.[47]

Substitution of (W.84) into (W.82) yields for $\beta^{(1)}$ the equation

$$D\,[(r\beta^{(1)\prime})' - r^{-1}\beta^{(1)}] + (rP\beta^{(1)} + C)(1 + PA_s) = 0\,, \tag{W.89}$$

with general solution

[47]This occurs generally in plate buckling problems *unless* coupling is reintroduced through boundary conditions. Eccentric ring supports can provide such a mechanism.

$$\beta^{(1)} = A_1 J_1(\lambda r) + A_2 Y_1(\lambda r) - r^{-1} C/P\ , \tag{W.90}$$

where $\lambda^2 = D^{-1}P(1+PA_s)$, J_1 and Y_1 are Bessel functions of the first order and of the first and second kinds, respectively, and the A's are constants. From (W.88), we find that

$$w = -\lambda^{-1}(1+PA_s)[A_1 J_0(\lambda r) + A_2 Y_0(\lambda r)] - (C/P)\ln(\lambda r/2) + A_3\ . \tag{W.91}$$

We need 4 boundary conditions on $\beta^{(1)}$, w, $M_\sigma^{(1)}$, and $V^{(1)}$ to determine the constants A_1, A_2, A_3, and C. The vanishing of the determinant of their coefficients yields the *buckling equation* for λ.

As an example, let the plate be clamped so that the boundary conditions are

$$\beta^{(1)}(0) = \lim_{r\to 0} rV^{(1)} = \beta^{(1)}(b) = w(b) = 0\ . \tag{W.92}$$

Hence, $A_2 = C = 0$ and the buckling equation reduces to $J_1(\lambda b) = 0$, the smallest positive root of which is $\lambda b = 3.832\cdots$. The buckling load is

$$P_{cr} = \frac{(1+4P_k A_s)^{1/2} - 1}{2A_s} \approx \frac{P_K}{1+P_K A_s}\ , \tag{W.93}$$

where $P_K = (14.68\cdots)(D/b^2)$ is the critical load under the Kirchhoff Hypothesis.

As a second example, we consider the clamped circular plate shown in Fig. 5.22 with an additional support at $r=0$; the condition on $rV^{(1)}$ is replaced by $w(0)=0$. As the support may exert a concentrated force at $r=0$, $C \neq 0$, in general.

To investigate the behavior of $\beta^{(1)}$ and w near $r=0$, we need the following properties of Bessel functions for small values of their arguments:

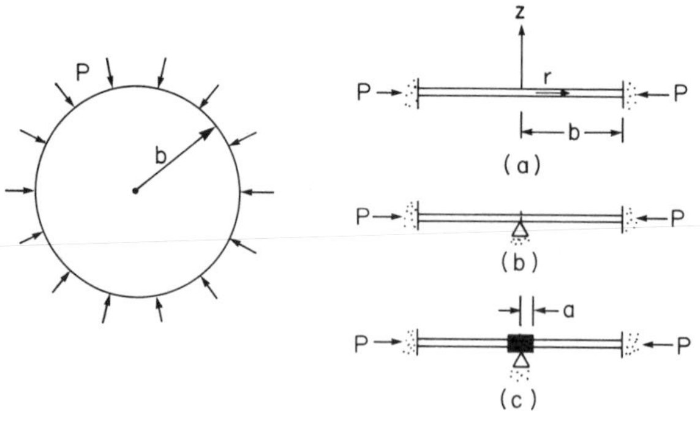

Fig. 5.22. Compressed, clamped circular plate: (a) without additional support; (b) with central support; (c) with central, immovable core.

$$J_0(z) = 1 - (z/2)^2 + O(z^4) \tag{W.94}$$

$$\tfrac{1}{2}\pi Y_0(z) = [\ln(z/2) + \gamma]J_0(z) + (z/2)^2 + O(z^4) \tag{W.95}$$

$$J_1 = (z/2) - \tfrac{1}{2}(z/2)^3 + O(z^5) \tag{W.96}$$

$$\pi Y_1(z) = -(2/z) + 2[\ln(z/2) + \gamma]J_1(z) - (z/2) + (5/4)(z/2)^3 + O(z^5)\,, \tag{W.97}$$

where $\gamma = 0.5772\cdots$ is Euler's constant. Hence, setting $\rho = \lambda r/2$ in (W.90) and (W.91), we have, near $r=0$,

$$\beta^{(1)}(\rho) = -\frac{1}{\rho}\left[\frac{A_2}{\pi} + \frac{C\lambda}{2P}\right] + \frac{2A_2}{\pi}\rho\ln\rho + O(\rho) \tag{W.98}$$

$$w(\rho) = -\left[\frac{2(1+PA_s)A_2}{\lambda\pi} + \frac{C}{P}\right]\ln\rho + \left[A_3 - \frac{1+PA_s}{\lambda}\left[A_1 + \frac{2\gamma}{\pi}A_2\right]\right] + O(\rho^2\ln\rho)\,. \tag{W.99}$$

Setting $\beta^{(1)}(0)=0$ yields

$$C = -\frac{2A_2P}{\lambda\pi} \tag{W.100}$$

$$w(\rho) = -\frac{2PA_sA_2}{\lambda\pi}\ln\rho + \left[A_3 - \frac{1+PA_s}{\lambda}\left[A_1 + \frac{2\gamma}{\pi}A_2\right]\right] + O(\rho^2\ln\rho)\,. \tag{W.101}$$

Turning, first, to *classical plate theory*, we set $A_s=0$. The logarithm term in (W.101) vanishes and $w(0)=0$ yields

$$A_3 = \lambda^{-1}[A_1 + (2\gamma/\pi)A_2]\,. \tag{W.102}$$

With these results, the boundary conditions $\beta^{(1)}(b)=w(b)=0$ yield the determinantal eigenvalue equation

$$J_1(x)[\ln(x/2)+\gamma] + (1/x)[J_0(x) - 2] - (\pi/2)Y_1(x) = 0\ ,\ x \equiv \lambda b\,, \tag{W.103}$$

whose lowest eigenvalue is $x = 6.65\cdots$; the associated *axisymmetric* buckling load is

$$P_{cr} = (44.2\cdots)(D/b^2)\,. \tag{W.104}$$

Turning to shearable plates ($A_s \neq 0$), we see that the vertical displacement w has a *logarithmic singularity* at $r=0$, so that the boundary condition $w(0)=0$ *cannot be satisfied.* In fact, removal of the singularity by setting $A_2=0$ and satisfaction of the boundary conditions at the edge leave a nonzero displacement at the origin.

To explain this phenomenon, we note that if $A_2 \neq 0$, the central support exerts a concentrated force of magnitude, say, F_0 at $r=0$. The occurrence of a logarithmic singularity in the vertical displacement under a concentrated force acting on a shearable plate has been noted and discussed before. For details, see Nordgren (1963) and Wilkinson & Kalnins (1966). The physical explanation is that the singularity in the transverse shear stress resultant Q (which, by linear bending theory, behaves as $F_0/2\pi r$ near $r=0$) produces, at the origin, an infinite shear strain and a vertical slope in the deformed reference surface. In *classical plate theory*, on the other hand, the shear strain is set to zero, the slope of the deformed reference surface is zero at the origin, and the vertical displacement under the concentrated force is *finite*.

To get a clue as to the "correct" displacement, we turn to the classical theory of linear elasticity. From Timoshenko & Goodier (1970, p. 402), we have the Boussinesq solution for the vertical deflection, u_z, in an elastic half-space under a concentrated vertical force acting at the origin of a set of circular cylindrical coordinates (r, θ, z):

$$u_z = \frac{F_0}{2\pi E}[(1+\nu)z^2(z^2+r^2)^{-3/2} + 2(1-\nu^2)(z^2+r^2)^{-1/2}] . \quad \text{(W.105)}$$

It follows that the *average* vertical displacement $\bar{u}_z \equiv d^{-1}\int_0^d u_z dz$ of a finite circular slab of depth d and radius b, supported along its circumference, has a *logarithmic singularity* at $r=0$, which gives us more confidence in the results for shearable plates.[48]

Returning to our plate, we can, practically, represent the effect of the central support by clamping the plate to an immovable hub of radius a ($\ll b$) and infinite bending rigidity.[49] The boundary conditions are now

$$\beta^{(1)}(a) = w(a) = \beta^{(1)}(b) = w(b) = 0 , \quad \text{(W.106)}$$

and these, with $x=\lambda b$ and $\rho_0 = xa/2b$, lead to the determinantal eigenvalue equation

$$J_1(x)[\gamma + \ln(x/2) + (1+PA_s)^{-1}(PA_s + 2\rho_0^2 \ln\rho_0)\ln(a/b)] - x^{-1}\{1 + [1-J_0(x)](1-2\rho_0^2\ln\rho_0)\} - \tfrac{1}{2}\pi Y_1(x) = 0 , \quad \text{(W.107)}$$

where PA_s can be extracted from

$$PA_s(1+PA_s) = DA_s x^2/b^2 . \quad \text{(W.108)}$$

In practical configurations, all the new terms that appear in going from (W.103)

---

[48]The integration produces $\bar{u}_z = -\dfrac{(3-2\nu)F_0}{4\pi G d}\ln r +$ regular terms. The shear coefficient here is $k=2/(3-2\nu)=0.857\cdots$ which compares very well with Cowper's (1966) result of $0.851\cdots$ for $\nu=1/3$.

[49]We assume that the extensional rigidities of the hub and the plate are equal. If $a/b \ll 1$, the effects of an extensionally stiffer hub are not important.

to (W.107) are of *comparable importance.* However, as a/b becomes small, only the shear term $(1+PA_s)^{-1}PA_sJ_1(x)\ln(a/b)$, which causes a reduction in the buckling load, retains significance. As $a/b \to 0$, the smallest positive solution of (W.107) approaches the smallest positive zero of $J_1(x)$ and the effects of the central support *vanish.* This result is not surprising, as a concentrated force of vanishingly small magnitude produces a nonzero deflection. We reemphasize here that the limit $a/b \to 0$ is impractical and that the validity of plate theory is questionable near a concentrated load. Nevertheless, the behavior of the plate for small values of a/b is of some theoretical interest. In Table 5.5 we list the smallest positive solutions of (W.107) for $DA_s/b^2=0.01$ and various values of a/b.

Table 5.5
Solutions of the eigenvalue equation (W.107)

a/b	0.1	0.05	0.01	10^{-3}	10^{-4}	10^{-5}	0
x	6.61	6.58	6.38	6.14	5.89	5.64	3.83

These numbers should be compared with the value of $x=6.65$ predicted by the classical theory of plates with a central support. We note that extremely small values of a/b are needed to produce a substantial drop in x.

5. The postbuckling of axishells: a short survey

There are 2 main reasons for investigating the postbuckling of shells.

(a) If the buckling is stable, then the shell can carry loads above the buckling load, thus permitting light-weight designs, a feature exploited widely in the aerospace industry.

(b) If the buckling is unstable and imperfection sensitive, then postbuckling analyses can estimate the reduction in the load-carrying capacity of the imperfect shell and the severity of the ensuing collapse.

Koiter (1945, 1963b, 1967a, 1967b, 1982) has developed powerful tools for determining the initial shape of the secondary equilibrium path at bifurcation and for estimating the effects of small imperfections. Although we touched on postbuckling theory in Section IV.R, we cannot do justice to this important topic within the scope of this book. Instead, we refer the reader to the references cited there; in particular, to the detailed and unified presentation of Koiter's theory given by Budiansky (1974) and to the useful reviews by Hutchinson & Koiter (1970) and Tvergaard (1976).

An alternative to a bifurcation/imperfection-sensitivity analysis is to attack the nonlinear shell equations directly. In a few cases, approximate methods such as Galerkin's or those of perturbation theory have been useful. More often, numerical methods are needed. For axishells, various general forms of the governing are listed in Section Q, but simplified versions of these equations are almost always used. The latter could be the (simplified) Reissner equations, (R.9) and (R.10) or (R.19) and (R.24); the equations of moderate rotation theory, (R.30) and (R.31); or the DVM equations, (W.70)-(W.74). For shallow geometries, (R.40) and (R.41) or (W.79) and (W.80) have been used extensively. Frequently, one sees the above approximate theories in displacement form.

The nonlinear load-deflection[50] behavior of a shell of revolution can take one of the following forms, depending on the geometry, the boundary conditions, the loading (uniform, partial, concentrated), and the imperfections:

(a) Monotonous increase in load up to a limit point where the shell collapses by axisymmetric snap buckling. See Fig. 4.13a for a qualitative picture. As we discuss later in this section, this can happen in shallow shells for certain geometries and loadings, but deep shells with imperfect shape or imperfect boundary conditions may also behave this way.

(b) Axisymmetric bifurcation of a perfect shell followed by a further increase in the load. Secondary bifurcation along the postbuckling path, if it occurs, is of minor importance. In a corresponding imperfect shell, the loading increases monotonously until failure. See Fig. 4.13d. Examples are found in circular and annular plates and in very shallow shells.

(c) A variant of (b), in which the bifurcation is unsymmetric, occurs in some edge-compressed, clamped, annular plates and in shallow spherical caps under concentrated loads.

(d) Axisymmetric or unsymmetric bifurcation of a perfect shell followed by a drastic reduction in the load along a secondary equilibrium path. The corresponding imperfect shell fails by snap buckling at a fraction of the bifurcation load. The many examples of such behavior include edge-loaded cylindrical, conical, and spherical shells, and uniformly loaded shallow spherical shells.

One approach to the analysis of the buckling and postbuckling of shells of revolution [which is embodied in several computer codes—see Bushnell (1976), for example] is to calculate the nonlinear (fundamental) equilibrium path, including boundary effects and, possibly, axisymmetric imperfections. One then searches for unsymmetric bifurcations from the axisymmetric path and notes their location relative to snap points. The reader is warned, however, that this approach is not always adequate: in some cases, a combination of axisymmetric and unsymmetric imperfections may prove to be more detrimental to the load-carrying capacity of a shell of revolution that axisymmetric imperfections alone. Moreover, as noted by Budiansky & Hutchinson (1972), sometimes transition from an unstable state to a stable state occurs in the postbuckling range, necessitating a thorough-going, unsymmetric analysis.

We conclude this section with a short review of some specific cases.

---

[50]The "deflection" is usually measured in some convenient norm.

(a) Axially compressed circular cylindrical shells. Because of their widespread utility and because of the great reduction in their theoretical buckling loads caused by small imperfections, circular cylindrical shells have been studied and tested extensively. Here, we summarize Koiter's (1963c) analysis of the effects of imperfections on the unsymmetric bifurcations. We start from the 2-dimensional Donnell equations for a stress function, F, and the outward radial displacement W. These equations consist of the strain-compatibility condition,

$$R^2 A \nabla^4 F - RW,_{zz} + W,_{\theta\theta}(W_0 + W),_{zz} - W,_{z\theta}^2 = 0\,, \qquad \text{(W.109)}$$

and the equilibrium condition in the radial direction,

$$R^2 D \nabla^4 W + RF,_{zz} - F,_{\theta\theta}(W_0 + W),_{zz} - F,_{zz} W,_{\theta\theta} + 2W,_{z\theta} F,_{z\theta} = 0\,. \qquad \text{(W.110)}$$

Here, R is the radius of the undeformed midsurface, $W_0(z)$ is the known axisymmetric imperfection, ∇^4 is the biharmonic operator in surface coordinates (z, θ), and a comma denotes partial differentiation. The solution strategy is as follows.

First, we seek a prebuckling solution of the form

$$W = W^*(z)\ ,\ F = -\tfrac{1}{2} PR^2\theta^2 + F^*(z)\,, \qquad \text{(W.111)}$$

where P is the axial compressive force per unit circumferential length. Substitution yields 2 ordinary differential equations for the axisymmetric unknowns W^* and F^* that can be solved in standard ways.

Next, we superimpose a small, θ-periodic, unsymmetric perturbation on the functions in (W.111) by setting

$$W = W^*(z) + W^{**}(z)\cos m\theta\,,\ F = F^*(z) + F^{**}(z)\cos m\theta - \tfrac{1}{2} PR^2\theta^2\,. \text{(W.112)}$$

Substituting these expressions into (W.109) and (W.110), we obtain ordinary differential equations for W^{**} and F^{**} which depend on $W_0(z)$, P, and m as parameters.

Third, we seek the smallest value of P which produces a nontrivial solution to the differential equations and boundary conditions—a standard eigenvalue problem—and then minimize this value over m. The resulting critical load, P_{cr}, depends on the given imperfection, $W_0(z)$.

Finally, we could attempt to minimize P_{cr} with respect to W_0, but in practice one usually fixes P_{cr} by assuming that W_0 is proportional to the buckling mode, $\overline{W}(z)$, of the *perfect cylinder*, i.e., one assumes that

$$W_0(z) = \mu h \overline{W}(z)\,, \qquad \text{(W.113)}$$

where μ is the relative amplitude of the imperfection and h is the thickness of the shell.

Koiter (1963c) has investigated the buckling of a *long* cylindrical shell with a sinusoidal imperfection. In his study, boundary conditions are ignored and the perturbed field equations solved approximately by a Rayleigh-Ritz procedure.[51] For $\mu = 0.3$, as much as a 60% reduction in load was obtained! See also Budiansky & Hutchinson (1972).

(b) Spherical shells. The bifurcation pressure of a complete spherical shell is given by (W.5).[52] The perfect shell is strongly unstable at the axisymmetric bifurcation point; the introduction of unsymmetrical imperfections or a *local* axisymmetric one, produces a large reduction in the buckling pressure. Recent investigations by Gräff *et al* (1985) and Drmota *et al* (1986) confirm this behavior.

The "locality" of the buckling modes of a complete, pressurized spherical shell has led investigators to focus on *shallow* spherical caps, which have also been examined for buckling under concentrated and nonuniform axisymmetric loads. Timoshenko & Gere (1961, pp. 512-519) summarize early work; for later developments, see Kaplan (1974), Tvergaard (1976), Kollar (1982), and Scheidl & Troger (1984).

We summarize here three specific studies.

Hutchinson (1967) investigated the imperfection sensitivity of spherical shells under uniform external pressure. Exploiting the shallowness of the buckling modes, he worked with the Cartesian form of (W.109) and (W.110) [modified for a spherical shell and with a pressure term added to (W.110)] and used Koiter's theory to study the interaction of 2 and 3 buckling modes on the postbuckling behavior. Hutchinson found substantially increased sensitivity due to mode interaction. He also performed an upper bound analysis using Koiter's (1963b) approach. Koiter's (1969) more refined *axisymmetric* analysis as well as numerical, axisymmetric snapping analyses by Bushnell (1967) and Koga & Hoff (1969) have yielded results very similar to those of Hutchinson. Fig. 5.23, which is copied from Fig. 5 of the review by Kollar (1982), summarizes some of these results.

Fitch & Budiansky (1970) have investigated the snap buckling and postbuckling behavior of shallow, clamped spherical caps subjected to a constant pressure over a circular region centered at the apex. Included are the limiting cases of uniform pressure and a concentrated force, analyized previously by Huang (1964), Weinitschke (1965), and Fitch (1968), respectively. The equations used by Fitch & Budiansky are essentially (W.109) and (W.110), written in plane polar coordinates with partial derivatives replaced by covariant derivatives and with a pressure term added. For each loading case, the axisymmetric snapping and the unsymmetric bifurcation were determined numerically. If bifurcation occurred before snapping, the "b" coefficient in (IV.R.4) was determined by an initial postbuckling analysis, $b < 0$ implying imperfection sensitivity. (Note

---

[51]This procedure ensures that the resulting bifurcation load is an upper bound on the true load (for the assumed imperfection).

[52]Zoelly (1915) and Schwerin (1922) obtained this result assuming axisymmetry; van der Neut (1932) included unsymmetric deformations but obtained the same result.

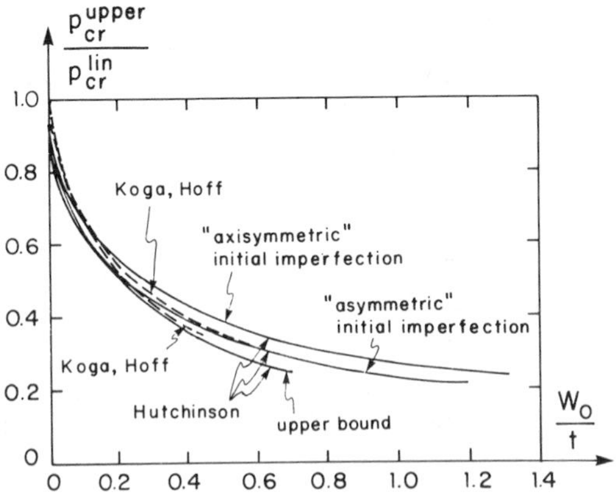

Fig. 5.23. Influence of the initial imperfection amplitude on the snapping load of a spherical shell. (From Fig. 5 of Kollar, 1982).

that "a" $=0$ for this class of problems.) Results were discussed in terms of the shell and loading parameters

$$\lambda, \bar{\lambda} = \left(\frac{r}{h}\right)\left[12(1-\nu^2)\frac{h^2}{R^2}\right]^{1/4}, \tag{W.114}$$

where, for the shell parameter λ, r is the planform radius, and for the loading parameter $\bar{\lambda}$, r is the radius of the loaded region. Some of their results are as follows:

For a typical λ (say $\lambda = 12$) and $\bar{\lambda} \leq 2$ (which approximates a concentrated load), the shell bifurcates first, with $b > 0$. For $2 < \bar{\lambda} < 4.8 \cdots$, the shell snaps; for $\bar{\lambda} > 4.8 \cdots$, it again bifurcates, but now with $b < 0$. Thus, as the size of the loading region is increased, the shell passes from stable bifurcation, to axisymmetric snapping, to unstable bifurcation. The results for $\lambda > 12$ differ little for those with $\lambda = 12$.

For uniform pressure over the entire shell, Huang's work shows that snapping occurs if $\lambda \lesssim 5.5$ and imperfection-sensitive bifurcation occurs for larger λ.

For concentrated loads and $\lambda \lesssim 7.8$, Fitch & Budiansky (1970) show that the spherical cap is stable and that for larger λ's, snapping occurs, though it is not severe. If $\lambda \gtrsim 9.2$, they show that unsymmetric bifurcation occurs, with $b > 0$. Their results are summarized by the formula

$$P_{cr} = \frac{\pi Q_c E h^3}{6(1-\nu^2)R}. \tag{W.115}$$

Values of Q_c, b, and the number of circumferential waves n are shown in Fig. 5.24, which is taken from their (1970) paper.

Dimpling, an essentially local, inextensional phenomenon, is another form of buckling which is often observed in thin shells. The remainder of the shell, which is practically undisturbed, is connected to the dimpled region by a narrow transition region where an elastic "hinge" forms. In a spherical shell, the dimple itself is an inverted spherical cap, to a first approximation. (See subsection S.4.c.) Wan (1980a, 1980b, 1984), Wan & Ascher (1980), and Wan & Yun (1985) have studied this phenomenon in shallow spherical caps by superimposing a meridionally varying external pressure on a uniform internal pressure and analyzing the transition layer using asymptotic methods. The original papers should be consulted for details.

Experimental results on the buckling and postbuckling of spherical shells, on the whole, agree with the tends discussed above. For details, the reader is referred to the extensive reviews by Kaplan (1974) and Kollar (1982).

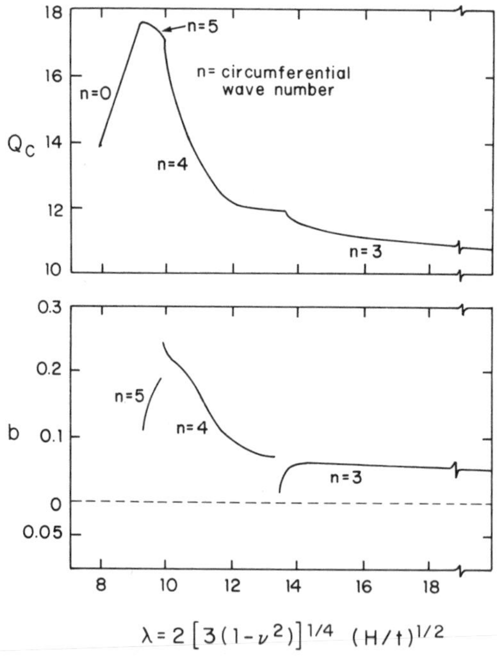

Fig. 5.24. Buckling and postbuckling of a clamped spherical cap under a concentrated load. (Fig. 3 of Fitch & Budiansky, 1970).

X. Thermodynamics

To emphasize conceptual similarities, we have made the body of this section virtually identical to Section T of Chapter IV on beamshells—only the details of the equations differ slightly. For introductory material, motivation, and references to relevant general treatments of thermodynamics, see the first two paragraphs of Section IV.T.

In an axishell, the rate of internal heat production, n, per unit undeformed volume is a function of σ and ζ only, while the cold flux vector in (II.C.2) has the form

$$\mathbf{c} = c^{\sigma}(\sigma, \zeta)\mathbf{x}_{,\sigma}(\sigma, \theta, \zeta) + c^{\zeta}(\sigma, \zeta)\mathbf{x}_{,\zeta}(\sigma, \theta, \zeta)\,. \tag{X.1}$$

Likewise, in (II.D.2), the entropy density η and the absolute temperature θ[53] are functions of σ and ζ only. As a consequence of these assumptions, it is sufficient in deriving the thermal equations for an axishell by descent from 3 dimensions to take the arbitrary volume V over which (II.C.1) and (II.D.1) apply to be the ring-like region

$$V = \{\sigma, \theta, \zeta \mid a \le \sigma \le b \;\;,\;\; 0 \le \theta < 2\pi \;\;,\;\; -H \le \zeta \le H\}\,. \tag{X.2}$$

Thus, the boundary, ∂V, now consists only of edges and faces on which, respectively,

$$\mathbf{c}\cdot\mathbf{n}dA = \pm c^{\sigma}\mu r d\theta d\zeta \;\;,\;\; \sigma = {}^{b}_{a} \;\text{ and }\; \mathbf{c}\cdot\mathbf{n}dA = \pm c^{\zeta}\mu r d\sigma d\zeta \;\;,\;\; \zeta = \pm H\,. \tag{X.3}$$

With the aid of (B.2), it follows that (II.C.1) and (II.D.2) take the forms

$$\mathcal{Q} = (rq)^{b}_{a} + \int_{a}^{b} s r d\sigma \tag{X.4}$$

$$\mathcal{H} = \int_{a}^{b} \iota r d\sigma \;\text{ and }\; \mathcal{J} = (rj)^{b}_{a} + \int_{a}^{b} \nu r d\sigma\,, \tag{X.5}$$

where

$$q = \int_{-}^{+} c^{\sigma}\mu d\zeta \;\;,\;\; s = c^{\zeta}\mu\,\big|^{+}_{-} + \int_{-}^{+} n\mu d\zeta \tag{X.6}$$

$$\iota = \int_{-}^{+} \eta\mu d\zeta \;\;,\;\; j = \int_{-}^{+} (c^{\sigma}/\theta)\mu d\zeta \tag{X.7}$$

$$\nu = (c^{\zeta}\mu/\theta)^{+}_{-} + \int_{-}^{+} (n/\theta)\mu d\zeta\,. \tag{X.8}$$

We call q and j the *heat and entropy influx resultants*, respectively, and ι the *entropy density resultant*; s and ν are, respectively, the *heating* and the *supply of*

---

[53]To avoid the introduction of an unconventional symbol, we have denoted by θ both the absolute temperature and the circumferential coordinate. As these distinct variables never appear in the same equation in what follows, there should be no chance of confusion.

entropy.

Observe that if we introduce a *mean reciprocal temperature* and a *transverse temperature gradient* by setting

$$\frac{1}{T} \equiv \frac{1}{2}\left[\frac{1}{\theta_-} + \frac{1}{\theta_+}\right] \ , \ \Delta \equiv \frac{T^2}{2H}\left[\frac{1}{\theta_-} - \frac{1}{\theta_+}\right] , \tag{X.9}$$

then the supply of entropy v, defined by (X.8), may be given the form

$$v = \frac{s + \Delta\chi}{T} , \tag{X.10}$$

where

$$\chi \equiv -\frac{H[(c^{\zeta}\mu)_+ + (c^{\zeta}\mu)_-]}{T} + \frac{1}{\Delta}\int_-^+ \left[\frac{T}{\theta} - 1\right] n\mu d\zeta . \tag{X.11}$$

The last term in (X.10) measures the supply of entropy due to the nonuniform distribution of temperature through the thickness and is thus analogous to the external distributed couple l, defined by (B.31) and (D.1).[54] Except in special cases, the distribution through the thickness of neither the temperature θ nor the deformed position $\bar{\mathbf{x}}$ can be inferred exactly from axishell variables, so that neither χ nor l can be specified precisely. However, in many practical problems, these terms are negligible.

Observe also that if $\theta = T$, then, by $(X.6)_1$ and $(X.7)_2$, $j = q/T$. This suggests that we work with a new unknown w by setting

$$j \equiv \frac{q}{T} + \Delta w . \tag{X.12}$$

Again, we emphasize that the equations of motion, (B.20) and (B.28), and the Clausius-Duhem inequality, (II.D.1), with $\mathcal{H}$ and $\mathcal{J}$ given by (X.5), are *exact;* the approximate nature of axishell theory enters when we take the apparent external mechanical power $\mathcal{W}$ to be the actual external mechanical power and when we approximate (or ignore) l and χ.

To obtain a complete thermodynamic theory of axishells, we *postulate* the following law of *Conservation of Energy* (1st Law of Thermodynamics):

$$\int_{t_1}^{t_2} (Q + \mathcal{W})dt = (\mathcal{K} + I)_{t_1}^{t_2} , \tag{X.13}$$

where I is the *internal energy*. By (H.1), (X.13) is equivalent to the *reduced equation of energy balance*

---

[54]If θ^{-1} is linear in ζ, then $-T\chi = H[(c^{\zeta}\mu)_+ + (c^{\zeta}\mu)_-] + \int_-^+ \zeta n\mu d\zeta$.

$$\int_{t_1}^{t_2}(Q+\mathcal{D})dt = I\Big|_{t_1}^{t_2}. \tag{X.14}$$

We call an axishell *thermoelastic* if there exists an *internal energy density* i, depending on

$$\Lambda \equiv (e_\sigma, e_\theta, g, k_\sigma, k_\theta, k_n, T, \Delta) \tag{X.15}$$

and its gradient $\Lambda' \equiv (e_\sigma{}', \cdots, \Delta')$, such that

$$I = \int_a^b ir\,d\sigma. \tag{X.16}$$

It is convenient to introduce the *free-energy density* Φ by setting

$$i = \Phi + \iota T + f\Delta, \tag{X.17}$$

where the unknown f may be called the *entropy density couple*. Both Φ and f are assumed to be functions of Λ and Λ'. Inserting (X.17) into (X.16), the resulting expression into (X.14), and using (L.5), (L.6), and (X.4), we have

$$\int_{t_1}^{t_2}[(rq)_a^b + \int_a^b(N_\sigma\dot{e}_\sigma + N_\theta\dot{e}_\theta + Q\dot{g} + M_\sigma\dot{k}_\sigma + M_\theta\dot{k}_\theta + M_n\dot{k}_n + s)r\,d\sigma]dt$$
$$= \int_a^b(\Phi + \iota T + f\Delta)r\,d\sigma\Big|_{t_1}^{t_2}. \tag{X.18}$$

For sufficiently smooth fields, (X.18) implies the *reduced differential equation of energy balance*

$$q' + (q/r)\cos\alpha + \Xi\cdot\dot{\Lambda} + s = \dot{\Phi} + \dot{\iota}T + \dot{f}\Delta, \tag{X.19}$$

where

$$\Xi \equiv (N_\sigma, N_\theta, Q, M_\sigma, M_\theta, M_n, -\iota, -f). \tag{X.20}$$

Likewise, for sufficiently smooth fields, (II.D.1) and (X.5) imply the *differential Clausius-Duhem inequality*

$$\dot{\iota} \geq j' + (j/r)\cos\alpha + v \tag{X.21}$$

which, with the aid of (X.10) and (X.12) and upon multiplication by T, may be given the form

$$T\dot{\iota} \geq q' + T^{-1}(-qT' + T^2w\Delta') + [(q + T\Delta w)/r]\cos\alpha + s + \Delta(Tw' + \chi). \tag{X.22}$$

Using the reduced equation of energy balance, (X.19), we obtain

$$\Xi\cdot\dot{\Lambda} - \dot{\Phi} - \dot{f}\Delta \geq T^{-1}(-qT' + T^2w\Delta') + \Delta\{T[w' + (w/r)\cos\alpha] + \chi\}. \tag{X.23}$$

We now assume that Φ, Ξ, q, and w are given by differentiable constitutive functions of the form

$$\Phi = \Phi(\Lambda, \Lambda'), \ldots, w = w(\Lambda, \Lambda'). \tag{X.24}$$

Thus, in particular,

$$\dot{\Phi} = \Phi,_{\Lambda} \cdot \dot{\Lambda} + \Phi,_{\Lambda'} \cdot \dot{\Lambda}' \,, \tag{X.25}$$

where $\Phi,_{\Lambda}$ is short for $(\Phi,_{e_\sigma}, \cdots, \Phi,_{\Delta})$, etc. (Though it is customary *not* to include strain gradients in the parameter list of the constitutive functions, it actually *simplifies* the form of the resulting equations to do so. As will be seen, the same final form for Φ emerges as if the strain gradients had not been included.)

We now wring a number of consequences from (X.23). First, since $\dot{f}$ may be expressed in a form similar to (X.25) and

$$w' = w,_{\Lambda} \cdot \Lambda' + w,_{\Lambda'} \cdot \Lambda'' \,, \tag{X.26}$$

(X.23) can be rewritten as

$$(\tilde{\Xi} - \tilde{\Phi},_{\Lambda}) \cdot \dot{\Lambda} - \tilde{\Phi},_{\Lambda'} \cdot \dot{\Lambda}' \\ \geq T^{-1}(-qT' + T^2 w \Delta') + \Delta\{T[w,_{\Lambda} \cdot \Lambda' + w,_{\Lambda'} \cdot \Lambda'' + (w/r)\cos\alpha] + \chi\} \,, \tag{X.27}$$

where $\tilde{\Xi} = (N_\sigma, N_\theta, Q, M_\sigma, M_\theta, M_n, -\iota, 0)$ and $\tilde{\Phi} = \Phi + f\Delta$ is a modified free-energy density.

At any σ and t,

$$\Upsilon \equiv (\Lambda, \dot{\Lambda}, \Lambda', \dot{\Lambda}', \Lambda'') \,, \tag{X.28}$$

can be assigned arbitrarily, provided that $\mathbf{p}$, l, and s are chosen so that the differential equations of momentum and reduced energy balance, (D.3), (D.5), and (X.19), are satisfied. By virtue of the *form* of the constitutive equations (X.24), it follows that a choice of Υ fixes every term in (X.27) *except* χ.[55] Thus, there must be a relation involving χ and at least some of the components of Υ; otherwise, by (X.11), we could give χ any value by prescribing suitable heat fluxes along the faces of the axishell and thus violate (X.27).

We write the relation involving χ in the form

$$T[w' + (w/r)\cos\alpha] + \chi = A \,. \tag{X.29}$$

If we regard A as a new unknown and adopt the *Principle of Equipresence* (Truesdell & Toupin, 1960, Sect. 293), then A must be a function of Λ and Λ'. With this understanding, the right side of (X.27) reduces to $T^{-1}(-qT' + T^2 w\Delta') + \Delta A$—a function of Λ and Λ' only. Because the coefficients of $\dot{\Lambda}$ and $\dot{\Lambda}'$ on the left side of (X.27) are independent of $\dot{\Lambda}$ and $\dot{\Lambda}'$, and because $\dot{\Lambda}$ and $\dot{\Lambda}'$ may be prescribed arbitrarily at any σ and t, we conclude that, to not violate the inequality, we must have

$$\tilde{\Xi} = \tilde{\Phi},_{\Lambda} \,, \quad \tilde{\Phi},_{\Lambda'} = 0 \,. \tag{X.30}$$

Thus, $\tilde{\Phi}$ depends on the strains and T only. But obviously, if we are to account for so simple a phenomenon as the build up of stress couples in a circular plate

---

[55]If $n=0$, then, from $(X.6)_2$ and (X.11), $s = (c^\zeta\mu)_+ - (c^\zeta\mu)_-$ and $\chi = -H[(c^\zeta\mu)_+ + (c^\zeta\mu)_-]$. Thus, satisfaction of (X.19) fixes the sum of the heat flow through the faces of the axishell, but leaves the difference at our disposal.

heated uniformly on one face and constrained to remain flat, we need constitutive equations for M_σ, M_θ, and M_n that depend on the transverse temperature gradient, Δ; i.e., $\tilde{\Phi}$ is *not* a suitable free-energy density. We therefore return to (X.23) and take the relation for χ to be of the form

$$\dot{f} + T[w' + (w/r)\cos\alpha] + \chi = B \ , \tag{X.31}$$

where

$$B = B(\Lambda, \Lambda') . \tag{X.32}$$

By arguments similar to those used before, we conclude that (X.27) will remain inviolate if and only if $\Phi = \hat{\Phi}(\Lambda)$ and $\Xi = \Phi,_\Lambda$, i.e., if and only if

$$\Phi = \Phi(e_\sigma, e_\theta, g, k_\sigma, k_\theta, k_n, T, \Delta) \tag{X.33}$$

and

$$\begin{gathered} N_\sigma = \Phi,_{e_\sigma} \ , \ N_\theta = \Phi,_{e_\theta} \ , \ Q = \Phi,_g \\ M_\sigma = \Phi,_{k_\sigma} \ , \ M_\theta = \Phi,_{k_\theta} \ , \ M_n = \Phi,_{k_n} \ , \ \iota = -\Phi,_T \ , \ f = -\Phi,_\Delta \ , \end{gathered} \tag{X.34}$$

where the unknown Lagrange multiplier that appears in (O.3), the isothermal form of this relation, has been set to zero. [See the discussion following (O.5)] With these relations, (X.19) and (X.23) reduce to

$$q' + (q/r)\cos\alpha + s = \dot{\iota}T + \dot{f}\Delta \tag{X.35}$$

$$-qT' + T^2 w\Delta' + T\Delta B \leq 0 . \tag{X.36}$$

By elaborating on the approach in Section O on strain-energy densities in isothermal axishells, we could, with the aid of Table 5.1 (p.189) and the last 5 columns of Table 4.2 (p.149), begin to delimit the functional form of the free-energy density Φ by laying down, in addition to items (1)-(10) in subsection V.2, new restrictions and assumptions suggested by the presence of thermal variables. The first two of these might be:

(11) The constitutive relations for q, w, and B do not depend on the spatial derivatives of the strain.

(12) In the undeformed (reference) state, the thermal variables $\iota, f, \tau, \Delta, B, \tau'$, and Δ' are zero, where $T \equiv T_0 + \tau$ and T_0 is a constant.

However, for simplicity and because of the paucity of literature on nonlinear stress-strain-temperature relations for axishells, we simply record here the analogues of (IV.T.44) and (IV.T.45), namely, the quadratic free-energy density, Φ, and the linear constitutive relations for the heat influx resultant, q, the entropy-flux deviation, w, and the auxiliary unknown, B:

$$\begin{aligned} \Phi = \tfrac{1}{2}C\{&e_\sigma^2 + e_\theta^2 + 2\nu e_\sigma e_\theta + (h^2/12)(k_\sigma^2 + k_\theta^2 + 2\nu k_\sigma k_\theta) \\ &- 2\tilde{\alpha}[(e_\sigma + e_\theta)\tau - (h^2/12)(k_\sigma + k_\theta)\Delta]\} - (cm/2T_0)[\tau^2 + (h^2/12)\Delta^2] \end{aligned} \tag{X.37}$$

$$q = hq_0\tau' \ , \ w = -h^3 w_0 \Delta' \ , \ B = -hB_0\Delta \ , \qquad \text{(X.38)}$$

where, as in (O.10), $(1-\nu^2)C = Eh$, $\tilde{\alpha}$ is the *coefficient of thermal expansion*, c is the *specific heat at constant volume*, and q_0 is the *thermal conductivity*.

If ι, f, q, w, and B are computed, respectively, from $(X.34)_7$, $(X.34)_8$, and (X.38), and the resulting expressions substituted into (X.35) and (X.31), the equations we obtain, *if linearized*, read

$$(hq_0\tau')' + (hq_0\tau'/r)\cos\alpha + s = [cm\tau + T_0 C\tilde{\alpha}(e_\sigma + e_\theta)]^{\bullet} \qquad \text{(X.39)}$$

$$[h^2/(12T_0)][cm\Delta - T_0C\tilde{\alpha}(k_\sigma + k_\theta)]^{\bullet} - T_0[(h^3w_0\Delta')' + (h^3w_0\Delta'/r)\cos\alpha] + \chi = -hB_0\Delta . \qquad \text{(X.40)}$$

These equations are valid if

$$\| e \|^2 + T_0^{-2}(\tau^2 + h^2\Delta^2) \ll 1 \ , \qquad \text{(X.41)}$$

where e is the strain vector defined following (O.20). They apply to transversely homogeneous axishells with an additional relative error of $O(h\sqrt{b_\sigma^2 + b_\theta^2})$ coming from the neglect of curvature effects.

For a discussion of the slight differences between (X.40) and equation (10.14) in Green & Naghdi (1979), see the paragraph following (IV.T.47).

References

[The letter in brackets indicates the section where the reference appears.]

Adkins, J. E., and Rivlin, R. S. (1952). Large elastic deformations of isotropic materials IX. The deformation of thin shells. *Phil. Trans. Roy. Soc. Lond.* A244, 505-531. [S]

Ahmadi, G., and Glockner, P. G. (1983). Collapse by ponding of pneumatic elastic spherical caps under distributed loads. *Canadian J. Civ. Engr.* **40,** 740-747. [T]

Antman, S. S. (1979). The eversion of thick spherical shells. *Arch. Rat. Mech. Anal.* **70,** 113-123. [S]

Antman, S. S., and Calderer, M. C. (1985a). Asymptotic shapes of inflated spheroidal nonlinearly elastic shells. *Math. Proc. Camb. Phil. Soc.* **97,** 541-549. [T]

Antman, S. S., and Calderer, M. C. (1985b). Asymptotic shapes of inflated non-circular elastic rings. *Math. Proc. Camb. Phil. Soc.* **97,** 357-379. [T]

Antman, S. S., and Osborn, J. E. (1979). The principle of virtual work and integral laws of motion. *Arch. Rat. Mech. Anal.* **69,** 232-261. [G]

Ashwell, D. G. (1960). On the large deflexion of a spherical shell with an inward point load. *In* "The Theory of Thin Elastic Shells," Proc. I.U.T.A.M. Sympos. Delft, 1969 (W.T. Koiter, ed.), pp. 43-63. North-Holland, Amsterdam. [S]

Axelrad, E. L. (1983). "Schalentheorie." B. G. Teubner, Stuttgart. [S]

Benedict, R., Wineman, A., and Yang, W. -H. (1979). The determination of limiting pressure in simultaneous elongation and inflation of nonlinear elastic tubes. *Int. J. Solids Struct.* **15,** 241-249. [T]

Berger, H. M. (1955). A new approach to the analysis of large deflections of plates. *J. Appl. Mech.* **22,** 465-472. [U]

Berger, M. S. (1967). An application of the calculus of variations in the large to the equations of nonlinear elasticity. *Bull. Am. Math. Soc.* **73,** 520-525. [S]

Bert, C. H. (1975). Analysis of plates (Chapter 4) and Analysis of shells (Chapter 5). *In* "Composite Materials—Structural Design and Analysis" (C. C. Chamis, ed.), pt. I, vol. 7. Academic Press, New York. [O]

Blatz, P. J., and Ko, W. L. (1962). Application of finite elastic theory to the deformation of rubbery materials. *Trans. Soc. Rheology* **6,** 223-251. [O, T]

Brodland, G. W., and Cohen, H. (1986). Deflection and snapping of spherical caps. (Submitted for publication.) [S]

Bromberg, E., and Stoker, J. J. (1945). Non-linear theory of curved elastic sheets. *Q. Appl. Math.* **3,** 246-265. [S]

Brush, D. O., and Almroth, B. O. (1975). "Buckling of Bars, Plates, and Shells." McGraw-Hill, New York, etc. [W]

Budiansky, B. (1968). Notes on nonlinear shell theory. *J. Appl. Mech.* **35,** 393-401. [P, U]

Budiansky, B. (1974). Theory of buckling and post-buckling behavior of elastic structures. *Adv. Appl. Mech.* **14,** 1-65. [W]

Budiansky, B., and Hutchinson, J. W. (1972). Buckling of circular cylindrical shells under axial compression. *In* "Contributions to the Theory of Aircraft Structures" (van der Neut Anniversary Volume, W. T. Koiter, ed.), pp. 239-260. Delft Univ. Press, Delft. [W]

Budiansky, B., and Sanders, J. L., Jr. (1963). On the "best" first-order linear shell theory. *In* "Progress in Applied Mechanics" (Prager Anniversary Volume, D. C. Drucker, ed.), pp. 129-140. Macmillan, New York. [P]

Burt, P. B., and Reid, J. L. (1976). Exact solution to a nonlinear Klein-Gordon equation. *J. Math. Anal. Appl.* **55,** 43-45 [U]

Bushnell, D. (1967). Bifurcation phenomena in spherical shells under concentrated and ring loads. *AIAA J.* **5,** 2034-2040. [S, W]

Bushnell, D. (1976). BOSOR5-program for buckling of elastic-plastic shells of revolution including large deflections and creep. *Comp. Struct.* **6,** 221-239. [W]

Bushnell, D. (1981). Computerized buckling analysis of shells. Air Force Wright Aeronautical Laboratories Rept. No. AFWAL-TR-81-3049. [W]

Bushnell, D. (1985). "Computerized Buckling Analysis of Shells." Martinus Nijhoff, Dordrecht. [W]

Calladine, C. R. (1983). "Theory of Shell Structures." Cambridge Univ. Press, Cambridge, etc. [Intro]

Cheo, L. S., and Reiss, E. L. (1974). Secondary buckling of circular plates. *SIAM J. Appl. Math.* **26,** 490-495. [S]

Chernykh, K. F. (1980). Nonlinear theory of isotropically elastic thin shells. *Iz. AN SSSR. Mekhanika Tverdogo Tela* **14,** 148-159. [Q]

Chien, W. -Z., and Hu, H. -C (1957). On the snapping of a thin spherical cap. *In*

"Proc. 9th Int. Cong. Appl. Mech., Brussels," vol. 6, pp. 309-320. [S]

Chia, C.-Y. (1980). "Nonlinear Analysis of Plates." McGraw-Hill, New York [S]

Clark, R. A. (1950). On the theory of thin elastic toroidal shells. *J. Math. & Phys.* **29,** 146-178. [S]

Clark, R. A., and Narayanaswamy, O.S. (1967). Nonlinear membrane problems for elastic shells of revolution. *In* Proceedings of the Symposium on the Theory of Shells to Honor L. H. Donnell (D. Muster, ed.), pp. 81-110. Univ. of Houston Press, Houston. [R, S]

Cline, G. B. (1963). Effect of internal pressure on the influence coefficients of spherical shells. *J. Appl. Mech.* **30,** 91-97. [S]

Comer, R. L., and Levy, S. (1963). Deflection of an inflated circular cylinder cantilever beam. *AIAA J.* **1,** 1652-1655. [T]

Courant, R., and Hilbert, D. (1962). "Methods of Mathematical Physics," vol. 2. Interscience, New York and London. [U]

Cowper, G. R. (1966). The shear coefficient in Timoshenko's beam theory. *J. Appl. Mech.* **33,** 335-340. [W]

Croll, J. G. A. (1985). A tension field solution for non-linear circular plates. *In* "Aspects of the Analysis of Plate Structures" (A volume in honor of W. H. Wittrick, Dawe *et al*, eds.), pp. 309-323. Clarendon Press, Oxford. [T]

Dean, W. R. (1939). The distortion of a curved tube due to internal pressure. *Phil. Mag.* **28,** 452-464. [S]

Dickey, R. W. (1973). Inflation of the toroidal membrane. *Q. Appl. Math.* **32,** 11-26. [S]

Dickey, R. W. (1976). Nonlinear bending of circular plates. *SIAM J. Appl. Math.* **30,** 1-9. [S]

Dickey, R. W. (1987). Membrane caps. *Q. Appl. Math.* (to appear) [T]

Drmota, M., Scheidl, R., Troger, H., and Weinmüller, E. (1986). On the imperfection sensitivity of complete spherical shells. *Computational Mech.* **2,** 63-74. [W]

Durban, D., and Baruch, M. (1974). Behaviour of an incrementally elastic thick walled sphere under internal and external pressure. *Int. J. Non-Lin. Mech.* **9,** 105-119. [T]

Dym, C. L. (1974). "Stability Theory and its Application to Structural Mechanics." Noordhoff, Leyden. [W]

Erdélyi, A. (1956). "Asymptotic Expansions." Dover, New York. [S]

Erdélyi, A., Magnus, W., Oberhettinger, F., and Tricome, F. G. (1953). "Higher Transcendental Functions", vol. 1. McGraw-Hill, New York, etc. [S]

Evan-Iwanowski, R. M., Cheng, H. S., and Loo, T. C. (1962). Experimental investigation of deformations and stability of spherical shells subject to concentrated load at the apex. *In* "Proc. 4th U. S. Nat. Cong. Appl. Mech.", vol. 1, pp. 563-575. [S]

Evans, A. E., and Skalak, R. (1980). "Mechanics and Thermodynamics of Biomembranes." CRC Press, Boca Raton, Florida. [T]

Feodosev, V. I. (1968). On equilibrium modes of a rubber spherical shell under internal pressure. *PMM* **32,** 335-341. [T]

Feng, W. W., and Yang, W. H. (1973). On the contact problem of an inflated spherical nonlinear membrane. *J. Appl. Mech.* **40,** 209-214. [T]

Fitch, J. R. (1968). The buckling and post-buckling behavior of spherical caps under concentrated load. *Int. J. Solids Struct.* **4,** 421-446. [S]

Fitch, J. R., and Budiansky, B. (1970). The buckling and postbuckling of spherical caps under axisymmetric load. *AIAA J.* **8,** 686-693. [W]

Flügge, W. (1960). "Stresses in Shells." Springer-Verlag, Berlin, etc. [S]

Flügge, W. (1973). "Stresses in Shells," 2nd ed. Springer-Verlag, New York, etc. [S]

Flügge, W., and Riplog, P. M. (1956). A large deformation theory of shell membranes. Tech. Rept. No. 102, Div. Engr. Mech., Stanford Univ. [R, S, T].

Föppl, A. (1905). "Vorlesungen über technische Mechanik," vol. 3. Teubner, Leipzig, pp. 298-299. [S, T]

Föppl, A. (1922). "Vorlesungen über technische Mechanik," vol. 3, 9th ed. Teubner, Leipzig. [S]

Foster, H. O. (1967). Very large deformations of axially symmetric membranes made of neo-Hookean materials. *Int. J. Engr. Sci.* **5,** 95-117. [T]

Friedrichs, K. O., and Stoker, J. J. (1941). The non-linear boundary value problem of the buckled plate. *Am. J. Math.* **63,** 839-887. [S]

Friedrichs, K. O., and Stoker, J. J. (1942). Buckling of the circular plate beyond the critical thrust. *J. Appl. Mech.* **9,** 7-13. [S]

Fulton, J. P. (1988). "Rubber-like spherical shells and circular plates under vertical point loads." Dissertation, Univ. of Virginia. [S]

Fulton, J. P., and Simmonds, J. G. (1986). Large deformations under vertical edge loads of annular membranes with various strain energy densities. *Int. J. Non-Lin. Mech.* **21,** 257-267. [T]

Fung, Y. C. (1979). Inversion of a class of nonlinear stress-strain relationships of biological soft tissues. *J. Biomed. Engr., Trans. ASME* **101,** 23-27. [T]

Fung, Y. C. (1980). On pseudo-elasticity of living tissue. *Mechanics Today* **5,** 49-66. [T]

Gelfand, I. M., and Fomin, S. V. (1963). "Calculus of Variations" (translated from the Russian by R. A. Silverman). Prentice-Hall, Englewood Cliffs, New Jersey. [V]

Gräff, R., Scheidel, R., Troger, H., and Weinmüller, E. (1985). An investigation of the complete post-buckling behavior of axisymmetric spherical shells. *ZAMP* **36,** 803-821. [W]

Grabmüller, H., and Novak, E. (1987a). Nonlinear boundary value problems for the annular membrane: new results on existence of positive solutions. *Math. Meth. Appl. Sci.* (to appear). [S]

Grabmüller, H., and Novak, E. (1987b). Nonlinear boundary value problems for the annular membrane: a note on uniqueness of positive solutions. *J. Elasticity* **17,** 279-284. [S]

Grabmüller, H., and Pirner, R. (1987). Positive solutions of annular elastic membrane problems with finite rotations. *Studies Appl. Math.* (to appear). [S]

Grabmüller, H., and Weinitschke, H. J. (1986). Finite displacements of annular membranes. *J. Elasticity* **16,** 135-147. [S]

Green, A. E., and Adkins, J. E. (1970). "Large Elastic Deformations," 2nd ed. Chapter IV. Clarendon Press, Oxford. [T]

Green, A. E., and Naghdi, P. M. (1979). On thermal effects in the theory of shells. *Proc. Roy. Soc. London* **A365,** 161-190. [X]

Hart, V. G., and Evans, D. J. (1964). Non-linear bending of an annular plate bent by transverse edge forces. *J. Math. & Phys.* **43,** 275-303. [S]

Hart-Smith, L. J., and Crisp, J. D. C. (1967). Large elastic deformations of thin rubber membranes. *Int. J. Engr. Sci.* **5,** 1-24. [T]

Hill, J. M. (1980). The finite inflation of a thick-walled elastic torus. *Q. J. Mech. Appl. Math.* **33,** 471-490. [S]

Hill, J. M. (1982). Finite deformation of thick-walled inner tubes and tyres under inflation and rotation. *In* "Finite Elasticity," Proc. I.U.T.A.M. Sympos. Lehigh, 1980. (D.E Carlson & R. T. Shield, eds.), pp. 211-236. Martinus Nijhoff, The Hague, etc. [S]

Hoff, N. J. (1969). Some recent studies of the buckling of thin shells. *Aero. J.* **73,** 1057-1070. [W]

Huang, N. C. (1964). Unsymmetrical buckling of thin spherical shells. *J. Appl. Mech.* **31,** 447-457. [W]

Hutchinson, J. W. (1967). Imperfection sensitivity of externally pressurized spherical shells. *J. Appl. Mech.* **34,** 49-55. [W]

Hutchinson, J. W., and Koiter, W. T. (1970). Postbuckling theory. *Appl. Mech. Rev.* **23,** 1353-1366. [W]

Isaacson, E. (1965). The shape of a balloon. *Comm. Pure Appl. Math.* **18,** 163-166. [T]

Jafari, A. H., Abeyaratne, R., and Horgan, C. O. (1984). The finite deformation of a pressurized circular tube for a class of compressible materials. *ZAMP* **35,** 227-246. [T]

Jenssen, O. (1986). Asymptotic integration applied to the differential equation for thin elastic toroidal shells. *J. Appl. Mech.* **53,** 461-462. [S]

Jordan, P. F. (1962). Stresses and deformations of the thin-walled pressurized torus. *J. Aerospace Sci.* **29,** 213-225. [S]

Kaplan, A. (1974). Buckling of spherical shells. "Thin Shell Structures" (The Sechler Anniversary Volume, Y. C. Fung & E. E. Sechler, eds.), pp. 247-288. Prentice-Hall, Englewood Cliffs, New Jersey. [W]

Keppel, W. J. (1984). Finite axisymmetric deformation of a thin shell of revolution. Thesis, Univ. of Arizona. [Q]

Klingbeil, W. W., and Shield, R. T. (1964). Some numerical investigations on empirical strain-energy functions in the large axisymmetric extensions of rubber membranes. *ZAMP* **15,** 609-629. [T]

Knowles, J. K. (1960). Large amplitude oscillations of a tube of incompressible elastic material. *Q. Appl. Math.* **18,** 71-77. [U]

Knowles, J. K. (1962). 0n a class of oscillations in the finite deformation theory of elasticity. *J. Appl. Mech.* **29,** 283-287. [U]

Koga, T. (1972). Bending rigidity of an inflated circular cylindrical membrane of rubbery material. *AIAA J.* **10,** 1485-1489. [T]

Koga, T., and Hoff, N. J. (1969). The axisymmetric buckling of initially imperfect complete spherical shells. *Int. J. Solids Struct.* **5,** 679-697. [W]

Koiter, W. T. (1945). "Over de Stabiliteit van het Elastisch Evenwicht." Thesis, Delft. Amsterdam. [W]

Koiter, W. T. (1960). A consistent first approximation in the general theory of thin elastic shells. "The Theory of Thin Elastic Shells," Proc. I.U.T.A.M. Sympos. Delft, 1959 (W. T. Koiter, ed.), pp. 12-33. North-Holland, Amsterdam. [P]

Koiter, W. T. (1963a). A spherical shell under point loads at its poles. *In* "Progress in Applied Mechanics" (The Prager Anniversary Volume, D. C. Drucker, ed.). Macmillan, New York, pp. 211-236. [S]

Koiter, W. T. (1963b). Elastic stability and postbuckling behavior. *In* "Nonlinear Problems" (R. E Langer, ed.), pp. 257-275. Univ. Wisconsin Press. [W]

Koiter, W. T. (1963c). The effect of axisymmetric imperfections on the buckling of cylindrical shells under axial compression. *Proc. Kon. Ned. Ak. Wet.* **B66,** 265-279. [W]

Koiter, W. T. (1964). Couple-stresses in the theory of elasticity. *Proc. Kon. Ned. Ak. Wet.* **B67,** 17-44. [Intro]

Koiter, W. T. (1966). On the nonlinear theory of thin elastic shells, I-III. *Proc. Kon. Ned. Ak. Wet.* **B69,** 1-54. [W]

Koiter, W. T. (1967a). General equations of elastic stability for thin shells. *In* "Proceedings of the Symposium on the Theory of Shells to Honor L. H. Donnell" (D. Muster, ed.), pp. 187-228. Univ. of Houston Press, Houston. [W]

Koiter, W. T. (1967b), "On the Stability of Equilibrium." NASA TT-F-10. [W]

Koiter, W. T. (1969). The nonlinear buckling problem of a complete spherical shell under uniform external pressure. *Proc. Kon. Ned. Ak. Wet.* **B72,** 40-123. [W]

Koiter, W. T. (1980). The intrinsic equations of shell theory with some applications. *Mechanics Today* **5,** 139-154. [O, R]

Koiter, W. T. (1982). The application of the initial postbuckling analysis to shells. *In* "Buckling of Shells" (E. Ramm, ed.), pp. 3-17. Springer-Verlag, Berlin. [W]

Kollár, L. (1982). Buckling of complete spherical shells and spherical caps. *In* "Buckling of Shells" (E. Ramm, ed.), pp. 401-425. Springer-Verlag, Berlin, etc. [W]

Kollár, L., and Dulácska, E. (1984). "Buckling of Shells for Engineers." Wiley, Chichester, etc. [W]

Kraus, H. (1967). "Thin Elastic Shells." Wiley, New York, etc. [S]

Kriegsman, G. A., and Lange, C. G. (1980). On large axisymmetrical deflection states of spherical shells. *J. Elasticity* **10,** 179-192. [S]

Kydoniefs, A. D. (1966). Finite deformation of an elastic torus under rotation and inflation. *Int. J. Engng. Sci.* **4,** 125-154. [S]

Kydoniefs, A. D. (1967). The finite inflation of an elastic toroidal membrane. *Int. J. Engng. Sci.* **5,** 477-494. [S]

Kydoniefs, A. D., and Spencer, A. J. M. (1965). The finite inflation of an elastic torus. *Int. J. Engng. Sci.* **3,** 173-195. [S]

Kydoniefs, A. D., and Spencer, A. J. M. (1967). The finite inflation of an elastic toroidal membrane of circular cross section. *Int. J. Engng. Sci.* **5,** 367-391. [S]

Kydoniefs, A. D., and Spencer, A. J. M. (1969). Finite axisymmetric deformation of an initially cylindrical elastic membrane. *Q. J. Mech. Appl. Math.* **22,** 87-95. [T]

Langer, R. E. (1931). On the asymptotic solution of ordinary differential equations. *Trans. Am. Math. Soc.* 23-64. [S]

Langer, R. E. (1935). On the asymptotic solutions of ordinary differential equations, with reference to the Stokes' phenomenon about a singular point. *Trans. Am. Math. Soc.* **37,** 397-416. [S]

Lardner, T. J., and Pujara, P. (1980). Compression of spherical cells. *Mechanics Today* **5,** 161-176. [T]

Leckie, F. A. (1961). Localized loads applied to spherical shells. *J. Mech. Engr. Sci.* **3,** 111-118. [S]

Leissa, A. W. (1969). Vibration of Plates. NASA SP-160, Washington, D. C. [U]

Libai, A. (1962). On the nonlinear elastokinetics of shells and beams. *J. Aerospace Sci.* **29,** 1190-1195, 1209. [W]

Libai, A. (1981). On the nonlinear intrinsic dynamics of doubly curved shells. *J. Appl. Mech.* **48,** 909-914. [U]

Libai, A. (1983). Nonlinear shell dynamics—intrinsic and semi-intrinsic approaches. *J. Appl. Mech.* **50,** 531-536. [U]

Libai, A., and Simmonds, J. G. (1983). Nonlinear elastic shell theory. *Adv. Appl. Mech.* **23,** 271-371. [R, T]

Librescu, L. (1987). Refined geometrically nonlinear theories of anisotropic laminated shells. *Q. Appl. Math.* **45,** 1-22. [O]

Lin, Y., and Wan, F. Y. M. (1985). Asymptotic solutions of steadily spinning shells of revolution under uniform pressure. *Int. J. Solids Struct.* **21,** 27-53. [S]

Lukasiewicz, S. A., and Glockner, P. G. (1983). Collapse by ponding of shells. *Int. J. Solids Struct.* **19,** 251-261. [T]

Malcolm, D. J, and Glockner, P. G. (1981). Collapse by ponding of air-supported spherical caps. *Proc. ASCE* **107,** 1731-1742. [T]

Mansfield, E. H. (1969). Tension field theory. *In* "Proc. 12th Int. Cong. Appl. Mech." (M. Hetényi & W. G. Vincenti, eds.). Springer-Verlag, Berlin & New York, pp. 305-320. [T]

Mansfield, E. H. (1970). Load transfer via a wrinkled membrane. *Proc. Roy. Soc. London* **A316,** 269-289. [T]

Marguerre, K. (1938). Zur Theorie der gekrümmten Platte grosser Formänderung. *In* "Proc. 5th Int. Cong. Appl. Mech.," pp. 93-101. [S]

Massonet, C. H. (1948). Le voilment des plaques planes sollicitées dans leur plan. *In* "Int. Assoc. Bridge Struct. Engr. Third Congress, Liege, Belgium."

Meissner, E. (1913). Das Elastizitätsproblem für dünne Schalen von

Ringflächen-, Kugel- oder Kegelform. *Physikalische Zeitschrift,* **14,** 343-349. [R]
Mescall, J. F. (1964). On the numerical analysis of the nonlinear axisymmetric equations for shells of revolution. U. S. Army Materials Res. Agency, Rept. TR 64-20. [S]
Mescall, J. F. (1965). Large deflections of spherical shells under concentrated loads. *J. Appl. Mech.* **32,** 936-938. [S]
Naruoka, M. (1981). "Bibliography on Theory of Plates." Gihodo Publ. Co., Tokyo. [S]
Mukherjee, K., and Chakraborty, S. K. (1985). Exact solutions for large amplitude free and forced oscillations of a thin spherical shell. *J. Sound Vib.* **100,** 339-342. [U]
Needleman, A. (1976). Necking of pressurized spherical membranes. *J. Mech. Phys. Solids* **24,** 339-359. [T]
Needleman, A. (1977). Inflation of spherical rubber balloons. *Int. J. Solids Struct.* **13,** 409-421. [T]
Niordson, F. I. (1985). "Shell Theory" (Chapter 16). North Holland, Amsterdam, etc. [U]
Nordgren, R. P. (1963). On the method of Green's function of the thermoelastic theory of shallow shells. *Int. J. Engr. Sci.* **1,** 279-308. [W]
Novozhilov, V. V. (1970). "Thin Shell Theory," 2nd ed. Wolters-Noordhoff, Groningen, the Netherlands. [S]
Ogden, R. W. (1972). Large deformation isotropic elasticity: on the correlation of theory and experiment for incompressible rubberlike solids. *Proc. Roy. Soc. London* **A326,** 565-584. [T]
Olver, F. W. J. (1974). "Asymptotics and Special Functions." Academic Press, New York, etc. [S]
Parker, D. F., and Wan, F. Y. M. (1984). Finite polar dimpling of shallow caps under sub-buckling pressure loading. *SIAM J. Appl. Math.* **44,** 301-326. [S]
Pipkin, A. D. (1968). Integration of an equation in membrane theory. *ZAMP* **19,** 818-819. [T, V]
Pujara, P., and Lardner, T. J. (1978). Deformations of elastic membranes: effect of different constitutive relations. *ZAMP* **29,** 315-327. [T]
Ranjan, G. V., and Steele, C. R. (1977). Large deflection of deep spherical shells under concentrated load. *In* "Proceedings, AIAA/ASME 18th Structures, Structural Dynamics, and Materials Conference," San Diego, pp. 269-278. [S]
Ranjan, G. V., and Steele, C. R. (1980). Nonlinear corrections for edge bending of shells. *J. Appl. Mech.* **47,** 861-865. [S]
Reissner, E. (1938). On tension field theory. *In* "Proc. 5th Int. Cong. Appl. Mech.," pp. 359-361. [T]
Reissner, E. (1947). Stresses and small displacements of shallow spherical shells, I, II. *J. Math. & Phys.* **25,** 80-85, 279-300. [S]
Reissner. E. (1949). On finite deflections of circular plates. *Proc. Sympos. Appl. Math.* **1,** 213-219. [R]

Reissner, E. (1950). On axisymmetrical deformations of thin shells of revolution. *Proc. Sympos. Appl. Math* **3,** 27-52. [Q, R, S]

Reissner, E. (1959). On influence coefficients and nonlinearity for thin elastic shells. *J. Appl. Mech.* **26,** 69-72. [S]

Reissner, E. (1963a). On the equations for finite symmetrical deflections of thin shells of revolution. *In* "Progress in Applied Mechanics" (The Prager Anniversary Volume, D. C. Drucker, ed.), pp. 171-178. Macmillan, New York. [M, R, V]

Reissner, E. (1963b). On stresses and deformations in toroidal shells of circular cross sections which are acted upon by uniform normal pressure. *Q. Appl. Math.* **21,** 177-188. [S]

Reissner, E. (1969). On finite symmetrical deflections of thin shells of revolution. *J. Appl. Mech.* **36,** 267-270. [Intro, M, P]

Reissner, E. (1972). On finite symmetrical strain in this shells of revolution. *J. Appl. Mech.* **39,** 1137. [P, R]

Reissner, E. (1974). Linear and nonlinear theory of shells. *In* "Thin-Shell Structures" (The Sechler Anniversary Volume, Y. C. Fung & E. E. Sechler, eds.), pp. 29-44. Prentice-Hall, Englewood Cliffs, New Jersey. [M]

Reissner, E. (1978). A note on finite deflections of circular ring plates. *ZAMP* **29,** 698-703. [S]

Reissner, E., and Wan, F. Y. M. (1965). Rotating shallow elastic shells of revolution. *J. Soc. Indust. Appl. Math.* **13,** 333-352. [R, S]

Reissner, H. (1912). Spannungen in Kugelschalen (Kuppeln). *H. Müller-Breslau Festschrift,* pp. 181-193. Kröner, Leipzig. [R]

Rogers, C., and Baker, J. A. (1980). The finite elastodynamics of hyperelastic thin tubes. *Int. J. Non-lin. Mech.* **15,** 225-233. [U]

Rosen, A., and Libai, A. (1975). Transverse vibrations of compressed annular plates. *J. Sound Vib.* **40,** 149-153. [U]

Rossettos, J. N., and Sanders, J. L. Jr. (1965). Toroidal shells under internal pressure in the transition range. *AIAA J.* **3,** 1901-1909. [S]

Sanders, J. L., Jr. (1959). An improved first-approximation theory for thin shells. NASA Rept. No. 24. [P]

Sanders, J. L. Jr., and Liepins, A. A. (1963). Toroidal membrane under internal pressure. *AIAA J.* **1,** 2105-2110. [S]

Scheidl, R., and Troger, H. (1986). On the buckling and postbuckling of spherical shells. "Flexible Shells" (E. L. Axelrad & F. A. Emmerling, eds.), pp. 146-162. Springer-Verlag, Heidelberg, etc. [W]

Schwerin, E. (1922). Zur stabilitat der dünnwandingen Hohlkugel unter gleichmäszigem Aussendruck. *ZAMM* **2,** 81-91. [W]

Schwerin, E. (1929). Über Spannungen und Formänderungen kreisringförminger Membranen. *Zeit. f. Tech. Physik* **12,** 651-659. [S, T]

Seide, P. (1956). Axisymmetrical buckling of circular cones under axial compression." *J. Appl. Mech.* **23,** 625-628. [S]

Seide, P. (1975). "Small Elastic Deformations of Thin Shells." Noordhoff, Leyden. [S]

Shahinpoor, I. M., and Nowinsky, J. L. (1971). Exact solution to the problem of forced large amplitude radial oscillations of a thin hyperelastic tube. *Int. J. Non-lin. Mech.* **6,** 193-207. [U]

Shield, R. T. (1972). On the stability of finitely deformed elastic membranes: part 2, stability of inflated cylindrical and spherical membranes. *ZAMP* **23,** 16-34. [T]

Simmonds, J. G. (1961). The general equations of equilibrium of rotationally symmetric membranes and some strain solutions for uniform centrifugal loading. NASA Tech. Note D-816. [R]

Simmonds, J. G. (1962). The finite deflection of a normally loaded, spinning, elastic membrane. *J. Aerospace Sci.* **29,** 1180-1189. [S]

Simmonds, J. G. (1975). Rigorous expunction of Poisson's ratio from the Reissner-Meissner equations. *Int. J. Solids Struct.* **11,** 1051-1056. [R, S]

Simmonds, J. G. (1985). A new displacement form for the nonlinear equations of motion of shells of revolution. *J. Appl. Mech.* **52,** 507-509. [P, Q, U]

Simmonds, J. G. (1986). The strain-energy density of rubber-like shells of revolution undergoing torsionless, axisymmetric deformation (axishells). *J. Appl. Mech.* **53,** 593-596. [O, Q]

Simmonds, J. G. (1987). The strain-energy density of compressible, rubber-like axishells. *J. Appl. Mech.* **54,** 453-454. [O, Q, T]

Simmonds, J. G., and Danielson, D. A. (1972). Nonlinear shell theory with finite rotation and stress-function vectors. *J. Appl. Mech.* **39,** 1084-1090. [R]

Simmonds, J. G., and Libai, A. (1987a). A simplified version of Reissner's non-linear equations for a first-approximation theory of shells of revolution. *Computational Mech.* **2,** 1-5. [R]

Simmonds, J. G., and Libai, A. (1987b). Asymptotic forms of a simplified version of the non-linear Reissner equations for clamped elastic spherical caps under outward pressure. *Computational Mech.* **2,** 231-244. [S]

Singer, J. (1982). Buckling experiments on shells—a review of recent developments. *Solid Mech. Arch.* **7,** 213-313. [W]

Stavsky, Y., and Hoff, N. J. (1969). Mechanics of composite structures. *In* "Composite Engineering Laminates" (A. G. H. Dietz, ed.), Chapt. 1. M.I.T Press, Cambridge. [O]

Steele, C. R. (1965). Toroidal pressure vessels. *J. Spacecraft* **2,** 937-943. [S]

Stein, M., and Hedgepath, J. M. (1961). Analysis of partially wrinkled membranes. NASA TN D-813. [T]

Stoker, J. J. (1963). Elastic deformations in thin cylindrical sheets. *In* "Progress in Applied Mechanics" (The Prager Anniversary Volume, D. C. Drucker, ed.). Macmillan, New York, pp. 179-188. [T]

Struik, D. J. (1961). "Differential Geometry," 2nd Ed. Addison-Wesley, Reading, Mass. [Intro]

Szegö, G. (1934). Über einige Asymptotische Entwicklungen der Legendreschen Funktionen. *Proc. London Math. Soc.* **36** (2nd Series), 427-450. [S]

Szyszkowski, W., and Glockner, P. G. (1984a). Finite deformation and stability behaviour of spherical inflatables under axi-symmetric concentrated loads. *Int. J. Non-Lin. Mech.* **19,** 489-496. [T]

Szyszkowski, W., and Glockner, P. G. (1984b). Finite deformation and stability behaviour of spherical inflatables subjected to axi-symmetric hydrostatic loading. *Int. J. Solids Struct.* **20,** 1021-1036. [T]

Taber, L. A. (1982). Large deflection of a fluid-filled spherical shell under a point load. *J. Appl. Mech.* **49,** 121-128. [S]

Taber, L. A. (1983). Compression of fluid-filled spherical shells by rigid indenters. *J. Appl. Mech.* **50,** 717-722. [S]

Taber, L. A. (1984). Large deformation mechanics of the enucleated eyeball. *J. Biomechanical Engr.* **106,** 229-234. [S, T]

Taber, L. A. (1985). On approximate large strain relations for a shell of revolution. *Int. J. Non-Lin. Mech.* **20,** 27-39. [Q]

Taber, L. A. (1987). On boundary layers in a pressurized Mooney cylinder. *J. Appl. Mech.* **54,** 280-286. [Q]

Tielking, J. T., and Feng, W. W. (1974). The application of the minimum potential energy principle to nonlinear axisymmetric membrane problems. *J. Appl. Mech.* **41,** 491-496. [T]

Timoshenko, S. P., and Gere, J. M. (1961). "Theory of Elastic Stability," 2nd ed. McGraw-Hill, New York, etc. [W]

Timoshenko, S. P., and Goodier, J. N. (1970). "Theory of Elasticity." 3rd ed., McGraw-Hill, New York, etc. [W]

Timoshenko, S., and Woinowsky-Krieger, S. (1959). "Theory of Plates and Shells," 2nd ed. McGraw-Hill, New York, etc. [S]

Treloar, L. R. G. (1944). Stress-strain data for vulcanized rubber under various types of deformation. *Trans. Faraday Soc.* **40,** 59-70. [T]

Truesdell, C. A., and Toupin, R. A. (1960). The classical field theories. *In* "Encyclopedia of Physics" (S. Flügge, ed.), vol. III/1. Springer-Verlag, Berlin and New York. [X]

Tsui, E. Y. W. (1968). "Stresses in Shells of Revolution." Pacific Coast Publishers, Menlo Park, California. [S]

Tvergaard, V. (1976). Buckling behaviour of plate and shell structures. *In* "Theoretical and Applied Mechanics" (W. T. Koiter, ed.), pp. 233-247. North-Holland. [W]

Updike, D. P., and Kalnins, A. (1970). Axisymmetric behavior of an elastic spherical compressed between rigid plates. *J. Appl. Mech.* **37,** 635-640. [S]

Updike, D. P., and Kalnins, A. (1972a). Axisymmetric postbuckling and nonsymmetric buckling of a spherical shell compressed between rigid plates. *J. Appl. Mech.* **39,** 172-178. [S]

Updike, D. P., and Kalnins, A. (1972b). Contact pressure between an elastic spherical shell and a rigid plate. *J. Appl. Mech.* **39,** 1110-1114. [S]

van der Neut, A. (1932). "Der Elastische Stabiliteit van den dunwandigen Bol." Thesis, Delft. [W]

VanDyke, P. (1966). Nonlinear influence coefficients for a spherical shell and pressure loading. *AIAA J.* **4,** 2045-2047. [S]

Vaughan, H. (1980). Pressurizing a prestressed membrane to form a paraboloid. *Int. J. Engr. Sci.* **18,** 99-107. [T]

Vinson, J. R., and Sierakowski, R. L. (1986). "The Behavior of Structures

Composed of Composite Materials." Martinus Nijhoff, Dordrecht. [O]
Wagner, H. (1929). Flat sheet metal girders with a very thin metal web. *Z. Flugtech. Motorluftsschiffahrt* **20,** 200-314. (Translated as NASA TM 604-606). [T]
Wah, T. (1963). Vibration of circular plates at large amplitudes. *ASCE J. Engr. Mech.* **5,** 1-15. [U]
Wan, F. Y. M. (1980a). Polar dimpling of complete spherical shells. *In* "Theory of Shells," Proc. 3rd I.U.T.A.M. Shell Sympos. Tbilisi, 1978 (W. T. Koiter & G. K. Mikhailov, eds.), pp. 589-605. North-Holland, Amsterdam, etc. [S, W]
Wan, F. Y. M. (1980b). The dimpling of spherical caps. *Mechanics Today* **5,** 495-508. [S, W]
Wan, F. Y. M. (1984). Shallow caps with a localized pressure distribution centered at the apex. "Flexible Shells" (E. Axelrad & F. Emmerling, eds.), pp. 124-145. Springer-Verlag, Berlin, etc. [S, W]
Wan, F. Y. M., and Ascher, U. (1980). Horizontal and flat points in shallow shell dimpling. *In* "Proc. BAIL I. Conf." Dublin (J. Miller, ed.), pp. 415-419. [S, W]
Wan, F. Y. M., and Yun, T. -q. (1985). Axisymmetric dynamic stability of polar dimpling. *In* "Proc. Int. Conf. Nonlin. Mech." (W. -Z. Chien, ed.), pp. 340-346. Science Press, Beijing. [S, W]
Wang, C. C. (1965). On the radial oscillations of a spherical thin shell in the finite elasticity theory. *Q. Appl. Math.* **23,** 270-274. [U]
Washizu, K. (1980). A note on the principle of stationary complementary energy in non-linear elasticity. *Mechanics Today* **5,** 509-522. [U]
Weinberg, I. J. (1962). Symmetric finite deflections of circular plates subjected compressive edge forces. *J. Math. & Phys.* **41,** 104-115. [S]
Weinitschke, H. J. (1960). On the stability problem of shallow spherical shells. *J. Math. & Phys.* **38,** 209-231. [W]
Weinitschke, H. J. (1965). On asymmetric buckling of shallow spherical shells. *J. Math. & Phys.* **44,** 141-163. [W]
Weinitschke, H. J. (1970). Existenz- und eindeutigkeitssätze für gleichungen der kreisförmigen membran. *Meth. u. Verf. d. Math. Physik* **3,** 117-139. [S]
Weinitschke, H. J. (1980). On axisymmetric deformations of nonlinear elastic membranes. *Mechanics Today* **5,** 523-542. [S]
Weinitschke, H. J. (1986). On finite displacement of circular membranes. *Math. Methods Appl. Sci.* (to appear). [S]
Weinitschke, H. J. (1987a). On uniqueness of axisymmetric deformations of elastic plates and shells. *SIAM J. Math. Anal.* (to appear). [R]
Weinitschke, H. J. (1987b). On finite displacements of circular elastic membranes. *Math. Meth. Appl. Sci.* **9,** 76-98. [S]
Weinitschke, H. J., and Lange, C. G. (1987). Asymptotic solutions for finite deformations of thin shells of revolution with a small circular hole. *Q. Appl. Math.* (to appear). [S]
Wempner, G. A. (1958). The conical disk spring. *In* Proceedings of the 3rd Nat. Cong. Appl. Mech., pp. 473-478. [S]

Wempner, G. A. (1964). Axisymmetrical deflections of shallow conical shells. *ASCE, J. Engr. Mech. Div.* **90,** 181-193. [S]

Wilkinson, J. P., and Kalnins, A. (1966). Deformation of open spherical shells under arbitrarily located concentrated loads. *J. Appl. Mech.* **33,** 305-312. [W]

Wu, C. -H. (1970a). Tube to annulus—an exact nonlinear membrane solution. *Q. Appl. Math.* **27,** 489-496. [T]

Wu, C. -H. (1970b). On certain integrable nonlinear membrane solutions. *Q. Appl. Math.* **28,** 81-90. [T]

Wu, C. -H. (1974a). The wrinkled axisymmetric air bags made of inextensible membrane. *J. Appl. Mech.* **41,** 963-968. [S, T]

Wu, C. -H. (1974b). Infinitely stretched Mooney surfaces of revolution are uniformly stretched catenoids. *Q. Appl. Math.* **33,** 273-284. [T]

Wu, C. -H. (1978). Nonlinear wrinkling of nonlinear membranes of revolution. *J. Appl. Mech.* **45,** 533-538. [S, T]

Wu, C. -H. (1979). Large finite strain membrane problems. *Q. Appl. Math.* **36,** 347-359. [T]

Yamaki, N. (1961). Influence of large amplitudes on flexural vibrations of elastic plates. *ZAMM* **41,** 501-510. [U]

Yamaki, N. (1984). "Elastic Stability of Circular Cylindrical Shells." North-Holland, Amsterdam. [S, W]

Yang, W. H., and Feng, W. W. (1970). On axisymmetrical deformations of nonlinear membranes. *J. Appl. Mech.* **37,** 102-1011. [T]

Yanowitch, M. (1956). Non-linear buckling of circular elastic plates. *Comm. Pure Appl. Math.* **9,** 661-672. [S]

Zak, M. (1982). Statics of wrinkling films. *J. Elasticity* **12,** 51-63. [T]

Zoelly, R. (1915). "Über ein Knickungproblem an der Kugelschale." Thesis, Zurich. [W]

Chapter VI

Shells Suffering 1-Dimensional Strains (Unishells)

In the preceding chapters on birods, beamshells, and axishells, we studied 3 important classes of structures that, under special loads and boundary/initial conditions, suffer strains, rotations, and displacements that are functions of one spatial variable only. The resulting simplifications of the general governing equations are enormous and in static problems, partial differential equations reduce to ordinary ones. Because strains, unlike rotations or displacements, are unaffected by rigid-body motions, it seems natural and possibly fruitful to seek other shells that, subject to suitable loads and boundary/initial conditions, suffer deformations in which the strains depend on one spatial variable only but in which the rotations and displacements depend (possibly) on a second one. The simplest and most important example is the bending of a pressurized tube of circular planform and arbitrary cross section—the subject of Sections A and B.

Specifically, in Section A we consider tubes subjected to a net end moment and, if pressurized, to an equilibrating net end force as well. We formulate the governing equations, discuss various approximations, and survey the literature on the collapse of such tubes. In Section B, we develop variational principles for curved tubes of either closed or open cross section, allowing for loads that act on the boundaries of the latter.

Beginning with Section C, we extend our analysis to *unishells*, i.e., to any shell with a set of loads and boundary/initial conditions (which must include the trivial ones that produce no distortion) such that there is a system of coordinates (σ, τ) on the reference surface in which the components of the deformed metric and curvature tensors are independent of τ. Simmonds (1979) has shown that any unishell must have a reference surface that is a general helicoid, i.e., a reference surface that is generated by translating an arbitrary plane curve along a circular helix.[1] We shall further demand that, in an unishell, the transverse shearing strain(s) be independent of τ so that in static problems—to which we devote the

[1]By regarding a straight line as a circular helix of zero pitch and infinite radius, we may include general cylinders in the class of general helicoids and thus regard beamshells as a special type of unishell.

remainder of the chapter—the governing differential equations are ordinary.[2] Thus, while from an *intrinsic* viewpoint, a unishell is 1-dimensional in its constitution and in its static behavior, from an *extrinsic* viewpoint, it exhibits *repeatability* or *periodicity* in its structure and in its response.

In Sections D-H we derive kinematic, equilibrium, compatibility, and constitutive relations for unishells. In Section I—the last of this last chapter—we consider several special cases of these general equations, including the *torsion, extension, and inflation* of a straight tube; the *extension and twist* of a right helicoidal shell; and the *inextensional deformation* of a general unishell.

A. In-Plane Bending of Pressurized Curved Tubes

1. Introduction

We shall be concerned in this section with the nonlinear behavior of curved tubes of arbitrary, closed cross section, subject to end couples in their planes of curvature and under internal or external pressure. The thicknesses of the tubes is taken constant. Straight tubes fall out as a special case. In what follows, we shall make the sweeping assumption that the problem can be modeled by a 1-dimensional, θ-independent state of stress and strain. Disturbances that might affect the validity of the 1-dimensional model are assumed to be confined to regions near the ends of the tube and will not be considered: their analysis is beyond the scope of this book.

The nonlinear bending of pressurized straight pipes and the linear and nonlinear bending of pressurized curved tubes have important biologic and engineering applications. Arteries and veins carry the nutrients of life while pipelines carry liquids and gases throughout the civilized world and are major components in many machines and industrial plants.

Often pipes burst under internal pressure or collapse under external pressure. These *modes of failure* are fairly well understood. Our concern in this section is with the more complex phenomenon of *failure in bending* (with or without coexisting pressure). In a circular pipe, the major effect of bending is an "ovalization" or flattening of the cross section accompanied by: increased overall flexibility, stress concentrations, an appearance of shell-like bending moments, and, finally, collapse at a bending couple below that predicted by elementary theory.

A common engineering approach to pipe bending is to apply "correction factors" to standard formulas of beam theory. (See the ASME Boiler and Pressure Vessel Code, Part III, the British ESDU data series No. 74073, or several commercially available computer codes for piping analysis.) However, the aerospace, nuclear, and petroleum industries have demanded refined numerical and laboratory studies. Several special and all-purpose computer codes now

[2]The nonlinear elastodynamics of general unishells awaits development.

include "pipe elbow" subroutines as well as procedures for handling inelastic behavior and 2-dimensional effects, such as end disturbances. For more information see the reviews by Sobel & Newman (1980) and Bushnell (1981).

The first *linear* study of the bending of curved tubes goes back to von Kármán (1911); the existing linear theory was reviewed and extended by Clark & Reissner (1951). Nonlinear effects were introduced by Reissner (1959), and Axelrad (1962a).[3] More recent investigations and reviews of the nonlinear bending of tubes include those of Fabian (1977), Boyle (1981), Reissner (1981), Emmerling (1982, 1984), Axelrad & Emmerling (1983, 1984), and Axelrad (1985a), who summarizes the approaches to the bending of finite tubes. Axelrad (1987) also devotes a substantial part of his recent book to the theory of curved tubes.

In the following, we develop the field equations for the nonlinear bending of shearable, curved tubes, subject to pressure and end loads, as a natural extension of the equilibrium and compatibility conditions for axishells. No assumption beyond 1-dimensionality will be made. At the end of this section, we reduce the equations, in stages, to their simplest possible approximate forms.

2. *Geometry*

Consider a surface $\mathcal{T}$ generated by rotating a simple, *closed* plane curve $\mathcal{L}: r(\sigma), z(\sigma)$ through an angle Θ about the z-axis of a fixed, circular cylindrical coordinate system (r, θ, z). Here, $\sigma \in [0, L]$ denotes meridional arc length and $0 \le \theta \le \Theta \le 2\pi$. We note that $\mathcal{T}$ is a *sector of a torus*. A shell that has this toroidal segment as its reference surface is called a *curved tube* for short.[4] Straight tubes will be included in the analysis as limiting cases of curved tubes. For simplicity, our curved tubes will be assumed to be homogeneous, elastically isotropic, and of constant thickness h, but polar, orthotropic materials may also be included.

In practice, the most common tube is the *circular pipe*

$$\mathcal{L}: \; r = a + b\sin(\sigma/b) \;, \quad z = -b\cos(\sigma/b) \;, \quad 0 \le \sigma < 2\pi b \;. \tag{A.1}$$

Here, a is the constant radius of the centerline of the pipe and b is the constant radius of the circular meridians.

3. *External loads*

We consider two types of loads:

(a) A pure couple $M\mathbf{e}_z$ applied to both edges ($\theta = 0, \Theta$) of the shell. This couple equilibrates the net moment produced at the edges by the stress resultant N_θ and stress couples M_θ and M_n, and is applied to the shell so that certain kinematical constraints, to be discussed later, are satisfied.

---

[3]This is a translation of a paper in Russian that appeared in 1960.

[4]A more precise name would be a *plane, circularly bent tube* since a "curved tube" could refer to a helicoidal tube or to a planar tube with a non-circular centerline.

(b) The pure couple of the preceding paragraph, plus a constant fluid pressure p, plus a force $P\overline{\mathbf{e}}_\theta$ at the edges which equilibrates the resultant of the pressure.[5] Here, P must equal the net force produced by the stress resultant N_θ at the edge. The force and couple at the edge are applied so that the kinematical constraints are satisfied. In this way, the effects of pressure on the bending of curved tubes can be studied.

4. Semi-inverse approach

We shall postulate a certain 1-dimensional state of stress and strain that is independent of θ. A solution of the resulting field equations (if it exists), periodic in σ and satisfying *integral* edge conditions on the stress resultant and couples, will be assumed to approximate an exact solution of the full shell equations at a sufficient distance from the edges. Of course, it is possible (but not likely in practice) that the actual boundary conditions on stresses and displacements will be met by the semi-inverse solution; otherwise, we must make a vague and perhaps unjustified appeal to St. Venant's Principle. In the linear theory of semi-infinite shells of revolution under self-equilibrated end stresses, Simmonds (1977) has shown that some of the stresses decay over a distance of the order of $\sqrt{R/h}$ times the radius R of the loaded edge. This result may be used to establish the unique role of axisymmetric, semi-inverse, St. Venant-like solutions. Intuitively, we expect a similar story for the weak, nonlinear bending of a straight tube. The question of uniqueness and the decay of θ-dependent edge disturbances in the bending of *curved* tubes is still open. For further discussion, see Axelrad & Emmerling (1983, 1984, Sections 4 and 5). Many tubes are designed too conservatively because the stiffening effect of rigid boundaries, which may influence a major portion of the tube, is not taken into account.

5. Kinematics of deformation

A curved tube undergoing bending, as shown in Fig. 6.1, behaves like a beam having no out-of-plane distortion of its meridional cross section. That is, material meridional planes of the undeformed shell become material meridional planes of the deformed shell, which is also toroidal. This is an important geometrical property because it allows us to decompose the deformation into two parts: an in-plane deformation of meridional cross sections, *plus* a rigid rotation of each of these meridional planes about the z-axis by a linearly varying angle, $\overline{k}\theta$. The two deformations obviously commute.

If $\mathbf{y}(r(\sigma), z(\sigma), \theta)$ is the position of a point on the initial surface $\mathcal{T}$, where (r, z, θ) are circular cylindrical coordinates in a fixed Cartesian frame $Oxyz$, then

$$\overline{\mathbf{y}} = \overline{\mathbf{y}}(\overline{r}(\sigma), \overline{z}(\sigma), \overline{\theta}(\theta)) \;,\; \overline{\theta} \equiv (1 + \overline{k})\theta \tag{A.2}$$

is the (averaged) deformed position of the point. Here, $(\overline{r}, \overline{z}, \overline{\theta})$ are the circular

[5] $\overline{\mathbf{e}}_r$ and $\overline{\mathbf{e}}_\theta$ are unit vectors in the *deformed* configuration.

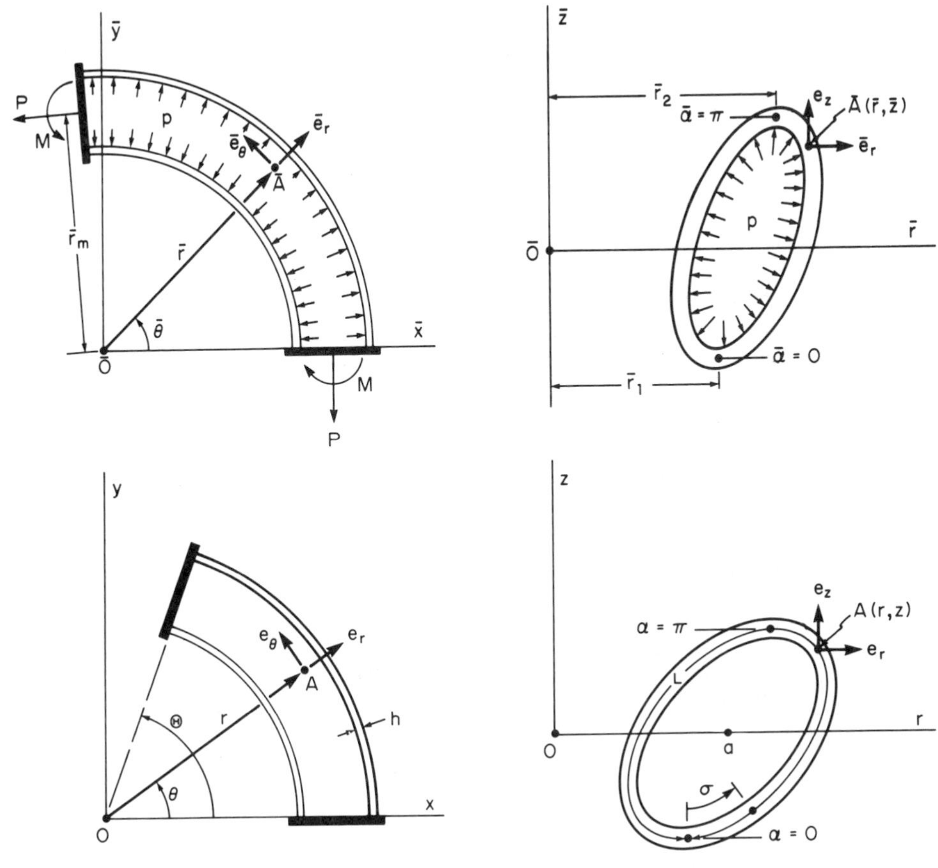

Fig. 6.1. Planform and cross-sectional views of an undeformed and deformed tube.

cylindrical coordinates of the point in a (generally) new Cartesian frame $\overline{Oxyz}$ whose $\overline{xy}$-plane has been translated and rotated parallel to the xy-plane of the original Cartesian frame; (σ, θ) are Lagrangian coordinates which serve as Gaussian coordinates on the deformed, toroidal reference surface $\overline{\mathcal{T}}$. We note that the σ- and θ-coordinate curves are lines of principal curvature on both $\mathcal{T}$ and $\overline{\mathcal{T}}$. It follows that the in-surface, $\sigma\theta$-shearing strain vanishes.

The two-part decomposition of the deformation of a curved tube implies that the equations for axishells hold for all in-surface quantities except those affected by the circumferential rotation of cross sections. Thus, we have

$$d\bar{s}_\theta \equiv \bar{r}d\bar{\theta} = (1 + \bar{k})\bar{r}d\theta \, , \tag{A.3}$$

so that the hoop strain is given by

$$e_\theta = \frac{d\bar{s}_\theta}{ds_\theta} - 1 = (1+\bar{k})\frac{\bar{r}}{r} - 1\,, \tag{A.4}$$

while the expressions for the bending strains take the form

$$rk_\theta = (1+\bar{k})\sin\bar{\alpha}_T - \sin\alpha \tag{A.5}$$

$$rk_n = (1+\bar{k})\cos\bar{\alpha}_T - \cos\alpha \tag{A.6}$$

$$k_\sigma = \beta'\,, \tag{A.7}$$

where, as before, $\bar{\alpha}_T = \alpha + \beta$.

An applied, constant fluid pressure is also affected by the rotation. To see how, let $\mathbf{p} = \bar{p}_H \bar{\mathbf{e}}_r + \bar{p}_V \mathbf{e}_z$ be the distributed surface load on the deformed shell reckoned per unit *undeformed* area. Then,

$$\mathbf{p} = -p(1+e_\theta)\mathbf{m} \tag{A.8}$$[6]

$$\bar{p}_H = -p(1+e_\theta)\mathbf{m}\cdot\bar{\mathbf{e}}_r \tag{A.9}$$

$$\bar{p}_V = -p(1+e_\theta)\mathbf{m}\cdot\mathbf{e}_z\,, \tag{A.10}$$

where p is the magnitude of the pressure reckoned per unit *deformed* area, and

$$\bar{\mathbf{e}}_r = \cos\bar{\theta}\mathbf{e}_x + \sin\bar{\theta}\mathbf{e}_y \quad \text{and} \quad \bar{\mathbf{e}}_\theta = -\sin\bar{\theta}\mathbf{e}_x + \cos\bar{\theta}\mathbf{e}_y \tag{A.11}$$

are unit vectors associated with the rotated meridians. They must be distinguished carefully from the non-rotated basis vectors $\mathbf{e}_r$ and $\mathbf{e}_\theta$. From $(A.2)_2$ and (A.11) we have

$$\bar{\mathbf{e}}_{r,\theta} = (1+\bar{k})\bar{\mathbf{e}}_\theta \;\;,\;\; \bar{\mathbf{e}}_{\theta,\theta} = -(1+\bar{k})\bar{\mathbf{e}}_r\,. \tag{A.12}$$

In view of (A.4), (V.L.2), and (V.Q.7), (A.9) and (A.10) reduce to

$$r\bar{p}_H = pr(1+\underline{e_\theta})[(1+\underline{e_\sigma})\sin\bar{\alpha}_T + \underline{g}\cos\bar{\alpha}_T] = (1+\bar{k})\bar{r}\bar{z}'p \tag{A.13}$$

$$r\bar{p}_V = -pr(1+\underline{e_\theta})[(1+\underline{e_\sigma})\cos\bar{\alpha}_T - \underline{g}\sin\bar{\alpha}_T] = -(1+\bar{k})\bar{r}\bar{r}'p\,, \tag{A.14}$$

where the underlined terms represent *strain effects* that we shall discuss later.

The compatibility condition for an axishell also has to be modified for a curved tube because of the factor $1+\bar{k}$ in (A.4). From the relation

$$\bar{r}' = \lambda_\sigma \cos\bar{\alpha}\,, \tag{A.15}$$

which holds for axishells as well as curved tubes, we have, with the aid of (V.L.1) and (A.4),

---

[6]As we deal in this chapter with constant pressure only, we have dropped the subscript 0 on the scalar pressure.

$$[r(1+e_\theta)]' = (1+\bar{k})[(1+e_\sigma)\cos\bar{\alpha}_T - g\sin\bar{\alpha}_T]\,. \tag{A.16}$$

6. Equilibrium equations

Deviations from the equilibrium equations of axishells, (V.D.3) and (V.D.5) (with the accelerations terms deleted), result from differentiating the vectors $N_\theta \bar{\mathbf{e}}_\theta$ and $\mathbf{M}_\theta$ (which involves $\bar{\mathbf{e}}_r$) with respect to θ, as required by (V.C.1) and (V.C.2). As a consequence of (A.12), the factor $1+\bar{k}$ has to be attached to N_θ, M_θ, and M_n in the equilibrium equations and in any equations derived from these. In particular, (V.L.15), (V.Q.2), and (V.Q.3) are modified for the elastostatics of curved tubes as follows:

$$(rM_\sigma)' - (1+\bar{k})(M_\theta\cos\bar{\alpha}_T - M_n\sin\bar{\alpha}_T) + r\,[(1+e_\sigma)Q - gN_\sigma] = 0 \tag{A.17}$$

$$(rH)' - (1+\bar{k})N_\theta + r\bar{p}_H = 0 \tag{A.18}$$

$$(rV)' + r\bar{p}_V = 0\,. \tag{A.19}$$

For fluid pressure, $r\bar{p}_H$ and $r\bar{p}_V$ are given by (A.13) and (A.14). These equations are supplemented by (V.Q.10) and (V.Q.11), taken in the form

$$N_\sigma = H\cos\bar{\alpha}_T + V\sin\bar{\alpha}_T\ ,\quad Q = -H\sin\bar{\alpha}_T + V\cos\bar{\alpha}_T\,. \tag{A.20}$$

In view of (A.14), equation (A.19) can be expressed in the integrated form

$$rV = C + \tfrac{1}{2}(1+\bar{k})p\bar{r}^2 = C + \frac{pr^2}{2(1+\bar{k})}\underline{(1+e_\theta)^2}\,, \tag{A.21}$$

where

$$C = r(0)V(0) - \tfrac{1}{2}(1+\bar{k})p\bar{r}^2(0) = r(L)V(L) - \tfrac{1}{2}(1+\bar{k})p\bar{r}^2(L) \tag{A.22}$$

is a constant of integration and, again, the underlined term in (A.21) represents a strain effect. The appearance of a (possibly) non-zero constant, even if $p=0$, reflects the fact that the deformed meridians are closed. We shall take (A.17), (A.18), and (A.21) as the basic equilibrium equations, with (A.13) and (A.20) as auxiliary.

7. Constitutive relations

The constitutive relations are not affected by the transition from axishell to curved tube. Preferably, they are expressed in terms of the mixed-energy density $\Psi(N_\sigma, N_\theta, Q, k_\sigma, k_\theta, k_n)$ and may have the general form of (V.O.5) or the widely-used linear form (V.R.2) and (V.R.3). The kinematic relations (A.5)-(A.7) allow the arguments of Ψ to be reduced to $N_\sigma, N_\theta, Q, \beta$, and β'. Substitution into the basic equilibrium equations and the compatibility condition produces four equations for N_σ, N_θ, Q, and β. If Ψ is taken to be quadratic and independent of k_n, we obtain the field equations of *shearable tubes.* Note that we are retaining the nonlinear strain effects in the equilibrium conditions and in the expressions for $\bar{p}_H$ and V.

8. Boundary and end conditions

In the classical theory of axishells, boundary conditions at $\sigma=0, L$ require the prescription of $\mathbf{N}_\sigma$ and M_σ or their kinematic duals, $\mathbf{u}$ and β. Here, because we deal with closed tubes, the boundary conditions reduce to the requirement that $M_\sigma, H, V, \beta, \bar{r}$ and $\bar{z}$ be L-periodic functions of σ, i.e., $M_\sigma(\sigma+L)=M_\sigma(\sigma)$, etc. All these requirements, except that on $\bar{z}$, are met by demanding that β, β', H and H' be L-periodic. [The periodicity of V is guaranteed automatically by the right side of (A.21).] The unknown constant C is determined by the periodicity requirement on $\bar{z}$, which is written most conveniently in the form

$$\oint_{L} \lambda_\sigma \sin(\alpha+\beta)d\sigma = \oint_{L} [(1+e_\sigma)\sin\bar{\alpha}_T + g\cos\bar{\alpha}_T]d\sigma = 0\,, \tag{A.23}$$

where $(\alpha+\beta) \equiv \bar{\alpha}_T + \gamma = \alpha+\beta+\gamma$. To avoid the occurrence of terms of different orders of magnitude, we rewrite (A.23) in the alternative form

$$\oint_{L} (\sin\bar{\alpha}_T - \sin\alpha + e_\sigma \sin\bar{\alpha}_T + g\cos\bar{\alpha}_T)d\sigma = 0\,. \tag{A.24}$$

The analytical difficulties in toroidal shells associated with enforcing axial continuity have an interesting history, some of which was given in subsection V.S.5. Usually, most of the axial deformation is confined to narrow zones near $\alpha=0$ and $\alpha=\pi$. These effects are especially strong if the meridional cross section is *not* symmetrical. For further details, see subsection V.S.5.

Next, we consider the conditions at $\theta=0, \Theta$. Here, we need to verify that the net force and moment are of the form $M\mathbf{e}_z$ and $P\bar{\mathbf{e}}_\theta$, the latter being necessary to equilibrate the net pressure loads. The end result is a formula relating the gross moment M to the gross curvature $\bar{k}$.

Starting with pure bending ($p=0$), we have for the end force

$$P = \oint_{L} N_\theta d\sigma = (1+\bar{k})^{-1}\oint_{L} (rH)'d\sigma \equiv 0\,. \tag{A.25}$$

The moment along the $\bar{r}$-axis at the ends is given by

$$\begin{aligned} M_r &= \oint_{L} (N_\theta \bar{z} - M_\theta \cos\bar{\alpha}_T + M_n \sin\bar{\alpha}_T)d\sigma \\ &= (1+\bar{k})^{-1}\oint_{L} [(rH)'\bar{z} - (1+\bar{k})(M_\theta\cos\bar{\alpha}_T - M_n\sin\bar{\alpha}_T)]d\sigma\,. \end{aligned} \tag{A.26}$$

Integrating by parts and expressing $\bar{z}'$ in terms of strains, we get

$$\begin{aligned} (1+\bar{k})M_r = \oint_{L} \{&-rH[(1+e_\sigma)\sin\bar{\alpha}_T + g\cos\bar{\alpha}_T] \\ &- (1+\bar{k})(M_\theta\cos\bar{\alpha}_T - M_n\sin\bar{\alpha}_T)\}d\sigma + (rH\bar{z})_0^L\,. \end{aligned} \tag{A.27}$$

Using (A.17) and (A.20), we can reduce (A.27) to

$$(1+\bar{k})M_r = -\oint_{\mathcal{L}} rV\bar{r}'d\sigma + (rH\bar{z} - M_\sigma)_0^L . \tag{A.28}$$

But for pure bending, $rV = C$; hence,

$$(1+\bar{k})M_r = (rH\bar{z} - rM_\sigma - C\bar{r})_0^L \equiv 0 . \tag{A.29}$$

The vanishing of M_r is thus seen to result from the periodicity (continuity) condition on $\bar{r}$, $\bar{z}$, H, and M_σ[7].

So far, we have found that the net end force and the moment along $\bar{\mathbf{e}}_r$ vanish identically, as they should. We now compute the moment along $\mathbf{e}_z$:

$$\begin{aligned} T_z = M &= \oint_{\mathcal{L}} (\bar{r}N_\theta + M_\theta \sin\bar{\alpha}_T + M_n \cos\bar{\alpha}_T) d\sigma \\ &= \oint_{\mathcal{L}} \{-(1+\bar{k})^{-1} rH[(1+e_\sigma)\cos\bar{\alpha}_T - g\sin\bar{\alpha}_T] + (M_\theta \sin\bar{\alpha}_T + M_n \cos\bar{\alpha}_T)\} d\sigma . \end{aligned} \tag{A.30}$$

Using (A.20), we find that

$$M = \oint_{\mathcal{L}} \{-(1+\bar{k})^{-1} r[(1+e_\sigma)N_\sigma + gQ] + (M_\theta \sin\bar{\alpha}_T + M_n \cos\bar{\alpha}_T)\} d\sigma . \tag{A.31}$$

This formula establishes the relationship between the applied moment, M, and $\bar{k}$.

We conclude that our equilibrium equations, with $p = 0$, the periodicity conditions, and (A.31), should possess solutions that satisfy the integral conditions for force and moment at the ends. In addition, the solutions of this system will produce rigid body motions of the ends with no out-of-plane distortion. The semi-inverse theory we have outlined can accommodate neither different nor additional conditions. In particular, *we can place no constraints on the distortions of the end cross sections in their own planes*. It is this requirement that potentially causes the greatest discrepancy between theory and practice in short tubes. However, in practice, tubes are usually connected to adjacent structures or to other tubes by means of flanges or other stiff devices that deform little. In such cases, our theory provides an approximate solution to the problem "far" from the ends. (See the discussion in subsection A.4).

We now consider the combined effects of pressure and end couples. Here, the net force on the ends of the tube does not vanish but is given by

$$P = (1+\bar{k})^{-1} \oint_{\mathcal{L}} r\bar{p}_H d\sigma = p \oint_{\mathcal{L}} \bar{r}[(1+e_\sigma)\sin\bar{\alpha}_T + g\cos\bar{\alpha}_T] d\sigma = p \oint_{\mathcal{L}} \bar{r} d\bar{z} , \tag{A.32}$$

which comes from substituting (A.18) into (A.25). The quantity on the far right of (A.32) is the radial resultant of the pressure on the deformed tube, per radian

[7] It is instructive to note that satisfaction of the equilibrium and periodicity conditions are sufficient, in this case, to ensure the vanishing of M_r for tubes of *arbitrary* cross section. Still, the vanishing of M_r can replace one of the periodicity conditions if the two are equivalent in the sense of (A.29)

in the $\bar{\theta}$-direction, as we expect from elementary considerations. It also equals the resultant of the pressure on the end closures of the tube.

To compute the moment turning about $\bar{\mathbf{e}}_r$, we substitute (A.18) into (A.26) and use (A.13) to obtain, instead of (A.29),

$$M_r(1+\bar{k}) = [r(H\bar{z} - M_\sigma) - C\bar{r} - (1/6)(1+\bar{k})p\bar{r}^3]_0^L + \oint_{\mathcal{L}} r\bar{p}_H\bar{z}\,d\sigma \,. \tag{A.33}$$

The bracketed terms vanish due to periodicity, while the last term above reduces, in view of (A.21), to

$$M_r = p\oint_{\mathcal{L}} \bar{r}\,\bar{z}\,d\bar{z} = -\tfrac{1}{2}p\oint_{\mathcal{L}} \bar{z}^2\,d\bar{r} \,. \tag{A.34}$$

This last expression is evidently the moment of P with respect to the $\bar{r}$-axis. It can be made to vanish by an appropriate choice of the xy base plane. In particular, we see from (A.34) that M_r vanishes automatically for tubes of symmetric cross section if the xy-plane coincides with a plane of symmetry.

Finally, we calculate the edge moment turning about $\mathbf{e}_z$. From (A.30)

$$\begin{aligned} T_z &= \oint_{\mathcal{L}} \{(1+\bar{k})^{-1}\bar{r}[(rH)' + r\bar{p}_H] + M_\theta \sin\bar{\alpha}_T + M_n\cos\bar{\alpha}_T\}\,d\sigma \\ &= \oint_{\mathcal{L}} \{-(1+\bar{k})^{-1} r[(1+\underline{e_\sigma})N_\sigma + \underline{gQ}] + (M_\theta\sin\bar{\alpha}_T + M_n\cos\bar{\alpha}_T) + (3/2)p\bar{r}^2\bar{z}'\}\,d\sigma \,, \end{aligned} \tag{A.35}$$

where, as before, the underlined terms represent strain effects. This equation relates the gross bending strain $\bar{k}$ to the *total edge moment* T_z. This moment should be decomposed as

$$T_z = M + P\bar{r}_m \,, \tag{A.36}$$

where $\bar{r}_m$ is the moment arm of P. Unfortunately, this decomposition depends on the user because pressure and couple effects are intermingled and only the total moment T_z can be calculated for a given $\bar{k}$ (and vice versa). *Our method* of decomposition is based on the observation that the rotation of the ends is caused by the end couple only; without the couple the pressure effect is the same as if the curved tube were a complete toroidal axishell and corresponds mathematically to setting $\bar{k}=0$ everywhere. For a given pressure we may therefore *define* the moment arm of P by

$$\bar{r}_m \equiv [P^{-1}\oint_{\mathcal{L}} (\bar{r}N_\theta + M_\theta\sin\bar{\alpha}_T + M_n\cos\bar{\alpha}_T)\,d\sigma]_{\bar{k}=0} \,, \tag{A.37}$$

from which we obtain

$$\begin{aligned} M = \oint_{\mathcal{L}} [&(\bar{r}N_\theta + M_\theta\sin\bar{\alpha}_T + M_n\cos\bar{\alpha}_T) \\ &- (\bar{r}N_\theta + M_\theta\sin\bar{\alpha}_T + M_n\cos\bar{\alpha}_T)_{\bar{k}=0}]\,d\sigma \,. \end{aligned} \tag{A.38}$$

This equation relates $\bar{k}$ to M *for a given level of pressure*. In practice the stress couples are small in toroidal axishells so membrane theory can be used if $\bar{k}=0$.

Another way of separating pressure and couple effects is based on the observation that $r\bar{p}_H$ is the single term responsible for producing a force at the ends of the curved tube. By omitting this term in (A.30), we cancel the moment due to this force and only "pure" couple effects are left. (This approximation implies that

$$\bar{r}_m \equiv \oint_{\mathcal{L}} r\bar{p}_H \bar{r}\, d\sigma \Big/ \oint_{\mathcal{L}} r\bar{p}_H\, d\sigma \,.) \tag{A.39}$$

We prefer our first method since we regard the shifting of the location of P as an effect produced by the couple. Other decompositions may be used so long as their implications are understood. For a variational approach to the definition of M, see subsection B.2.

9. Reductions and approximations

Let us assume that Ψ is quadratic as in (V.R.1) and does not depend on k_n. Inserting the strains and moments of (V.R.2) and (V.R.3) into the compatibility condition (A.16) and the moment equilibrium expression (A.17), we obtain the following 2, coupled, 2nd order differential equations for H and $\bar{\alpha}_T$:[8]

$$\begin{aligned}(1+\bar{k})^{-1}[r(rH)']' - \{(1+\bar{k})[\cos^2\bar{\alpha}_T + (A_s/A_e)\sin^2\bar{\alpha}_T] + \nu r(\cos\bar{\alpha}_T)'\}H \\ + A_e^{-1}[\cos\alpha - (1+\bar{k})\cos\bar{\alpha}_T] + (1+\bar{k})^{-1}(r^2\bar{p}_H)' + \underline{\nu rp(1+e_\theta)g} \\ - [(1-A_s/A_e)(1+\bar{k})\sin\bar{\alpha}_T\cos\bar{\alpha}_T + \nu r(\sin\bar{\alpha}_T)']V = 0\end{aligned} \tag{A.40}$$

$$\begin{aligned}D\{[r(\bar{\alpha}_T-\alpha)']' - (1+\bar{k})r^{-1}[(1+\bar{k})\sin\bar{\alpha}_T - \sin\alpha]\cos\bar{\alpha}_T + \nu\alpha'[(1+\bar{k})\cos\bar{\alpha}_T - \cos\alpha]\} \\ + r(-H\sin\bar{\alpha}_T + V\cos\bar{\alpha}_T) \\ \times\underline{\{1+(A_e-A_s)(H\cos\bar{\alpha}_T + V\sin\bar{\alpha}_T) - \nu A_e(1+\bar{k})^{-1}[(rH)'+r\bar{p}_H]\}} = 0,\end{aligned} \tag{A.41}$$

where $\bar{p}_H$ is given by (A.13) and V by (A.21). All these equations are complicated by the strain effects indicated by underlines. In (A.13), (A.21), and (A.40), ignoring the underlined terms is equivalent to making a slight change in the pressure. This shall be our first approximation, so that, in particular, (A.13) and (A.21) are to be replaced by

$$V \approx \tilde{V} \equiv \frac{C}{r} + \frac{pr}{2(1+\bar{k})} \tag{A.42}$$

$$\bar{p}_H \approx \tilde{p}_H \equiv p\sin\bar{\alpha}_T\,. \tag{A.43}$$

(We could weaken the above approximation by retaining *linear* strain effects and still have two equations in H and $\bar{\alpha}_T$. However, as such equations seem

[8] In Section V.S on axishells, we set $\beta = \bar{\alpha}_T - \alpha$ because we often linearized in β; in this section we do not.

nearly as complicated as their parents, there seems to be little motive for pursuing this further.)

A further approximation, consistent with the quadratic form of Ψ, is to delete all remaining nonlinear extensional and shear strain effects while retaining terms nonlinear in β. Thus, we drop the underlined term in (A.41). The resulting two differential equations together with the periodicity conditions on β, β', H, and H' plus the integral conditions (A.23) and (A.30) or (A.35) constitute the *small strain-finite rotation theory* for the nonlinear bending of shearable tubes. The classical form of the theory is obtained by setting $A_s/A_e = 0$.

Because there are good computer codes for the numerical solution of nonlinear shell equations, we need not simplify the governing equations further. Indeed, one might ask: "Why make any simplifications at all if numerical methods are to be used?" However, we can gain insight and clarity if we pare away all but the essential terms in our equations, especially if (approximate) analytic methods are to be used. This has been a traditional and fruitful approach in shell theory.

With this in mind, let us consider slender tubes such that $b/r \ll 1$, where b is a typical dimension of the cross section. In slender circular tubes, $b/a \ll 1$. [See (A.1).] Noting that differentiation with respect to σ will normally have the scale of b, we conclude that, compared to the first term in (A.41), all other bending terms are $O(b^2/r^2)$.[9] Thus, (A.41) takes the simplified form

$$D\,[r\,(\bar{\alpha}_T - \alpha)']' - r\,(H \sin\bar{\alpha}_T - \tilde{V}\cos\bar{\alpha}_T) = 0\,, \tag{A.44}$$

which is what we would have obtained exactly had we neglected M_θ and taken $M_\sigma = Dk_\sigma$ in deriving (A.41). Similar arguments applied to (A.40) yield the simplified compatibility condition

$$A_e[r\,(rH)']' + (1+\bar{k})[\cos\alpha - (1+\bar{k})\cos\bar{\alpha}_T] + A_e(r^2\tilde{p}_H)' = 0\,, \tag{A.45}$$

which is what we would have obtained exactly had we neglected e_σ (meridional inextensionality) and taken $e_\theta = A_e N_\theta$ in deriving (A.40). The preservation of the static-geometric duality is obvious. Similar approximations have been made by Reissner (1981) [especially see his equations (31) and (32)] and Axelrad (1962a, 1965). For further discussion, see Axelrad & Emmerling (1984) and Axelrad (1985b).

In obtaining (A.45) we neglected the terms $\nu\bar{\alpha}_T{}'rH$ and $\nu\bar{\alpha}_T{}'rV$. Because of the $\bar{\alpha}_T{}'$ factor, such terms (which come from the term $\nu e_\sigma{}'$) might be larger than some of the neglected terms. Had we assumed meridional inextensionality so that $e_\sigma \equiv 0$ and $e_\theta = (1-\nu^2)A_e N_\theta$, these terms would have been zero automatically. To preserve the static-geometric duality, we should also replace D by $(1-\nu^2)D$ [which is equivalent to setting $M_\theta = 0$ in (A.41)]. Some authors prefer

[9]Actually, the scale of differentiation is even smaller than b, since most of the strong bending tends to concentrate in *narrow interior zones* around $\alpha = 0, \pi$, as we saw in subsection V.S.5. This probably explains the success of this approximation in less-than-slender tubes.

this *constitutive approximation*. Its justification remains an open problem.

As for the integral conditions, the stress resultants in (A.35) can be omitted and possibly the strain effects (indicated by the underlines) as well. But, unless we revert to classical theory, we *should not* neglect the strain terms in (A.24), especially the shear-strain term which might become important if the cross section of the curved tube is not symmetrical, especially if the material is weak in shear.

Finally, we note that although

$$(1+\bar{k})u_r = r\{(1-\nu^2)A_e[(rH)' + r\tilde{p}_H] + \bar{k}\} \tag{A.46}$$

is a first integral of (A.45), it does not reduce the order of our system, but only serves as a way of computing the additional unknown u_r.

10. The collapse of tubes in bending

As discussed in the introduction to this section, one of the important applications of the nonlinear theory of straight and curved tubes is to estimate the maximum (limit point) value of the end couple, M_s, that a tube can sustain, with or without pressure. In practice, the actual value of M_s may fall below its ideal value because either the tube exceeds its elastic limit at some point (necessitating a more elaborate elastic-plastic analysis) or else the tube buckles first in a local mode. Fortunately, in most slender tubes, the actual maximum bending moment is not far below what the simplified theory predicts. The latter therefore becomes a convenient and practical upper bound. The possible coexistence of a large external pressure, which might cause collapse if acting alone, is excluded. This case is beyond the scope of this book. For further discussion, see Axelrad (1978).

Our present discussion is limited to *long tubes* such that the boundary conditions at the ends of the tube do not affect the behavior of its central portion[10]. The problem for medium and short tubes will be discussed at the end of this section.

Seide & Weingarten (1961) analyzed the bifurcation of a circular cylindrical shell under pure bending, assuming that the prebuckled state could be described by linear membrane theory. They found that the *critical maximum membrane compressive stress* is only slightly greater than that of the uniformly, axially loaded shell. Libai & Durban (1973, 1977) proved that the two coincide as $h \to 0$. The result can be expressed as

$$M_{cr}^L \approx 1.81(1-\nu^2)^{-1/2}Ebh^2 , \tag{A.47}$$

where M_{cr}^L denotes the buckling end-couple of the linear prebuckled state.

[10]As noted before, the relevant length parameter for the decay of some of the edge effects is $b^{3/2}h^{-1/2}$. For long circular cylinders, Axelrad (1965) found that the length of the tube should be at least $2.5b^{3/2}h^{-1/2}$ for the theory to hold. For more details on the effects of the length of the tube, see Fig. 5 of Axelrad & Emmerling (1983), where the tube length is $5.7b^{3/2}h^{-1/2}$

Using an initial postbuckling analysis for oval shells, Kempner & Chen (1974) showed that this buckling point is unstable.

Brazier (1927) was the first to apply a *nonlinear* analysis to compute the limit point collapse moment, M_s, for a long circular cylindrical tube. Using an energy approach, Brazier obtained

$$M_s \approx 0.987(1-\nu^2)^{-1/2}Ebh^2 , \qquad \text{(A.48)}$$

which is close to half the linear result! The phenomenon of collapse due to increasing ovalization of tube cross sections still carries the name *Brazier effect*. Later studies on the collapse of long cylindrical shells include those of Wood (1958), who introduced internal pressure effects; Reissner (1959), who obtained approximate results using the first two terms in a power series expansion in β and who also included pressure effects; and Thurston (1977), who refined earlier calculations by Reissner & Weinitschke (1963). Thurston's results were corroborated by Emmerling (1982), who used a Fourier expansion to treat the more general case of pressurized tubes of *elliptic* cross section. For the pure bending of an unpressurized circular tube, Emmerling gives

$$M_s = 0.959(1-\nu^2)^{-1/2}Ebh^2 . \qquad \text{(A.49)}$$

As this is only slightly below (A.48), Brazier's formula can be accepted.

The appearance of axial wrinkles at the intrados of strongly bent tubes has been observed in experiments. This bifurcation phenomenon occurs below the limit point and can cause the shell to collapse. Axelrad (1965) was the first to investigate the bifurcation of the tube using the *nonlinear* prebuckling state derived from a bending analysis. Because each buckling wave is confined to a small, locally shallow region of the tube, Axelrad introduced a "local buckling" approximation by taking the stresses and curvatures of the prebuckling nonlinear analysis to be *constant* over these regions. The buckling equations thereby simplify considerably. Assuming that $\overline{R}_\theta \ll \overline{R}_\sigma$ and $N_\sigma \ll N_\theta$ in the buckled region, Axelrad derived the approximate formula

$$N_{\theta_{cr}} \approx h[3(1-\nu^2)]^{-1/2}Eh(\alpha+\beta)' , \qquad \text{(A.50)}$$

where $(\alpha+\beta)'$ is, for small strains, the local meridional curvature at the intrados point closest to the $\overline{z}$-axis of the deformed cylindrical tube. Further calculation yields

$$M_{cr} \approx 0.921Ebh^2(1-\nu^2)^{-1/2} , \qquad \text{(A.51)}$$

which, as noted before, is slightly less than the corresponding value for M_s.

Fabian (1977) studied the interaction of bifurcation and limit point failures in long cylindrical tubes under bending and pressure by using a nonlinear bending theory to obtain the prebuckled state and then assuming that shallow shell theory governed any subsequent deformation. He used finite difference methods to obtain numerical results which show for large b/h that the "local buckling" provides a good approximation to the bifurcation moment. Whether the bifurcation actually brings about collapse is not always clear.

Stephens *et al* (1975), who did a numerical analysis with the STAGS program, indicated that in long shells the Brazier effect dominates the behavior, despite the appearance of wrinkles.

Up to now, our discussion has been limited to *initially straight tubes*. The collapse of long, initially curved tubes has been studied less extensively. Reissner (1959) and Axelrad (1962b) obtained approximate results. For more recent work, see Boyle (1981), Emmerling (1982), and Axelrad & Emmerling (1983); the last two include the effects of superimposed bifurcation. For pure bending, approximate formulas such as (A.50) can be used, but, in general, the complexity of the problem requires the use of computer codes that incorporate plastic effects and which can predict the subcritical response of the tube and either the collapse moment or points of bifurcation. Some of the codes use a 1-dimensional approach and are restricted to long tubes; others include edge effects. The state of the art is reviewed by Bushnell (1981).

The discussion so far has been limited to *long* tubes. In short and intermediate length tubes, the stiffening effects of the edges, which tend to decrease cross-sectional distortion, cannot be ignored. In short tubes (length/radius of order 1), the distortion can be ignored and a linear prebuckling analysis used; if the tube is circular, the critical moment is given by (A.47). In tubes of intermediate length, the situation is more complex and calls for the use of computer codes or analytic approximations. Thus, Stephens *et al* (1975) used the code STAGS to study the effects of length on the collapse and bifurcation of circular cylindrical shells while Emmerling (1982) and Axelrad & Emmerling (1983) used semi-membrane theory in their analyses of tubes of intermediate length. A discussion of the details of this work is beyond the scope of this book.

B. Variational Principles for Curved Tubes

1. General Remarks

We shall restrict our attention in this section to the development of static variational principles for shearable, curved tubes under axisymmetric surface loads and end moments in the planes of curvature of the tubes. This is one of the more important cases in which variational principles can be useful. Dynamic effects may be incorporated merely by adding the kinetic energy to the total static potential. The extension of our analysis to more general unishells should be straightforward.

To generalize the considerations of Section A, we consider tubes with both open and closed cross sections and let $0 \le \sigma \le L$ and $0 \le \theta \le \Theta$, as in Fig. 6.2. We indicate the *circumferential boundaries* of the tube, $\sigma = 0, L$, by the subscript σ and the *ends* of the tube, $\theta = 0, \Theta$, by the subscript θ. Thus, ∂U_σ and ∂N_σ are those parts of the circumferential boundaries where kinematical constraints or loads are imposed, respectively, with similar definitions of ∂U_θ and ∂N_θ. Axisymmetric stress resultants $\hat{\mathbf{N}}_\sigma$ and stress couples $-\hat{M}_\sigma \bar{\mathbf{e}}_\theta$ are assumed to act on ∂N_σ; a *net* moment $\hat{T}_z \mathbf{e}_z$ and a *net* force $\hat{P} \bar{\mathbf{e}}_\theta$ are assumed to act on ∂N_θ. The net end force $\hat{P}$ cannot be prescribed arbitrarily for it must equilibrate the

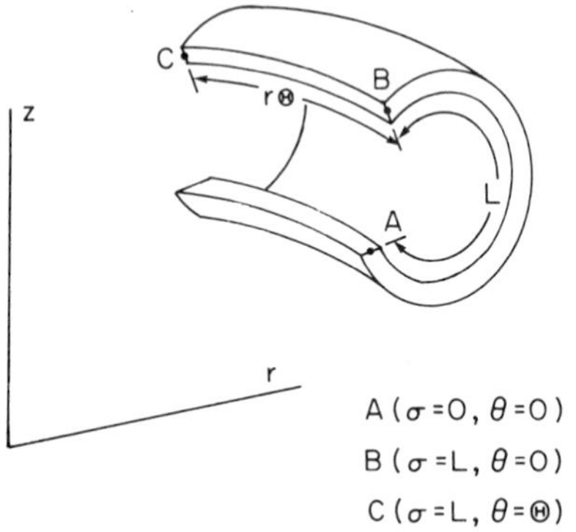

Fig. 6.2. A meridionally-incomplete tube.

surface load $\mathbf{p}$ and the boundary stress resultant, $\hat{\mathbf{N}}_\sigma$, as we found in Section A. The loads on the tube must also be compatible with the basic kinematic assumptions.

Compared with axishells, the main new feature in the development of variational principles for curved tubes is that the displacements and rotations of the tube vary with σ *and* θ, whereas the stress resultants and couples remain axisymmetric. Accordingly, we shall use 2-dimensional forms of the equilibrium and kinematic equations. After each variational principle has been derived, it will be reduced to 1-dimensional form.

2. *Principles of Virtual Work and Stationary Total Potential*

In subsection A.5, we presented an analysis and discussion of the kinematics of deformation of curved tubes. The following development is based on these results.

Recall that position of a point on the deformed reference is given by

$$\overline{\mathbf{y}} = \overline{r}\,\overline{\mathbf{e}}_r + \overline{z}\mathbf{e}_z \,, \tag{B.1}$$

where $\overline{\mathbf{e}}_r$ and $\overline{\mathbf{e}}_\theta = \mathbf{e}_z \times \overline{\mathbf{e}}_r$ are given by (A.11). The velocity of the point is

$$\dot{\overline{\mathbf{y}}} = \dot{\overline{r}}\,\overline{\mathbf{e}}_r + \dot{\overline{z}}\mathbf{e}_z + \overline{r}\theta\dot{\overline{k}}\overline{\mathbf{e}}_\theta \,, \tag{B.2}$$

where $\overline{k}$ is defined in (A.2). The spin $\boldsymbol{\omega}$ has two components: an angular velocity $-\dot{\beta}$ about $\overline{\mathbf{e}}_\theta$ and an angular velocity $\theta\dot{\overline{k}}$ about $\mathbf{e}_z$ due to the rotation of the cross section of the tube. Thus,

$$\boldsymbol{\omega} = -\dot{\beta}\overline{\mathbf{e}}_\theta + \theta\dot{\overline{k}}\mathbf{e}_z \,. \tag{B.3}$$

The form of the finite rotation vector $\boldsymbol{\beta}$ itself is not needed for now. Replacing linear and angular scalar velocities by variations, we obtain the following expressions for the variations of deformed position and rotation:

$$\delta\bar{\mathbf{y}} = \delta\bar{r}\,\bar{\mathbf{e}}_r + \delta\bar{z}\mathbf{e}_z + \bar{r}\theta\delta\bar{k}\,\bar{\mathbf{e}}_\theta \tag{B.4}$$

$$\delta\boldsymbol{\beta} = -\,\delta\beta\bar{\mathbf{e}}_\theta + \theta\delta\bar{k}\mathbf{e}_z\,. \tag{B.5}$$

Without loss of generality, we have assumed that the z-axis remains fixed and that the cross section $\theta = 0$ does not rotate during the deformation.

Turning to the equations of motion, (V.C.1) and (V.C.2), with the surface couple and the inertia terms deleted, we take the dot product of (V.C.1) with $\delta\mathbf{y}$ and (V.C.2) with $\delta\boldsymbol{\beta}$, add, and integrate by parts to obtain

$$\begin{aligned} -\int_0^L\int_0^\Theta (&\mathbf{N}_\sigma\cdot\delta\bar{\mathbf{y}},_\sigma + r^{-1}\mathbf{N}_\theta\cdot\delta\bar{\mathbf{y}},_\theta - \mathbf{p}\cdot\delta\mathbf{y} \\ &+ \mathbf{M}_\sigma\cdot\delta\boldsymbol{\beta},_\sigma + r^{-1}\mathbf{M}_\theta\cdot\delta\boldsymbol{\beta},_\theta + \mathbf{m}\cdot\mathbf{N}_\sigma\bar{\mathbf{e}}_\theta\cdot\delta\boldsymbol{\beta})rd\theta d\sigma \\ &+ \int_0^\Theta (r\mathbf{N}_\sigma\cdot\delta\bar{\mathbf{y}} + r\mathbf{M}_\sigma\cdot\delta\boldsymbol{\beta})_0^L d\theta + \int_0^L (\mathbf{N}_\theta\cdot\delta\bar{\mathbf{y}} + \mathbf{M}_\theta\cdot\delta\boldsymbol{\beta})_0^\Theta d\sigma = 0\,. \end{aligned} \tag{B.6}$$

We now insert into this equation the special forms of the stress resultants and couples given by (V.D.1) and (V.D.2), the variations of position and rotation given by (B.4) and (B.5), and the strains given by (A.4)-(A.7) and (V.L.1). After some manipulations, we obtain

$$\begin{aligned} -\int_0^L\int_0^\Theta (&N_\sigma\delta e_\sigma + N_\theta\delta e_\theta + Q\delta g + M_\sigma\delta k_\sigma + M_\theta\delta k_\theta + M_n\delta k_n - \mathbf{p}\cdot\delta\bar{\mathbf{y}})rd\theta d\sigma \\ &+ \int_0^\Theta (r\hat{\mathbf{N}}_\sigma\cdot\delta\bar{\mathbf{y}} + r\hat{M}_\sigma\delta\beta)_{\partial N_\sigma} d\theta + \hat{T}_z\theta\delta\bar{k}\,|_{\,\partial N_\theta} = 0\,, \end{aligned} \tag{B.7}$$

where the *total cross-sectional moment* T_z turning about $\mathbf{e}_z$ is defined by

$$T_z \equiv \int_0^L (\bar{r}N_\theta + \mathbf{M}_\theta\cdot\mathbf{e}_z)d\sigma\,. \tag{B.8}$$

[Also see (A.30).] In the above, ∂N_σ and ∂N_θ are the parts of the boundary that are loaded, as explained before, and $\hat{N}_\sigma$, $\hat{M}_\sigma$, and $\hat{T}_z$ are the specified loads at these boundaries; it is also assumed that $\delta\mathbf{y}$ and $\delta\boldsymbol{\beta}$ satisfy any kinematic constraints on ∂U_σ or ∂U_θ.

Equation (B.7) is a statement of the Principle of Virtual Work for a tube, valid for the restricted kinematic and static fields assumed in this section. In a meridionally complete tube, the ∂N_σ term vanishes because of periodicity.

As none of the terms in (B.7) depends on θ, we can integrate from $\theta = 0$ to $\theta = \Theta$ and then divide by Θ to obtain the 1-dimensional principle

$$\begin{aligned} -\int_0^L (&N_\sigma\delta e_\sigma + N_\theta\delta e_\theta + Q\delta g + M_\sigma\delta k_\sigma + M_\theta\delta k_\theta + M_n\delta k_n - \mathbf{p}\cdot\delta\bar{\mathbf{y}})rd\sigma \\ &+ (r\hat{\mathbf{N}}_\sigma\cdot\delta\bar{\mathbf{y}} + r\hat{M}_\sigma\delta\beta)_{\partial N_\sigma} + \hat{T}_z\delta\bar{k}\,|_{\,\partial N_\theta} = 0\,. \end{aligned} \tag{B.9}$$

We now assume that the tube is elastic. The integrand in (B.9) is then equal to the variation of the strain-energy density Φ. We also assume that the surface and boundary loads possess a potential $\mathcal{P}$ such that

$$\delta\mathcal{P} = -\int_0^L \mathbf{p}\cdot\delta\bar{\mathbf{y}} r d\sigma - [r(\hat{\mathbf{N}}_\sigma\cdot\delta\bar{\mathbf{y}} + \hat{M}_\sigma\delta\beta)]_{\partial N_\sigma} - \hat{T}_z\delta\bar{k}|_{\partial N_\theta}\,. \tag{B.10}$$

Under these conditions, (B.9) reduces to

$$\delta(\mathcal{E} + \mathcal{P}) = 0\,, \tag{B.11}$$

where $\mathcal{E} \equiv \int_0^L \Phi r d\sigma$ is the total strain energy (per unit radian) of the tube. Equation (B.11) expresses the Principle of Stationary Total Potential Energy for curved tubes. It states that among all admissible deformed positions $\bar{\mathbf{y}}$ and rotations (as represented by β and $\bar{k}$) which are sufficiently smooth on $0 < \sigma < L$ and satisfy any kinematic constraints on the boundaries, those which satisfy the equations of equilibrium (including any prescribed loads on ∂N_σ and ∂N_θ) render $\mathcal{E} + \mathcal{P}$ stationary.

Note that the Principle is the same as for axishells *except* for the additional boundary term $\hat{T}_z\delta\bar{k}$ at the ends of the tube. Note also that, even though both a force and a moment act at these ends, their effects are combined into a single work term. This is in accord with the mechanics which, as we noted earlier, implies that the end force is determined by the internal pressure.

In meridionally-closed tubes, it is convenient to apply a specified rotation $\bar{k}$ rather than a specified end moment $\hat{T}_z$, in which case the boundary terms in (B.10) vanish. These terms also vanish in incomplete tubes with specified end rotations and *free* (i.e., unloaded) boundaries, ∂N_σ.

The potential of a closed tube under uniform internal pressure (say fluid pressure) merits special attention. Here, the surface loading is given by (A.8)-(A.10). If the end rotation is specified, we obtain

$$\delta\mathcal{P} = -p(1+\bar{k})\oint \mathbf{m}\cdot(\delta\bar{r}\,\bar{\mathbf{e}}_r + \delta\bar{z}\mathbf{e}_z)\bar{r}d\sigma = p(1+\bar{k})\oint(\bar{z}'\delta\bar{r} - \bar{r}'\delta\bar{z})\bar{r}d\sigma\,. \tag{B.12}$$

Consider now the functional

$$\mathcal{F} \equiv -p(1+\bar{k})\oint \bar{r}'\bar{r}\bar{z}d\sigma\,, \tag{B.13}$$

where $\mathcal{F}/p$ is the toroidal volume (per unit radian) bounded by the deformed reference surface and its ends. Taking variations, we have

$$\delta\mathcal{F} = -p\delta\bar{k}\oint \bar{r}'\bar{r}\bar{z}d\sigma + p(1+\bar{k})\oint(\bar{z}'\delta\bar{r} - \bar{r}'\delta\bar{z})\bar{r}d\sigma\,. \tag{B.14}$$

The first term on the right vanishes because the end rotation is specified, so that $\delta F = \delta\mathcal{P}$. Thus, we can take

$$\mathcal{P} = p\oint[r'rz - (1+\bar{k})\bar{r}'\bar{r}\bar{z}]d\sigma\,. \tag{B.15}$$

(The term $r'rz$ has been added to make $\mathcal{P}$ vanish if the tube is undeformed.)

If the end rotations are not specified, we may group the first term of (B.14) with $\hat{T}_z$ of (B.10) and require that $-T\delta\bar{k}$ have a potential, where

$$T \equiv \hat{T}_z - p\oint \bar{r}'\bar{r}\bar{z}d\sigma\,. \tag{B.16}$$

This last equation is the *variational solution* to the problem of separating moment and pressure effects at the ends of a tube. (See our discussion in Section A.) In (B.16), T_z is the total moment, the integral term is the pressure effect[11], and $\hat{T}$ is the end couple which is left after pressure effects have been removed. The use of (B.15) for $\mathcal{P}$ and the replacement of T_z by $\hat{T}$ in (B.10) must go together.

3. Extended variational principles

The reader will appreciate by now that variational principles for curved tubes are quite similar to those for axishells, the main difference being in the end terms involving T_z, $\delta\bar{k}$, and in the pressure term, as we saw before. Also, if the variational equations are to be used to derive stationary conditions, then the 2-dimensional form might be more useful.

The Hu-Washizu Principle for curved tubes is

$$\delta\Pi_1[e_\sigma, e_\theta, g, k_\sigma, k_\theta, k_n, N_\sigma, N_\theta, Q, M_\sigma, M_\theta, M_n, \bar{\mathbf{y}}, \beta, \bar{k}, \tilde{\mathbf{N}}_\sigma, \tilde{M}_\sigma, \tilde{M}_z] = 0\,, \quad \text{(B.17)}$$

where

$$\begin{aligned}\Pi_1 = \mathcal{E} + \mathcal{P} - \int_0^L \{&\lambda_T N_\sigma + \lambda_\theta N_\theta + gQ + (k_\sigma - \beta')M_\sigma + (k_\theta + r^{-1}\sin\alpha)M_\theta + \\ &+ (k_n + r^{-1}\cos\alpha)M_n - [\mathbf{N}_\sigma\cdot\bar{\mathbf{y}}' + r^{-1}(1+\bar{k})(\bar{r}N_\theta + \mathbf{M}_\theta\cdot\mathbf{e}_z)]\}\, r\, d\sigma \quad \text{(B.18)} \\ &- [r\tilde{\mathbf{N}}_\sigma\cdot(\bar{\mathbf{y}} - \hat{\bar{\mathbf{y}}}) + r\tilde{M}_\sigma(\beta - \hat{\beta})]_{\partial U_\sigma} - \tilde{T}_z(\bar{k} - \hat{\bar{k}})_{\partial U_\theta}\,.\end{aligned}$$

In the above, $\lambda_T \equiv 1 + e_\sigma$, $\lambda_\theta \equiv 1 + e_\theta$, the hat ($\hat{\ }$) denotes a prescribed quantity, and the tilde ($\tilde{\ }$) denotes a Lagrange multiplier.[12] The fields in Π_1 are subject to no constraints, save for sufficient smoothness for the integral in (B.18) to exist.

To obtain stationary conditions for this principle, we write (B.18) in its 2-dimensional form and then take variations. After somewhat lengthy calculations, we obtain an expression for $\delta\Pi_1$ which is virtually identical with (V.V.19-20), *except* for the additional terms

$$[(T_z - \hat{T}_z)\delta\bar{k}]_{\partial N_\theta} + [(T_z - \tilde{T}_z)\delta\bar{k} + (\hat{\bar{k}} - \bar{k})\delta\tilde{T}_z]_{\partial U_\theta}$$

which reflect conditions at the end of the tube, and for the $(1+\bar{k})$ coefficient which modifies the various equilibrium and kinematic quantities. Setting $\delta\Pi_1 = 0$ yields necessary stationary conditions—the constitutive laws, the strain-kinematic relations, the equilibrium equations, all the boundary conditions, and the equality of the Lagrange Multipliers with corresponding forces and moments on the boundary.

---

[11]The integral itself is the moment of the area enclosed by the deformed meridian with respect to the z-axis.

[12]If $g = 0$, $\lambda_T = \lambda_\sigma$, where λ_σ is defined in (V.P.2)

Proceeding as in previous chapters, we now derive variational principles in terms of the mixed-energy density Ψ, defined in (V.O.4), and its modified form

$$\Omega(N_\sigma, N_\theta, Q, k_\sigma, k_\theta, k_n) \equiv \Phi - \lambda_T N_\sigma - \lambda_\theta N_\theta = \Psi - N_\sigma - N_\theta . \quad \text{(B.19)}$$

By *defining* the stretches and shearing strain as

$$\lambda_T \equiv -\Omega,_{N_\sigma} \ , \ \lambda_\theta \equiv -\Omega,_{N_\theta} \ , \ g = -\Omega,_Q \ , \quad \text{(B.20)}$$

we eliminate them from Π_1, so obtaining the new principle

$$\delta\Pi_2(N_\sigma, N_\theta, Q, M_\sigma, M_\theta, M_n, k_\sigma, k_\theta, k_n, \bar{\mathbf{y}}, \beta, \bar{k}, \tilde{\mathbf{N}}_\sigma, \tilde{M}_\sigma, \tilde{T}_z) = 0 \ , \quad \text{(B.21)}$$

where

$$\begin{aligned}\Pi_2 = \mathcal{P} + \int_0^L \{&\Omega - (k_\sigma - \beta')M_\sigma - (k_\theta + r^{-1}\sin\alpha)M_\theta - (k_n + r^{-1}\cos\alpha)M_n \\ &+ [\mathbf{N}_\sigma \cdot \bar{\mathbf{y}}' + r^{-1}(1+\bar{k})(\bar{r}N_\theta + \mathbf{M}_\theta \cdot \mathbf{e}_z)]\} r d\sigma \qquad \text{(B.22)} \\ &- [r\tilde{\mathbf{N}}_\sigma \cdot (\bar{\mathbf{y}} - \hat{\bar{\mathbf{y}}}) + r\tilde{M}_\sigma(\beta - \hat{\beta})]_{\partial U_\sigma} - [\tilde{T}_z(\bar{k} - \hat{\bar{k}})]_{\partial U_\theta} \ .\end{aligned}$$

In the next variational principle, we further eliminate the stress couples and the bending strains by inserting the *definitions*

$$M_\sigma \equiv \Omega,_{k_\sigma} \ , \ M_\theta \equiv \Omega,_{k_\theta} \ , \ M_n \equiv \Omega,_{k_n} \ , \quad \text{(B.23)}$$

and (A.5)-(A.7) into (B.22) to obtain the new principle

$$\delta\Pi_3(N_\sigma, N_\theta, Q, \bar{\mathbf{y}}, \beta, \bar{k}, \tilde{\mathbf{N}}_\sigma, \tilde{T}_\sigma, \tilde{M}_z) = 0 \ , \quad \text{(B.24)}$$

where

$$\begin{aligned}\Pi_3 \equiv \mathcal{P} + \int_0^L &[\Omega + \mathbf{N}_\sigma \cdot \bar{\mathbf{y}}' + r^{-1}(1+\bar{k})\bar{r}N_\theta] r d\sigma \\ &- [r\tilde{\mathbf{N}}_\sigma \cdot (\bar{\mathbf{y}} - \hat{\bar{\mathbf{y}}}) + r\tilde{M}_\sigma(\beta - \hat{\beta})]_{\partial U_\sigma} - [\tilde{T}_z(\bar{k} - \hat{\bar{k}})]_{\partial U_\theta} \ . \qquad \text{(B.25)}\end{aligned}$$

The fields in Π_3 are subject to no constraints, save the requisite smoothness. The stationary conditions are the stretch-strain-kinematic relations [with $\lambda_T, \lambda_\theta$, and g defined by (B.20)], the equilibrium equations, and the same boundary conditions that go with Π_2. For details, see (V.V.25) where $\delta\Pi_3$ for an axishell is calculated. The extension to curved tubes is obvious.

Next, we eliminate the deformed position $\bar{\mathbf{y}}$ as a field variable to arrive at a variational principle in terms of the stress resultants N_σ, N_θ, and Q, and the rotations β and $\bar{k}$ only. As we shall see, this will put some restrictions on the stress resultant arguments in the resulting functionals. For this purpose, we assume that the surface load $\mathbf{p}$ is a given function of $\bar{\mathbf{y}}$ which can be inverted to yield $\bar{\mathbf{y}}(\mathbf{p})$. We assume that an analogous situation holds regarding any prescribed boundary load, i.e., we assume that the deformed position on the loaded parts of the circumferential boundaries of the tube can be expressed as

$\bar{\mathbf{y}}(\hat{\mathbf{N}}_\sigma)_{\partial N_\sigma}$. Under these conditions, a *mixed* load potential can be constructed of the form

$$\mathcal{P}[\mathbf{p}, \hat{\mathbf{N}}_\sigma, \beta, \bar{k}] \equiv \mathcal{P} + \int_0^L \mathbf{p}\cdot\bar{\mathbf{y}}(\mathbf{p}) r d\sigma + r\hat{\mathbf{N}}_\sigma\cdot\bar{\mathbf{y}}(\hat{\mathbf{N}}_\sigma)|_{\partial N_\sigma}, \tag{B.26}$$

such that

$$\delta\mathcal{P} = \int_0^L \bar{\mathbf{y}}(\mathbf{p})\cdot\delta\mathbf{p} r d\sigma + [r\bar{\mathbf{y}}(\hat{\mathbf{N}}_\sigma)\cdot\delta\hat{\mathbf{N}}_\sigma - r\hat{M}_\sigma\delta\beta]_{\partial N_\sigma} - (\hat{T}_z\delta\bar{k})_{\partial N_\theta}. \tag{B.27}$$

We now integrate Π_3 by parts, add and subtract the term $\int_0^L \mathbf{p}\cdot\bar{\mathbf{y}} r d\sigma$, and identify $\mathbf{N}_\sigma$ on ∂U_σ with $\tilde{\mathbf{N}}_\sigma$. Furthermore, we restrict $\mathbf{N}_\sigma$, $\mathbf{N}_\theta$, and $\mathbf{p}$ to satisfy, *a priori*, the force equilibrium equation, (V.C.1). With these restrictions and operations, we obtain the variational principle

$$\delta\Pi_4(N_\sigma, N_\theta, Q, \beta, \bar{k}, \tilde{M}_\sigma, \tilde{T}_z, \hat{\mathbf{N}}_\sigma, \mathbf{p}) = 0, \tag{B.28}$$

where

$$\Pi_4 \equiv \int_0^L \Omega r d\sigma + \mathcal{P} + [r\mathbf{N}_\sigma\cdot\hat{\bar{\mathbf{y}}} - r\tilde{M}_\sigma(\beta - \hat{\beta})]_{\partial U_\sigma} - [\tilde{T}_z(\bar{k} - \hat{\bar{k}})]_{\partial U_\theta}. \tag{B.29}$$

The arguments of Π_4 are constrained to satisfy the boundary conditions $\mathbf{N}_\sigma = \hat{\mathbf{N}}_\sigma$ on ∂N_σ as well as (V.C.1).

To derive our final principle, we impose the additional constraint that β and $\bar{k}$ satisfy the boundary conditions $\beta = \hat{\beta}$ on ∂U_σ and $\bar{k} = \hat{\bar{k}}$ on ∂U_θ.[13] The terms in (B.29) involving $\tilde{M}_\sigma$ and $\tilde{T}_z$ then drop out of Π_4 and we obtain

$$\delta\Pi_5(N_\sigma, N_\theta, Q, \beta, \bar{k}, \hat{\mathbf{N}}_\sigma, \mathbf{p}) = 0, \tag{B.30}$$

where

$$\Pi_5 \equiv \int_0^L \Omega r d\sigma + \mathcal{P} + (r\mathbf{N}_\sigma\cdot\hat{\bar{\mathbf{y}}})_{\partial U_\sigma}. \tag{B.31}$$

The constraints on the fields to be varied here are the equilibrium equations and boundary conditions on the forces, as with Π_4, and any boundary conditions on the rotations. The stationary conditions are the force-kinematic relations, the moment equilibrium equations, the boundary conditions on M_σ and T_z, and any kinematic boundary conditions on $\bar{\mathbf{y}}$. We observe that Π_5 has exactly the same form as in previous chapters.

Again, we emphasize that in a meridionally complete (i.e., closed) tube, all terms evaluated on ∂U_σ or ∂N_σ drop out of all of the principles, to be replaced by periodicity conditions on the argument functions.

Dead loading, i.e., surface and boundary loading that is independent of $\bar{\mathbf{y}}$ and β, is worth mentioning as it has many practical applications and sometimes can be used to approximate more complex loading. In this case, both $\mathbf{p}$ and $\hat{\mathbf{N}}_\sigma$

---

[13] These conditions may be vacuous.

cancel out of $\mathcal{P}$ and $\delta\mathcal{P}$ and we are left with

$$\delta\overline{\mathcal{P}}_{DL} = -(r\hat{M}_{\sigma}\delta\beta)_{\partial N_{\sigma}} - (\hat{T}_{z}\delta\bar{k})_{\partial N_{\theta}} \,. \qquad \text{(B.32)}$$

For further remarks and applications, see Section V.V.

C. Helicoidal Shells

In studying the in-plane bending of curved tubes, we encountered displacements that were functions of two variables (σ and θ); the associated strains, however, were functions of but one (σ). This suggests the question, "What is the most general class of materially isotropic shells that, under suitable loads and boundary conditions, admits physical strain fields that depend on one spatial variable only (if such solutions exist)?" As mentioned in the introduction to this chapter, Simmonds' work (1979) provided the answer: "General helicoidal shells." Later (1984), he reduced the field equations of such shells to a system of ordinary differential equations.[14] Fig. 6.3 shows a typical case in which an elastically isotropic helical tube, with a thickness variation that is identical but arbitrary in every section through its longitudinal axis, is under a constant internal pressure p, an axial force P, and an axial torque T. The tube may be considered to have an infinite number of turns or else, as shown schematically, to be sealed by rigid end plates that exert forces on the ends of the tube that react the internal pressure on the plates. (The latter reactions are not shown in Fig. 6.3.)

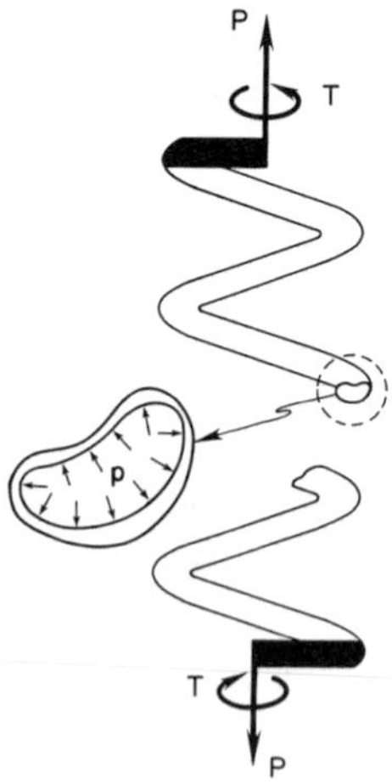

Fig. 6.3. A helicoidal shell with a 1-dimensional strain field.

---

[14]In what follows, we correct several typographical errors in Simmonds (1984), usually without mentioning that we have done so.

In the 3 subsections to follow, we first discuss the kinematics of deformation of an arbitrary shell. Next, we summarize Simmonds' (1979) results. Finally, we state a few elementary facts about the geometry of a helicoid.

1. Deformation of an arbitrary shell

Let $\mathbf{y}(\sigma, \tau)$ denote the parametric equation of the reference surface, $\mathcal{S}$, of an arbitrary shell. The triad $\{\mathbf{a}_\sigma, \mathbf{a}_\tau, \mathbf{n}\}$, where $\mathbf{a}_\sigma \equiv \partial \mathbf{y}/\partial\sigma \equiv \mathbf{y},_\sigma$, $\mathbf{a}_\tau \equiv \partial \mathbf{y}/\partial\tau \equiv \mathbf{y},_\tau$, and $\mathbf{n} \equiv \mathbf{a}_\sigma \times \mathbf{a}_\tau / |\mathbf{a}_\sigma \times \mathbf{a}_\tau|$, is a basis in 3-dimensional Euclidean space at every smooth point of $\mathcal{S}$ where $\mathbf{a}_\sigma$ and $\mathbf{a}_\tau$ are not parallel. A sufficiently smooth deformation of the shell carries this basis into the deformed basis $\{\bar{\mathbf{a}}_\sigma, \bar{\mathbf{a}}_\tau, \bar{\mathbf{n}}\}$, where $\bar{\mathbf{a}}_\sigma \equiv \bar{\mathbf{y}},_\sigma, \bar{\mathbf{a}}_\tau \equiv \bar{\mathbf{y}},_\tau$, and $\bar{\mathbf{n}} \equiv \bar{\mathbf{a}}_\sigma \times \bar{\mathbf{a}}_\tau / |\bar{\mathbf{a}}_\sigma \times \bar{\mathbf{a}}_\tau|$. This deformation may be decomposed into a *rigid body rotation*,

$$\{\mathbf{A}_\sigma, \mathbf{A}_\tau, \mathbf{N}\} = \mathbf{Q}\cdot\{\mathbf{a}_\sigma, \mathbf{a}_\tau, \mathbf{n}\} , \tag{C.1}$$

followed by a *distortion*,

$$\{\bar{\mathbf{a}}_\sigma, \bar{\mathbf{a}}_\tau\} = \{\mathbf{A}_\sigma, \mathbf{A}_\tau\} + \{\mathbf{E}_\sigma, \mathbf{E}_\tau\} . \tag{C.2}$$

Here, $\mathbf{Q}$ is a *rotator* (or proper orthogonal tensor) and $\{\mathbf{E}_\sigma, \mathbf{E}_\tau\}$ is a set of *extensional strain vectors*. The rotator has the representation (e.g., Beatty, 1977)

$$\mathbf{Q} = (2\mu^2 - 1)\mathbf{1} + 2\boldsymbol{\beta}\boldsymbol{\beta} + 2\mu\boldsymbol{\beta}\times , \tag{C.3}$$

where

$$\mu \equiv \cos(\beta/2) \ \text{ and } \ \boldsymbol{\beta} \equiv \sin(\beta/2)\,\mathbf{e} \tag{C.4}$$

are *Euler parameters*. Here, β is an angle of rotation, $\mathbf{e}$ is a unit vector along the axis of rotation, and $\boldsymbol{\beta}\,\boldsymbol{\beta}$ denotes the direct product of $\boldsymbol{\beta}$ with itself. The Euler parameters are not independent because $\mu^2 + \boldsymbol{\beta}\cdot\boldsymbol{\beta} = 1$. The angles that determine the Euler parameters are the primary kinematic unknowns in what follows. (Alternatively, we could take $\boldsymbol{\beta}$ and μ as the kinematic unknowns, thereby introducing algebraic functions in place of trigonometric ones.)

2. Dependence of the metric and curvature components on the same, single surface coordinate implies a general helicoid

We now assume that, in the coordinate system (σ, τ), the coefficients of the first fundamental form of $\bar{\mathcal{S}}$, $\bar{\mathbf{a}}_\sigma\cdot\bar{\mathbf{a}}_\sigma$, $\bar{\mathbf{a}}_\sigma\cdot\bar{\mathbf{a}}_\tau$, and $\bar{\mathbf{a}}_\tau\cdot\bar{\mathbf{a}}_\tau$ (i.e., the covariant components of the metric tensor), are functions of σ only. Likewise, with $\bar{\mathbf{n}} \equiv \bar{\mathbf{a}}_\sigma \times \bar{\mathbf{a}}_\tau / |\bar{\mathbf{a}}_\sigma \times \bar{\mathbf{a}}_\tau|$, we assume that the coefficients of the second fundamental form of $\bar{\mathcal{S}}$, $\bar{\mathbf{n}}\cdot\bar{\mathbf{a}}_\sigma,_\sigma$, $\bar{\mathbf{n}}\cdot\bar{\mathbf{a}}_\sigma,_\tau$, and $\bar{\mathbf{n}}\cdot\bar{\mathbf{a}}_\tau,_\tau$ (i.e., the covariant components of the curvature tensor), are functions of σ only. If we then apply the well-known Frenet formulas for a curve on a surface (Struik, 1961) to the τ-coordinate curve on $\bar{\mathcal{S}}$ (along which the differential element of arc length is $\sqrt{\bar{\mathbf{a}}_\tau\cdot\bar{\mathbf{a}}_\tau}\, d\tau$), we obtain a set of 3 coupled *vector* differential equations that, *formally*, resemble constant coefficient *ordinary* differential equations in τ, inasmuch as σ enters these equations only as a parameter. Using this observation, Simmonds (1979) was able to integrate these equations. Under a change of variable, he showed (in our notation) that the vector-parametric form of $\mathfrak{Z}$ is that of a general helicoid:

$$\bar{S}:\ \bar{\mathbf{y}}(s,\theta)=\bar{r}(s)\mathbf{e}_r(\theta+\bar{k}\theta-\kappa(s))+[\bar{z}(s)+\bar{b}\theta]\mathbf{e}_z\,. \tag{C.5}$$

Here, s and θ are surface coordinates; $\bar{r}$, κ, and $\bar{z}$ are arbitrary, sufficiently smooth functions of s; $\bar{k}$ and $\bar{b}$ are constants; and $\mathbf{e}_r(\theta)=\mathbf{e}_x\cos\theta+\mathbf{e}_y\sin\theta$. This family of helicoids includes the undeformed reference surface $\mathcal{S}$ which, without loss of generality, we distinguish by taking $\bar{r}=r$, $\bar{z}=z$, $\bar{b}=b$, and $\kappa=\bar{k}=0$. That is,

$$\mathcal{S}:\ \mathbf{y}(s,\theta)=r(s)\mathbf{e}_r(\theta)+[z(s)+b\theta]\mathbf{e}_z\,. \tag{C.6}$$

For convenience, we shall henceforth take s to measure arc length along a meridian of $\mathcal{S}$, i.e., arc length along a section cut from $\mathcal{S}$ by a plane through the z-axis.

3. The geometry of a general helicoid

On $\mathcal{S}$, the (covariant) base vectors associated with the coordinates s and θ are denoted and defined by

$$\mathbf{a}_s\equiv\mathbf{y}'=\cos\alpha\,\mathbf{e}_r+\sin\alpha\,\mathbf{e}_z\ ,\quad \mathbf{a}_\theta\equiv\dot{\mathbf{y}}=r\mathbf{e}_\theta+b\mathbf{e}_z\,, \tag{C.7}$$

where primes and dots henceforth denote differentiation with respect to s and θ, respectively, and

$$\cos\alpha=r'(s)\ ,\quad \sin\alpha=z'(s)\,. \tag{C.8}$$

With

$$a=|\mathbf{a}_s\times\mathbf{a}_\theta|^2=r^2+b^2\cos^2\alpha\,, \tag{C.9}$$

the expression for the unit normal to $\mathcal{S}$ reads

$$\mathbf{n}=a^{-1/2}\mathbf{a}_s\times\mathbf{a}_\theta=a^{-1/2}(-r\sin\alpha\,\mathbf{e}_r-b\cos\alpha\,\mathbf{e}_\theta+r\cos\alpha\,\mathbf{e}_z)\,. \tag{C.10}$$

The triad $\{\mathbf{a}_s,\mathbf{a}_\theta,\mathbf{n}\}\equiv\{\mathbf{a}_i\}$, $i=1,2,3$, forms a *nonorthonormal* frame at each point of $\mathcal{S}$.

D. Force and Moment Equilibrium

We consider a shell-like, helicoidal body whose reference shape has the vector-parametric form $\mathbf{x}=\hat{\mathbf{x}}(\xi,\eta,\zeta)$, where (ξ,η) belongs to some rectangular domain and $-H\le\zeta\le H$. By generalizing the approaches of Chapters IV and V, we may, in some rational way, associate a reference surface $\mathcal{S}$ with this body. No matter how $\mathcal{S}$ is chosen—as the geometric midsurface, as a centroidal surface, or via a dynamic consistency condition—we demand that $\mathcal{S}$ be a general helicoid. In analogy to what we did in Chapters IV and V, we assume that there is a change of variables such that the reference shape of the body has a parametric equation of the form $\mathbf{x}=\tilde{\mathbf{x}}(s,\theta,\tilde{\zeta})$, where $0\le s\le L$, $-\Theta\le\theta\le\Theta$, $-H\le\tilde{\zeta}\le H$, and, in (C.6), $\mathbf{y}=\tilde{\mathbf{x}}(s,\theta,0)$.

Let s and θ represent a fixed but arbitrary point on S and consider the differential strip swept out by $\mathbf{x}_{,\theta}d\theta$ as $\tilde{\zeta}$ goes from $-H$ to H. We denote by $\sqrt{a}\mathbf{N}^s d\theta$ the force exerted on the particles composing the deformed image of this strip by the particles on its outward side. See Fig. 6.4 (where we have used a common convention and denoted vectors that represent couples by double-headed arrows). Likewise, we denote by $\sqrt{a}\mathbf{N}^\theta ds$ the force exerted from the outside on the the particles that compose the deformed image of the differential strip swept out by $\mathbf{x}_{,s}ds$ as $\tilde{\zeta}$ goes from $-H$ to H. Finally, let $\mathbf{p}\sqrt{a}\,dsd\theta$ denote the external force acting on the particles composing the deformed image of the differential element of volume dV swept out by the (oriented) differential element of surface area $\mathbf{x}_{,s}\times\mathbf{x}_{,\theta}dsd\theta$ as, again, $\tilde{\zeta}$ goes from $-H$ to H. Then, static force equilibrium of the deformed image of the skewed panel given by $0\le s_1 < s < s_2 \le L, -\Theta \le \theta_1 < \theta < \theta_2 \le \Theta, -H \le \tilde{\zeta} \le H$ requires that

$$\int_{\theta_1}^{\theta_2} \sqrt{a}\mathbf{N}^s \Big|_{s_1}^{s_2} d\theta + \int_{s_1}^{s_2} \sqrt{a}\mathbf{N}^\theta \Big|_{\theta_1}^{\theta_2} ds + \int_{s_1}^{s_2}\int_{\theta_1}^{\theta_2} \mathbf{p}\sqrt{a}dsd\theta = \mathbf{0}\,. \quad \text{(D.1)}$$

To derive overall moment equilibrium, let $\sqrt{a}\mathbf{M}^s d\theta$ and $\sqrt{a}\mathbf{M}^\theta ds$ denote, respectively, the moments (at the particle with with position $\bar{\mathbf{y}}$) exerted from the

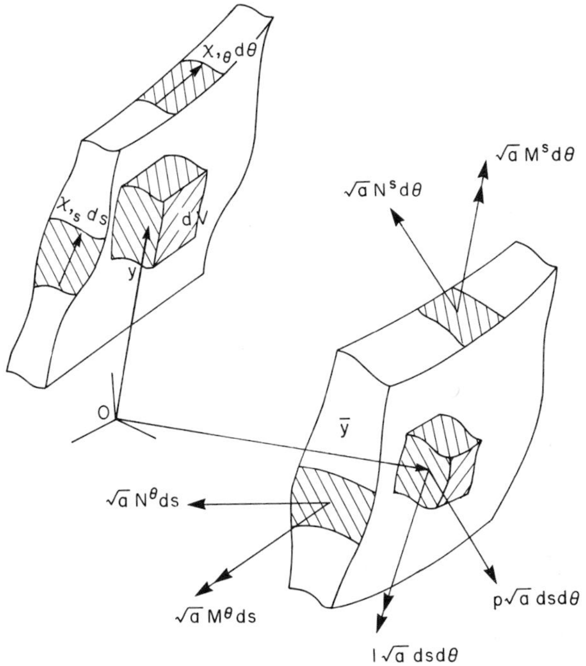

Fig. 6.4. Geometry and static conventions for a helicoidal shell.

outside on the particles that compose the deformed images of the differential strips swept out by $\mathbf{x},_\theta d\theta$ and $\mathbf{x},_s ds$, as ζ goes from $-H$ to H. Furthermore, let $\mathbf{l}\sqrt{a}\,dsd\theta$ denote the external moment (at the particle with position $\bar{\mathbf{y}}$) exerted on the deformed image of dV. Then, taking moments about the origin, we have

$$\int_{\theta_1}^{\theta_2} \sqrt{a}(\bar{\mathbf{y}}\times\mathbf{N}^s+\mathbf{M}^s)\Big|_{s_1}^{s_2} d\theta + \int_{s_1}^{s_2} \sqrt{a}(\bar{\mathbf{y}}\times\mathbf{N}^\theta+\mathbf{M}^\theta)\Big|_{\theta_1}^{\theta_2} ds + \int_{s_1}^{s_2}\int_{\theta_1}^{\theta_2}(\bar{\mathbf{y}}\times\mathbf{p}+\mathbf{l})\sqrt{a}\,dsd\theta = \mathbf{0}\,. \tag{D.2}$$

If the stress resultants $\mathbf{N}^s$ and $\mathbf{N}^\theta$ are sufficiently smooth for the Divergence Theorem to hold, then (D.1) may be reduced to the form $\int\int(\cdots)\sqrt{a}dsd\theta=\mathbf{0}$, and, if the integrand is continuous, it follows from the arbitrariness of s_1, s_2, θ_1, and θ_2 that the following differential equation of force equilibrium holds at each point of $\mathcal{S}$:

$$(\sqrt{a}\mathbf{N}^s)' + (\sqrt{a}\mathbf{N}^\theta)^\bullet + \sqrt{a}\,\mathbf{p} = \mathbf{0}\,. \tag{D.3}$$

Here, as noted earlier, a prime and a dot denote, respectively, ordinary or partial differentiation with respect to s and θ.

By similar arguments, we obtain from (D.2) the differential equation of moment equilibrium

$$[\sqrt{a}(\bar{\mathbf{y}}\times\mathbf{N}^s+\mathbf{M}^s)]' + [\sqrt{a}(\bar{\mathbf{y}}\times\mathbf{N}^\theta+\mathbf{M}^\theta)]^\bullet + \sqrt{a}(\bar{\mathbf{y}}\times\mathbf{p}+\mathbf{l}) = \mathbf{0}\,, \tag{D.4}$$

assuming, of course, that the various fields in (D.2) are sufficiently smooth. With the aid of (D.3), this differential equation may be reduced to

$$(\sqrt{a}\mathbf{M}^s)' + (\sqrt{a}\mathbf{M}^\theta)^\bullet + \sqrt{a}(\bar{\mathbf{a}}_s\times\mathbf{N}^s+\bar{\mathbf{a}}_\theta\times\mathbf{N}^\theta+\mathbf{l}) = \mathbf{0}\,, \tag{D.5}$$

where

$$\bar{\mathbf{a}}_s \equiv \bar{\mathbf{y}}' \text{ and } \bar{\mathbf{a}}_\theta \equiv \bar{\mathbf{y}}^\bullet\,. \tag{D.6}$$

The differential equations of equilibrium have first integrals. Thus, if we take the dot product of (D.3) with $\mathbf{e}_z$ and note that, by hypothesis, the components of the stress resultant vectors are functions of s only, we have $(\sqrt{a}V)' + \sqrt{a}p_V = 0$, where

$$V \equiv \mathbf{N}^s\cdot\mathbf{e}_z\ ,\ p_V = \mathbf{p}\cdot\mathbf{e}_z = p_V(s)\,. \tag{D.7}$$

Integration yields

$$\sqrt{a}\,V + \int_0^s p_V\sqrt{a}\,d\bar{s} = \sqrt{a_0}\,V_0 \equiv (1+\bar{k})C\,, \tag{D.8}$$

where $2\pi C$ is is the axial force per turn of the *deformed* shell along the deformed coordinate curve s=constant. [For an axishell, $b=0$, $\sqrt{a}=r$, $s=\sigma$, and (D.8) reduces to (V.Q.5).] Likewise, if we dot (D.4) with $\mathbf{e}_z$ and assume that $(\mathbf{M}^s, \mathbf{M}^\theta)\cdot\mathbf{e}_z$ as well as $(\mathbf{N}^s, \mathbf{N}^\theta)\cdot(\mathbf{e}_z\times\bar{\mathbf{y}})$—the latter, by (C.5), being equal to $(\mathbf{N}^s, \mathbf{N}^\theta)\cdot\mathbf{e}_\theta(\theta+\bar{k}\theta-\kappa(s))$—are functions of s only, then the resulting differential equation may be integrated from 0 to s to yield

$$[\sqrt{a}(\bar{\mathbf{y}}\times\mathbf{N}^s+\mathbf{M}^s)+\int_0^s\sqrt{a}(\bar{\mathbf{y}}\times\mathbf{p}+\mathbf{l})d\bar{s}]\cdot\mathbf{e}_z=\sqrt{a}(\bar{\mathbf{y}}\times\mathbf{N}^s+\mathbf{M}^s)|_0 \equiv(1+\bar{k})T\,, \tag{D.9}$$

where T is the net axial moment per turn of the *deformed* shell acting over the deformed coordinate curve $s=$constant. This equation has no counterpart in the theory of axishells because there we did not consider pure torsion.

Note that we may satisfy (D.3) identically in terms of a stress function $\mathbf{F}$ by setting

$$\sqrt{a}\mathbf{N}^s=\mathbf{F}^{\bullet}-\int\sqrt{a}\mathbf{p}ds\ ,\ \ \sqrt{a}\mathbf{N}^{\theta}=-\mathbf{F}'\,. \tag{D.10}$$

In view of (D.8) and the fact that the components of $\mathbf{N}^s$ and $\mathbf{N}^{\theta}$ must depend on s only, we can conclude that the vertical component of the stress function must be of the form

$$\mathbf{F}\cdot\mathbf{e}_z=F_z(s)+\sqrt{a_0}\,V_0\theta\,. \tag{D.11}$$

We can also satisfy (D.5) identically in terms of $\mathbf{F}$ and another stress function $\mathbf{G}$. However, such a representation will not be used in what follows.

E. Virtual Work and Strains

The appropriate strain measures conjugate to $\mathbf{N}^s, \mathbf{N}^{\theta}, \mathbf{M}^s$, and $\mathbf{M}^{\theta}$ follow from the Principle of Virtual Work. To derive this principle for helicoidal shells, we first take the dot product of (D.3) with $\delta\bar{\mathbf{y}}$, the variation of the deformed position, and integrate the resulting scalar equation over an arbitrary panel $0\le s_1<s<s_2\le L$, $-\Theta\le\theta_1<\theta<\theta_2\le\Theta$. Applying the Divergence Theorem to take derivatives from $\mathbf{N}^s$ and $\mathbf{N}^{\theta}$ onto $\bar{\mathbf{y}}$, and noting (D.5), we have

$$\int_{\theta_1}^{\theta_2}\sqrt{a}\mathbf{N}^s\cdot\delta\bar{\mathbf{y}}\,\Big|_{s_1}^{s_2}d\theta+\int_{s_1}^{s_2}\sqrt{a}\mathbf{N}^{\theta}\cdot\delta\bar{\mathbf{y}}\,\Big|_{\theta_1}^{\theta_2}ds+\int_{s_1}^{s_2}\int_{\theta_1}^{\theta_2}\sqrt{a}\mathbf{p}\cdot\delta\bar{\mathbf{y}}dsd\theta =\int_{s_1}^{s_2}\int_{\theta_1}^{\theta_2}\sqrt{a}(\mathbf{N}^s\cdot\delta\bar{\mathbf{a}}_s+\mathbf{N}^{\theta}\cdot\delta\bar{\mathbf{a}}_{\theta})dsd\theta\,. \tag{E.1}$$

We next introduce extensional strain vectors in the coordinate system (s,θ) by setting

$$\{\bar{\mathbf{a}}_s,\bar{\mathbf{a}}_{\theta}\}=\{\mathbf{A}_s,\mathbf{A}_{\theta}\}+\{\mathbf{E}_s,\mathbf{E}_{\theta}\}\,, \tag{E.2}$$

where

$$\{\mathbf{A}_s,\mathbf{A}_{\theta},\mathbf{N}\}=\mathbf{Q}\cdot\{\mathbf{a}_s,\mathbf{a}_{\theta},\mathbf{n}\} \tag{E.3}$$

and $\mathbf{Q}$ is the rotator given in (C.3) in terms of the unknown rotation vector $\boldsymbol{\beta}$. From (E.3),

$$\delta\{\mathbf{A}_s, \mathbf{A}_\theta, \mathbf{N}\} = \delta\mathbf{Q}\cdot\{\mathbf{a}_s, \mathbf{a}_\theta, \mathbf{n}\} = \delta\mathbf{Q}\cdot\mathbf{Q}^T\cdot\{\mathbf{A}_s, \mathbf{A}_\theta, \mathbf{N}\}\,. \tag{E.4}$$

But $\mathbf{Q}\cdot\mathbf{Q}^T = \mathbf{1}$, so that $\delta\mathbf{Q}\cdot\mathbf{Q}^T + \mathbf{Q}\cdot\delta\mathbf{Q}^T = \mathbf{0}$, i.e., $\delta\mathbf{Q}\cdot\mathbf{Q}^T$ is skew. Hence, there exists a vector $\delta\boldsymbol{\omega}$ such that $\delta\mathbf{Q}\cdot\mathbf{Q}^T\cdot\mathbf{v} = \delta\boldsymbol{\omega}\times\mathbf{v}$, for any vector $\mathbf{v}$. Furthermore, if $\mathbf{v}$ is represented in the basis $\{\mathbf{A}_s, \mathbf{A}_\theta, \mathbf{N}\}$, then

$$\delta\mathbf{v} = \delta^*\mathbf{v} + \boldsymbol{\omega}\times\mathbf{v}\,, \tag{E.5}$$

where the asterisk (*) indicates that we are to take the variation holding $\{\mathbf{A}_s, \mathbf{A}_\theta, \mathbf{N}\}$ fixed. Thus, we have

$$\delta\{\bar{\mathbf{a}}_s, \bar{\mathbf{a}}_\theta\} = \delta^*\{\mathbf{E}_s, \mathbf{E}_\theta\} + \delta\boldsymbol{\omega}\times\{\bar{\mathbf{a}}_s, \bar{\mathbf{a}}_\theta\}\,, \tag{E.6}$$

Inserting (E.6) into the right side of (E.1) and making use of moment equilibrium, (D.5), we may, after an integration by parts, rewrite (D.2) in the form

$$\mathrm{EVW} = \mathrm{IVW}\,, \tag{E.7}$$

where

$$\mathrm{EVW} \equiv \int_{\theta_1}^{\theta_2}\sqrt{a}(\mathbf{N}^s\cdot\delta\bar{\mathbf{y}} + \mathbf{M}^s\cdot\delta\boldsymbol{\omega})\Big|_{s_1}^{s_2}d\theta + \int_{s_1}^{s_2}\sqrt{a}(\mathbf{N}^\theta\cdot\delta\bar{\mathbf{y}} + \mathbf{M}^\theta\cdot\delta\boldsymbol{\omega})\Big|_{\theta_1}^{\theta_2}ds$$
$$+\int_{s_1}^{s_2}\int_{\theta_1}^{\theta_2}\sqrt{a}(\mathbf{p}\cdot\delta\bar{\mathbf{y}} + \mathbf{l}\cdot\delta\boldsymbol{\omega})\,ds\,d\theta \tag{E.8}$$

is the *External Virtual Work* and

$$\mathrm{IVW} \equiv \int_{s_1}^{s_2}\int_{\theta_1}^{\theta_2}\sqrt{a}(\mathbf{N}^s\cdot\delta^*\mathbf{E}_s + \mathbf{N}^\theta\cdot\delta^*\mathbf{E}_\theta + \mathbf{M}^s\cdot\delta\boldsymbol{\omega}' + \mathbf{M}^\theta\cdot\delta\dot{\boldsymbol{\omega}})\,ds\,d\theta \tag{E.9}$$

is the *Internal Virtual Work*. [As we noted in the second footnote in Chapter I, the work of Antman & Osborn (1979) and Carey & Dinh (1986) leaves little doubt that (E.7) is implied by the integral equations of force and moment equilibrium, (D.1) and (D.2); i.e., the fields in (E.8) and (E.9) need only be smooth enough for the integrals in (D.1) and (D.2) to make sense.]

To introduce bending strain vectors, we note that differentiation of the identity $\mathbf{Q}\cdot\mathbf{Q}^T = \mathbf{1}$ with respect to s or θ shows that $\mathbf{Q}'\cdot\mathbf{Q}^T$ and $\dot{\mathbf{Q}}\cdot\mathbf{Q}^T$ are skew. Thus, there exist vectors $\mathbf{K}_s$ and $\mathbf{K}_\theta$ such that

$$\mathbf{Q}'\cdot\mathbf{Q}^T\cdot\mathbf{v} = \mathbf{K}_s\times\mathbf{v}\ ,\ \ \dot{\mathbf{Q}}\cdot\mathbf{Q}^T\cdot\mathbf{v} = \mathbf{K}_\theta\times\mathbf{v}\ ,\ \ \forall\,\mathbf{v}\,. \tag{E.10}$$

To express the derivatives of $\boldsymbol{\omega}$ in terms of $\mathbf{K}_s$ and $\mathbf{K}_\theta$, note that if $\mathbf{c}$ is an arbitrary *constant* vector and if we set $\mathbf{u} \equiv \mathbf{Q}\cdot\mathbf{c}$, then $\delta\mathbf{u} = \delta\boldsymbol{\omega}\times\mathbf{u}$, $\mathbf{u}' = \mathbf{K}_s\times\mathbf{u}$, and $\dot{\mathbf{u}} = \mathbf{K}_\theta\times\mathbf{u}$. Now consider the relation $(\delta\mathbf{u})' = \delta\mathbf{u}'$. If both sides are expanded, we have $(\delta\boldsymbol{\omega}\times\mathbf{u})' = \delta(\mathbf{K}_s\times\mathbf{u})$ or, expanding still further,

$$\delta\boldsymbol{\omega}'\times\mathbf{u} + \delta\boldsymbol{\omega}\times(\mathbf{K}_s\times\mathbf{u}) = \delta\mathbf{K}_s\times\mathbf{u} + \mathbf{K}_s\times(\delta\boldsymbol{\omega}\times\mathbf{u})\,. \tag{E.11}$$

But if $\mathbf{a}, \mathbf{b}$, and $\mathbf{c}$ are any three vectors, $\mathbf{a}\times(\mathbf{b}\times\mathbf{c}) - \mathbf{b}\times(\mathbf{a}\times\mathbf{c})$, $= \mathbf{c}\times(\mathbf{b}\times\mathbf{a})$. Hence, (E.11) may be reduced to the form

$$(\delta\dot{\boldsymbol{\omega}}' - \delta\mathbf{K}_s + \boldsymbol{\omega}\times\mathbf{K}_s)\times\mathbf{u} = \mathbf{0}\ . \tag{E.12}$$

The vector in parentheses multiplying $\mathbf{u}$ must vanish everywhere, for if there is a point with coordinates (s^*, θ^*) where it does not, then, by taking $\mathbf{u}^*$ to be *any* constant vector such that the left side of (E.12) is not zero at (s^*, θ^*), we have, with $\mathbf{c} = \mathbf{Q}^{*T}\cdot\mathbf{u}^*$, that the right side of (E.12) *is* zero, a contradiction. Thus,

$$\delta\boldsymbol{\omega}' = \delta\mathbf{K}_s - \boldsymbol{\omega}\times\mathbf{K}_s = \delta^*\mathbf{K}_s\ . \tag{E.13}$$

A similar argument shows that

$$\delta\dot{\boldsymbol{\omega}} = \delta\mathbf{K}_\theta - \boldsymbol{\omega}\times\mathbf{K}_\theta = \delta^*\mathbf{K}_\theta\ . \tag{E.14}$$

F. The Rotation Vector for 1-Dimensional Strains

We seek to reduce the field equations governing the deformation of helicoidal shells suffering 1-dimensional strains (i.e., unishells) to a system of ordinary differential equations. Our basic unknowns will be the components of the stress function $\mathbf{F}$, introduced in (D.10) to satisfy identically the force equilibrium equations (D.3), and the components of the rotation vector $\boldsymbol{\beta}$, in terms of which we can express the bending strains, via (C.3) and (E.10). (In linear theory, $\mathbf{F}$ and $\boldsymbol{\beta}$ are static-geometric duals.) In terms of the Euler parameters $(\mu, \boldsymbol{\beta})$ defined by (C.4), we have from equation (3.70) of Libai & Simmonds (1983) and equations (3.7) and (3.8) of Simmonds (1984),

$$\mathbf{K}_s = 2(\mu\boldsymbol{\beta}' - \mu'\boldsymbol{\beta} + \boldsymbol{\beta}\times\boldsymbol{\beta}') \tag{F.1}$$

$$\mathbf{K}_\theta = 2(\mu\dot{\boldsymbol{\beta}} - \dot{\mu}\boldsymbol{\beta} + \boldsymbol{\beta}\times\dot{\boldsymbol{\beta}})\ . \tag{F.2}$$

Simmonds (1984) has shown that if we assume that the components of the bending strain vectors in the basis $\{\mathbf{A}_s, \mathbf{A}_\theta, \mathbf{N}\}$ are functions of s only, then (F.1) and (F.2) may be used to determine the explicit θ-dependence of $\mathbf{Q}$ and its associated Euler parameters, μ and $\boldsymbol{\beta}$. Furthermore, from his equations (3.41)-(3.44),[15] we infer that $\mathbf{Q} = \mathbf{Q}_{III}\cdot\mathbf{Q}_{II}\cdot\mathbf{Q}_I$, where the three elementary rotators on the right have the following associated Euler parameters:

$$\mathbf{Q}_I:\ \boldsymbol{\beta}_I = \sin(\bar{k}\theta/2 - \gamma/2)\mathbf{e}_z \tag{F.3}$$

$$\mathbf{Q}_{II}:\ \boldsymbol{\beta}_{II} = -\sin(\psi/2)\bar{\mathbf{e}}_\theta\ ,\ \ \bar{\mathbf{e}}_\theta \equiv \mathbf{Q}_I\cdot\mathbf{e}_\theta \tag{F.4}$$

$$\mathbf{Q}_{III}:\ \boldsymbol{\beta}_{III} = -\sin(\phi/2)\mathbf{E}_z\ ,\ \ \mathbf{E}_z \equiv \mathbf{Q}_{II}\cdot\mathbf{e}_z\ . \tag{F.5}$$

Here, $\bar{k}$ is the same constant that appears in (C.5) and the unknown angles γ, ψ, and ϕ are functions of s only. In words: The rigid body part of the local deformation of a neighborhood of a particle on the reference surface $\mathcal{S}$ consists, first,

---

[15]In (3.44), $\mathbf{e}_r$, $\mathbf{e}_\theta$ should be replaced by $\bar{\mathbf{e}}_r$, $\bar{\mathbf{e}}_\theta$.

of a rotation about the (fixed) unit vector $\mathbf{e}_z$ through an angle of $\bar{k}\theta-\gamma$ radians that sends the frame $\{\mathbf{e}_r, \mathbf{e}_\theta, \mathbf{e}_z\}$ into the new frame

$$\{\overline{\mathbf{e}}_r, \overline{\mathbf{e}}_\theta, \mathbf{e}_z\} \equiv \mathbf{Q}_I \cdot \{\mathbf{e}_r, \mathbf{e}_\theta, \mathbf{e}_z\} \,. \tag{F.6}$$

This is followed by a rotation about the (changing) unit vector $\overline{\mathbf{e}}_\theta$ through a negative angle of ψ radians. Finally, there is a rotations about the (changing) unit vector $\mathbf{E}_z$ through a negative angle of ϕ radians. In terms of these unit vectors and angles, the bending strain vectors turn out to be

$$\begin{aligned}\mathbf{K}_\theta &= (1+\bar{k})\mathbf{e}_z - \mathbf{E}_z \\ &= \sin\psi\overline{\mathbf{e}}_r + (1+\bar{k}-\cos\psi)\mathbf{e}_z\end{aligned} \tag{F.7}$$

$$\begin{aligned}\mathbf{K}_s &= -(\phi'\mathbf{E}_z + \psi'\overline{\mathbf{e}}_\theta + \gamma'\mathbf{e}_z) \\ &= \phi'\sin\psi\,\overline{\mathbf{e}}_r - \psi'\overline{\mathbf{e}}_\theta - (\gamma' + \phi'\cos\psi)\mathbf{e}_z \,.\end{aligned} \tag{F.8}$$

G. Strain Compatibility

Because $\bar{\mathbf{a}}_s = \bar{\mathbf{y}}'$ and $\bar{\mathbf{a}}_\theta = \dot{\bar{\mathbf{y}}}$, we have $\dot{\bar{\mathbf{a}}}_s = \bar{\mathbf{a}}_\theta'$, or, by (E.2),

$$(\mathbf{A}_s + \mathbf{E}_s)^\bullet = (\mathbf{A}_\theta + \mathbf{E}_\theta)' \,. \tag{G.1}$$

A first integral of these equations follows immediately from (C.5):

$$(\mathbf{A}_\theta + \mathbf{E}_\theta)\cdot\mathbf{e}_z = \bar{b} \,. \tag{G.2}$$

In linear theory, this equation is the geometric dual of (D.9) with the distributed loads absent.

H. Component Form of the Field Equations

As with the simpler case of axishells, different subsets of the field equations for unishells take relatively simple component forms in different bases. Clearly, $\mathbf{e}_z$ is a preferred direction before and after deformation—the first integrals (D.8), (D.9), and (G.2) show this. In problems in which the rotations are large but the extensional strains small, it seems advantageous to use rotated frames as this helps to separate the kinematic effects of bending from those of stretching. The frame of reference that preserves the special role of $\mathbf{e}_z$ is that introduced in (F.6), namely, $\{\overline{\mathbf{e}}_r, \overline{\mathbf{e}}_\theta, \mathbf{e}_z\} = \mathbf{Q}_I \cdot \{\mathbf{e}_r, \mathbf{e}_\theta, \mathbf{e}_z\}$. On the other hand, regardless of whether the extensional strains are small or large, the components of the stress resultants tangent to $\bar{S}$ generally dominate the transverse shear stress resultants; likewise, the components of the extensional strain vectors tangent to $\bar{S}$ generally dominate the transverse shearing strains. If the extensional strains are small, these considerations suggest using the rotated basis $\{\mathbf{A}_s, \mathbf{A}_\theta, \mathbf{N}\} = \mathbf{Q}\cdot\{\mathbf{a}_s, \mathbf{a}_\theta, \mathbf{n}\}$ introduced in (E.3). To accommodate these conflicting demands, we shall strike a compromise, using $\{\overline{\mathbf{e}}_r, \overline{\mathbf{e}}_\theta, \mathbf{e}_z\}$ for the static and kinematic equations and $\{\mathbf{A}_s, \mathbf{A}_\theta, \mathbf{N}\}$ for the constitutive relations.

1. Compatibility conditions

In the frame $\{\bar{\mathbf{e}}_r, \bar{\mathbf{e}}_\theta, \mathbf{e}_z\}$, the $\mathbf{e}_z$-component of vector compatibility condition (G.1) has the first integral (G.2). Two more scalar conditions may be obtained from (G.1) by setting

$$\bar{\mathbf{a}}_s = \mathbf{A}_s + \mathbf{E}_s \equiv \bar{a}_{sr}\bar{\mathbf{e}}_r + \bar{a}_{s\theta}\bar{\mathbf{e}}_\theta + \bar{z}'\mathbf{e}_z \tag{H.1}$$

$$\bar{\mathbf{a}}_\theta = \mathbf{A}_\theta + \mathbf{E}_\theta \equiv \bar{a}_{\theta r}\bar{\mathbf{e}}_r + \bar{a}_{\theta\theta}\bar{\mathbf{e}}_\theta + \bar{b}\mathbf{e}_z\,. \tag{H.2}$$

To compute the $\bar{\mathbf{e}}_r$- and $\bar{\mathbf{e}}_\theta$-components of (G.1), we note from (F.3) and (F.6) that

$$(\bar{\mathbf{e}}_r, \bar{\mathbf{e}}_\theta)' = -\gamma'\mathbf{e}_z \times (\bar{\mathbf{e}}_r, \bar{\mathbf{e}}_\theta) \;\;,\;\; (\bar{\mathbf{e}}_r, \bar{\mathbf{e}}_\theta)^{\bullet} = (1+\bar{k})\mathbf{e}_z \times (\bar{\mathbf{e}}_r, \bar{\mathbf{e}}_\theta)\,. \tag{H.3}$$

As all components in (H.1) and (H.2) are functions of s only, it follows that

$$\bar{a}_{\theta r}{}' + \gamma'\bar{a}_{\theta\theta} + (1+\bar{k})\bar{a}_{s\theta} = 0 \tag{H.4}$$

$$\bar{a}_{\theta\theta}{}' - \gamma'\bar{a}_{\theta r} - (1+\bar{k})\bar{a}_{sr} = 0\,. \tag{H.5}$$

To express $\bar{a}_{sr}, \bar{a}_{s\theta}, \bar{a}_{\theta r}$, and $\bar{a}_{\theta\theta}$ in terms of the components of the extensional strain vectors, we make use of equations (4.4) and (5.10)-(5.12) of Simmonds (1984) to write[16]

$$\begin{bmatrix} \mathbf{A}_s\cdot\bar{\mathbf{e}}_r & \mathbf{A}_s\cdot\bar{\mathbf{e}}_\theta & \mathbf{A}_s\cdot\mathbf{e}_z \\ \mathbf{A}_\theta\cdot\bar{\mathbf{e}}_r & \mathbf{A}_\theta\cdot\bar{\mathbf{e}}_\theta & \mathbf{A}_\theta\cdot\mathbf{e}_z \\ \mathbf{N}\cdot\bar{\mathbf{e}}_r & \mathbf{N}\cdot\bar{\mathbf{e}}_\theta & \mathbf{N}\cdot\mathbf{e}_z \end{bmatrix} = \begin{bmatrix} \cos\Lambda & -\sin\phi\cos\alpha & \sin A \\ r\cos\psi\sin\phi - b\sin\psi & r\cos\phi & r\sin\psi\sin\phi + b\cos\psi \\ \cos\Omega & \cos Z & \cos\Xi \end{bmatrix}, \tag{H.6}$$

where

$$\cos\Lambda \equiv \cos\alpha\cos\psi\cos\phi - \sin\alpha\sin\psi \tag{H.7}$$

$$\cos\Omega \equiv -a^{-1/2}[r(\sin\alpha\cos\psi\cos\phi + \cos\alpha\sin\psi) + b\cos\alpha\cos\psi\sin\phi] \tag{H.8}$$

$$\cos Z \equiv a^{-1/2}(r\sin\alpha\sin\phi - b\cos\alpha\cos\phi) \tag{H.9}$$

$$\sin A \equiv \cos\alpha\sin\psi\cos\phi + \sin\alpha\cos\psi \tag{H.10}$$

$$\cos\Xi \equiv a^{-1/2}[r(\cos\alpha\cos\psi - \sin\alpha\sin\psi\cos\phi) - b\cos\alpha\sin\psi\sin\phi]\,. \tag{H.11}$$

---

[16]The notations α and ϕ of that reference are here *reversed*.

Using these relations and the component representations[17]

$$\mathbf{E}_s = E^{s}_{s\bullet}\mathbf{A}_s + E^{\theta}_{s\bullet}\mathbf{A}_\theta + E_s\mathbf{N} \tag{H.12}$$

$$\mathbf{E}_\theta = E^{s}_{\theta\bullet}\mathbf{A}_s + E^{\theta}_{\theta\bullet}\mathbf{A}_\theta + E_\theta\mathbf{N}\,, \tag{H.13}$$

we find that

$$\bar{a}_{sr} = (1 + E^{s}_{s\bullet})\cos\Lambda + E^{\theta}_{s\bullet}(r\cos\psi\sin\phi - b\sin\psi) + E_s\cos\Omega \tag{H.14}$$

$$\bar{a}_{\theta\theta} = -E^{s}_{\theta\bullet}\cos\alpha\sin\phi + (1 + E^{\theta}_{\theta\bullet})r\cos\phi + E_\theta\cos Z \tag{H.15}$$

$$\bar{a}_{s\theta} = -(1 + E^{s}_{s\bullet})\cos\alpha\sin\phi + E^{\theta}_{s\bullet}r\cos\phi + E_s\cos Z \tag{H.16}$$

$$\bar{a}_{\theta r} = E^{s}_{\theta\bullet}\cos\Lambda + (1 + E^{\theta}_{\theta\bullet})(r\cos\psi\sin\phi - b\sin\psi) + E_\theta\cos\Omega\,, \tag{H.17}$$

and (G.2) takes the more explicit form

$$E^{s}_{\theta\bullet}\sin A + (1 + E^{\theta}_{\theta\bullet})(r\sin\psi\sin\phi + b\cos\psi) + E_\theta\cos\Xi = \bar{b}\,. \tag{H.18}$$

Once angles and strains are known, $\bar{z}$, to within a rigid axial displacement, follows from (E.5), (H.1), and (H.6) as

$$\bar{z} = \int[(1 + E^{s}_{s\bullet})\sin A + E^{\theta}_{s\bullet}(r\sin\psi\sin\phi + b\cos\psi) + E_s\cos\Xi]ds\,. \tag{H.19}$$

To express $\bar{\mathbf{y}}$ in terms of angles and strains, we use (D.6), (E.2), (H.2), and (H.3) to write

$$\dot{\bar{\mathbf{y}}} = (1 + \bar{k})^{-1}(\bar{a}_{\theta\theta}\dot{\bar{\mathbf{e}}}_r - \bar{a}_{\theta r}\dot{\bar{\mathbf{e}}}_\theta) + \bar{b}\mathbf{e}_z\,. \tag{H.20}$$

But, as $\bar{a}_{\theta\theta}$ and $\bar{a}_{\theta r}$ are functions of s only,

$$\bar{\mathbf{y}} = (1 + \bar{k})^{-1}(\bar{a}_{\theta\theta}\bar{\mathbf{e}}_r - \bar{a}_{\theta r}\bar{\mathbf{e}}_\theta) + \bar{b}\theta\mathbf{e}_z + \mathbf{f}(s)\,, \tag{H.21}$$

where $\mathbf{f}$ is an arbitrary vector function of integration. Differentiating this expression with respect to s and making use of (H.1) and (H.3)-(H.5), we find that $\mathbf{f}' = \bar{z}'\mathbf{e}_z$. Hence, to within a rigid body movement which we discard,

$$\bar{\mathbf{y}} = (1 + \bar{k})^{-1}(\bar{a}_{\theta\theta}\bar{\mathbf{e}}_r - \bar{a}_{\theta r}\bar{\mathbf{e}}_\theta) + (\bar{z} + \bar{b}\theta)\mathbf{e}_z\,. \tag{H.22}$$

Comparing this expression with (C.5), we have

$$(1 + \bar{k})\bar{r} = (\bar{a}^2_{\theta\theta} + \bar{a}^2_{\theta r})^{1/2}\,. \tag{H.23}$$

Furthermore, (C.5), (F.3), and (F.6) imply that

---

[17]Equations (H.12) and (H.13) represent the tensor expression $\mathbf{E}_\alpha = E^{\beta}_{\alpha\bullet}\mathbf{A}_\beta + E_\alpha\mathbf{N}$, $\alpha = 1, 2$, written in extended form. Because $\{\mathbf{A}_s, \mathbf{A}_\theta, \mathbf{N}\}$ is a nonorthonormal basis, the components of strain in (H.12) and (H.13) are *nonphysical*. An alternative way of expressing extensional strain vectors is in the *contravariant form* $\mathbf{E}^\beta = E^{\beta}_{\bullet\alpha}\mathbf{A}^\alpha + E^\beta\mathbf{N}$. Because, in general, $E^{\beta}_{\alpha\bullet} \neq E^{\beta}_{\bullet\alpha}$, we cannot drop the dots.

$$\mathbf{e}_r(\theta + \bar{k}\theta - \kappa) = \cos(\gamma - \kappa)\bar{\mathbf{e}}_r + \sin(\gamma - \kappa)\bar{\mathbf{e}}_\theta \,. \tag{H.24}$$

Hence,

$$\kappa = \gamma + \tan^{-1}(\bar{a}_{\theta r}/\bar{a}_{\theta\theta}) \,. \tag{H.25}$$

2. Force equilibrium

The force equilibrium equations are to be satisfied identically by using the stress function representation (D.10). The presence of the factor $1+\bar{k}$ in the second of these equations and the distinguished role of $\mathbf{e}_z$ suggests that we set $\sqrt{a_0}\,V_0 = (1+\bar{k})C$ and decompose the external distributed load as follows:

$$\mathbf{p} = (1+\bar{k})(q_r\bar{\mathbf{e}}_r + q_\theta\bar{\mathbf{e}}_\theta + q_z\mathbf{e}_z) \,. \tag{H.26}$$

If then—noting (D.11)—we represent the stress function in the component form

$$\mathbf{F} = F_r(s)\bar{\mathbf{e}}_r + F_\theta(s)\bar{\mathbf{e}}_\theta + [F_z(s) + (1+\bar{k})C\theta]\mathbf{e}_z \,, \tag{H.27}$$

(D.10) expands into

$$\sqrt{a}\,\mathbf{N}^s = (1+\bar{k})[-F_\theta\bar{\mathbf{e}}_r + F_r\bar{\mathbf{e}}_\theta + (C - \textstyle\int_0^s \sqrt{a}\,q_z d\bar{s})\mathbf{e}_z] \tag{H.28}$$

$$\sqrt{a}\,\mathbf{N}^\theta = -(F_r' + \gamma' F_\theta + \sqrt{a}\,q_\theta)\bar{\mathbf{e}}_r - (F_\theta' - \gamma' F_r - \sqrt{a}\,q_r)\bar{\mathbf{e}}_\theta - F_z'\mathbf{e}_z \,. \tag{H.29}$$

3. Moment equilibrium

To express (D.5) in component form, we set, in analogy with (H.1), (H.2), (H.12), and (H.13),

$$\begin{aligned}\sqrt{a}\,\mathbf{M}^s &= M^{s\bullet}_{\theta}\mathbf{A}_s - M^{s\bullet}_{s}\mathbf{A}_\theta + \sqrt{a}\,M^s\mathbf{N} \\ &\equiv \bar{M}_{sr}\bar{\mathbf{e}}_r + \bar{M}_{s\theta}\bar{\mathbf{e}}_\theta + \bar{M}_s\mathbf{e}_z\end{aligned} \tag{H.30}$$

$$\begin{aligned}\sqrt{a}\,\mathbf{M}^\theta &= M^{\theta\bullet}_{\theta}\mathbf{A}_s - M^{\theta\bullet}_{s}\mathbf{A}_\theta + \sqrt{a}\,M^\theta\mathbf{N} \\ &\equiv \bar{M}_{\theta r}\bar{\mathbf{e}}_r + \bar{M}_{\theta\theta}\bar{\mathbf{e}}_\theta + \bar{M}_\theta\mathbf{e}_z \,.\end{aligned} \tag{H.31}$$

Resolving $\bar{\mathbf{a}}_s \times \mathbf{N}^s$ and $\bar{\mathbf{a}}_\theta \times \mathbf{N}^\theta$ into components in the basis $\{\bar{\mathbf{e}}_r, \bar{\mathbf{e}}_\theta, \mathbf{e}_z\}$ with the aid of (H.1), (H.2), (H.28), and (H.29), and using the differentiation formulas (H.3), we obtain from (D.5) the following equations in the $\bar{\mathbf{e}}_r$- and $\bar{\mathbf{e}}_\theta$-directions:

$$\begin{aligned}&\bar{M}_{sr}' + \gamma'\bar{M}_{s\theta} - (1+\bar{k})[\bar{M}_{\theta\theta} + F_r\bar{z}' + \bar{a}_{s\theta}(\textstyle\int_0^s \sqrt{a}\,q_z d\bar{s} - C)] \\ &\qquad + (F_\theta' - \gamma' F_r - \sqrt{a}\,q_r)\bar{b} - \bar{a}_{\theta\theta}F_z' + \sqrt{a}\,\mathbf{l}\cdot\bar{\mathbf{e}}_r = 0\end{aligned} \tag{H.32}$$

$$\begin{aligned}&\bar{M}_{s\theta}' - \gamma'\bar{M}_{sr} + (1+\bar{k})[\bar{M}_{\theta r} - F_\theta\bar{z}' + \bar{a}_{sr}(\textstyle\int_0^s \sqrt{a}\,q_z d\bar{s} - C)] \\ &\qquad + \bar{a}_{\theta r}F_z' - (F_r' + \gamma' F_\theta + \sqrt{a}\,q_\theta)\bar{b} + \mathbf{l}\cdot\bar{\mathbf{e}}_\theta = 0 \,.\end{aligned} \tag{H.33}$$

In view of (H.6), (H.30) and (H.31) yield

$$\overline{M}_{sr} = M^{s\bullet}_{\theta}\cos\Lambda - M^{s\bullet}_{s}(r\cos\psi\sin\phi - b\sin\psi) + \sqrt{a}M^{s}\cos\Omega \tag{H.34}$$

$$\overline{M}_{s\theta} = -M^{s\bullet}_{\theta}\sin\phi\cos\alpha - M^{s\bullet}_{s}r\cos\phi + \sqrt{a}M^{s}\cos Z \tag{H.35}$$

$$\overline{M}_{s} = M^{s\bullet}_{\theta}\sin A - M^{s\bullet}_{s}(r\sin\psi\sin\phi + b\cos\psi) + \sqrt{a}M^{s}\cos\Xi \tag{H.36}$$

$$\overline{M}_{\theta r} = M^{\theta\bullet}_{\theta}\cos\Lambda - M^{\theta\bullet}_{s}(r\cos\psi\sin\phi - b\sin\psi) + \sqrt{a}M^{\theta}\cos\Omega \tag{H.37}$$

$$\overline{M}_{\theta\theta} = -M^{\theta\bullet}_{\theta}\sin\phi\cos\alpha - M^{\theta\bullet}_{s}r\cos\phi + \sqrt{a}M^{\theta}\cos Z \tag{H.38}$$

$$\overline{M}_{\theta} = M^{\theta\bullet}_{\theta}\sin A - M^{\theta\bullet}_{s}(r\sin\psi\sin\phi + b\cos\psi) + \sqrt{a}M^{\theta}\cos\Xi\,. \tag{H.39}$$

Further, the above equations, together with (H.22), give the first integral (D.9) the more explicit form

$$\begin{aligned} &M^{s\bullet}_{\theta}\sin A - M^{s\bullet}_{s}(r\sin\psi\sin\phi + b\cos\psi) + \sqrt{a}M^{s}\cos\Xi \\ &\quad + \overline{a}_{\theta\theta}F_{r} - \overline{a}_{\theta r}F_{\theta} + \int_0^s(\overline{a}_{\theta r}q_r + \overline{a}_{\theta\theta}q_{\theta})\sqrt{a}\,d\overline{s} = (1+\overline{k})T\,. \end{aligned} \tag{H.40}$$

4. Constitutive relations

To obtain a complete set of field equations, we must supplement the 3 compatibility conditions, (H.4), (H.5), and (H.18), and the 3 conditions of moment equilibrium, (H.32), (H.33), and (H.40), with constitutive relations. Obviously, if the stresses and strains are to depend on s only, so must any inhomogeneities.

If the shell is elastic, then there exists a strain-energy density Φ, reckoned per unit area of the undeformed reference helicoid S, such that the Internal Virtual Work defined by (E.9) takes the form

$$IVW = \int_{S_1}^{S_2}\int_{\theta_1}^{\theta_2}\sqrt{a}\,\delta\Phi\, ds\, d\theta\,. \tag{H.41}$$

If we insert the component representations (H.12), (H.13), (H.30), (H.31) into the the right side of (E.9), use (E.13) and (E.14) to express variations of the derivatives of the rotation in terms of variations of the bending strain vectors, and assume that the resulting integrand as well as Φ are continuous in s and θ, then, because the right sides of (E.9) and (H.35) apply to an arbitrary panel, we have

$$\begin{aligned} \delta\Phi &= N^{s\bullet}_{s}\delta E^{\,s}_{s\bullet} + N^{s\bullet}_{\theta}\delta E^{\,\theta}_{s\bullet} + N^{\theta\bullet}_{s}\delta E^{\,s}_{\theta\bullet} + Q^{s}\delta E_{s} + Q^{\theta}\delta E_{\theta} \\ &\quad + M^{s\bullet}_{s}\delta K^{s}_{s\bullet} + M^{s\bullet}_{\theta}\delta K^{\theta}_{s\bullet} + M^{\theta\bullet}_{s}\delta K^{s}_{\theta\bullet} + M^{\theta\bullet}_{\theta}\delta K^{\theta}_{\theta\bullet} + M^{s}\delta K_{s} + M^{\theta}\delta K_{\theta}\,, \end{aligned} \tag{H.42}$$

where, from (F.7), (F.8), (H.6), (H.28), and (H.29),

$$N_s^{s\bullet} = \mathbf{N}^s \cdot \mathbf{A}_s = -(1+\bar{k})a^{-1/2}[F_\theta \cos\Lambda + F_r \sin\phi\cos\alpha - (C - \int_0^s \sqrt{a}\, q_z d\bar{s})\sin A] \tag{H.43}$$

$$N_\theta^{s\bullet} = \mathbf{N}^s \cdot \mathbf{A}_\theta = (1+\bar{k})a^{-1/2}[F_\theta(b\sin\psi - r\cos\psi\sin\phi) + rF_r\cos\phi + (C - \int_0^s \sqrt{a}\, q_z d\bar{s})(r\sin\psi\sin\phi + b\cos\psi)] \tag{H.44}$$

$$N_s^{\theta\bullet} = \mathbf{N}^\theta \cdot \mathbf{A}_s = -a^{-1/2}[(F_r{}' + \gamma' F_\theta + \sqrt{a}\, q_\theta)\cos\Lambda + (-F_\theta{}' + \gamma' F_r + \sqrt{a}\, q_r)\sin\phi\cos\alpha + F_z{}'\sin A] \tag{H.45}$$

$$N_\theta^{\theta\bullet} = \mathbf{N}^\theta \cdot \mathbf{A}_\theta = -a^{-1/2}[(F_r{}' + \gamma' F_\theta + \sqrt{a}\, q_\theta)(r\cos\psi\sin\phi - b\sin\psi) + (F_\theta{}' - \gamma' F_r - \sqrt{a}\, q_r)r\cos\phi + F_z{}'(r\sin\psi\sin\phi + b\cos\psi)] \tag{H.46}$$

$$Q^s = \mathbf{N}^s \cdot \mathbf{N} = (1+\bar{k})a^{-1/2}[-F_\theta\cos\Omega + F_r\cos Z + (C - \int_0^s \sqrt{a}\, q_z d\bar{s})\cos\Xi] \tag{H.47}$$

$$Q^\theta = \mathbf{N}^\theta \cdot \mathbf{N} = -a^{-1/2}[(F_r{}' + \gamma' F_\theta + \sqrt{a}\, q_\theta)\cos\Omega + (F_\theta{}' - \gamma' F_r - \sqrt{a}\, q_r)\cos Z + F_z{}'\cos\Xi] \tag{H.48}$$

$$\mathbf{K}_s \cdot \mathbf{A}_s = \sqrt{a}\, K_{s\bullet}^{\theta} = -\gamma'\sin A - \phi'\sin\alpha + \psi'\sin\phi\cos\alpha \tag{H.49}$$

$$-\mathbf{K}_s \cdot \mathbf{A}_\theta = \sqrt{a}\, K_{s\bullet}^{s} = \gamma'(r\sin\psi\sin\phi + b\cos\psi) + \phi' b + \psi' r\cos\phi \tag{H.50}$$

$$\mathbf{K}_\theta \cdot \mathbf{A}_s = \sqrt{a}\, K_{\theta\bullet}^{\theta} = (1+\bar{k})\sin A - \sin\alpha \tag{H.51}$$

$$-\mathbf{K}_\theta \cdot \mathbf{A}_\theta = \sqrt{a}\, K_{\theta\bullet}^{s} = b - (1+\bar{k})(r\sin\psi\sin\phi + b\cos\psi) \tag{H.52}$$

$$\mathbf{K}_s \cdot \mathbf{N} = K_s = -\gamma'\cos\Xi - a^{-1/2}\phi' r\cos\alpha - \psi'\cos Z \tag{H.53}$$

$$\mathbf{K}_\theta \cdot \mathbf{N} = K_\theta = (1+\bar{k})\cos\Xi - a^{-1/2} r\cos\alpha\,. \tag{H.54}$$

The strain energy per unit area of the undeformed midsurface has the form

$$\Phi = \hat{\Phi}(E_{s\bullet}^{s}, \cdots, K_\theta) \tag{H.55}$$

so that

$$N_s^{s\bullet} = \frac{\partial\Phi}{\partial E_{s\bullet}^{s}}\,, \cdots, M^\theta = \frac{\partial\Phi}{\partial K_\theta}\,. \tag{H.56}$$

As the components of $\boldsymbol{\beta}$ and $\mathbf{F}$ are the basic unknowns, we assume that (H.56) can be solved for the extensional and transverse shearing strains as functions of the bending strains and stress resultants. Then, by a Legendre transformation, we may introduce the *mixed*-energy density

$$\Psi = \Phi - (N_s^{s\bullet} E_{s\bullet}^{s} + \cdots + Q^{\theta} E_{\theta}) \tag{H.57}$$

so that

$$E_{s\bullet}^{s} = -\frac{\partial \Psi}{\partial N_s^{s\bullet}} \;, \cdots , \; M^{\theta} = \frac{\partial \Psi}{\partial K_{\theta}} \; . \tag{H.58}$$

When (H.58) is substituted into the compatibility and moment equilibrium conditions, we obtain 6 quasilinear ordinary differential equations for $\phi, \gamma, \psi, F_r, F_\theta$, and F_z.

I. Special Cases

In several important special cases, including two we have already discussed—axishells and curved tubes—the equations of the preceding section simplify considerably, as we now show.

1. Axishells

Here,

$$\begin{gathered} s = \sigma \;, \; \psi = \beta \;, \; F_\theta = -rH \;, q_r = p_H \;, \; \mathbf{l}\cdot\bar{\mathbf{e}}_\theta = -l \\ \gamma = \phi = F_r = F_z = \bar{k} = b = \bar{b} = q_\theta = 0 = \mathbf{l}\cdot\bar{\mathbf{e}}_r = 0. \end{gathered} \tag{I.1}$$

Of the 3 compatibility equations, (H.4) and (H.18) are satisfied trivially while (H.5) reduces to (V.M.2) with

$$E_{s\bullet}^{s} = e_\sigma \;, \; E_{\theta\bullet}^{\theta} = e_\theta \;, \; E_s = g \; . \tag{I.2}$$

Of the 3 moment equilibrium equations, (H.32) and (H.40) are satisfied trivially while the negative of (H.33) reduces to (V.Q.12) with

$$M_s^{s\bullet} = M_\sigma \;, \; M_\theta^{\theta\bullet} = M_\theta \;, \; rM^\theta = M_n \; . \tag{I.3}$$

2. Pure bending of pressurized curved tubes

Here (assuming the external distributed surface couple is absent),

$$\begin{gathered} s = \sigma \;, \; \psi = \beta \;, \; (1+\bar{k})F_\theta = -rH \\ \gamma = \phi = F_r = F_z = b = \bar{b} = q_\theta = 0 \; . \end{gathered} \tag{I.4}$$

The non-trivial compatibility and moment equilibrium equations are (H.5) and (H.33) which, if we make the same identification as in (I.3), reduce, respectively, to (A.16) and (A.17) [plus the auxiliary equations (A.20)-(A.22) that accompany the latter].

3. Torsion, inflation, and extension of a tube

We consider a circular cylindrical shell with an undeformed midsurface of radius R, acted upon by an internal pressure p, a net axial end force P, and a net end torque T. Assuming the shell to be sufficiently long to ignore end effects, we make the simplest static and kinematic assumptions consistent with overall equilibrium. Thus, with $\alpha=\pi/2$, $s=z$, and a superior zero denoting a constant, we take

$$\begin{gathered} F_\theta=\overset{0}{F}_\theta\ ,\ F_z=(z/R)\overset{0}{F}_z\ ,\ C=P/2\pi\ ,\ \gamma=\pi/2-(z/R)\overset{0}{\chi} \\ \phi=-\pi/2\ ,\ \psi=\overset{0}{\psi}\ ,\ R\mathbf{p}=-p\mathbf{a}_s\times\mathbf{a}_\theta\ ,\ F_r=\bar{k}=b=\bar{b}=0\ . \end{gathered} \tag{I.5}$$

From (H.6)-(H.11) there follows

$$\cos\Lambda=-\sin\overset{0}{\psi}\ ,\ \cos Z=-1\ ,\ \sin A=\cos\overset{0}{\psi}\ ,\ \cos\Omega=\cos\Xi=0 \tag{I.6}$$

and

$$\mathbf{A}_s=-\sin\overset{0}{\psi}\bar{\mathbf{e}}_r+\cos\overset{0}{\psi}\mathbf{e}_z\ ,\ R^{-1}\mathbf{A}_\theta\equiv\tilde{\mathbf{A}}_\theta=-(\cos\overset{0}{\psi}\bar{\mathbf{e}}_r+\sin\overset{0}{\psi}\mathbf{e}_z)\ ,\ \mathbf{N}=-\bar{\mathbf{e}}_\theta\ . \tag{I.7}$$

From (F.7), (F.8), and (H.49)-(H.54), we have for the bending strains

$$\mathbf{K}_s\equiv k_{z\theta}\mathbf{A}_s-k_{zz}\tilde{\mathbf{A}}_\theta+k_z\mathbf{N}=R^{-1}\overset{0}{\chi}\mathbf{e}_z \tag{I.8}$$

$$R^{-1}\mathbf{K}_\theta\equiv k_{\theta\theta}\mathbf{A}_s-k_{\theta z}\tilde{\mathbf{A}}_\theta+k_\theta\mathbf{N}=R^{-1}[\sin\overset{0}{\psi}\bar{\mathbf{e}}_r+(1-\cos\overset{0}{\psi})\mathbf{e}_z]\ , \tag{I.9}$$

where $k_{z\theta}, \cdots$ are physical components. Thus,

$$Rk_{z\theta}=\overset{0}{\chi}\cos\overset{0}{\psi}\ ,\ Rk_{zz}=\overset{0}{\chi}\sin\overset{0}{\psi}\ ,\ Rk_{\theta\theta}=\cos\overset{0}{\psi}-1\ ,\ Rk_{\theta z}=\sin\overset{0}{\psi} \tag{I.10}$$

and $k_z=k_\theta=0$.

Likewise, from (H.28), (H.29), and (H.43)-(H.48), we have

$$\mathbf{N}^s\equiv n_{zz}\mathbf{A}_s+n_{z\theta}\tilde{\mathbf{A}}_\theta+n_z\mathbf{N}=R^{-1}[-\overset{0}{F}_\theta\,\bar{\mathbf{e}}_r+(P/2\pi)\mathbf{e}_z] \tag{I.11}$$

$$R\mathbf{N}^\theta\equiv n_{\theta z}\mathbf{A}_s+n_{\theta\theta}\tilde{\mathbf{A}}_\theta+n_\theta\mathbf{N}=(R^{-1}\overset{0}{\chi}\overset{0}{F}_\theta-Rq_\theta)\bar{\mathbf{e}}_r-R^{-1}\overset{0}{F}_z\mathbf{e}_z \tag{I.12}$$

where $n_{zz}, \cdots$ are physical components. Thus,

$$\begin{gathered} Rn_{zz}=\overset{0}{F}_\theta\sin\overset{0}{\psi}+(P/2\pi)\cos\overset{0}{\psi}\ ,\ Rn_{z\theta}=\overset{0}{F}_\theta\cos\overset{0}{\psi}-(P/2\pi)\sin\overset{0}{\psi} \\ Rn_{\theta z}=-[(\overset{0}{\chi}\overset{0}{F}_\theta-R^2q_\theta)\sin\overset{0}{\psi}+\overset{0}{F}_z\cos\overset{0}{\psi}]\ ,\ Rn_{\theta\theta}=-(\overset{0}{\chi}\overset{0}{F}_\theta-R^2q_\theta)\cos\overset{0}{\psi}+\overset{0}{F}_z\sin\overset{0}{\psi} \end{gathered} \tag{I.13}$$

and $n_z=n_\theta=0$.

Because the 2 normal bending strains, k_z, k_θ, and the 2 transverse shear stress resultants, n_z and n_θ, are zero, we assume that the mixed-energy density defined in (H.57) is such that the normal components of the stress couples and the transverse shearing strains are also zero. Thus, if we introduce physical components by setting

$$\mathbf{E}_s\equiv e_{zz}\mathbf{A}_s+e_{z\theta}\tilde{\mathbf{A}}_\theta\ ,\ R^{-1}\mathbf{E}_\theta\equiv e_{\theta z}\mathbf{A}_s+e_{\theta\theta}\tilde{\mathbf{A}}_\theta \tag{I.14}$$

and

$$\mathbf{M}^s \equiv m_{z\theta}\mathbf{A}_s - m_{zz}\tilde{\mathbf{A}}_\theta \ , \ R\mathbf{M}^\theta \equiv m_{\theta\theta}\mathbf{A}_s - m_{\theta z}\tilde{\mathbf{A}}_\theta \ , \tag{I.15}$$

we find that the compatibility condition (H.4) is satisfied identically, while (H.5) and (H.18) reduce to

$$-\overset{0}{\chi}[e_{\theta z}\sin\overset{0}{\psi} + (1+e_{\theta\theta})\cos\overset{0}{\psi}] + (1+e_{zz})\sin\overset{0}{\psi} + e_{z\theta}\cos\overset{0}{\psi} = 0 \tag{I.16}$$

$$e_{\theta z}\cos\overset{0}{\psi} - (1+e_{\theta\theta})\sin\overset{0}{\psi} = 0 \, . \tag{I.17}$$

Likewise, the moment equilibrium condition (H.32) is satisfied identically while (H.33) and (H.40) reduce to

$$\begin{aligned}&\overset{0}{\chi}(-m_{z\theta}\sin\overset{0}{\psi} + m_{zz}\cos\overset{0}{\psi}) - m_{\theta\theta}\sin\overset{0}{\psi} + m_{\theta z}\cos\overset{0}{\psi} - [(1+e_{zz})\cos\overset{0}{\psi} - e_{z\theta}\sin\overset{0}{\psi}]\overset{0}{F}_\theta \\ &- [(1+e_{\theta\theta})\cos\overset{0}{\psi} + e_{\theta z}\sin\overset{0}{\psi}]\overset{0}{F}_z + [(1+e_{zz})\sin\overset{0}{\psi} + e_{z\theta}\cos\overset{0}{\psi}](P/2\pi) = 0\end{aligned} \tag{I.18}$$

$$m_{z\theta}\cos\overset{0}{\psi} + m_{zz}\sin\overset{0}{\psi} + [e_{\theta z}\sin\overset{0}{\psi} + (1+e_{\theta\theta})\cos\overset{0}{\psi}]\overset{0}{F}_\theta = T \, . \tag{I.19}$$

Furthermore, from (H.26) and (I.5), we have

$$q_\theta = p[(1+e_{zz})^2 e_{\theta\theta}^2 + e_{z\theta}^2 e_{\theta z}^2]^{1/2} \tag{I.20}$$

and from (H.19),

$$\bar{z} = [(1+e_{zz})\cos\overset{0}{\psi} - e_{z\theta}\sin\overset{0}{\psi}]z \, . \tag{I.21}$$

Once a mixed-energy density is prescribed[18] we may express the extensional strain components and the stress couple components in terms of the components of the stress resultants and bending strains and hence, via (I.10) and (I.13), in terms of $\overset{0}{\chi}, \overset{0}{\psi}, \overset{0}{F}_\theta$, and $\overset{0}{F}_z$. These expressions, substituted into (I.16)-(I.19), yield 4 *algebraic* equations for these 4 unknown constants. These equations contain, implicitly, torque-twist-extension and force-extension-twist relations, with the pressure entering as a parameter.

4. Extension and twist of a right helicoidal shell

Here, $r = r_0 + s$ and $\alpha = 0$. Following Wan (1982), we use a semi-inverse method to obtain the kinematic and static equations governing the gross extension and twist of a right helicoidal shell, stress free along its radial edges $s = 0$ and $s = L$. That is, we retain only enough freedom among the dependent variables to allow for the imposition of a net axial force P and a net axial torque T over any edge $\theta =$ constant. We assume that the constitutive laws are consistent with our *a priori* assumptions, but do not attempt to give them explicit form. [For a linear analysis using the Kirchhoff Hypothesis, see Reissner and Wan

---

[18]So far as we know, none has been proposed in the literature that would allow us to include the effects of $k_{z\theta} - k_{\theta z}$ and $n_{z\theta} - n_{\theta z}$, i.e., to go beyond a first-approximation shell theory.

(1968)].

Thus, we set

$$\phi = \pi/2 \ , \ \gamma = -\pi/2 \ , \ F_\theta = C = 0 \tag{I.22}$$

and assume that the external distributed surface forces and couples are zero. With

$$r = \sqrt{a}\cos\chi \ , \ b = \sqrt{a}\sin\chi \ , \tag{I.23}$$

it follows from (H.6)-(H.11) that

$$\sin A = \cos Z = \cos\Lambda = 0 \ , \ \cos\Xi = \cos X \ , \ \cos\Omega = -\sin X \ , \tag{I.24}$$

and

$$\mathbf{A}_s = -\bar{\mathbf{e}}_\theta \ , \ a^{-1/2}\mathbf{A}_\theta \equiv \tilde{\mathbf{A}}_\theta = \cos X\bar{\mathbf{e}}_r + \sin X\mathbf{e}_z \ , \ \mathbf{N} = -\sin X\,\bar{\mathbf{e}}_r + \cos X\,\mathbf{e}_z \ , \tag{I.25}$$

where

$$X = \chi + \psi \ . \tag{I.26}$$

From (F.7), (F.8), and (H.49)-(H.54), we have for the bending strains

$$\mathbf{K}_s \equiv k_{r\theta}\mathbf{A}_s - k_{rr}\tilde{\mathbf{A}}_\theta + k_r\mathbf{N} = -\psi'\,\bar{\mathbf{e}}_\theta \tag{I.27}$$

$$a^{-1/2}\mathbf{K}_\theta \equiv k_{\theta\theta}\mathbf{A}_s - k_{\theta r}\tilde{\mathbf{A}}_\theta + k_\theta\mathbf{N} = a^{-1/2}[\sin\psi\,\bar{\mathbf{e}}_r + (1 + \bar{k} - \cos\psi)\mathbf{e}_z] \ , \tag{I.28}$$

where k_{rr}, $\cdots$ are physical components. Thus, with the aid of (I.25),

$$\begin{gathered} k_{r\theta} = \psi' \ , \ \sqrt{a}\,k_{\theta r} = \sin\chi - (1 + \bar{k})\sin X \ , \ \sqrt{a}\,k_\theta = (1 + \bar{k})\cos X - \cos\chi \\ k_{rr} = k_r = k_{\theta\theta} = 0 \ . \end{gathered} \tag{I.29}$$

Likewise, from (H.28), (H.29), and (H.43)-(H.48), we have

$$\mathbf{N}^s \equiv n_{rr}\mathbf{A}_s + n_{r\theta}\tilde{\mathbf{A}}_\theta + n_r\mathbf{N} = a^{-1/2}(1 + \bar{k})F_r\bar{\mathbf{e}}_\theta \tag{I.30}$$

$$\sqrt{a}\mathbf{N}^\theta \equiv n_{\theta r}\mathbf{A}_s + n_{\theta\theta}\tilde{\mathbf{A}}_\theta + n_\theta\mathbf{N} = -(F_r{}'\bar{\mathbf{e}}_r + F_z{}'\mathbf{e}_z) \ , \tag{I.31}$$

where n_{rr}, $\cdots$ are physical components. Thus, with the aid of (I.25),

$$\begin{gathered} \sqrt{a}\,n_{rr} = -(1 + \bar{k})F_r \ , \ n_{\theta\theta} = -(F_r{}'\cos X + F_z{}'\sin X) \ , \ n_\theta = F_r{}'\sin X - F_z{}'\cos X \\ n_{r\theta} = n_{\theta r} = n_r = 0 \ . \end{gathered} \tag{I.32}$$

We assume that the mixed strain-energy density given by (H.57) is such that the second lines of (I.29) and (I.32) imply that the corresponding components of the stress couple and extensional strain vectors are zero. Thus, in terms of physical components, we have

$$\mathbf{M}^s = m_{r\theta}\mathbf{A}_s \ , \ \sqrt{a}\mathbf{M}^\theta = -m_{\theta r}\tilde{\mathbf{A}}_\theta + m_\theta\mathbf{N} \tag{I.33}$$

$$\mathbf{E}_s = e_{rr}\mathbf{A}_s \ , \ a^{-1/2}\mathbf{E}_\theta = e_{\theta\theta}\tilde{\mathbf{A}}_\theta + e_\theta\mathbf{N} \ . \tag{I.34}$$

From these last expressions and (H.1), (H.2), and (I.25) follows

$$\bar{a}_{s\theta} = -(1+e_{rr}) \ , \ \bar{a}_{\theta r} = \sqrt{a}\,[(1+e_{\theta\theta})\cos X - e_\theta \sin X] \\ \bar{a}_{sr} = \bar{a}_{\theta\theta} = 0 \ . \tag{I.35}$$

so that the compatibility condition (H.5) is satisfied identically while (H.4) and (H.18) reduce to

$$\{\sqrt{a}\,[(1+e_{\theta\theta})\cos X - e_\theta \sin X]\}' - (1+\bar{k})(1+e_{rr}) = 0 \tag{I.36}$$

$$\sqrt{a}\,[(1+e_{\theta\theta})\sin X + e_\theta \cos X] = \bar{b} \ . \tag{I.37}$$

For future reference, we note that these compatibility conditions together with (I.29), imply that

$$\sqrt{a}\,[1 + \tfrac{1}{2}(e_{rr}+e_{\theta\theta})](k_{r\theta} - k_{\theta r}) \\ = (e_{rr} - e_{\theta\theta})[\sqrt{a}\,\tfrac{1}{2}(k_{r\theta}+k_{\theta r}) - \sin\chi] - (\sqrt{a}\,e_\theta)' \ . \tag{I.38}$$

(This equation also follows directly from $\bar{\mathbf{a}}_\theta{}' \cdot \mathbf{N} = \dot{\bar{\mathbf{a}}}_s \cdot \mathbf{N}$.)

The moment equilibrium equations (H.32) and (H.40) are satisfied identically while the negative of (H.33) reduces to

$$(\sqrt{a}\,m_{r\theta})' + \sqrt{a}\,(1+\bar{k})(m_{\theta r}\cos X + m_\theta \sin X) \ . \\ + \sqrt{a}\,[(1+e_{\theta\theta})(F_r{}'\sin X - F_z{}'\cos X) + e_\theta(F_r{}'\cos X + F_z{}'\sin X) = 0 \tag{I.39}$$

Overall force equilibrium along any edge $\theta =$ constant requires that

$$\int_0^L \sqrt{a}\,\mathbf{N}^\theta ds = P\mathbf{e}_z \ , \tag{I.40}$$

which, by (H.29) and (I.22), implies

$$F_r(L) - F_r(0) = 0 \ , \ F_z(L) - F_z(0) = -P \ . \tag{I.41}$$

Overall moment equilibrium along any edge $\theta =$ constant requires that

$$\int_0^L \sqrt{a}\,(\mathbf{M}^\theta + \bar{\mathbf{y}} \times \mathbf{N}^\theta) ds = T\mathbf{e}_z \ , \tag{I.42}$$

which, with the aid of (H.22), $(I.33)_2$, and (I.35), leads to the two scalar conditions

$$\int_0^L \{(1+\bar{k})(m_{\theta r}\cos X + m_\theta \sin X) - \sqrt{a}\,[(1+e_{\theta\theta})\cos X - e_\theta \sin X]F_z{}'\}ds = 0 \tag{I.43}$$

$$\int_0^L \{(1+\bar{k})(m_{\theta r}\sin X - m_\theta \cos X) + \sqrt{a}\,[(1+e_{\theta\theta})\cos X - e_\theta \sin X]F_r{}'\}ds \\ = -(1+\bar{k})T \ , \tag{I.44}$$

a third condition being identical to $(I.41)_1$.

Equations (I.30) and (I.33)$_1$ imply that the edges $s=0, L$ will be stress free if

$$F_r(0) = F_r(L) = m_{r\theta}(0) = m_{r\theta}(L) = 0 \,. \tag{I.45}$$

A complete set of field equations follows upon specifying the mixed-energy density

$$\Psi = \hat{\Psi}(n_{rr}, n_{\theta\theta}, n_\theta, k_{r\theta}, k_{\theta r}, k_\theta) = \tilde{\Psi}(F_r, F_z, \psi, \bar{k}) \,. \tag{I.46}$$

If the basic assumptions of first-approximation shell theory are invoked, namely that Ψ is independent of the transverse shear stress resultant n_θ and the normal bending strain k_θ, then $e_\theta = m_\theta = 0$ and the governing equations simplify in an obvious way. We further note that (I.38) shows that, *if the extensional strains are also small*, we can ignore the difference between $k_{r\theta}$ and $k_{\theta r}$. In this case, with $\tau = \frac{1}{2}(k_{r\theta} + k_{\theta r})$, the mixed strain-energy density may be taken to be a function of n_{rr}, $n_{\theta\theta}$, and τ only, and we may further simplify our equations by setting $m_{r\theta} = m_{\theta r} = \Psi_{,\tau}$. The reader should consult Wan (1982) for more details.

5. Inextensional deformation

We extend to general helicoidal shells the results of Reissner (1968) for the inextensional deformation of split shells of revolution. Interestingly, it is of no help to determine the rotation field first. Rather, we obtain $\bar{\mathbf{r}}$, κ, and $\bar{z}$ directly, starting from (C.5).

With $\lambda \equiv 1 + \bar{k}$ and

$$\hat{\mathbf{e}}_r = \cos(\lambda\theta - \kappa)\mathbf{e}_x + \sin(\lambda\theta - \kappa)\mathbf{e}_y \tag{I.47}$$

$$\hat{\mathbf{e}}_\theta = -\sin(\lambda\theta - \kappa)\mathbf{e}_x + \cos(\lambda\theta - \kappa)\mathbf{e}_y \,, \tag{I.48}$$

we have

$$\mathbf{A}_s = \bar{\mathbf{a}}_s = \bar{\mathbf{y}}' = \bar{r}'\hat{\mathbf{e}}_r - \bar{r}\kappa'\hat{\mathbf{e}}_\theta + \sin A\,\mathbf{e}_z \tag{I.49}$$

$$\mathbf{A}_\theta = \bar{\mathbf{a}}_\theta = \dot{\bar{\mathbf{y}}} = \lambda\bar{r}\hat{\mathbf{e}}_\theta + \bar{b}\mathbf{e}_z \,, \tag{I.50}$$

where, from (H.10),

$$\bar{z}' = \mathbf{A}_s \cdot \mathbf{e}_z = \sin A \,. \tag{I.51}$$

The deformation being inextensional, we have, by (C.28),

$$a_{ss} \equiv \mathbf{a}_s \cdot \mathbf{a}_s = 1 = \bar{\mathbf{a}}_s \cdot \bar{\mathbf{a}}_s = \bar{r}'^2 + (\bar{r}\kappa')^2 + \sin^2 A \tag{I.52}$$

$$a_{s\theta} \equiv \mathbf{a}_s \cdot \mathbf{a}_\theta = b\sin\alpha = \bar{\mathbf{a}}_s \cdot \bar{\mathbf{a}}_\theta = -\lambda\bar{r}^2\kappa' + \bar{b}\sin A \tag{I.53}$$

$$a_{\theta\theta} \equiv \mathbf{a}_\theta \cdot \mathbf{a}_\theta = r^2 + b^2 = \bar{\mathbf{a}}_\theta \cdot \bar{\mathbf{a}}_\theta = (\lambda\bar{r})^2 + \bar{b}^2 \,. \tag{I.54}$$

These are 3 nonlinear equations for $\bar{r}$, A, and κ.

From (I.54),

$$\lambda\bar{r} = (r^2 + b^2 - \bar{b}^2)^{1/2} . \tag{I.55}$$

Thence, by (I.53),

$$\bar{r}\kappa' = \frac{\bar{b}\sin A - b\sin\alpha}{(r^2 + b^2 - \bar{b}^2)^{1/2}} . \tag{I.56}$$

To get $\sin A$, we first differentiate (I.55):

$$\bar{r}' = \frac{r\cos\alpha}{\lambda(r^2 + b^2 - \bar{b}^2)^{1/2}} . \tag{I.57}$$

Then, substituting (I.56) and (I.57) into (I.52), we get a quadratic equation for $\sin A$ whose solution is

$$\sin A = \frac{b\bar{b}\sin\alpha}{r^2 + b^2} \pm \left[\frac{a(r^2 + b^2 - \bar{b}^2)}{(r^2 + b^2)^2} - \frac{r^2\cos^2\alpha}{\lambda^2(r^2 + b^2)} \right]^{1/2} , \tag{I.58}$$

where a is given by (C.9).

Equations (I.55) and (I.58) are explicit formulas for $\bar{r}$ and $\sin A$. Substituted into (I.51) and (I.56), they reduce the determination of $\bar{z}$ and κ to quadratures.

The bending strains may be computed as the differences in the curvature tensors of the deformed and undeformed midsurface. However, it is simpler to note, first, that (H.18), (H.51), and (H.52) give

$$\sqrt{a}K^{\theta}_{\theta\bullet} = \lambda\sin A - \sin\alpha \tag{I.59}$$

$$\sqrt{a}K^{s}_{\theta\bullet} = b - \lambda\bar{b} . \tag{I.60}$$

Then,

$$\begin{aligned} K_{s\theta} = K_{\theta s} &= K^{s}_{\theta\bullet}a_{ss} + K^{\theta}_{\theta\bullet}a_{s\theta} \\ &= a^{-1/2}[b(\cos^2\alpha + \lambda\sin A\sin\alpha) - \lambda\bar{b}] \end{aligned} \tag{I.61}$$

$$\begin{aligned} K_{\theta\theta} &= K^{s}_{\theta\bullet}a_{s\theta} + K^{\theta}_{\theta\bullet}a_{\theta\theta} \\ &= a^{-1/2}[(b - \lambda\bar{b})b\sin\alpha + (\lambda\sin A - \sin\alpha)(r^2 + b^2)] . \end{aligned} \tag{I.62}$$

To compute K_{ss} (and thence $K^{s}_{s\bullet}$), we follow Reissner (1968) and note that, as the Gaussian curvature G is unchanged in an inextensional deformation,

$$aG = b_{ss}b_{\theta\theta} - b^2_{s\theta} = (b_{ss} + K_{ss})(b_{\theta\theta} + K_{\theta\theta}) - (b_{s\theta} + K_{s\theta})^2 , \tag{I.63}$$

i.e.,

$$K_{ss} = \frac{(2b_{s\theta} + K_{s\theta})K_{s\theta} - b_{ss}K_{\theta\theta}}{b_{\theta\theta} + K_{\theta\theta}} , \tag{I.64}$$

where, from (C.7) and (C.10),

$$b_{ss} = \mathbf{n}\cdot\mathbf{a}_s{}' = a^{-1/2} r\alpha' \tag{I.65}$$

$$b_{s\theta} = \mathbf{n}\cdot\dot{\mathbf{a}}_s = \mathbf{n}\cdot\mathbf{a}_\theta{}' = -\,a^{-1/2} b\cos^2\alpha \tag{I.66}$$

$$b_{\theta\theta} = \mathbf{n}\cdot\dot{\mathbf{a}} = a^{-1/2} r^2 \sin\alpha\,. \tag{I.67}$$

Finally,

$$K^s_{s\bullet} = K_{ss}a^{ss} + K_{s\theta}a^{s\theta} \quad , \quad K^\theta_{s\bullet} = K_{ss}a^{s\theta} + K_{s\theta}a^{\theta\theta} \ , \tag{I.68}$$

where

$$a^{ss} = a^{-1}(r^2+b^2)\,,\ a^{s\theta} = -\,a^{-1} b\sin\alpha\,,\ a^{\theta\theta} = a^{-1}\,. \tag{I.69}$$

Substituting (I.61), (I.62), and (I.65)-(I.67) into (I.64) and the resulting equation along with (I.61) into (I.68), we obtain

$$K^s_{s\bullet} = a^{-3/2}\left\{ \frac{a\,[a(r^2+b^2)G + \lambda^2\bar{b}(\bar{b} - b\sin A\sin\alpha)]}{\lambda[(r^2+b^2)\sin A - b\bar{b}\sin\alpha]} - (r^2+b^2)r\alpha' - b^2\sin\alpha\cos^2\alpha \right\} \tag{I.70}$$

$$K^s_{\theta\bullet} = -\,a^{-3/2}\left\{ \frac{a\,[abG\sin\alpha - \lambda^2\sin A(b\sin A\sin\alpha - \bar{b})]}{\lambda[(r^2+b^2)\sin A - b\bar{b}\sin\alpha]} - b(r\alpha'\sin\alpha + \cos^2\alpha) \right\}, \tag{I.71}$$

where

$$a^2 G = r^3\alpha'\sin\alpha - b^2\cos^4\alpha\,. \tag{I.72}$$

If we set $b = 0$, our equations reduce to those of Reissner (1968) for a slit shell of revolution. If we set $\alpha = \pi/2$ (in which case r is a constant), we obtain equations for a cylindrical helicoidal shell, which has been studied by Mansfield (1980). If we set $\alpha = 0$ and $r = r_0 + s$, we obtain the following displacement and bending fields for the inextensional deformation of a right helicoidal shell:

$$\bar{r} = \lambda^{-1}(r^2 + b^2 - \bar{b}^2)^{1/2} \tag{I.73}$$

$$\sin A = \pm\lambda^{-1}a^{-1/2}[\lambda^2(r^2 + b^2 - \bar{b}^2) - r^2]^{1/2} \tag{I.74}$$

$$\kappa = \lambda\bar{b}\int (r^2 + b^2 - \bar{b}^2)^{-1}\sin A\,dr \tag{I.75}$$

$$\bar{z} = \int \sin A\,dr \tag{I.76}$$

$$K^{\theta}_{\theta\bullet} = \frac{\lambda \sin A}{(r^2+b^2)^{1/2}} \; , \; K^{s}_{\theta\bullet} = \frac{b-\lambda\bar{b}}{(r^2+b^2)^{1/2}} \tag{I.77}$$

$$K^{s}_{s\bullet} = \frac{\lambda^2\bar{b}^2 - b^2}{\lambda(r^2+b^2)^{3/2}} \; , \; K^{s}_{\theta\bullet} = \frac{\lambda\bar{b} - b\sin A}{(r^2+b^2)^{3/2}} \; . \tag{I.78}$$

References

[A letter in brackets indicates the section where the reference appears]

Antman, S. S., and Osborn, J. E. (1979). The principle of virtual work and integral laws of motion. *Arch. Rat. Mech. Anal.* **69**, 232-261. [E]

Axelrad, E. L. (1962a). Equations of deformation for shells of revolution and for bending of thin-walled tubes subjected to large elastic displacements. (Translated from a 1960 article in Russian.) *Am. Rocket Soc. J.* **32**, 1147-1151. [A]

Axelrad, E. L. (1962b). Flexure and stability of thin walled tubes under hydrostatic pressure (in Russian). *Izv. Akad. Nauk SSSR, OTN Mekhanika i Mashinostroenie* **1**, 98-144. [A]

Axelrad, E. L. (1965). Refinement of buckling load analysis for tube flexure by way of considering precritical deformation (in Russian). *Izvestiya Akademii Nauk SSSR, Mekhanika* **4**, 133-139. [A]

Axelrad, E. L. (1978). Flexible shell theory and buckling of toroidal shells and tubes. *Ing. Arch.* **47**, 95-104. [A]

Axelrad, E. L. (1985a). Elastic tubes—assumptions, equations, edge conditions. *Thin Walled Structures* **3**, 193-215. [A]

Axelrad, E. L. (1985b). On local buckling of thin shells. *Int. J. Non-Lin. Mech.* **20**, 249-259. [A]

Axelrad, E. L. (1987). "Theory of Flexible Shells." North Holland, Amsterdam, etc. [A]

Axelrad, E. L., and Emmerling, F. A. (1983). Grosse Verformungen und Traglasten elastischer Rohre unter Biegung und Aussendruck. *Ing. Arch.* **53**, 41-52. [A]

Axelrad, E. L., and Emmerling, F. A. (1984). Elastic tubes. *Appl. Mech. Rev.* **37**, 891-897. [A]

Beatty, M. F. (1977). Vector analysis of finite rigid rotations. *J. Appl. Mech.* **44**, 501-502. [C]

Boyle, J. T. (1981). The finite bending of curved pipes. *Int. J. Solids Struct.* **17**, 515-529. [A]

Brazier, L. G. (1927). On the flexure of thin cylindrical shells and other "thin" sections. *Proc. Roy. Soc. London* **A116**, 104-114. [A]

Bushnell, D. (1981). Elastic-plastic bending and buckling of pipes and elbows. *Computers and Structures* **13**, 241-248. [A]

Carey, G. F., and Dinh, H. T. (1986). Conservation principles and variational problems. *Acta Mech.* **58**, 93-97. [E]

Clark, R. A., and Reissner, E. (1951). Bending of curved tubes. *Adv. Appl.*

Mech. **2**, 93-122.
Emmerling, F. A. (1982). Nichtlinerae Biegung und Beulen von Zylindern und krummen Rohren bei Normaldruck. *Ing. Arch.* **52**, 1-16. [A]
Emmerling, F. A. (1984). Nonlinear bending of curved tubes. *In* "Flexible Shells, Theory and Applications" (E. L. Axelrad and F. A. Emmerling, eds.), pp. 175-191. Springer-Verlag, Berlin, etc. [A]
Fabian, O. (1977). Collapse of cylindrical, elastic tubes under combined bending, pressure and axial loads. *Int. J. Solids Struct.* **13**, 1257-1270. [A]
Kempner, J., and Chen, Y. -N. (1974). Buckling and initial postbuckling of oval cylindrical shells under combined axial compression and bending. *Trans. New York Acad. Sci.* **36**, Series II, 171-191. [A]
Libai, A., and Durban, D. (1973). A method for approximate stability analysis and its application to circular cylindrical shells under circumferentially varying edge loads. *J. Appl. Mech.* **40**, 971-976. [A]
Libai, A., and Durban, D. (1977). Buckling of cylindrical shells subjected to nonuniform axial loads. *J. Appl. Mech.* **44**, 714-718. [A]
Libai, A., and Simmonds, J. G. (1983). Nonlinear elastic shell theory. *Adv. Appl. Mech.* **23**, 271-371. [F]
Mansfield, E. H. (1980). On finite inextensional deformation of a helical strip. *Int. J. Non-lin. Mech.* **15**, 459-467. [I]
Reissner, E. (1959). On finite bending of pressurized tubes. *J. Appl. Mech.* **26**, 386-392. [A]
Reissner, E. (1968). Finite inextensional pure bending and twisting of thin shells of revolution. *Q. J. Mech. Appl. Math.* **21**, 293-306. [I]
Reissner, E. (1981). On finite pure bending of curved tubes. *Int. J. Solids Struct.* **17**, 839-844. [A]
Reissner, E., and Wan, F. Y. M. (1968). On axial extension and torsion of helicoidal shells. *J. Math. & Phys.* **47**, 1-31. [I]
Reissner, E., and Weinitschke, H. J. (1963). Finite pure bending of circular cylindrical tubes. *Q. Appl. Math.* **20**, 305-319. [A]
Seide, P., and Weingarten, V. I. (1961). On the buckling of cylindrical shells under pure bending. *J. Appl. Mech.* **28**, 112-116. [A]
Simmonds, J. G. (1977). St. Venant's principle for semi-infinite shells of revolution. *Recent Advances in Engineering Science* **8**, 367-374, Scientific Publishers, Boston. [A]
Simmonds, J. G. (1979). Surfaces with metric and curvature tensors that depend on one coordinate only are general helicoids. *Q. App. Math.* **37**, 82-85. [C]
Simmonds, J. G. (1984). General helicoidal shells undergoing large, one-dimensional strains or large inextensional deformations. *Int. J. Solids Struct.* **20**, 13-30. [C]
Sobel, L. H., and Newman, S. Z. (1980). Comparison of experimental and simplified analytical results for the in-plane plastic bending and buckling of an elbow. *J. Press. Vess. Tech.* **102**, 400-409. [A]
Stephens, W. B., Starnes, J. H., and Almroth, B. O (1975). Collapse of long cylindrical shells under combined bending and pressure loads. *AIAA J* **13**, 20-25. [A]

Struik, D. J. (1961). "Differential Geometry," 2nd ed. Addison-Wesley, Reading Mass. [C]

Thurston, G. A. (1977). Critical bending moment of circular cylindrical tubes. *J. Appl. Mech.* **44**, 173-175. [A]

von Kármán, T. (1911). Über die Formänderung dünnwandiger Rohre, insbesondere federnder Ausgleichrohre. *Z. Vereines Deutscher Ingenieure* **55**, 1889-1895.

Wan, F. Y. M. (1982). Finite axial extension and torsion of elastic helicoidal shells. Technical Rep. No. 82-15, The Institute of Applied Mathematics and Statistics, The Univ. of British Columbia. [I]

Wood, J. D. (1958). The flexure of a uniformly pressurized circular cylindrical shell. *J. Appl. Mech.* **25**, 453-458. [A]

Appendices

A. Guide to Notation

Shell theory uses many symbols. Some have become standard (e.g., N's for stress resultants, M's for stress couples, Q's for transverse shear stress resultants), but others vary from writer to writer. Below, we give a few *guidelines* to our scheme of notation and list some of the more important *global* symbols. We have not attempted to list *local* symbols, i.e., symbols peculiar to a section or subsection. However, we warn the reader that some global symbols have a different meaning locally. In part, this was forced on us because of the limited number of alphabets and fonts available on our laser printer, and, sometimes, we had to use other than optimal symbols. Thus, our capital "script" ABC's are $\mathcal{ABC}$ and our few bold Greek fonts, such as $\boldsymbol{\beta}$ in Chapter VI, were "jerry-rigged" by backspacing. (We find it frustrating and amusing that the the set of fonts available to us had no bold Greek and no italic Greek, yet included hundreds of dingbats such as ✂, ✏, and ❁.)

1. General scheme of notation

Bold letters (e.g., **F**) denote vectors.

Script letters (e.g, $\mathcal{F}$) usually denote lines, regions, matrices, column vectors, functionals, and operators.

A comma followed by a subscript (as in $F_{,\theta}$) denotes differentiation with respect to the subscript.

A prime (as in F') denotes differentiation with respect to undeformed arc length σ.

A dot [as in $\dot{F}$ or $(F+G)^{\bullet}$] denotes differentiation with respect to time t.

An overbar (as in $\bar{F}$) denotes a quantity associated with the deformed configuration. The same quantity, without the overbar, is associated with the undeformed configuration.

A hat (as in $\hat{F}$) denotes a quantity prescribed or specified on the boundary and, sometimes (as in $[F = \hat{F}(\sigma, t)]$, denotes the arguments of a function.

2. Global notations

Some of the following symbols come with or without subscripts. Subscripts signify either direction (e.g., e_θ is *circumferential* strain) or else qualify a quantity (e.g., A_e is *extensional* compliance).

A, A_i	constraint matrix, compliances
b_σ, b_θ	meridional and circumferential curvatures on a surface of revolution
$\mathbf{b}$	unit normal to a reference line
$\mathbf{B}$	rotated unit normal (through an angle β)[1]
$\mathcal{B}$	3-dimensional body, edge-constraint vector
C, C_i	stiffness coefficients in the strain-energy density
d	baseline deviation (distance between base and reference curves)
D	bending rigidity
$\mathcal{D}$	deformation power density
e, e_i	extensional (engineering) strains
$\mathbf{e}_i$	unit vector in the i-direction
E, E_i	Young's or related moduli
$\mathcal{E}$	strain energy in a shell
$\mathbf{f}$	3-dimensional body force
F	stress function
$\mathbf{F}$	force vector in beamshell
$\mathcal{F}$	functional
g	transverse shear strain
h	thickness of the undeformed shell
H	bound on ζ-coordinate, x or radial component of stress resultant vector
$\mathcal{H}$	unique solution of $A\mathcal{H}=\mathcal{B}$
I	moment of inertia
k, k_i	bending strains
$\mathcal{K}$	kinetic energy
$l, \mathbf{l}$	scalar and vector external distributed couples
L	length of undeformed reference curve or meridian
$\mathcal{L}$	Lagrange multiplier vector
m	mass/length or mass/area
$\mathbf{m}$	rational normal to the deformed reference line or meridian
$M, M_i, \mathbf{M}_i$	stress couples (bending moments), stress couple vector
$\mathcal{M}_k$	null vectors of constraint matrix A
$N, N_i, \mathbf{N}_i$	stress resultants, stress resultant vector
$\mathcal{N}$	edge-load vector
$p, \mathbf{p}, p_i$	external surface normal load, load vector, load vector components
$\mathcal{P}, \bar{\mathcal{P}}$	load potential, complementary load potential
Q	transverse shear stress resultant
$\mathcal{Q}$	rate of heating
r	radial distance from axis to surface of revolution

[1] $\mathbf{B}=\bar{\mathbf{b}}$ under the Kirchhoff Hypothesis.

$\mathbf{S}, \mathbf{S}_i, \mathbf{S}^i$	stress vectors
$\mathcal{S}$	shock line, reference surface
t	time
$\mathbf{t}$	unit tangent to reference line and reference meridian
T	1- or 2-dimensional temperature field
$\mathbf{T}$	rotated unit tangent (through an angle β)[2]
$\mathcal{T}$	set of test functions
u	displacement in the longitudinal, x-, or r-direction
$\mathbf{u}$	displacement
$\mathbf{v}, v_i$	velocity, velocity components
V	component of stress resultant vector in the y- or axial direction
$\mathbf{V}$	velocity test function
$\mathcal{v}$	edge velocity vector
$\mathcal{V}$	edge test-function vector
w	displacement in the vertical or z-direction
$\mathcal{W}$	external mechanical power
x, y, z	Cartesian coordinates
$\mathbf{x}$	position in space
$\mathbf{y}$	position of reference line or surface
$\mathcal{Y}$	edge kinematic vector
z	longitudinal coordinate in birod
$\mathbf{z}$	vector from reference line or surface to a generic point in a shell-like body ($=\mathbf{x}-\mathbf{y}$)
α	angle between tangent to reference line or meridian and xy-plane
$\bar{\alpha}_T$	$\alpha+\beta$
β	rotational deformation (angle between $\mathbf{T}$ and $\mathbf{t}$)
γ	transverse shear angle (between $\mathbf{T}$ and $\mathbf{t}$)
δ	variation operator
ε	Lagrangian strain, perturbation parameter
ζ	transverse coordinate in a shell-like body
θ	3-dimensional temperature field, circumferential coordinate
ι	entropy resultant
κ, κ_i	modified bending strains
λ, λ_i	stretches ($\lambda_i = 1 + e_i$)
ν, ν_i	Poisson ratios
Π_i	variational functionals
σ	arc length and Lagrangian coordinate along reference line or meridian
Φ, ϕ	dimensional and nondimensional strain-energy densities
Ψ, ψ	dimensional and nondimensional mixed-energy densities
ω	spin in a beamshell or axishell ($=\dot{\beta}$)
$\boldsymbol{\omega}$	spin in a unishell
Ω	angular test function, rotational test function, modified, mixed energy density

---

[2] $\mathbf{T} = \mathbf{t}$ under the Kirchhoff Hypothesis.

B. Some Isotropic, 3-Dimensional Strain-Energy Densities

We list here some of the isotropic, 3-dimensional strain-energy densities that have been proposed in the literature.[3] Many are for incompressible, rubber-like materials, two are for compressible rubbers, and one is for a biologic material. Most of the formulas may be found in either Alexander (1968) or Ogden (1972), both of whom discuss experiments and give typical values for the constants that appear below. The notation Mooney [1940;A] means that the formula beside the name may be found in a paper written by Mooney in 1940 that is discussed and referenced in Alexander (1968). The I_i are the standard, 3-dimensional strain invariants (Green & Zerna, 1970), the λ_i are principal stretches, and the C_i are constants. Other symbols are defined as they occur.

Mooney [1940;A][4]	$C_1(I_1-3)+C_2(I_2-3)$
Treloar [1943;A][5]	$C_1(I_1-3)$
Rivlin-Saunders [1951;A][6]	$C_1(I_1-3)+f(I_2-3)$
Thomas [1955;A][7]	$\dfrac{C_1}{\lambda_1\lambda_2\sqrt{C}}F(k,\alpha)$
Hart-Smith [1966;A]	$C_1\int \exp[C_2(I_1-3)^2]dI_1+C_3\ln(I_2/3)$
Alexander (1968)	$C_1(I_1-3)+C_3(I_2-3)+C_4\ln[C_5(I_2-3)+1]$
Alexander (1968)	$C_1\int \exp[C_2(I_1-3)^2]dI_1+C_3(I_2-3)+C_4\ln[C_5(I_2-3)+1]$
Ogden (1972)[8]	$\sum_i C_i(\lambda_1^{\alpha_i}+\lambda_2^{\alpha_i}+\lambda_3^{\alpha_i}-3)$
Blatz-Ko (1962)[9]	$C_1\left[I_1+\dfrac{1-2\nu}{\nu}I_3^{-\nu/(1-2\nu)}-\dfrac{1+\nu}{\nu}\right]$
Blatz-Ko (1962)	$C_1(I_2/I_3+2\sqrt{I_3}-5)$
Fung *et al* (1978)	$C_1\exp(C_2I_1^2+C_3I_2)$

[3]See Wu (1979) for a list of some 2-dimensional strain-energy densities.

[4]Sometimes called a "Mooney-Rivlin" material.

[5]Commonly called a "neo-Hookean" material.

[6]f is an arbitrary function.

[7]$C=1-\lambda_3^2/\lambda_1^2$, $D=1-\lambda_3^2/\lambda_2^2$, $k^2=D/C$, $\sin\alpha=\sqrt{C}$, and F is the elliptic integral of the first kind.

[8]The α_i are constants.

[9]ν is a constant which may be identified with Poisson's ratio as the strains grow small.

References

Alexander, H. (1968). A constitutive relation for rubber-like materials. *Int. J. Engr. Sci.* **6,** 549-563.

Blatz, P. J., and Ko, W. L. (1962). Application of finite elastic theory to the deformation of rubbery materials. *Trans. Soc. Rheology* **6,** 233-251.

Fung, Y. C., Tong, P., and Patitucci, P. (1978). Stress and strain in the lung. *J. Engr. Mech. Div., ASCE* **104,** 201-223.

Ogden, R. W. (1972). Large deformation isotropic elasticity—on the correlation of theory and experiment for incompressible rubberlike solids. *Proc. R. Soc. London* **A326,** 565-584.

Wu, C. H. (1979). Large finite strain membrane problems. *Q. Appl. Math.* **39,** 347-359.

Index

C

F

N

O

P

T

U

V

W

Y